The American Railroad Passenger Car

THE JOHNS HOPKINS UNIVERSITY PRESS   BALTIMORE AND LONDON

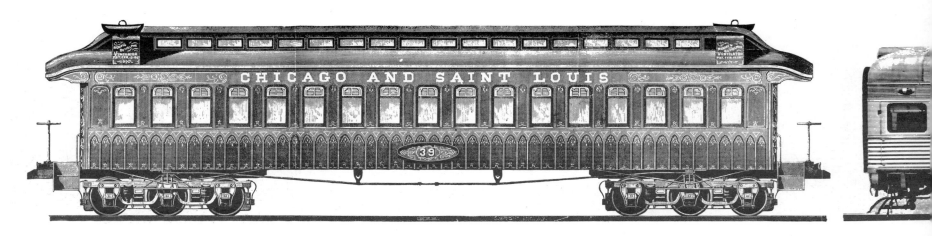

The American Railroad Passenger Car

JOHN H. WHITE, JR.

Part 2

© 1978 The Johns Hopkins University Press
All rights reserved
Printed in the United States of America on acid-free paper

Originally published in one hardcover volume, 1978
Johns Hopkins Paperbacks edition, published in two softcover volumes, 1985
9 8 7 6

The Johns Hopkins University Press
2715 North Charles Street
Baltimore, Maryland 21218-4363
www.press.jhu.edu

Library of Congress Cataloging in Publication Data

White, John H.
The American railroad passenger car.

"Originally published in one hardcover volume, 1978"—
Vol. 1, t.p. verso.
Bibliography: v. 2, p. 677
Includes index.
1. Railroads—United States—Passenger-cars—History.
I. Title.
[TF455.W45   1985]   625.2´3´0973   84–26161
ISBN 0–8018–2743–4 (pbk.: set: alk. paper)
ISBN 0–8018–2722–1 (pbk.: v. 1: alk. paper)
ISBN 0–8018–2747–7 (pbk.: v. 2: alk. paper)

A catalog record for this book is available from the British Library.

CONTENTS

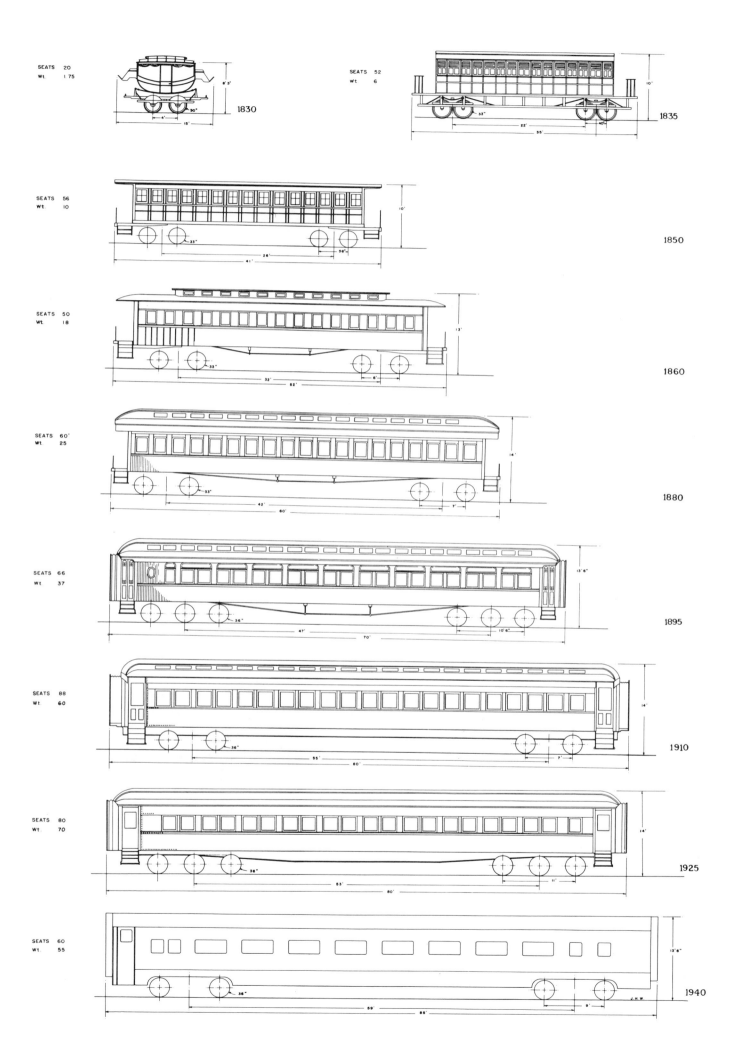

SEATS 20
Wt 1.75
1830

SEATS 52
Wt 6
1835

SEATS 56
Wt 10
1850

SEATS 50
Wt 18
1860

SEATS 60'
Wt 25
1880

SEATS 66
Wt 37
1895

SEATS 88
Wt 60
1910

SEATS 80
Wt 70
1925

SEATS 60
Wt 55
1940

CHAPTER FIVE

Passenger Car Accommodations

FROM DECORATION TO VESTIBULES

OF ALL THE COMMODITIES CARRIED BY RAILWAYS, none is more fragile, sensitive, or quarrelsome than human cargo. A lump of coal or a stick of lumber can be trundled about in the simplest sort of car; it requires no shelter, heat, seating, or washrooms. Schedules, rough track, and rude trainmen are of no consequence. But these and countless other matters are of paramount concern to the railway passenger. The degree of heat, the mechanics of the sunshade, the angle of the seat back, the texture of the cushions, and even their color form the basis for endless complaints, comments, and recommendations. To answer the demand for greater comfort, the railroad passenger car became an increasingly complicated structure with a profusion of subsystems and fancy auxiliary apparatus. Year by year the addition of some new improvement for convenience, sanitation, safety, or decorative effect increased weight and cost. The trend was irreversible and explains more than any other single factor the car builders' failure to reduce dead weight.

Thus the history of the passenger car cannot be told simply by tracing its increase in size or the shift from wood to steel construction. The story must also cover the progressive development of its fittings and the evolution of a simple shelter on flanged wheels into a complex vehicle interwoven with pipes, ducts, wiring, and gadgets.

Seating

To the first railway passengers, land travel was a hardship to be borne. They expected little more than a hard seat, a leather curtain, and perhaps a hand strap. Lighting came from the sun and ventilation from chance breezes, while heating was self-generated. The ride was wonderfully smooth compared with highway travel, and since the journey was short, pioneer lines gave minimal attention to most interior fixtures.

Seating was an exception, however. A good seat is a basic comfort and was recognized as such from the beginning. The earliest four-wheel cars had high-backed upholstered seats arranged face-to-face, highway coach fashion. When the large, open, eight-wheel cars came into fashion, this time-honored plan was abandoned, because Americans had developed a phobia about riding backwards. Apparently the syndrome was peculiar to this country; British travelers seemed content with the old system, and indeed it is universally followed today on the Continent. Of necessity Pullman reinstated the stationary-back, face-to-face seat in his sleeping cars, but it was considered a drawback. Coach passengers would accept nothing less than a reversible seat back, and until very recent times the swingback or walkover seat was the standard for American coaches.

Conduce Gatch, car builder for the B & O, claimed to have introduced the reversible seat around 1833.[1] He said that it was used on several experimental eight-wheel cars built at the time. The idea was an old one; in fact little was new in the area of furniture design by the opening of the last century. (An attractive bench with a reversible back appears in a fifteenth-century painting of Saint Barbara by Robert Campin.) The Camden and Amboy adopted the design for its eight-wheel coaches of 1836. The B & O itself employed both metal-frame swingback seats, as recorded in the Austrian Karl Gegha's 1844 report, and a high-backed wooden Empire-style seat which seems to have had a platform rocker base. The Empire chair appears in a B & O drawing (1835c., shown in Figure 1.70) and the engraving of a B & O coach in Douglas Galton's 1857 report on American railways (Figure 1.82).

British visitors were well pleased with American seating. A correspondent for the *Illustrated London News* wrote glowingly of the standard of comfort available to travelers in North America and the absence of the mean wooden benches in third-class carriages.[2] Except for emigrants, all passengers could expect a seat that was "well stuffed, and covered with a fine plush, the arms of the seats being made of polished black walnut or mahogany." The reporter was equally impressed by the generous seat size—18 inches deep by 39 to 42 inches long—and the deep, comfortable cushions.

American travelers of the period were not so well pleased. Compliments were few, complaints legion. An advertisement for a luxury seat in the 1851 *American Railway Guide* claimed that even after a short time in a common car seat, the passenger felt as if his bones were pounded in a bag. Another critic wrote that the experienced traveler "will breathe dust and sparks, and submit to any amount of jolting with entire serenity," but that his language would become irreligious when faced with a car seat that lacked such conveniences as a simple footboard.[3]

A dozen solutions were devised to resolve the car seat dilemma. Because no one design could comfortably accommodate all passenger sizes, shapes, and weights, some form of adjustable seat was regarded as essential. August Mencken and other historians have claimed that the first adjustable or reclining railroad car seat was introduced in 1852. This is the year when the first U.S. patent was issued for such a device, but as Chapter 4 described, similar seats were being tested on the Camden and Amboy more than a decade earlier. Reclining seats were also used by the Eastern and P W & B railroads in the mid-1840s. It is true, however, that no wide-scale interest in reclining seats developed until the middle of the following decade. Although the inventions which flooded the Patent Office at that time aimed generally at a more comfortable seat, their specific purpose was to provide a recumbent sleeping-car seat. Bailey, Kasson, Smith, and Hammitt seats were rushed onto the market (see Figures 5.1 and 5.2). The usual exaggerated claims were made for them. Hammitt announced that his seat "will render railroad traveling more pleasant than staying at home."[4] A number of railroads adopted these new designs for at least some of their better cars, and the introduction of parlor and chair cars encouraged a new wave of reclining-seat inventions. Adjustable seats were expensive and heavy, however. They required more space and hence reduced the total seating capacity, and a fee had to be paid for the patent. More important, they failed to supplant the sleeping car. In testimony during the Pullman-Wagner legal battle of

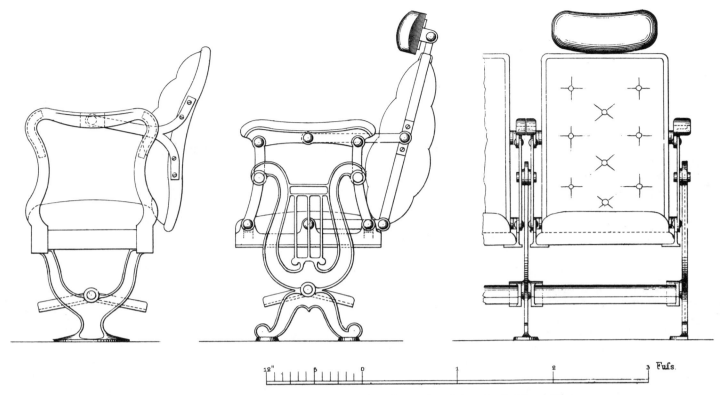

Figure 5.1   American passenger car seats typical of the 1850–1870 period. The reclining seat on the left was patented by C. P. Bailey in 1854. (A. Bendel, Aufsätze Eisenbahn-wesen in Nord-Amerika, Berlin, 1862, plate E)

ALBERT M. SMITH'S
Patent Premium Reclining and Self-Adjusting
CAR SEAT.

For a Night and Day High or Low Back Seat combined in One.
PATENTED AUGUST 21st, 1855.

It was awarded two first premiums, a Gold Medal, at the great "Fair of the American Institute," held at New York, and a Diploma at the State Fair, held at Elmira, N. Y., 1855.

This valuable improvement is adapted and can be applied at a very trifling expense, to the ordinary Seat now in use, without impairing its present qualifications as a day Seat, and a new Seat provided with it, and made in single seats, costs no more than the ordinary seat.

By an arrangement that is very simple and not liable to get out of order, the back is so hung at points, varying from the centre, that it can be converted into a High Back Night Seat, by pushing against the upper part of the Back, which disconnects the lower part, and allows the sides to be reversed, the outside placed in, which is the natural form and shape of the person, and raised high enough to support the head, this slides the seat forward on a curve, so as to be in conformity with the angle of the back, and it is then self-adjusting to any position of the person, and cannot be moved from it by the sudden motion of the Cars, making a seat as perfectly adapted to its intended use, (as a High or Low Back), as if made specially for it, and no other.

The Seat can be seen and examined, and orders will be received for the improvement to apply to old Seats or for new Seats at the office of

ALBERT M. SMITH,
Patentee and Manufacturer,
13 NORTH St. PAUL STREET, ROCHESTER, N. Y

or **TAULMAN & LOW,** *Agents,*
157 BROADWAY, NEW YORK.

Figure 5.2   Smith's reclining seat, from an advertisement in the Railroad Advocate *of December 15, 1855.*

Figure 5.3 Numerous inventors attempted to produce headrests to convert ordinary coach seats for sleeping purposes. (American Railway Review, *April 16, 1860*)

1881, W. N. Bragg, who had patented a reclining seat in 1857, admitted that he could never sleep in one of his own chairs.[5]

An elaborate car seat would be worth the money if it eliminated the need for sleeping cars, but because it could not, most railroads were content to give this trade to Pullman. The less affluent traveler was forced to make do with a low-backed, one-position coach seat. American inventors offered the coach passenger a wonderful variety of headrests, stirrups, and hammocks (see Figure 5.3). Their advertisements claimed that each individual willing to buy and carry his own apparatus might convert an ordinary seat into a commodious chair or bed. The earliest device of this nature seems to be a headrest patented in 1860 by Amos Gore.[6] One of the wildest, patented in 1889 by Herbert M. Small, was a small canvas hammock, fitted with hooks, to be mounted between two seat backs. The sleeper's legs hung down between the supporting straps, and the feet were held by a pair of looped rope stirrups. The hammock must have had approximately the stability of a trapeze in a hurricane. It required the cooperation of the people in the forward seat, since three of Small's patented hooks protruded over their seat backs. The experienced traveler avoided such gadgets and paid the extra fare to go Pullman.

For almost the entire history of railway passenger service in America, the typical car seat was a small, relatively uncomfortable two-passenger bench (Figures 5.4 and 5.5). The low one-piece back was reversible. The seat cushion was flat or convex. To withstand heavy use, both the back and seat cushions were hard, with stiff, unyielding springs. The armrests were often unpadded; they were commonly of wood, though in some instances nickel-plated cast iron was considered good enough. No adjustment was possible except that the position of the seat back could be reversed. The tall, the short, the fat, and the lean were expected to adapt themselves to the shape of the seat. Footboards were often dispensed with because they made cleaning more difficult. (A fast sweeping was a necessity at many terminals, where the cars were sent out on another train almost as soon as they were emptied.) The typical coach seat was not really a full sofa, though in size it appears to duplicate a love seat. A pair of legs was commonly omitted, for one end of the seat was attached to the inside wall of the car. This saved weight and the cost of the legs. It also simplified cleaning under the seat.

In the universal American-style coach, double passenger seats were placed on either side of the aisle. A rare exception was the rows of double and single seats used by narrow-gauge lines since the 1870s and adopted many years later for some deluxe standard-gauge cars. Suburban cars were occasionally equipped with three-abreast seats on one side of the aisle and conventional two-abreast seats on the other. The Southern Pacific adopted this plan as early as 1912 on some of its Oakland commuter cars, and the New Haven followed suit several years later.[7] The scheme was never popular, however, and when it was proposed in recent years for the new Erie-Lackawanna suburban service, it was bitterly opposed by the riders.

Before 1850, car seat frames appeared to be fabricated almost entirely of wood. After that time metal began to play a greater role. Cast-iron end frames in fact became very popular by mid-century. This style of seat framing has been credited to George Buntin, foreman of the Eastern Railroad's paint shop.[8] Although the claim cannot be verified, Buntin did patent a peculiar style of reversible seat that employed a cast-iron frame. The principal feature of the device was its cradle-like body; the seat and the back were identical, so that the seat was the back when it was tipped one way, and vice versa. It was called the poor man's reclining chair, because its central pivot did offer a variation in angle. Of course, both occupants of the seat had to agree on the same position. A number of New England railroads used the Buntin seat, and it was being produced at least until the late 1880s. A variation on the design is shown in Figure 5.6.

Others tried to improve the standard car seat at a realistic cost for the average coach patron. Reclining seats were simply too expensive for wide application; a coach seat mechanism had to be simple. One of the best of these schemes was the rocker seat patented in May 1879 by C. C. Mason of Altoona, Pennsylvania.[9] The loose seat cushion rested on an elliptical cradle and could thus be shifted and inclined fore or aft according to the position of the seat back (see Figure 5.7). This simple design made it possible to slope the seat cushion so that the passenger slid back against the seat back. In 1883 the inventor claimed that over 20,000 of his seats were in use. They were a great favorite on the Pennsylvania Railroad.

The comfort of the day coach rider was the abiding concern of Mathias N. Forney, locomotive designer, editor of the *Railroad Gazette*, and longtime secretary of the Master Car Builders Association.[10] Forney was appalled at the lack of science displayed in car seat design. The car seat manufacturer's aim was to produce the cheapest form of chair, add a few fistfuls of stuffing, and let the human shape conform as best it could. Forney understood the need for a cheap, simple seat, but felt that economy should be carried just so far. In analyzing the problem, he concluded that no one had bothered to study human anatomy or the basic elements of physiology. He set out to design an organic seat that suited the human form.

Forney's study led him to stipulate five conditions that were necessary for a comfortable seat:

1. A large bearing or seating surface for maximum support.
2. A concave seat cushion to receive and support the human posterior.

THE WAKEFIELD RATTAN CO., BOSTON,

Gardner's New Reversible Car Seat, No. 8,
Patented December 6th, 1881.

PATENTED FEB. 20, 1877.

Figure 5.4   The variety of seat styles available in 1870–1890 is suggested by this selection of advertising cuts. The Gardner bench had a pierced wood veneer seat and back rather than upholstered cushions. (Miscellaneous trade journals)

3. A seat cushion of the proper depth and height, with a soft front edge that would not dig into the legs.
4. A seat cushion inclined toward the back, to relieve strain and cradle the passenger more securely in his chair.
5. A high and properly shaped seat back that supported both the shoulder blades and the small of the back.

The common car seat was deficient on all five points. On August 25, 1885, Forney patented a seat (No. 324,825) that satisfied his conditions. A clever arrangement of levers, which he called cross links, permitted the seat back to reverse (Figures 5.8 and 5.9). Forney promoted his design in the *Railroad Gazette* and by personal appearances before such groups as the New England Railroad Club. Several railroads, including the New York Central and the Long Island, adopted his seat. How widely it was used cannot be determined, but it did enjoy a long life (see Figure 5.10). The Scarritt Car Seat Company of St. Louis came to control the patents, and as late as 1928 its successor, Hale and

Kilburn, continued to manufacture the Forney seat under the trade name of Scarritt.[11]

Except for the Forney seat and the relief offered by an occasional chair car, the coach passenger seemed condemned to suffer forever in the walkover seat (Figures 5.11 and 5.12). By 1930, however, luxurious long-haul motor buses unexpectedly began to cut into railroad passenger traffic.[12] Since a notable feature of the buses was their comfortable bucket seats, the venerable arguments of railroad managements about maximum capacity and the undesirability of squandering cash on fancy coach seating were silenced. More legroom and soft, well-formed seats became the new order. Of course, it was not until the lightweight era that coach passengers escaped the old walkovers, but the railroads were at least committed now to improving the lot of the coach patron (see Figure 5.13).

The bucket seats introduced in 1930 had a double set of springs. The upper decks were encased in nearly airtight upholstery to form an air cushion. In 1936 molded foam-rubber

Figure 5.5   *Reversible or walkover seat backs were standard in American coaches. Both wooden and cast-iron seat ends were common. (Pullman Neg. 2776)*

Figure 5.6   *This cast-iron seat, patented by T. J. Close in 1860, is similar to the popular Buntin seat except for the wooden-slat seat and back.*

cushions were introduced.[13] Gone was the walkover back; the seats had individual reclining backs (Figure 5.14). To reverse, the entire seat rotated on its base. On some cars it was possible to turn a seat 90 degrees and view the scenery in comfort. Some manufacturers used a simple pedestal base, but one of the largest car seat makers, Heywood-Wakefield, devised a platform-mounted center plate. The seat could thus be placed tight against the interior wall, ensuring maximum aisle width. It was pulled away from the wall along the track of its platform and then swung around on the center plate. Even with such an ingenious mechanism, more generous spacing was needed for the seat. It was possible to keep it within a few inches of the old 38-inch center spacing, but most roads were willing to give their patrons even more legroom. The Illinois Central changed to a 52-inch spacing that cut a standard-sized car's capacity to a mere 48 passengers.[14]

The success of the deluxe bucket seat inspired further investigation into this subject by the Association of American Railroads' Mechanical Division. In 1945 with a grant from the Heywood-Wakefield Company, the group hired Professor E. A. Hooton of the Harvard anthropology department to conduct research on car seat design.[15] Hooton, an expert in the field of human anatomy, set up test chairs in major Boston and Chicago stations to gather the needed measurements. Data collected from the 3,867 adults who volunteered for measurement were used to prepare a standard profile guaranteed to satisfy the greatest number of coach passengers (Figures 5.15 and 5.16). The most critical single dimension was the seat height of 16¾ inches.

Hooton's recommendations led to the development of the famous Sleepy Hollow car seat that became very nearly the industry standard in the years after World War II. Heywood-Wakefield offered the seat in three basic styles: a night chair with reclining leg rest, a day chair without the leg rest, and a lightweight model with a tubular frame. By the time the Sleepy Hollow line appeared, its maker all but controlled the car seat business. (Through a predecessor firm it claimed an establishment date of 1826. In 1897 the Heywood Brothers and the Wakefield Rattan Company had combined, and during the next half century they absorbed or outlived their major competitors, including their old rival, the Hale and Kilburn Car Seat Company of Philadelphia.)

Deluxe car seating was expensive not only because the seats themselves cost $3,000 to $4,000 but also because they added dead weight.[16] For generations seat designers had tried to trim weight. Early in the century pressed steel was substituted for cast-iron parts, but with the introduction of oversized deluxe seats the weight per seat jumped to 175 to 200 pounds each. Even if a car had only thirty seats, they weighed a total of 3 tons. Designers hoped to cut that figure in half, and one manufacturer actually succeeded in producing a 105-pound coach seat by using a welded tubular frame.[17] However, the overwhelming popularity of the Sleepy Hollow seat ended all hope of cutting overall car weight by creating lightweight furniture.

A great variety of materials was used for seat coverings. Bare wooden seats appeared only in the poorest class of cars. Emigrant cars, commuter cars, and special-accommodation cars for railroad shop employees might be outfitted with slat-bottomed seats. So far as comfort was concerned, the perforated veneer seats manufactured by Gardner and Company of New York probably ranked one level above the slat seat. In them thin sheets of veneer, steam-molded to the desired contour, were perforated with a geometric pattern of holes for ventilation and extra springiness. Even so they made a poor seat and had only

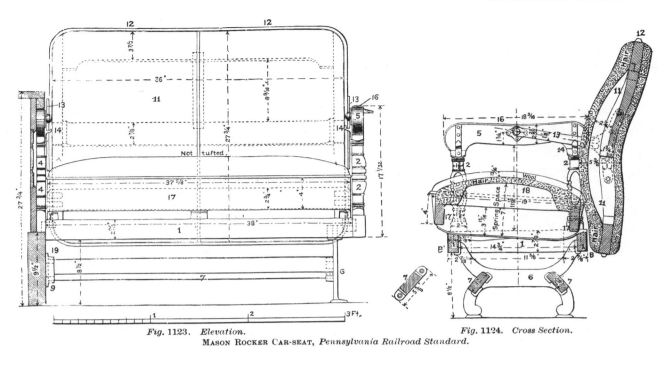

Fig. 1123.   *Elevation.*   Fig. 1124.   *Cross Section.*
MASON ROCKER CAR-SEAT, *Pennsylvania Railroad Standard.*

Fig. 1125.   Fig. 1126.
With lock.   Without lock.
SEAT-STOP.   (4 5-16x⅝ in)

Fig. 1127.   Fig. 1128.
MASON ROCKER CAR-SEAT.
Perspective Views.

Fig. 1129.
SEAT-ARM PLATE.
(2 7-16x1⅞ in.)

Fig. 1130.
SEAT-ARM PIVOT.
(Head, ¾ in. diameter.)

Figure 5.7   It was possible to tilt the seat cushion of the Mason seat by adjusting its position on the rocker-like base. The design was patented in 1879. (Car Builders' Dictionary, 1888, figs. 1123–1130)

limited use in steam railroad coaches. Gardner's major market was in the rapid-transit field.

From the beginning, coach passengers could expect some form of upholstered seat. Before 1880 the covering might be cloth, leather, rattan, velvet, or even woven wire.[18] Of these options leather appeared to be the best choice, since it is handsome, durable, stain-proof, and readily available. But it is expensive and slippery, and in the winter months passengers complained that it was cold. Even so, leather has had a steady if not universal application throughout the history of passenger cars.

Almost no information exists on seat coverings in the early period, though it seems that velvet was favored for the superior class of cars. In 1849 the Hudson River Railroad's best cars were upholstered with the finest plush velvet.[19] Several years later the Erie draped its cars with 300 feet of the same material at $5 a yard.[20] Red hues, from crimson to dark maroon, were most often chosen, though in the mid-1870s Pullman was said to favor Utrecht velvet, whose olive bronze was such a subtle shade that only an expert could identify it.[21]

Around the mid-century a miraculous material was introduced

that was to become the standard railroad car seat covering: bristly, bright mohair. It could endure heavy wear and retain an attractive appearance after years of use. Two and one-half times as strong as wool, mohair was the longest-wearing natural fiber known.[22] It could be counted on to last twelve to fifteen years, and some seats were still good after twenty-five years. Its stiff pile tended to hold the passenger in place, unlike a slick, soft-pile velvet. Mohair was also a sanitary covering, since it did not attract dirt and would not hold it. Because the oil was inside the individual hairs rather than on the outside, as with wool, normal processing did not destroy its silky luster.

Mohair reportedly came to the United States in 1839. At that time the first Angora goats were sent to a South Carolina planter by the Turkish sultan in gratitude for the American's help in showing the Turks how to grow cotton.[23] For more than forty years, however, plush made from the fleece of the Angora goats was imported. It was an expensive handmade textile until 1882 when, after three years of experimentation, the Sanford Mills in Maine perfected a practical mohair loom.[24] The domestic machine-made textile soon captured the market. It had few seri-

Figure 5.8   M. N. Forney, editor of the Railroad Gazette, *patented a seat in 1885 which became widely used on many American railroads.* (Railroad Gazette, *June 4, 1886*)

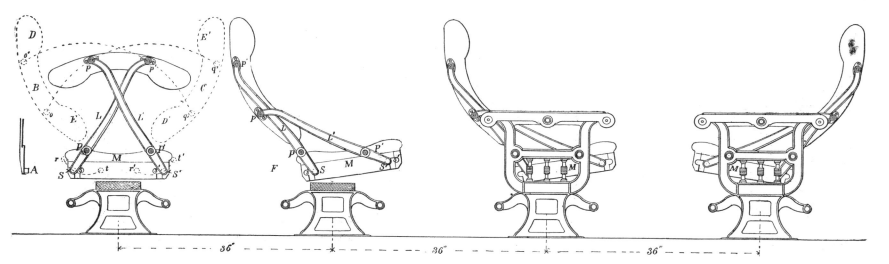

Figure 5.9   Forney designed his seat so that the back and seat conformed to the human anatomy in either their forward or their reverse positions. (Railroad Gazette, *June 4, 1886*)

ous rivals for railroad car seat upholstery until the introduction of cheap, soil-resistant fabrics during the 1940s.

Heating

During most of the nineteenth century, passenger cars were heated in the winter months by the common iron stove. It was the least expensive and simplest form of artificial heat available at the time. In a land of cheap firewood and coal these fuel-hungry ovens were not considered inefficient, and the brakeman's work in stoking them was regarded as time well spent. By modern standards, however, the stove is second only to the open fire in inefficient heating, because it warms by direct radiation. This is adequate for a small enclosure, but in the long tube of an American passenger car, a stove heated only a portion of the structure. To equalize the effect, it became common during the heating season to remove several seats on one side of the aisle and place the stove in the middle of the car. After about 1870 this practice seems to have been abandoned and a permanently fixed stove was installed at both ends. But neither method was really effective; passengers next to the stove roasted, while those far away from it froze. The car was difficult to heat not just because of its shape, but because the side walls were thin and generally uninsulated, while the windows and doors were loose-fitting. The doors seemed to be open more than closed as trainmen walked through the car, passengers boarded, and porters propped the door open to load luggage. Added to the inefficiency of the iron stove was the very real fire hazard. On every count the car stove was unsatisfactory, but for many generations it was accepted as preferable to no heat at all.

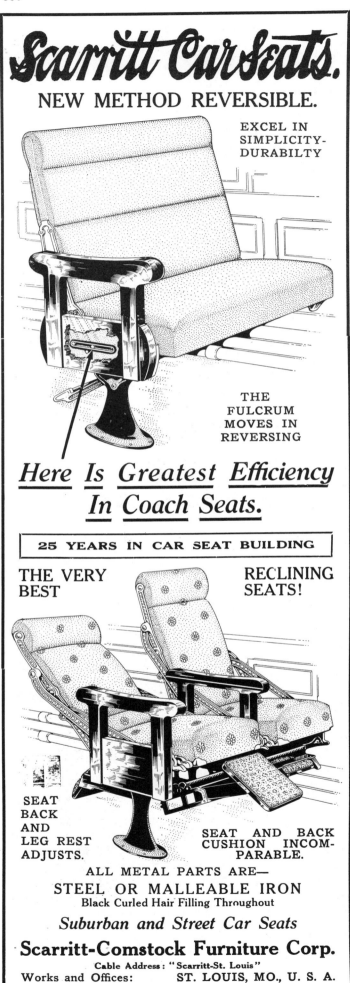

Figure 5.10   A Forney-style seat was still being produced by Scarritt-Comstock in 1915. (Railway Mechanical Engineer, March 1915)

Besides the stove, there were other major car-heating systems over the years. Hot-air systems used a stove as the source of heat but broadcast a stream of warm air through the car, usually by baseboard registers. Hot-water systems again used the stove but gained better distribution through a system of pipe radiators. Steam, usually drawn from the locomotive, was common in this country from about 1890 to 1910. Vacuum, a low-pressure steam system, was seldom chosen for car heating. Vapor, a popular extremely low-pressure steam system, became standard after about 1910.

There were, of course, countless experimental systems. An idea that was regularly proposed was the storage heater, for which cannon balls, bricks, or iron slugs would be heated in a central plant and deposited in the cars. Such schemes may have been plausible for city transit cars but not for long-distance trains. Electric, alcohol, and even gasoline heaters were proposed at an early date. The most original system, developed by a Swede in 1870, aimed at heating the train with a central furnace in the baggage car—not by circulating hot water or steam, but by propelling heated sand from car to car.[25] The inventor felt confident that he could make sand circulate through a labyrinth of iron pipes, flexible couplings, elbows, and tees the length of the train and then return to the head-end furnace for reheating. Such an invention could only convince practical railroad managers that stoves must be the best system after all.

In the very first trains there were no car heaters, and the passengers complained not about the poor heating but about the lack of all heat. Travelers bundled up and tried a variety of foot and lap warmers fueled by hot bricks, steaming water, or live coals. These methods were a holdover from coaching days, and as long as railway cars were divided into small coach-like compartments, the era of self-heating prevailed. No railway could be expected to install a stove in each compartment. But the early emergence of the large single-compartment eight-wheel car quickly changed the heating possibilities. First came the single space heater, a modest enough accommodation. In December 1835 a Boston newspaper reported the advent of anthracite coal-burning stoves on the lines then running out of the city.[26] A month later it was announced that the B & O was using a small, under-the-floor heater patented by Alexander McWilliams of New York (Figure 5.17).[27] Other Northern railroads began to install stoves at about the same time, according to reports in accounts by David Stevenson and Von Gerstner.[28] The Austrian engineer said that the Utica and Schenectady had mounted a wood-fired stove outside of the car body and run a pipe under the seats to distribute the heat.[29] A large Stephenson-built, double-truck car purchased by the Auburn and Syracuse had a similar system, according to a note in the Rochester Democrat of August 28, 1839. But a few roads, such as the New Haven–Hartford line, were running unheated cars as late as the winter of 1839.[30]

Almost no specific information appeared about the early railway heaters; it can only be assumed that the stoves used were so ordinary that no description was thought necessary (see Figure

Figure 5.11   Cast-iron-frame, mohair-covered seats such as these were the industry standard throughout the heavyweight era. (Standard Steel Car Company)

Figure 5.12   The standard walkover seat was given modern ends and high individual seat backs, but basically it was unchanged from the traditional design. (American Car and Foundry Company)

Figure 5.14   An aluminum-frame reclining seat on a swivel base made by Karpen and Brother in 1937.

Figure 5.13   Lightweight tubular frame seats made for the New Haven in about 1937 by Heywood-Wakefield. (Railroadians of America)

5.18). One exception was the Littlefield car heater, a double-walled coal burner patented on March 4, 1856.[31] Manufactured by Erastus Corning of Albany, it was put into use on the New York Central Railroad, of which Corning happened to be president. The self-feeding central hopper, filled with some 50 pounds of fuel, was good for the entire run between Albany and Buffalo. It operated at one-tenth the cost of a wood-fired stove and was the only heater on the train that could keep frost from the windows when the temperature dropped to zero (Figure 5.19). Passengers, particularly ladies and children, were said to crowd into the cars fitted with Littlefield's miracle, leaving the other coaches nearly empty. Even the trainmen, who rarely had a civilized word to offer on any subject, eloquently praised the heater because it required no attention throughout the long trip. No record exists on how widespread the Littlefield stove became. Its introduction suggests, however, that a general conversion from wood to coal firing began at about the same time that a similar shift in locomotive fuels took place. The shift from wood to coal represents a significant advance in railroad car heating.

Another special form of car heater that gained some attention was the cone disc stove manufactured by Gardner Chilson of Mansfield, Massachusetts (Figure 5.20). It was a common cast-iron stove of the 1870s except for the large-diameter disc that deflected the heat, preventing the immediate rise of hot air to the clerestory and circulating some heat to the floor where it was most needed. As a safety feature in the event of a wreck, long

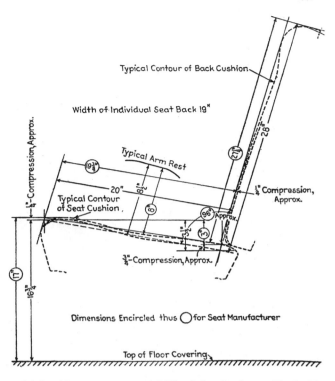

Figure 5.15 *After measuring 3,867 adults, Professor E. A. Hooton of Harvard University proposed this seat contour as a standard for railroad car seating. The design was completed in 1945.* (Railway Mechanical Engineer, *August 1945*)

Figure 5.16 *The Sleepy Hollow seat, based on Hooton's profile, provided luxury seating for coach passengers.* (Burlington Lines)

Figure 5.17 *McWilliams's coach heater, originally proposed for carriages, was used by the Baltimore and Ohio in the 1830s. C is the stove; d the heat pipe; b the register, and e the exhaust.* (U.S. Patent Office)

bolts secured the stove to the floor, and the fire doors could be locked as another precaution against the scattering of live coals.

The basic defects of the car stove, however, were not solved by Littlefield, Chilson, or a host of other inventors. The most prevalent complaint was that the stove was too hot for passengers unlucky enough to be sitting near it. The experienced traveler made a dash for the intermediate seats—not too close, not too distant. But for the people pressed near its simmering iron walls, the atmosphere was dry and baking as a kiln, so that they dozed and then woke sodden with perspiration. A giant 350-pound heater crammed with a quarter cord of wood could "burn the air and blister the varnish."[32] The villain here was the oafish brakeman whose job it was to tend the stove and who invariably went to extremes of overstoking or understoking. Overstoking was the more common; it seemed to be the pleasure of this barbaric race to burn up as much of the company's fuel as possible.

During the dark age of the car stove, many mechanics were busy devising improved heating systems. The most promising, the hot-air heater, was both simple and cheap enough for large-scale application. The basis of all such plans was to direct a stream of air over the stove to be heated and then to distribute it through the car by registers or ducts. The forward motion of the train created the air current, which was gathered by a roof-mounted hood or scoop. Axle-driven fans were also proposed, but most inventors wisely chose to avoid this complication. One of the first forced-air systems was patented on July 24, 1846 (No. 4654) by Tappen Townsend of Albany. It was a wind scoop mounted on the locomotive. After being heated in a furnace, the air was directed back along the train by ducts, with flexible joints between the cars. As far as is known, Townsend's plan was never tested. A few years later a Canadian, Henry Jones Ruttan, began to promote a forced-air system that received many trials during the 1850s and 1860s. Ruttan's chief concern was the purification of air for car ventilation (covered in the following section), but he incorporated a hot-air heater as part of the system.

In 1862 William Westlake (1831–1900), better known in later years as a partner in the giant Chicago railway supply house of Adams and Westlake, devised a forced-air heating system.[33] As well as the usual roof-mounted scoops, it had an under-the-floor heater, and registers running along the floor on both sides of the car assured a more even distribution of hot air than other systems provided. Westlake's heater was used by the La Crosse and Mil-

Figure 5.18   The common wood or coal stove is shown in this coach interior engraving.
(Illustrated London News, *April 6, 1861*)

waukee Railroad. Pullman was an early champion of the West-lake design but abandoned it within a few years in favor of the Baker hot-water heater.

The most popular of all hot-air stoves, patented on July 7, 1857 (No. 17,756), by James Spear of Philadelphia, was a cast-iron potbelly enclosed in a sheet-metal circular casing (Figures 5.21 and 5.22). The radiating surface was increased by a similar casing around the smoke pipe. Spear mounted a wind scoop on the roof that directed the air to the base of the stove and thence out along the length of the car through a baseboard register. The heater was so effective that it was adopted as standard by the Pennsylvania Railroad. This in itself guaranteed the success of Spear's patent, and soon such imitators as A. P. Winslow of Cleveland and F. S. Bissell of Pittsburgh were competing for a portion of the trade. A major attribute of the Spear plan was its low first cost—$45, about one-tenth the cost of a hot-water system.[34] However, the wind scoop heaters were not faultless. When the train was stationary, the air flow was reversed and the heat was exhausted out of the stovepipe.[35] The clerestory vents were too large, creating an imbalance that tended to draw cold air in and around the windows, doors, and other openings. The same imbalance was said to produce crosscurrents that inter-

fered with the lamps, causing them to smoke and blacken the chimneys.

Another class of car heater closely related to these hot-air systems was the suspended, or pendant, stove. It was mounted outside the car under the body, thereby leaving valuable space for more passengers. Suspended stoves were said to pay for themselves by the revenue from the extra seats. Optimistic supporters believed that the stove would be less dangerous in a collision because it would be thrown clear of the train.[36] And since the stove was not in the car, passengers were not burned, bruised, and blistered by bumping against it when the train lurched or abruptly stopped. Also, the stove's dust, ash, and smoke were not produced inside the car. The scheme had so many apparent advantages that it was tried at an early date. (McWilliams's patented heater of 1832 shown in Figure 5.17 was of this type.) The Old Colony tried some form of under-the-floor furnace in 1846.[37] Westlake pushed the suspended stove during the 1860s.

The chief advocate of the pendant heater was the Reading Railroad. Its partiality for this system was explained by an unusual circumstance. The road was double-tracked at such an early time that the clearances between the tracks, tunnels, and cuts were made extremely narrow.[38] Although adequate for their

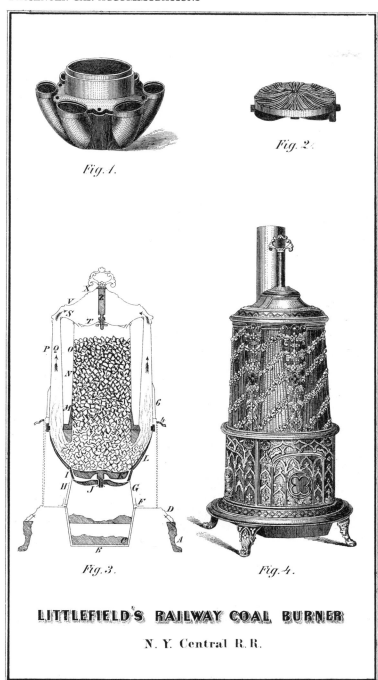

Figure 5.19   Littlefield's self-feeding car stove was patented in 1856. Coal-fired car stoves gradually replaced wood stoves after 1860. (New York State Railroad Report, 1856)

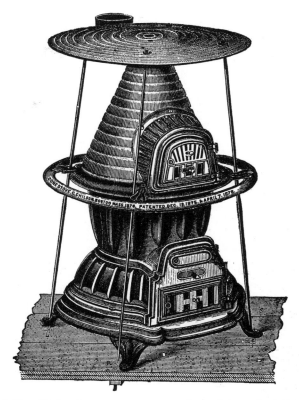

Figure 5.20   Chilson's car stove, patented in 1876 and 1878, featured a large cast-iron disc top that was intended to deflect the heat to the floor. Note the tie rods securely mounting the stove to the floor. (Poor's Manual of Railroads, 1875–1876)

day, they proved burdensome as larger and wider rolling stock came into favor. The cars passed so closely that passengers indiscreet enough to poke a head or hand out of the window were in danger of being hit. Enough damage claims resulted that wire grills were mounted over the windows. And now that the passengers were effectively caged inside the cars, the management felt compelled to do as much as possible to lessen the danger of a human barbecue. Gaslights and under-the-floor heaters were installed beginning around 1860.

The ascendancy of the pendant stove was assured on the Reading when John E. Wootten, the chief mechanical officer, took an interest in developing the idea. On September 29, 1868, Wootten was granted Patent No. 82,672 for a pendant-style heater. Since the general scheme was an ancient one, Wootten could claim little more in his patent application than the invention of a

reversible air vane, or scoop, mounted on the side of the furnace. Within ten years it was reported that all the Reading's cars would soon be outfitted with under-the-floor units, despite the fact that successive rebuilding of the line had corrected its clearance problems and the window screens might soon be phased out.[39] A better understanding of the Wootten stove, its radiator, and the stovepipe arrangement can be gained from Figure 5.23. These drawings appeared in the 1888 *Car Builders' Dictionary*. The major parts are designated by the letters A through H. A, B, and C are not shown, but the furnace is clearly represented. D is the cold-air supply pipe, E the hot-air flue, F the smoke pipe, G the hot-air delivery pipe, and H the coal box.

Soon after Wootten's retirement in 1886, the Reading's enchantment with the suspended heater evaporated. It was referred to as "old-fashioned," and though 400 cars were still so equipped, no new cars had been outfitted with it for some time.[40] The Wootten stoves were said to consume enormous quantities of fuel. Because they were hung out in the cold, much of the heat was lost simply in heating the car's underside rather than its interior. They required special attendants and fueling carts at every major stop. And of course, it was impossible to tend their fires while the train was in motion. The pendant idea was not entirely rejected in the industry, however; both W. C. Baker and George Westinghouse promoted a steam heater of this type.

Hot water was an attractive alternative to the methods just described. It is recognized as one of the best heating systems for small and intermediate-sized structures like railway cars. It distributes heat evenly from a central source and retains its warmth after the fire dies. It is relatively safe—certainly when compared with steam. A major drawback, however, is its high first cost: it requires a boiler, an expansion tank, piping, and radiators, which are much more costly than an equivalent hot-air system.

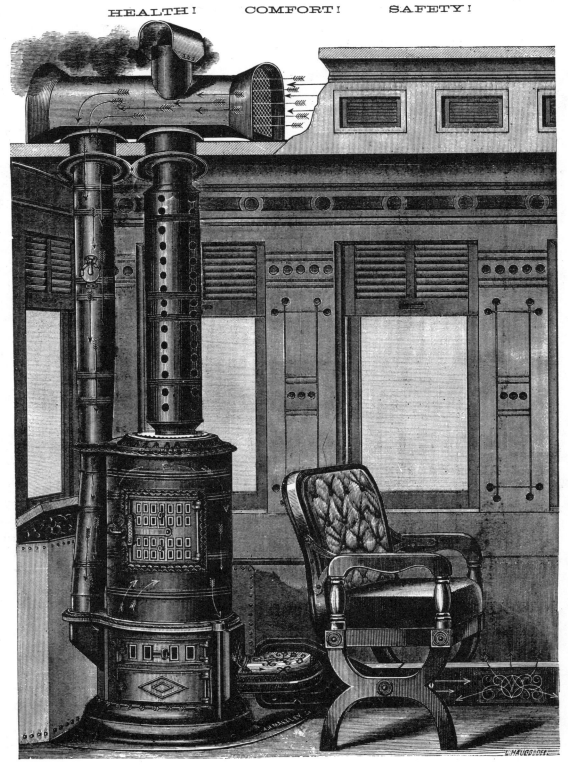

HEALTH!   COMFORT!   SAFETY!

Figure 5.21   James Spear manufactured a popular style of hot-air stove based on his patent of 1857. (National Car Builder, October 1881)

Curiously, hot-water heating for railroad cars seems to have received little attention until the middle 1860s. And even then it was pushed by a single man, William C. Baker (1828–1901) of Dexter, Maine. The Baker heater became the best-known, if not the most widely used, single make of car stove, and its inventor's name is one of the few that can be recognized by those with only a casual knowledge of railroad car history. Baker's interest in central heating began soon after he moved to New York City around 1850. Apparently he was drawn to the subject through a lecture delivered by Professor Benjamin Silliman, Jr., a popular scientific speaker and engineering gadfly of that day. In 1851

Baker helped Silliman to install the first steam-heating system in New York City.[41] Baker's interest gradually turned from heating buildings to heating railroad cars. Around 1865 he was ready to test the car heating system he had developed; during the next year he received the first of forty patents he would eventually acquire. The first Baker heater was demonstrated aboard the private car of the New Haven Railroad's president. It proved so successful that it was tried on some of the road's coaches. Then the New York–Boston Through Line adopted it for sleeping and parlor cars. In an advertisement of 1868 Baker listed seven railroads, including the Erie and the Michigan Southern, that were

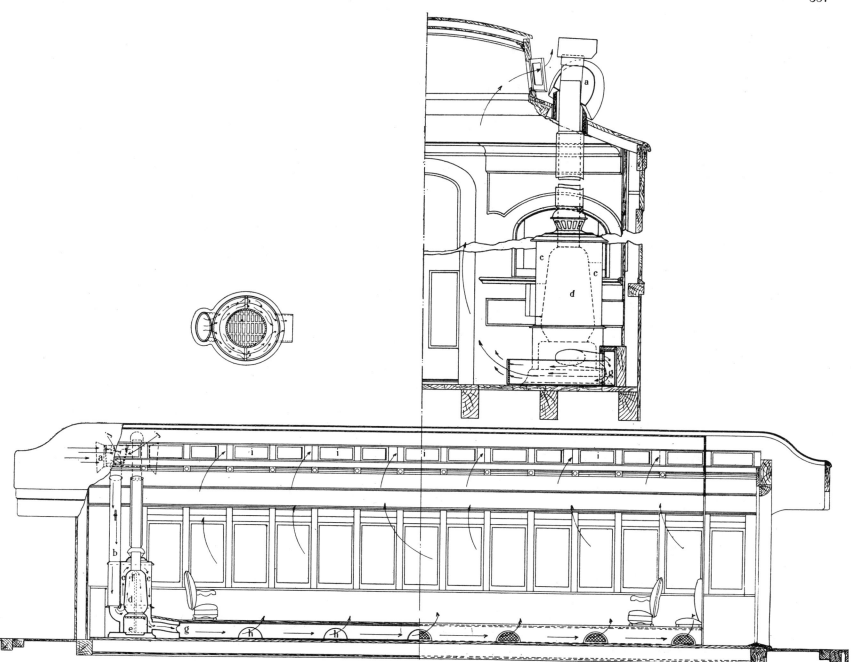

Figure 5.22   Spear's hot-air stove had floor-level registers and a roof-mounted air scoop.
(Journal of the Franklin Institute, 1897)

using his apparatus.[42] Within a few years the Baker heater was found on nearly every first-class car in this country. But few roads felt that they could afford to spend $500 each to install the heaters in their coaches when $2 to $10 coal stoves did an adequate job.[43]

Baker's invention was not an original contribution; in the main he was simply the first person to apply hot-water heating to railroad cars. However, his system was well arranged and executed. The source of heat was a small coal-burning stove shrouded with a sheet-metal jacket placed in one corner of the car (Figures 5.24 and 5.25). An iron pipe coil was mounted inside the stove.[44] The lower end of the coil passed out of the stove and looped its way around the floor under each seat, thus forming a continuous radiator. The pipe returned to the area of the heater and rose to terminate in the cistern, or expansion tank, mounted on the car roof. The upper end of the coil ended in the same tank. A filler plug and a safety valve were attached to the

expansion tank, which was a hollow, melon-shaped iron casting. The pipe used was 1- and ¼-inch, with 11 running feet allowed for each passenger. No pump was necessary; the natural currents set up by the fire at one end of the pipe were enough to circulate the water.

A basic problem was that the water would freeze in cold weather unless the fire was kept burning, but Baker thought of an ingenious solution. He devised a simple antifreeze by saturating the water with as much fine, dry salt as it would dissolve. The strong brine solution, which any mechanic could make, would not freeze until the temperature dropped to one degree above zero. In these extreme conditions, of course, the system would have to be drained or the fire kept going. Because the system was airtight, the salt did not attack the pipes—a fact that Baker claimed was proved by long years of service.

The salt water in turn caused a problem that Baker confessed had been difficult to work out. If the safety valve were to open,

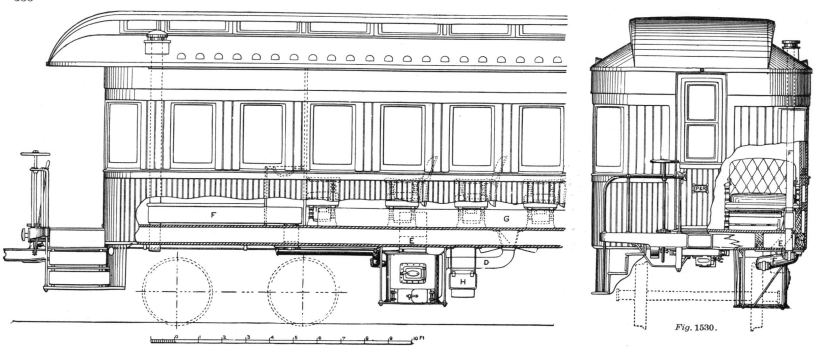

Fig. 1530.

Figure 5.23   Under-the-floor stoves were adopted by some lines to reduce the fire hazard. J. E. Wootten's patented design shown here was used on the Reading between 1870 and 1890. (Car Builders' Dictionary, 1888, figs. 1529–1532)

WARMING AND VENTILATING
Railroad Cars
BY HOT WATER.

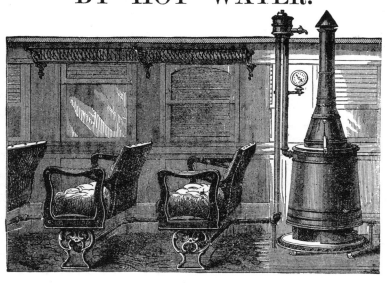

BAKER'S PATENT CAR WARMER.—One way of Applying it.

A very simple, safe and efficient plan for

Warming Railway Carriages !
—BY—
HOT WATER PIPES,

Which Radiates the Heat Directly at the Feet of Each Passenger without the Necessity of Going to the Stove to Get Warmed !

☞ All the finest Drawing-Room and Sleeping Cars in the United States have it, or are adopting it. Full descriptive Pamphlets furnished on application.

Baker, Smith & Co.,
Cor. Greene and Houston Sts., N. Y., and 127 Dearborn St., Chicago.

Figure 5.24   Baker's hot-water car heater was superior to ordinary car stoves because it could distribute heat more uniformly through the car. (Railroad Gazette, August 5, 1871)

salt crystals would form on the seat, preventing it from closing tightly once the pressure was relieved. His solution was a rubber-ball safety valve, which was used for many years until even Baker came to admit that it was a menace, for reasons that will be explained later. The valve was set for 150 pounds of pressure. Care was needed to fill the system and expel all the air. During this operation the system was subjected to a 400-pound hydrostatic test to expel the air and test all joints for absolute tightness. Theoretically, once this was done no more water need ever be added, but in practice an occasional topping off was found necessary. A pressure gauge attached above the heater, on one end of the coil, indicated a blockage in the system if it registered more than a trifling pressure. Once heated up, a dull fire could normally maintain a comfortable temperature. Two pecks of coal were sufficient for a twenty-four-hour period.

The Baker heater offered the great advantage of a steady, even warming of the entire car. It took up less floor space; one Baker could easily do the job of two stoves. It saved fuel. Unlike the Spear and most other hot-air systems, its operation was not dependent on the motion of the train. The brakeman, since he was not forever having to attend the heater, was free to "answer the call of the whistle." In a serious smashup, Baker cheerfully suggested the pipes would rupture and put out the fire. Apparently the scalding brine was not supposed to touch the passengers.

One of the Baker heater's failings has already been mentioned: the first cost discouraged its use for anything but palace cars. Also, it did not eliminate the fire peril; as long as live coals were carried inside a car, that danger was present. And it was said to be a noisy system, given to fits of pounding.[45] Far worse, however, was the danger of explosion. Like any closed system with a heat source, it could burst, with terrible results. During the 1890s several explosions were blamed on Baker heaters.[46] One of the worst, which occurred on the Santa Fe, wrecked one end of a smoking car and injured five passengers, two of them seriously. Baker blamed the safety valves; he said that trainmen would screw down the rubber ball to stop them from leaking. To solve

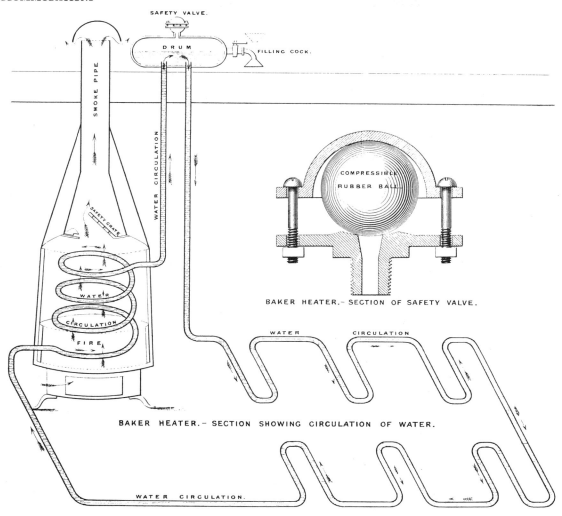

Figure 5.25 A schematic drawing of Baker's hot-water car-heating system. (Institution of Civil Engineers, *vol. 53, 1878, plate 4*)

the problem he devised a new throwaway style of cast-iron valve.

The popularity of Baker's apparatus encouraged others to produce variations on the same theme. A Cincinnati inventor, J. Q. C. Searle, introduced a hot-water system in 1870 that broke Baker's monopoly and was eventually manufactured by the Union Brass Company of Chicago. It was good enough for use on some of Pullman's deluxe cars. A second, less popular hot-water system was patented by J. Johnson in 1881. The Pennsylvania Railroad tested its own hot-water system in 1879.[47] It differed from the Baker system in that it had a single boiler, placed in a baggage car, for the entire train. A steam pump circulated the water. The device did not succeed, but it was one of the first instances of continuous train heating in this country.

In 1882 Baker, anxious to outdistance the competition, introduced an entirely new car heater: a pendant, low-pressure steam cooker, to be produced by a new company formed in Pittsburgh with the help of George Westinghouse.[48] The cup-shaped boiler held 8 gallons of water and measured roughly 2 feet in diameter and 1 foot deep. Two inclined chutes to the firebox from either side of the car served as hoppers which automatically fed in coal through a shaking action induced by the motion of the car. The hoppers carried 175 pounds of coal—enough for twenty-two hours of fire. Within two years the business was reorganized as the Standard Car Heating and Ventilating Company, with Westinghouse as president.

In a few years Baker was back in business for himself, again as a champion of hot water. Steadily growing car sizes called for larger heaters, and Baker increased the coil length from 16 to 30 linear feet.[49] The profile of his new model was squatter, and to lessen the danger of the stove's shattering in the event of an accident, the iron castings were replaced by pressed steel or malleable iron. A safe-like wrought-iron box offered further security against fire (Figure 5.26). This stove, if its fire was properly laid with 75 pounds of hard coal, could burn without attention for twenty-four hours. Baker introduced a heater of even greater capacity in 1892. It featured two coils, both made of tightly wound, tapered pipes for a maximum heating surface.[50]

In 1890 some 7,000 cars (or approximately one-quarter of all passenger cars) were heated with hot water; the majority of these were most likely of Baker's manufacture.[51] An 1891 Baker catalog in the Library of Congress claims that 30,000 Baker heaters had been sold since 1868. In 1893 Baker's business was so prosperous that he built a new plant in Hoboken, New Jersey, but after 1900 all forms of heating were superseded by steam for main-line passenger cars. Baker heaters gradually disappeared, although they continued in limited use for cars that required some independent form of heating. This was especially true for private cars that were set out at isolated stations which had no steam. Some self-propelled rail cars also used Baker-style heaters into the 1920s.

All the early forms of car heating depended on a live fire, and this meant that the danger of incineration was an everyday hazard. But the fire peril was an accepted fact of life at home or

lished. "Not long since," one correspondent said, "the writer was riding with a friend on a train which passed through a short tunnel. The register of the heater was in such a position as to allow the fire to shine out of it through two apertures about two inches apart. When the car became dark the fire glared out of these openings like the eyes of some demon. This caused the friend to remark: 'There is death, staring us in the face'."[53]

The combination of stoves and wooden cars was unquestionably dangerous. Tinder-dry wood heavily coated with varnish, oilcloth head linings, upholstered seats, lamp reservoirs sloshing with kerosene—it was a rich flammable mixture waiting only for the stove to topple over and spew out its cherry-red coals. As if there were not enough kindling aboard already, car builders discovered that wood shavings were a cheap insulating material and blithely packed the spaces between the upper and lower floorboards with this highly flammable material. In time the wood shavings were replaced by mineral wool, a material that was adopted by the Pennsylvania Railroad in 1882.[54]

Other measures were taken to lessen the likelihood of fire. Stoves were, of course, set on metal sheets, often zinc, to protect the floor from sparks. The doubtful precaution of pendant heaters has already been mentioned. Stoves were bolted to the floor sills in the hope that they would not fly loose and break open during a wreck. The B & O locked the stove doors to prevent the fire from spilling out in case of an upset. Each brakeman had to carry a key to open them. Often a water box was built into the stove's base so that if the stove tipped over, the water would spill and theoretically extinguish the fire (Figure 5.27). In 1868 Ruttan's hot-air stove was reported as having a water bottom containing five pails of water.[55] Many other stove manufacturers, such as Winslow and Bissell, offered similar features if desired. How effective these tanks were remains a question. If they were widely used, which is uncertain, they do not appear to have solved the fire problem.

More elaborate extinguishers were also developed. In 1887 the Pittsburgh, Chicago, and St. Louis tested a water sprinkler that would be automatically set off by a derailment.[56] Even if there was no fire, the passengers would receive a hosing. The Dyer Railway Automatic Fire Extinguishing Company of Chicago advertised a device that promised to kill any stove or lamp fire. Apparently it was a simple chemical (soda) extinguisher with tubes running to every open flame in the car. A more practical short-term solution was to install a better class of stove: cheap, brittle, cast-iron models were easily shattered, while as early as 1875 it was noted that wrought-iron stoves could withstand all but the most disastrous shocks.[57] Most railroads did not bother to make this simple substitution for car safety, however, and *The Railroad Gazette* announced that they had foolishly ignored the public interest. "And now the day of stove reform is past, and the people demand a root and branch revolution in [the] methods of heating. Who can blame them."[58]

Even if the railroads were not ardently interested in passenger safety, dozens of valuable cars were destroyed yearly by fire. The most publicized losses were the result of wrecks. Trains that were

abroad, since all heating and lighting depended on an open flame. Houses, barns, sheds, steamships, and churches regularly burned to the ground, and there were few communities that had not experienced at least one courthouse fire. Consequently there was no general outcry from critics of the railroads about the dangers of stove heating until rather late in the century. Beginning in the early 1880s, however, a storm of indignation suddenly seemed to explode. True, there had been a number of disasters in which passengers were burned in the wreckage, and some were surely burned alive. Although this was not a new development in railway travel, and although a minute number of people died by fire relative to the millions of passengers carried, it was an unspeakably gruesome prospect. Perhaps public wrath was stirred by the culmination of years of railroad accidents. Certainly the press also played a major role in the antistove crusade. The illustrated national weeklies reported all gory details of railway calamities. *Harper's* editorialized in a number of cartoons, the most stirring of which, entitled "The Modern Altar of Sacrifice," pictured a glowing car stove standing in the midst of a flaming wreck. Draped on top of the stove was a prostrate female figure. The wooden railway car and its demon heater were likened to the Prophet Elijah's fiery chariot.

The railroad trade press joined the cause. In June 1882 the *American Railroad Journal* pointed out that with the coming of warm weather, "the season of railroad barbecues is over, the annual sacrifice of baked and roasted victims has been offered up." Another journal described stove-heated cars as "perambulating crematories."[52] Dramatic letters to the editor were pub-

Figure 5.27 Water-bottom stoves were another way to lessen the fire peril from coal stoves. Bissell's stove was a product of the 1870s. (Poor's Manual of Railroads, 1879)

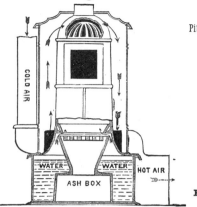

BISSELL SAFETY CAR STOVE,
WITH WATER BASE.
Fire Extinguished Instantly in case of Accident.

In use on following Railroads:
Pittsburgh, Ft. Wayne & Chicago Railway,
Pittsburgh, Cincinnati & St. Louis Railway,
Cleveland & Pittsburgh R. R.,
Pittsburgh & Lake Erie R. R.,
Evansville & Terre Haute R. R.
&c., &c.

BISSELL & CO.,
Office, 235 Liberty Street and 242 Penn Avenue,
PITTSBURGH, PA.
Works, Allegheny City, Pa.
[Sole Licensees under the Winslow Patents.]
Full Descriptive Circulars sent on Application.

not badly damaged by the accident itself could be swept by fire, and wrecks had a way of happening in isolated places where no fire equipment was handy. Cars standing idle between runs were also known to go up in smoke, frequently because of an unattended stove. Between 1887 and 1890, ninety passenger cars were lost by fire.[59] Claims against the railroads for lost baggage, express shipments, and mail were a secondary expense. In these same fires twenty passengers died, and while not every death can be attributed exclusively to fire, the claims that the roads normally paid amounted to $5,000 each. Hence the liability loss itself was over $100,000, much of which could have been saved by eliminating the stove menace. After generations of stubborn resistance, even the railroads were beginning to see the need for some better way to heat their cars.

STEAM HEATING

The answer to the fire peril was steam heating. It was a clean, wonderfully efficient method that had become standard in factories, hospitals, apartment buildings, and other large institutions following its introduction in the 1780s by James Watt. It was superior even to hot water as a medium of heat transfer. Like hot-water heating, steam also required elaborate plumbing that made it an expensive installation. The initial investment was justified by the uniform heating it provided and by the savings in fuel it made possible.

In addition to its inherent advantages, steam appeared to be a natural for railway service. The locomotive at the head of every train was a ready source of steam; the hot vapor had only to be piped back to the cars. No stoves, hence no fire in any car of the train, put an end to the danger of fire. The logic of the plan must have been obvious to the earliest car builders, yet nothing can be found to show that any form of train-board steam heating was tried before 1843. Around then Jacob Perkins (1766–1849), who was described as an American mechanical genius with a head that fairly rattled with ideas, settled in England and pursued various projects to employ high-pressure steam.[60] A sickly offspring of his studies was a steam heater that Perkins fastened beneath the Queen's private carriage on the London and Birmingham Railway. The oil-lamp–fired boiler fed steam to pipes running between the double floor. Like so many of Perkins's efforts, it was prophetic of what would one day be an important innovation.

The first reported attempt at steam heating in the United States was on the Old Colony in 1851.[61] A few years later a test was made on the New York Central.[62] Although details are lacking, it can be assumed that nothing practical was developed, because the next news in the field came from Europe. In 1860 several German railways began serious tests of steam car heating. Within eight years another four German lines were using steam heat. In 1865 the Prussian railways were trying a low-pressure steam system fed by a separate boiler placed in the baggage car.[63] Sweden began to experiment in the early 1870s, and the

Alsace-Lorraine railways attempted to heat their cars with steam drawn from the locomotive in 1877.[64]

News of these experiments seem to have inspired fresh interest in America. A system patented by H. R. Robbins in 1868, which involved a small vertical boiler placed in the baggage car, was tested on the B & O. In 1875 a low-pressure scheme invented by William M. Fuller was tested on an unnamed railroad in the New York City area.[65] Separate stoves fitted under the floor supplied the steam. (It will be recalled that Baker revived the same idea in 1882.) A few years after Fuller, the Camden and Atlantic Railroad tried out a design patented by Wood, Scull, and Snyder.[66] All these pioneering efforts were failures, owing generally to the resistance of the railroads, defects in the individual systems, and the lack of prestige of the inventors themselves.

In contrast, circumstances were favorable for William C. Baker, who unveiled a steam heating system on the New York Elevated late in March 1878.[67] Steam was drawn from the locomotive and distributed through the cars by 1¼-inch pipes. Rubber hoses made a flexible connection between the cars. Leakage at the hose connections eliminated the condensate without the need of elaborate traps. While it is generally thought that steam heating was not adopted on the Elevated until 1883, *Locomotive Engineering* claimed that the El's first 700 cars were fitted with Baker's system. This inventor's success can be ascribed to the utility of his design, the receptiveness of the Elevated, and his own reputation as a leader in his field. The Elevated's desperate need for a compact, efficient heating system may indeed have been the chief reason that Baker was given his opportunity. The cars were so crowded that there was little space for customers, much less stoves, and the crews were much too busy to tend fires. Stops were frequent—the doors were open as often as they were closed—and only the most efficient system could keep the cars even tolerably warm. Baker's direct use of the locomotive rather than an auxiliary boiler as the heat source was especially significant, because it prefigured the future of steam heating in America.

Main-line managers discounted Baker's system as a novelty suitable only for rapid transit. They disparaged the devices of other inventors as foolish gadgetry, and one official stated that any more tests would prove "a waste of money simply to gratify the whim of a lot of cranks."[68] Railroad managements maintained that steam heating would tax the locomotive's strength and delay schedules; that separate boilers were too expensive; that the cars would be too hot. They argued that the fire hazard

BION BRADBURY, President.   D. D. SEWALL, Vice-President and General Manager.   C. B. STROUT, Sec. and Treas.

THE SEWALL SAFETY CAR HEATING COMPANY,

No burning or scalding.

No coal dust, ashes or gases.

No freezing.

No complicated parts.

No difficult manipulations.

General view of system, showing pipes, patent valve-coupling, auxiliary boiler, etc.

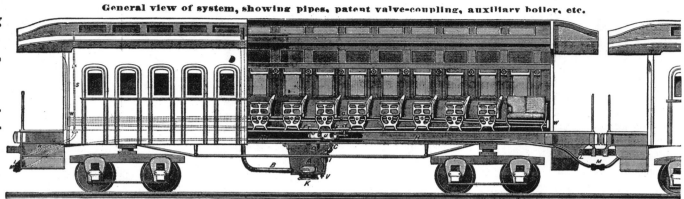

No fire in case of accident.

No waste or leakage of steam at high pressure

Saving in labor, fuel and seating capacity.

Supplies heat to cars side-tracked or separated from the engine

Coupling in like halves uncoupled.

Top view of coupling being coupled.

Controls steam from any source perfectly with entire safety, and the greatest economy.

A single valve in each car, and heat in each managed independently of any and all others!

Coupling as closed between cars.

Has been tested fully in temperatures from 30° below to 50° above zero F. Has continuously and perfectly heated five seven and twelve-car trains, and shown capacity for heating longer trains under test of railway experts.

IN ALL TESTS OF LONG TRAINS ENGINEMEN DECLARE DRAUGHT ON THEIR ENGINES TO BE IMPERCEPTIBLE.

This system has given PERFECT SATISFACTION throughout the winter in DAILY REGULAR USE ON MAINE CENTRAL R. R. Is being supplied to Michigan Central and many New England roads. Inquiry or inspection will satisfy the most skeptical that

IN SAFETY, SIMPLICITY, ECONOMY AND COMPLETENESS IT IS UNRIVALED.

Inspection of system on regular trains of Maine Central R. R. is solicited. Illustrated description. Estimates of equipment, etc., furnished on application to

SEWALL SAFETY CAR HEATING CO. PORTLAND, ME

Figure 5.28 *Steam heating was introduced in the 1880s to eliminate car heating stoves.*

could never be eliminated so long as there was a single open flame on the train, including cigars, lamps, and cooking ranges. And what of the locomotive itself? Had not the embers from the engine's firebox caused the Ashtabula holocaust?[69] Moreover, the wreck on the Maine Central had fried two mail clerks in a steam-heated car because the cooking stove had toppled over, setting the wreckage ablaze.[70] The bogeyman of scalding steam was another convenient argument against the new system; the railroad establishment claimed that the choice was between being roasted by the stove or parboiled by the steam pipe. In addition, the rear cars of steam-heated trains were so chilly that the danger of pneumonia was greater than the danger of fire. Should these statements fail to convince the traveler that steam heating was visionary nonsense, a final stopper was thrown out: "The locomotive is meant to haul the train, not heat it."[71]

The inventors persisted in the face of official hostility. Not all railway managers were unsympathetic; some felt that steam heating must be accepted in time and actually encouraged experi-

mentation. The problem was to perfect a workable system, and this could be done only in the field. The best laboratory was a train in everyday service, operated by everyday trainmen and patronized by everyday passengers. In the early 1880s half a dozen railroads cooperated in holding road tests that led to the real beginning of steam car heating in this country.

The proving ground was New England. The participants were the smaller trunk lines, such as the Connecticut River Railroad, which permitted James Emerson to outfit a train in 1881.[72] Emerson's separate boiler overheated, endangering the baggage car, so that he next tried taking steam from the locomotive with separate heaters under the cars to assist in warming the train. Since Emerson, like many others, believed that a separate boiler was preferable as the source of steam, he felt that an iron head-end car might solve the problem. Emerson's system worked well enough to remain in use for at least six years.

Lieutenant James W. Graydon of the U.S Navy was another proponent of the auxiliary baggage-car boiler. In the winter of

Figure 5.29 The Martin steam heating system, as advertised in the 1886 Railway Review.

1881–1882 he tested apparatus on the Troy and Boston Railroad with fair success, but the interchange of cars with another line that was not equipped for the new form of heating ended Graydon's work after only a single season. At the same time the Maine Central allowed D. D. and J. H. Sewall of Portland, Maine, to try out their patented design. A notable feature was their snap coupling head for the steam hoses (Figure 5.28). The Sewall system was eventually adopted by the Maine Central and received some use on other New England lines and the Michigan Central.

William Martin of Dunkirk, New York, was even more fortunate with his design than the Sewalls and has been credited with the first commercial success in this specialty.[73] His initial patent was granted in 1882, and following a few small test installations Martin broke into the big time when the Boston and Albany adopted his plan (Figure 5.29). In 1884–1885 Martin outfitted 200 of the road's cars. By the spring of 1887 three railroads were using his equipment; ten months later twenty-one roads had signed up.[74] By the summer of 1889 Martin apparatus had been installed in over 2,600 cars.[75]

Martin had many rivals: Wilder, McElroy, Westinghouse, and others. None, however, posed a major threat except Edward E. Gold (1847–1931) of New York. He was the nephew of Stephen J. Gold, an expert in the heating of large buildings.[76] Gold had been raised in the business and came to the problem with great experience in the intricacies of steam heating. He was also a friend of Russell Sage, a Wall Street raider who bought and sold railroads together with his ally, Jay Gould. Sage encouraged Gold to pursue his interest in car heating; there might be big money in it one day if the government forced its adoption. Gold obtained a few patents (the first in 1881) while Sage, a major shareholder in the New York Elevated, smoothed the way for a test. In 1883 Gold equipped the first cars with his heaters, and the El's general manager displayed an unusual enthusiasm for the system.[77] Within five years all Elevated stock was remodeled with Gold apparatus. Because the line's small engines had to make the frequent starts necessary to rapid-transit operations, they were not capable of supplying steam for the cars. Direct steam heating was therefore not practical. In the compromise worked out by Gold, steam was sent back to hot-water heaters in the cars after the train was underway and the need for maximum steam was temporarily at an end (Figure 5.30). Each car had four cylindrical, floor-mounted radiators.[78] These double-tube reservoirs measured 4 inches in diameter by 10 feet, 10 inches long. The close-fitting inner pipe was filled seven-eighths full with salt water, which was heated by a steam jacket formed by the outer tube. The radiators could retain enough heat to warm the car for two hours. Gold calculated that 32 pounds of coal would heat an eight-car train for one hour.

The dramatic success of the Gold system on the nation's largest single passenger-carrying railroad prompted main-line managers to consider its adoption. The Elevated had grown into a major enterprise since the time of Baker's steam trials, and it was now managed by experienced main-line railroad executives. If they

WM. MARTIN, President.   FRANK E. SHAW, Vice President.   C. A. CLUTE, Secy. & Treas.

THE MARTIN
ANTI-FIRE CAR HEATER CO.,
Dunkirk, N.Y.

Martin's Anti-Fire Car Heater.

Manufacture

THE ONLY SAFE CAR HEATER MADE.

No Fires in the Cars. No one Roasted Alive in Case of Wreck. No More Rio Holocausts.

Traveling rendered comparatively safe. The Martin Heater uses steam from the Locomotive Has been tested two Winters in low temperatures. Cars always warm and comfortable.

In use on the "Bee Line" (C., C., C. & I. R. R.) also the Boston & Albany, Dunkirk, Allegheny Valley & Pittsburgh, New York Central & Hudson River, Lake Shore & Michigan Southern and Chicago, Milwaukee & St. Paul Railroads.

Nothing but Good Results Wherever Used.

No leakage. No trouble with coupligs. Only takes four pounds pressure of Steam from the locomotive. THE ONLY SAFE HEATER. ☞Send for Circulars and Testimonials.

were satisfied with the Gold plan, it must be good. The Boston and Providence, the Long Island, and the Staten Island quickly fell into line. Next came the main-line roads like the B & O. By 1906 Gold boasted of being the largest car heating supplier in the nation, with apparatus on 40,000 cars.[79] By this time he was offering all types of heating apparatus—direct-steam, vapor, and electric—but his principal system was similar to that perfected for the Elevated. In addition to hot-water pipe radiators, he had devised a simple plan to convert existing Baker-type systems to steam. He replaced the original coil with a spiral pipe that had a smaller-diameter steam line inside. Thus the existing piping and heater could stay in place, while steam rather than a stove fire heated the water. In an emergency the fire could always be lighted.

By and large, government action to speed the cause of steam heating was the familiar story of too little, too late. Few laws

NO MORE PASSENGERS BURNED ALIVE.
THE DANGEROUS STOVE SUPERSEDED.
THE EDWARD E. GOLD SYSTEM.

Heating and Ventilating Steam Railway Cars, Horse Railway Cars, Cable Cars, Stations, Steamboats, Ferryboats, etc.   Safety, Economy, Comfort, Convenience, Uniformity, Pure Air for Breathing, and Increased Seating Capacity Secured by the Edward E. Gold System of Heating Cars without Stoves or Fire.

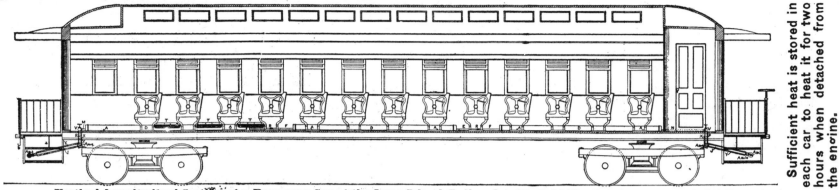

Vertical Longitudinal Section of a Passenger Car of the Long Island Railroad, with Gold Heating Apparatus.   (Scale 5-32″ = 1 foot.)

ADVANTAGES. —Perfectly safe.   Entirely efficient.   Takes up no room available for passengers.   Utilizes waste steam from locomotive for heating.   No fire is carried by cars.   Always ready.   Easily managed.   Very durable.   Floors of cars heated.   Uniform heat thoroughly diffused through the vehicle.   Excels anything yet devised for car heating.   Send for Illustrated Catalogue to

EDWARD E. GOLD & CO., N. E. Corner Frankfort and Cliff Streets, New York.
Proof of Value: 800 cars on Manhattan Elevated Railway are heated by this system.   Adopted also by Long Island R. R.; Prov., W. & B. R. R.; Suburban Rapid Transit R. R.; Staten Island R. R.; Hoboken Cable Cars and Ferryboats; Interstate Rap. Trans. Co. (Wyandotte, Kansas).

Sufficient heat is stored in each car to heat it for two hours when detached from the engine.

Figure 5.30   The Gold steam heating apparatus, introduced in 1883, was one of the first commercially successful systems in America. (Railroad Gazette, *March 4, 1887*)

were enacted until the late 1880s, by which time the railroads were voluntarily, albeit slowly, installing the necessary apparatus. Actually the legislators had been talking about the stove menace for decades before the first bill was passed. A railway safety bill was introduced in Congress in 1868 that specifically called for safer car heating and the conversion to iron cars.[80] More bills were introduced, but none passed during the next years. The first measure enacted that had any effect at all was a rider on the 1881 Post Office appropriation specifying a reduction of 10 percent in payments to any railroad that did not provide mail cars with safe lighting and heating. Apparently the language and enforcement of the bill was too weak to have any major influence, but it did give the railroads notice that Congress was concerned about the problem.

Ohio was one of the first state legislatures to act on the matter of railroad passenger safety. It passed a law in May 1869 specifying that all car heaters be self-extinguishing in the event of a derailment.[81] The act was apparently ineffective, however, because Ohio found it necessary to pass an antistove law in 1892.[82]

In 1887 a number of other states began to hold hearings on antistove bills. The fire-peril hysteria was at its zenith. Several workable, if not entirely perfect, steam systems were on the market, and antistove groups hoped that a little pressure from the state capitols might inspire a more expeditious conversion. Albany appears to have been the most aggressive in pushing through a steam law. The New York State legislature announced in August 1887 that stoves must be removed from trains by May 1, 1888.[83] An extension was granted until November 1888, but as late as 1894 the New Haven was still operating stove-heated coaches in New York.[84] In fairness it must be mentioned that the state railroad commissions took the case through the courts and a $7,000 fine was upheld. Other state antistove laws were passed, some as late as Maryland's in 1894, but in fact the railroads appear to have converted at their own pace.

There was some justification for the measured rate of the conversion. The industry was already obliged to finance the installation of air brakes and automatic couplers on several million freight cars in order to comply with the recently enacted Safety Appliance Act. Steam apparatus cost about $200 a car, so that 6.8 million dollars would be needed to change the whole fleet. To this must be added the cost of boilers at terminals to warm idle cars, as well as the investment lost by premature scrapping of the stoves. Yet one state legislator could not understand the fuss about expense.[85] Didn't stoves consume $35 to $50 worth of coal a season? Wouldn't this all be saved by using "free" steam from the locomotive? From his vantage point, steam heating would quickly pay for itself.

The progress of steam heating is difficult to outline precisely. Few overall figures appear to exist. Most early statistics are from the Master Car Builders reports, and the roads which answered the questionnaires on which they were based usually represented only a portion of the total passenger car fleet. The fragmentary information available reveals the following facts. In 1887, 150 railroads which owned one-third of the nation's passenger fleet (representing about 8,500 cars) reported that 392 were steam-heated.[86] The figure climbed to 800 in the following year, but only 24 roads reported. Individual manufacturers, such as Martin and Gold, claimed to have installed far more systems than that. In 1889 with only 45 railroads returning the questionnaire, they reported 1,572 steam-heated cars in service. In 1891 the first industry-wide total was offered.[87] According to the *Railroad Gazette*, 9,017 cars, or 26.7 percent of the total, were now steam-heated.

A better idea of the transition to steam can be gained from the reports of individual railroads. In 1893 the Pennsylvania had 41 percent of the cars on its Lines East under steam, and another 23 percent were being fitted out.[88] Its western lines reported that 56 percent of their cars were already converted. Early in 1895 the B & O had fitted all its 700 cars with steam, the majority of them with apparatus furnished by Gold.[89] It is probably safe to assume that by the turn of the century all first-class cars were heated by direct steam or by steam-actuated Baker heaters. However, some commuter and branch-line cars never gave up stove heating, and some railroads maintained stove-heated cars

Figure 5.31   Gibbs's steam hose coupling was marketed by the Safety Car Heating and Lighting Company. By 1892 it was used on more than 13,000 cars. (Railway Review, *December 5, 1892*)

for secondary or excursion trains years after the general conversion was in effect. This policy resulted in at least one major tragedy long after the fire peril was no longer an issue. On December 5, 1921, a head-end collision occurred on the Reading in a deep cut near Philadelphia.[90] The train, composed of aging wooden coaches, piled up and caught fire. Fire-fighting equipment could not reach the blaze because of the location, and twenty-seven people died. Most of the fatalities were blamed on the fire; once again the car stove had gathered in new victims.

Why weren't the railroads more enthusiastic about adopting steam in the middle 1880s? Some reasons have already been given, but a number of less obvious technical and operational problems were also involved. Temperature regulation was the major difficulty, and overheating was the chief complaint. The head-end cars received steam fresh from the boiler, and even if the pressure was reduced to 5 pounds or less, the temperature was well over 200 degrees. In mild weather the front cars were insufferably hot, like a fireless cooker. But because of radiation, steam leaks, and the insistence that train steam-line pressures be kept as low as possible for safety reasons, live pressure dropped to nothing by the time it reached the rear cars. No pressure, no heat. Automatic regulation could at least partially solve the problem, but the railroads complained about the expensive controls, so complicated that no ordinary trainman could learn to operate them in a lifetime.[91] Anyhow, the jarring of the trains upset most of the thermostats that were tried.[92] As early as 1888 the Milwaukee Road tested electric-pneumatic controls designed by George Gibbs. They had some success, but not enough to win many converts.

Traps were one automatic feature that could not be avoided. There had to be some way to discharge the steam once it condensed, or the system would fill with water. For maximum heat economy, the traps should be set to flush only after the water had cooled to, say, 120 degrees. But because traps froze with maddening regularity, it was common to adjust them well above this temperature. Once frozen, the system was blocked and all heating stopped at the point of blockage. Premature flushing made the direct steam system very uneconomic. In a test conducted in 1895, the difference between the costs of stove and steam heating were negligible, when theoretically steam heat should have been substantially cheaper.[93] Some lines attempted to avoid traps altogether, eliminating the condensate through leakage at the hose couplings. This was not only sloppy but self-defeating, because steam escaped as well as water.

The flexible connection between the cars was a fruitful source of mischief. Martin advocated a flexible pipe connection articulated through ball and sliding joints. Most heating suppliers favored rubber hoses as a cheaper method, and during the first half century of steam heating this was the standard connection. Manufacturing a durable rubber hose was not easy, however. Some lines received as little as two months' service from their hoses. Wire reinforcing actually shortened hose life, since it acted as a heating element. In 1894 the Peerless Rubber Company produced a durable hose from the finest Para rubber which was

guaranteed for at least one year.[94] In 1912 the Master Car Builders Association voted to issue very detailed specifications for steam hose manufacture in an effort to assure longer life.

The diversity in hose coupling heads was also a continuing problem. Each manufacturer had his own design, though all followed the basic air brake hose coupling devised years before by Westinghouse. This snap-lock coupling was made fast by a quarter turn and would automatically pull apart if the cars separated. Some modest advances in standardization were reached by 1892, when thirty-two railroads were said to be using the hose head patented in 1887 by George Gibbs (Figure 5.31).[95] These roads operated 13,301 cars, or about one-third of the entire fleet. But as late as 1912 the Master Car Builders were still struggling with the problem of nonstandard hose couplers.[96]

Steam heating placed a considerable drain on the locomotive boiler. In frigid weather the engine needed all the steam it could produce just to move the train. Snow and ice on the tracks, as well as journal oil that became stiff as grease, all impeded the train's progress. In 1887 it was calculated that a fourteen-car train consumed 22.6 horsepower of steam for heating, which was equal to 4.5 percent of the total engine power.[97] Computations made some years later showed that 100 pounds of coal an hour were needed to heat a ten-car train by steam.[98] Any way it was figured, steam heating robbed the locomotive of fuel and power. Larger locomotives were a necessity if schedules were to be maintained.

A possible solution to the steam-drain problem was an auxiliary head-end boiler. Many of the earliest steam heating promoters advocated this scheme, but few were able to sell it. Only while George Gibbs was on the Milwaukee Road's engineering staff did a heating engineer have any success in promoting the head-end system. He had initially favored drawing steam from the locomotive, but by 1890 he had come to believe that there should be an independent source of steam. On his recommendation the road built two 35-foot-long light and heat tenders— short, 32½-ton baggage cars that were encased with ¼-inch iron plate as a fire precaution.[99] A large horizontal boiler justified, because it supplied both the engine generator for lighting and the steam for heating needs. The locomotive thus entirely relieved of auxiliary requirements. The plan worked well, especially in the bitter cold of the northwest few other roads were ready to adopt Gibbs's scheme revived many years later.

A final difficulty in the adoption of steam transition, particularly for cars involved For Pullman this was a major obstacle changing cars were equipped for steam

SPECIALS.

		CAT. NO.
24	2 in. x ⅜ in. Lag Screws	179
2	1 in. Angle Valves	603a
2	1 in. Drain "	603b
1	1½ in. x 1 in. Cross	638
9	feet Covering for 1 in. Pipe	680
2	" " 1 in. Ells	681
2	" " 1 in. Tees	682
2	2 in. x 1 in. Tees (eccentric)	708
4	2 in. Return Bends	723a
1	Directions for Management (framed)	741a
24	Pipe Straps	749
2	2 in. x 1 in. Ells (eccentric left)	752
2	2 in. x 1 in. " (" right)	753
2	Automatic Traps	833
30	Pressed Steel Pipe Shields	

DIRECT STEAM HEATING SYSTEM

FOR

PASSENGER CARS

THE SAFETY CAR HEATING & LIGHTING CO.

160 BROADWAY, NEW YORK, N. Y.

SCALE

DRAWN BY 503 — CORRECT G.G.A. — APPROVED R.M. Dixon

JUNE 5TH 1900 — ENG.

TRAIN-PIPE SPECIALS

		CAT. NO.
60	feet Covering for 1½ in. Pipe	677
1	" " 1½ in. Cross	683
2	End Train-Pipe Valves	776
2	Extension Handles	777

PIPE AND FITTINGS (other than for Train-Pipe)

2	1 in. Ells	616
4	1 in. R. and L. Ells	616a
2	1 in. Tees	621
25	feet 1 in. Standard Pipe	625
200	" 2 in. " "	685
4	2 in. R. and L. Couplings	732

PIPE AND FITTINGS (Train-Pipe)

2	1½ in. R. and L. Couplings	618
60	feet 1½ in. Standard Pipe	628

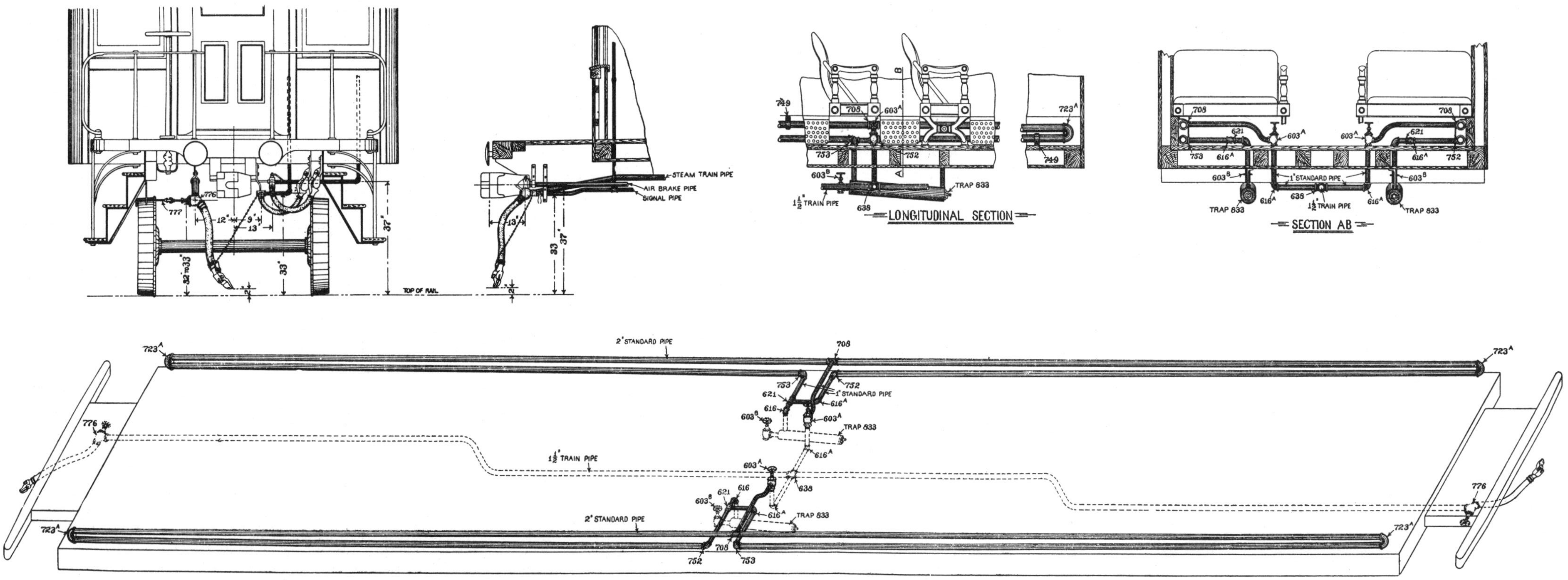

Figure 5.32 This direct steam heating system of 1901 used steam, at a reduced pressure, directly from the locomotive's boiler. (Safety Car Heating and Lighting Company catalog, 1901)

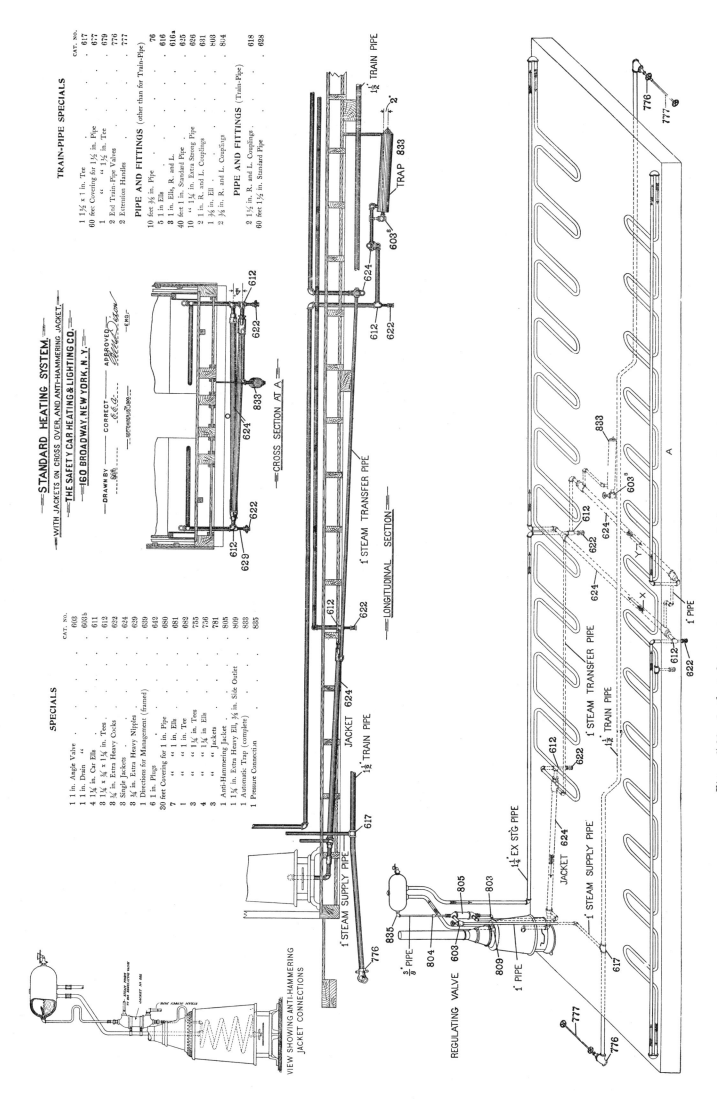

Figure 5.33 In the indirect system, steam from the locomotive heated water in the car heater. The water was then circulated through the internal heating system of the car. (Safety Car Heating and Lighting Company catalog, 1901)

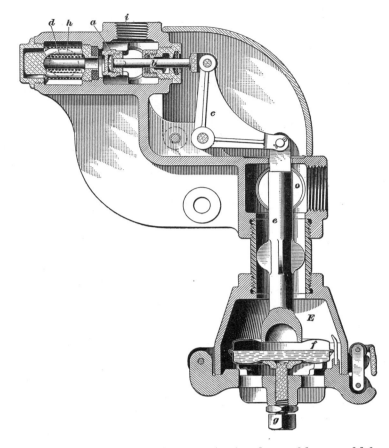

Figure 5.34   After 1910, vapor heating replaced both the direct and indirect steam systems shown in Figures 5.32 and 5.33. The reducing valve pictured here was the heart of the system. (International Text-book Company, 1912)

to worry about cold cars on nonsteam interchanges, or isolated terminals devoid of steam. It was also reassuring to have a fail-safe heating system for emergencies, like stalled or snowbound trains. If the locomotive's fuel ran out, the individual car heaters could warm the train. As late as 1905 the authority George W. Fowler, editor of the *Railroad Gazette*, said that the indirect steam system was the American standard.[100]

Yet for all its advantages, the indirect system was a makeshift that could be roughly compared to a plan for using a steam engine to pump water uphill to turn a waterwheel. Some more direct way to exploit steam must be found. The vacuum system was suggested, but it involved a supply-and-return line and a pump.[101] A far simpler scheme was introduced around 1902 by Egbert H. Gold of Chicago, who proposed using extremely low-pressure steam—no more than a few ounces above atmospheric pressure.[102] Heating engineers knew of this method as the vapor system, but no one had thought of applying it to railroad cars.[103] In Gold's modified version, pressure steam of 150 to 170 pounds filled the train line. At this pressure the rear cars would have plenty of steam, but its temperature would be 375 degrees, far too hot to introduce into the radiators. A reducing valve dropped the pressure to less than 1 pound, so that the steam entering the radiator pipes was at a manageable 212 degrees (Figures 5.34 and 5.35). Overheating of the cars was less likely, and regardless of the variations in the train-line steam pressure, the reducing valve kept the vapor pressure at a uniform level. Vapor also proved more economical than the old direct systems because steam was metered into the car as needed.

In 1904 the Chicago Car Heating Company was organized to exploit Gold's patents, and the business flourished as railroads switched over to vapor heating. Established suppliers like Edward E. Gold were forced to develop rival systems. Others fell by the wayside, while a few firms like the old Standard Heating and Ventilating Company were absorbed by the Chicago Car Heating Company, which in March 1917 became the Vapor Heating Company.[104]

Vapor was the greatest revolution in car heating since the beginning of steam in the 1880s, but like steam, vapor did not entirely solve the heating problem. The passenger car is inherently difficult to heat.[105] It is exposed on all six sides. Steel cars were said to require 20 percent more steam than wooden cars. Its rapid movement through cold air increases its radiation losses. It may pass through highly variable temperature extremes in a matter of hours—from boiling deserts to snow-covered mountains. It has a large window area, again causing radiation losses. Frequent opening of the end doors constantly admits cold air. The heating system must fit into a confined space yet remain accessible, and it must be able to perform for long periods with minimum servicing. Indeed, it is surprising that any system could be devised which partially meets these requirements.

Insulation, double-sash windows, and in later years Thermopane glass all helped to reduce radiation losses. But even a well-insulated car consumed 250 to 300 pounds of steam an hour.[106] Train pipe size and pressure were increased to satisfy the de-

one participant to convert (Figure 5.32). The problem could be partially solved by the use of dummy steam lines, so that the vapor could be passed the length of the train to the cars that were already fitted. A better solution was a combination steam and hot-water system, in which the stove could be lit if no steam connection was available. In reality this problem was not widespread, because the interchange of passenger cars was relatively small except for the Pullman fleet. Many lines ran trains that remained an integral package for weeks on end, moving back and forth between the same points. In these cases it was quite practical to have several different plans of steam heating on a single railroad, since the cars of the Golden Gate Special were rarely if ever mixed with, say, those of the Sacramento Limited.

A full solution to the problems of direct steam heating seemed impossible, and what developed instead was a major compromise: the union of steam and hot water. Edward Gold achieved this alliance in the early 1880s. Others, including Westinghouse, came to accept it as a blessing—the best of both schemes combined. Steam was introduced into a tube running inside the coil of a Baker-style heater. Hot water circulated through the pipe radiators. Because live steam was not flowing near the passengers, it was possible to boost pressure from 5 to 30 or even 60 pounds. The rear cars were thus guaranteed a lively supply of heat, and the danger of line freezes was reduced. Efficiency was materially improved, because the hot water fully utilized the latent heat rather than dumping it overboard, as in the direct steam system. More even temperatures were possible by regulating the hot-water flow. And overheating was less likely because no live steam was introduced into the radiators.

Railroads tended to favor the indirect method of heating, as it came to be called, because it solved what had seemed to be the unsolvable regulating problem (Figure 5.33). It also offered a way to salvage the existing investment in Baker heaters. A simple conversion kit would do the job. And it was still possible to operate the heating independently by stoking up a fire; no need

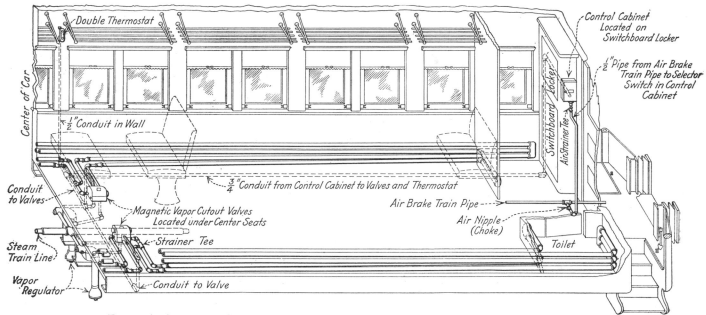

Figure 5.35   A typical vapor steam heating system of the 1920s. Electrical controls regulated the degree of heating. (Car Builders' Cyclopedia, 1928, fig. 1917)

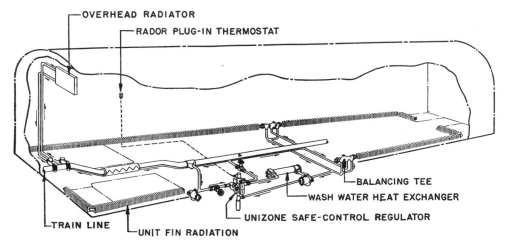

Figure 5.36   Fin-style radiators helped improve the efficiency of vapor heating. (Car Builders' Cyclopedia, 1953, fig. 918)

mand. For years 1½-inch pipe had been considered adequate, but in 1903 the Master Car Builders recommended an increase to 2 inches, and in 1911 the association made this size standard practice.[107] By 1948 the pipe size was increased by another ½ inch. Pressure followed a similar upward course, from 30 to 60 to 130 to full boiler pressures of 250 to 300 psi.[108] After pressures exceeded 130 psi, rubber hoses were no longer suitable. By the late 1920s flexible pipe connections between the cars were becoming more common. Steam hoses were made obsolete in January 1937 when they were barred from interchange service by the Association of American Railroads.

Other refinements were adopted in the 1930s. Electric thermostats had been reintroduced in 1921 and began to gain favor within a few years.[109] With the advent of air conditioning, thermostats became standard equipment. At this time the heating, ventilating, and cooling apparatus were integrated as a single system. Vapor continued to furnish the heat, but it was now distributed by overhead forced-air ducts and floor-mounted finned pipe radiators (Figure 5.36). Some cars had liquid-filled floor radiators reminiscent of the old indirect steam systems.[110]

The abandonment of steam locomotive operations did not bring an end to vapor heating. Small upright boilers stationed within the diesel-electric locomotive's body took over this duty. Similar oil-fired units had been used almost since the beginning of electric traction. Because main lines electrified only short portions of their tracks, through cars did most of their traveling behind a steamer. When it was clear that diesels were soon to be the dominant form of power, the design of steam heating generators was refined. Since the late 1920s the Otis style of boiler manufactured by the Vapor Heating Company was the best package unit available.[111] An even more efficient boiler was introduced around 1933 by an Englishman named Alic Clarkson. It was tried on the Union Pacific motor train streamliners with great success and soon became the favored style of steam unit. On long trains, Clarkson boilers were worked without respite at 100 percent of capacity. A typical steam generator burned diesel fuel and produced 4,500 pounds of steam an hour.

According to William D. Edson, the former manager of cars for Amtrak, head-end power plants—a modern version of Gibbs's 1890 boiler tender—began to replace small Clarkson-type gen-

Figure 5.37 Salisbury's dusters, a series of under-floor side curtains, created a tunnel under the train to carry the dust away from the cars. The scene was photogaphed on the Michigan Southern in 1868. (Pennsylvania State Railroad Museum)

erators in the late 1950s.[112] A head-end car could provide ample room for a heating boiler and generator capable of servicing the growing power needs of modern trains. The decline in passenger revenues at this time ended hopes for the new investment necessary to produce such cars. A few experimental units were created, however, such as the separate power car for the Pennsylvania's Keystone. According to Edson, vapor heating has definitely been abandoned in present practice. Electric heating was installed on most cars after about 1965. Diesel-electric locomotives are being fitted with alternators for head-end power. In some cases the main traction motor generator is given enough extra capacity to meet this requirement. Some older cars that have independent engine generator sets, like the Santa Fe hi-level cars, will be equipped with larger (60-kilowatt) units to handle electric heating.

Ventilating and Air Conditioning

Ventilating involves nothing more than drawing the good air in and forcing the bad air out. On first consideration it seems a simple matter of throwing open the windows and letting in all the fresh air in the universe. But in its wake come dust, smoke, cinders, and the partially condensed steam that billows along the train's smoky course. Just how to balance the conflicting requirements for a large volume of clean, pure air, without drafts, while maintaining a comfortable temperature on cold days, was a problem that was never resolved until the advent of air conditioning in the 1930s. Before that time ventilating was at best a sorry compromise.

At first the problem did not receive much attention; passengers seeking more air were advised to open the windows. For short journeys the inconveniences of this method could be tolerated, but as distances increased, it became a major hardship. The first concession to improved ventilation appears to have been a row of small openings above the top of the window sashes. The main windows could remain closed while the open transom sashes drew in fresh air above the passengers' heads. The oxygen supply

could be replenished without blowing the air and its disagreeable load of soot into people's faces. An early example of this arrangement is found in the 1836 Camden and Amboy coach shown in Chapter 1 (Figure 1.71). By the 1850s the scheme had been refined: very small square openings were cut into the letterboard, with either a simple wooden-center hinged, turnover sash, or with a more elaborate sheet-metal, clamshell closure.[113] Wire screen or pierced sheet-metal guards were sometimes placed over the openings to keep out the cinders. Of greater importance, the bad air was given an avenue of escape through roof-mounted ventilators, which were usually short stovepipes with raised caps to keep out the rain. Three or four of them were mounted along the center line of the roof. Examples of cars with this primitive form of ventilation are pictured in Chapter 1.

The combination of intake and exhaust vents provided a natural circulation absolutely essential to good ventilation. In theory the stale, hot air would rise and flow out of the roof vents. In practice, however, both the inlet and exit openings were too small to move much air, particularly in the summer, when sweltering passengers would have to throw open the windows and suffer smoke and cinders for a little cooling air.

Passengers rarely suffered in silence; travelers are a complaining race whose voices are amplified by the grumbling of journalists.[114] One early-day editor compared the dust storm that followed his train to a heavy Ohio River fog, as thick as gruel, but dry and choking rather than cool and moist. He said that it enveloped passengers and their garments with a heavy, dry coating. Early travelers on the Michigan Southern were sprinkled with a shower of fine sand along the lake shore that changed to a powdery clay dust near Cleveland—a dust so fine and dry that it stuck in the throat like flour. According to another traveler, the car interiors became so charged with dust that it was impossible to see from one end to the other through the twilight haze. Adding to the natural misery of a summer rail excursion was the fresh-air fiend, a first cousin to the notorious seat hog. The fresh-air addict would occupy a front position in the car with his window raised high to admit a hurricane of smoke and dust that blanketed everyone seated behind.[115] These worst of all health cranks had "the constitution of sawlogs and the manners of oxen."[116] Trainmen were too indifferent or timid to restrain them, and less hardy travelers could only insist at the journey's end that something must be done about the dust.

One possible solution was to suppress the dust at its source. Ballast was a rare commodity on early railways; most tracks were simply set on bare ground. In dry weather a passing train naturally raised a minor dust storm throughout its journey. Because most lines insisted that they could not afford stone ballast, the only alternative was to sprinkle the tracks. In 1851 the Hudson River Railroad ran a sprinkler car in conjunction with its passenger trains.[117] The improvement was so remarkable that the Boston and Providence and the Stonington railroads followed suit.[118] These lines abandoned the practice, however, after only three or four years; apparently the cost was too great to continue it. In fact the service may have been initiated only to silence agitation for a state law requiring gravel or sod banks along the tracks as a dust shield.[119] The general practice was probably revived from time to time, for as late as 1898 the Long Island Railroad was reported to be sprinkling its tracks with oil.[120] The procedure not only increased passenger comfort by cutting down the dust, but reduced running-gear wear and allowed the use of cheaper forms of ballast such as sand or cinders.

A simpler solution to the dust problem was offered in 1855 by Elam C. Salisbury. He proposed directing it under the train by forming a wind tunnel of side curtains hung along the underside of each car. Rather than raising a cloud around the side of the train, the dust would then be pushed harmlessly out behind the last car. The side curtains, also called petticoats, could be made of canvas, boards, or sheet metal. They would serve secondary functions in deadening the noise of the running gear and preventing pedestrians from falling under the moving train. The scheme was attractive because it was low in first cost and involved no complicated maintenance problems. A year after Salisbury's patent, the Michigan Central was testing his invention.[121] The New York and Harlem also gave it a trial, but Salisbury's plan received the warmest welcome in the Midwest. The Michigan Southern was a strong supporter of the petticoat duster. A photograph, dated 1868, in the collection of the late Walter Lucas shows a full train of cars equipped with it (Figure 5.37). Yet within two years the scheme was abandoned, for the road discovered that a disastrous side effect of concentrating the dust under the train was excessive wear on the brake gears and journal bearings.[122] The mechanical department, deciding that Salisbury's invention was a luxury it could not afford, once again let the passengers fend off dust for themselves.

At the time that the Michigan Southern was stripping off its petticoat dusters, the Missouri inventor W. M. K. Thornton offered an improvement on Salisbury's scheme.[123] Rather than side curtains, he proposed a horizontal canvas curtain fitted to the truck frames below the axles, just 6 inches off the rails. Thornton thus hoped to keep the dirt below the journals and the passenger compartment. A secondary curtain was suspended between the trucks to form a continuous shield. After a trial in the

West, nothing more was heard of Thornton's patent. A few years later, however, the Central Railroad of New Jersey tested a simpler form of duster that consisted of leather curtains stretched between and below the platforms. (The 1879 *Car Builders' Dictionary* shows the device in its figure numbered 216). Around 1880 the Pennsylvania Railroad put some fashion of duster on its Cape May express trains to help exclude the fine sand on that route.[124] Another style of duster was employed in 1930 by the Missouri Pacific on its observation cars.[125] A pipe frame covered with canvas extended out from below the rear platform and helped prevent the dust from billowing back to the passengers.

More elaborate antidust systems were proposed and tested during the 1850s. The most sophisticated involved air washers to cleanse the air. Some included ice cooling and might therefore be counted among the ancestors of modern air conditioning. In the various designs dust was collected by a water spray, a water bath, wet sponges, excelsior (which was often nothing more than wood shavings), or a combination of these methods.

One of the first air washers tried was patented in July 1854 by George F. Foote. A roof air scoop pulled air to a tank beneath the car, where a belt-driven rotary pump produced a spray. The air was then returned to the car's interior by floor ducts. August Mencken claims that Foote assigned his patent to the Michigan Southern shortly after it was issued.[126] The road installed the device at great expense in twenty-four new coaches received from Eaton and Gilbert that same summer.[127] A tank 18 feet long, 8 feet wide, and about 18 inches deep was fastened under each car, and water was added to a depth of 4 inches. In operation the system proved to be an effective dust killer—whenever it was in working order. Passengers did complain about the high humidity, but they rarely had to suffer long, because the pump or the spray jets quickly clogged with cinders, or else the belt slipped off one of the pulleys. And even if the system was working, the flow of air stopped when the train did. The windows were sealed, and passengers were left to endure mounting temperatures until the train got underway. During the following winter one of the Foote-equipped cars caught fire and was totally destroyed. The ventilating system was blamed for this calamity. A second series of trials was made during the next summer, and then the entire scheme was written off as a failure.

Foote's apparatus was also tested on the C C & C and the Erie. The Erie's mechanical personnel made some improvements to Foote's original designs.[128] The pump impeller was fitted directly to one of the axles or driven by a friction wheel. The water spray jets were attached to a standpipe reaching up into the intake air duct. A glass window was placed in one side of the duct so that the sprinkler action could be observed. Drawings of this apparatus were published in Douglas Galton's 1857 report on U.S. railways (see Figure 1.85). The service history of Foote's ventilator on the Erie and other lines where it may have been tested has not been uncovered. Presumably it was a short-lived phenomenon that promised much but delivered little.

The most talked-about air-washing system was introduced by

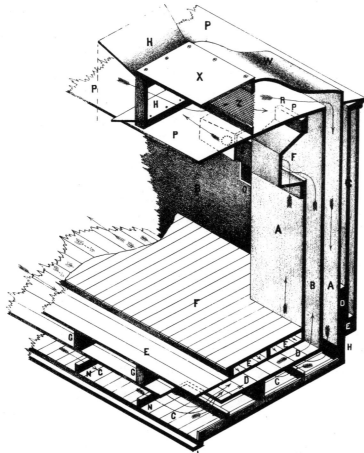

Figure 5.38   Henry Ruttan of Canada invented a ventilating and heating system that was used on a few U.S. lines between 1855 and 1875. Shallow water tanks below the floor helped trap cinders and dust. (Ventilation & Warming of Buildings, *1862*)

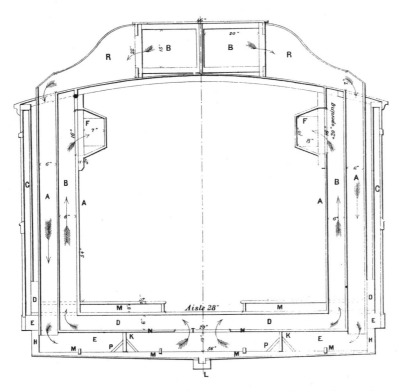

Figure 5.39   In Ruttan's ventilating system, B–R was the roof air scoops; A directed the air to the water pans, M–P; B directed the cleansed air to the interior of the car via the register, F. (Ventilation & Warming of Buildings, *1862*)

Henry Jones Ruttan (1792–1871) of Cobourg, Ontario.[129] Ruttan, a person of many occupations and interests, spent much of his life in the army, served as sheriff of his home town for thirty years, and became a representative in the House of Assembly. His avocation was heating and ventilating—an interest dating from the early 1840s that resulted in seven patents (1846–1858) and a major textbook on the subject published in 1862. In the course of his investigations Ruttan became interested in improving railroad car ventilating. In some respects his plan was very similar to Foote's, except that Ruttan dispensed with the pump and water spray and depended entirely on the water bath to collect dust and cinders (Figures 5.38 and 5.39). Tests of his system apparently began in May 1854 on the New York Central, and in the fall of the same year the Erie was ready to give his apparatus a trial.[130]

In Ruttan's device a 12-inch-deep tank under the floor provided 200 square feet of water surface. The bath was 3 inches deep. Baffles made the air take a meandering course to ensure maximum dirt collection. After passing over the bath, the clean air was directed into standing ducts that delivered it just above the passengers' heads. The intake was the familiar roof air scoop. No fans were employed; the current was dependent on the motion of the train. In winter one of the interior air ducts was replaced by a stove. The entire system, less the stove, cost a modest $100.

Ruttan was able to convince several railroads to install or at least test his apparatus during the following fifteen or so years. Few other inventors found so many openings, which says something for the system and its zealous originator. In 1857 the Grand Trunk was trying out an improved version.[131] Two years later the Boston and Lowell was also conducting road tests.[132] In 1863 the P W & B was expected to outfit thirty cars with Ruttan's ventilating system.[133] A year earlier all cars on the Michigan Central were reported to be so equipped.[134] By 1868 Ruttan announced that 300 cars were using his apparatus—a number far above the handful of cars that rival inventors could claim.[135]

After Ruttan's death, enthusiasm for the system declined. In 1873 the Burlington was still using it, but the road admitted that the apparatus had to be carefully managed in order to function properly, and that the task was too complex for the average brakeman.[136] Sometime before 1877 the Michigan Central, the largest single user, abandoned the Ruttan system.[137] It worked well enough on short trips, but after 50 miles the tank became choked with cinders, rendering it ineffective. Small leaks, which were unavoidable in the large sheet-metal tank, rotted out the floor sills. It is also likely that the car superintendent and his staff grew weary of attending to repairs on the complicated ventilator. Popular though it may have been with the traveling public, these men presented a good enough case about high maintenance costs to justify its removal.

The slow disappearance of Ruttan's system did not discourage other inventors from attempting to market similar devices. The railroad trade papers mentioned these efforts, but nothing of consequence occurred until the sleeping-car operator W. D.

Mann took an interest in the subject. Mann was seeking ways to discredit Pullman's service and style of car, and a major complaint against the open-section car was the poor ventilation. Passengers said that once they were closed inside the curtains, particularly in the upper berth, they felt as if they might suffocate. Mann thus found a ready-made issue and a ready-made answer in the ideas of Foote, Ruttan, and other early ventilation engineers. In Mann's version fresh air was drawn in through a low, wide hood that straddled nearly the full girth of the arched roof at one end of the car.[138] The air was directed into a closet, where it passed through a filter made of dampened mattress packing (which must have produced a conspicuous odor). In the summer the air was cooled by passing it over 350 to 600 pounds of block ice. In winter it was heated by steam pipes, though a small block of ice was still used to keep the excelsior moist with its melting water. In this way the need for a pump was avoided. The air was broadcast through the car by a register. Roof vents maintained the circulation.

With his usual bravado Mann made extravagant claims for the system, asserting that it achieved a full change of air every five minutes and that in summer it lowered temperature by 12 to 20 degrees. His claims were naturally challenged by the sober-minded *Railroad Gazette*, while the *Car Builders' Dictionary* conceded that the Mann system could effect temperature drops of about 10 degrees.[139] Opinions on the healthfulness of the air varied considerably between Colonel Mann and his critics.

Mann obtained a patent in September 1885, but he had already applied his system to some of his first North American cars in 1883. How many cars he equipped with it is uncertain, but as late as August 1887 the *American Railroad Journal* reported that the Mann sleeping car *Martha* had an ice-cooled ventilating system. Just how long he would have stood by his pet scheme will never be known, for the Pullman takeover in 1889 put an end to such unorthodox experiments for many years.

The dust problem simply remained unsolved and largely ignored. Some practical form of air-washer system could have been developed in the nineteenth century if the railroads had been willing to keep trying. The difficulty was not so much a lack of technology as a lack of commitment. An apologist for the industry might claim that the problem was indirectly solved by improved roadbeds, with crushed-rock ballast and mechanized ballast cleaners. Yet even today on European roads, despite their reputation for good track maintenance, major and minor clouds of dust are kicked up by trains. This powdery debris pours into non-air-conditioned compartments, coating passengers and their belongings with grime. One reason for the dust is that the leavings of the ballast cleaner are dumped close to the trackside, where passing trains are certain to stir it up. Whatever the cause, the dust problem lives on.

During the 1870s a new concern over ventilation arose in medical circles and spread to the general public. The purity, chemical composition, and quantity of the air people breathe became the subject of intense discussion. Railroad cars were singled out for particular criticism. As early as 1840, when stove heating had become common practice, the atmosphere of the tightly closed cars was so foul that it made passengers delirious.[140] As the general ventilation craze grew, everyone attacked the railroads. A lady passenger complained of insolent train crews who prevented heat loss by refusing to allow passengers to crack a window sash.[141] To avoid the ill effects of "second-hand breath," she said that she had to spend part of the trip standing on the platform.

The members of the Master Car Builders Association, while determined not to pamper the public, tended to agree that railroad cars were the most poorly ventilated of all public gathering places. Their air space was extremely limited; even a prisoner was furnished twenty times more air then a coach passenger had.[142] The total volume of air in a railway car was adequate for no more than 4 people rather than the 50 to 75 who were crammed into a full car. Moreover, the air was hazardously contaminated with carbonic acid. Fitch Adams, the crusty master car builder of the Boston and Albany, expressed a minority opinion: "This matter of ventilation is a great hobby with some people. I knew a man who said more people died every year from breathing impure air than were carried off by whisky. The anti-whisky people say whisky kills $\frac{7}{8}$ of the people in the world, and if impure air kills more, where are they to be found?"[143]

Various guesstimates on the quantity of fresh air required per railway passenger were projected. In 1840, 240 cubic feet was given as the amount needed per hour. This figure was repeated in the 1874 Master Car Builders Report; twenty years later, however, nothing less than 3,000 cubic feet per hour was considered safe for the health and comfort of each individual.[144] At the same time, it was contended that the real villain was the high level of carbonic acid, now commonly called carbon dioxide or CO_2. Samples from railroad cars showed that the air contained 0.23 percent CO_2, nearly double the amount measured in theaters and other poorly ventilated places.[145] Carbon dioxide, it was claimed, was the cause of the lassitude, delirium, uneasiness, and dull headaches associated with long railroad trips. This notion was countered in 1910. In a paper read before the American Public Health Association, Dr. Thomas R. Crowder admitted finding high levels of CO_2 caused by normal respiration of passengers in such confined spaces as railroad cars.[146] But he contended that the passenger discomfort was produced primarily by lack of oxygen, excessive heat, and too much or too little humidity.

Throughout the pure-air discussion another basic theme was the need to coordinate heating and ventilating. The problem was that while it was possible to pull in enough fresh air to satisfy the most ardent CO_2 crank, it was not possible to heat the car unless the exchange was limited. In the days of stove heating, only a very small amount of the cold outside air could be taken in. Even with steam heating, only about 1,000 cubic feet per passenger per hour could be admitted if a 70-degree temperature was to be maintained.[147] This was one-third of the volume needed for a fresh interior atmosphere. The 1,000-cubic-feet compromise having been accepted, it was now necessary to find the best way to

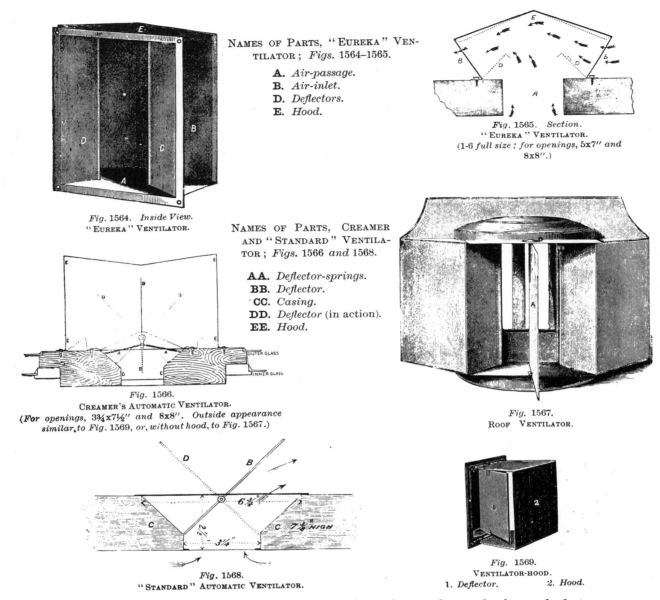

NAMES OF PARTS, "EUREKA" VEN-
TILATOR; *Figs. 1564–1565.*

A. *Air-passage.*
B. *Air-inlet.*
D. *Deflectors.*
E. *Hood.*

Fig. 1565. Section.
"EUREKA" VENTILATOR.
(1-6 *full size; for openings,* 5x7″ *and*
8x8″.)

Fig. 1564. Inside View.
"EUREKA" VENTILATOR.

NAMES OF PARTS, CREAMER
AND "STANDARD" VENTILA-
TOR; *Figs. 1566 and 1568.*

AA. *Deflector-springs.*
BB. *Deflector.*
CC. *Casing.*
DD. *Deflector* (in action).
EE. *Hood.*

Fig. 1566.
CREAMER'S AUTOMATIC VENTILATOR.
(*For openings,* 3¾x7½″ *and* 8x8″. *Outside appearance
similar to Fig.* 1569, *or, without hood, to Fig.* 1567.)

Fig. 1567.
ROOF VENTILATOR.

Fig. 1568.
"STANDARD" AUTOMATIC VENTILATOR.

Fig. 1569.
VENTILATOR-HOOD.
1. *Deflector.*   2. *Hood.*

*Figure 5.40 Creamer's clerestory-mounted ventilator and several other such devices
popular between about 1870 and 1890 are shown in this series of engravings. (Car Builders' Dictionary, 1888, fig. 1566)*

exchange the air. The exhaust was no problem; with the clerestory roof that had been introduced in the 1860s, warm, foul air
rose to the ceiling and flowed out the deck windows or vents. But
the clerestory was not efficient in introducing new air into the
car. Psychologically, open clerestory windows created a breezy
effect, but most of the air that blew in blew back out before
benefiting the passengers.[148] Only when the train was moving
slowly did much of the outside air entering the clerestory drop
down into the car. The crosscurrents set up by open deck windows often confounded the exhaust or made the lamps smoke,
fouling the atmosphere and blackening the ceiling.

One way to improve the clerestory's effectiveness was to hinge
the deck windows at the middle so that they could be swiveled
partially open *away* from the direction of travel. The air blowing
past the open end created a small vacuum, thus helping to suck
out the stale air inside. The idea worked well enough if the
brakeman positioned the deck windows correctly, but the
brakemen had become an indifferent lot since the introduction of
air brakes and steam heating.[149] In earlier days they had
manned the brakes, tended the stove, trimmed the lamps, reversed seats, and adjusted the clerestory windows, but now they
would do nothing more than call out the stations.

A railway supplier, William G. Creamer, decided to solve the
problem with a ventilator that automatically reversed itself. The
idea came to him about 1860 when he was riding in a new car on
the Hudson River Railroad, one of the first lines to adopt the
newfangled clerestory.[150] Cinders rained in upon him intermit-
tently because the slats in the deck window opening were facing
ahead, thus acting as a scoop to pull in the smoke as it wafted
over the top of the cars. The brakeman had failed to reverse the
slats at the end of the run. On the spot Creamer thought of a
simple solution: he would mount a series of small sheet-metal
boxes, open at both ends, along the side of the clerestory. A hole,
say 4 by 7 inches, would be cut through. A butterfly valve set
inside the box would swing to a position opposite the direction of
the train's travel because of the breeze acting on the other end of
the valve (Figure 5.40). Cinders and smoke would be deflected,
and no one need tend the valve. Light springs would hold it at a
center or fully open position when the train was stationary to
ensure an exhaust while standing at a station or siding. Naturally
the deck windows would be kept closed. Creamer's invention
was widely used and spawned many imitations.

Getting the bad air out was relatively easy; getting the good
air in was more difficult. The problem was not lack of power,

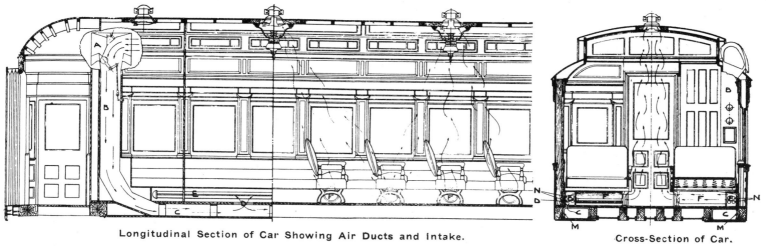

Longitudinal Section of Car Showing Air Ducts and Intake.   Cross-Section of Car.

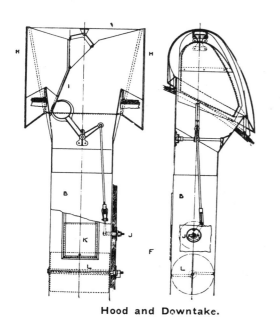

Hood and Downtake.

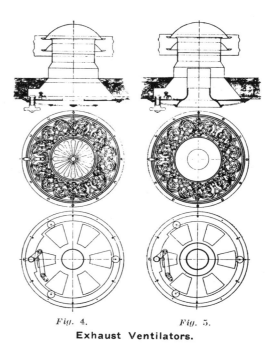

Fig. 4.   Fig. 5.

Exhaust Ventilators.

Figure 5.41   The Pennsylvania used this roof-mounted, air-scoop ventilating and heating system between 1900 and 1935. (Railroad Gazette, September 30, 1904)

since the train's motion created a draft strong enough to ventilate the deepest mine or the stalest dungeon, to paraphrase Creamer.[151] But where could the new air be drawn in without producing too strong a draft or admitting clouds of dust and smoke? The side windows were out on both counts, and the clerestory could not do the job. That left the ends of the car body, where the end windows were a logical choice except that the wind tended to blow back directly on the passengers. It occurred to someone to make an opening, called the transom or hood sash, above the doors at each end of the car. Opening out under the platform roof or hood, it was protected from the worst of the smoke and cinders, which ideally blew over the top. The transom was relatively small and high, so that it was not likely to admit drafts or a great rush of air even when it was open. And it brought the air in at about the right level—even with the passengers' heads. Evidence on the origins of this scheme is tenuous, but Figure 1.79 in Chapter 1 shows that some of the earliest eight-wheel cars had small transom vents over the end windows, if not over the doors. The Boston and Albany had adopted the hood transom sometime before 1876. The sash win-

dow had a ground glass lettered to read "If You Want to Breathe Pure Air Keep This Ventilator Open."[152]

The history of early railroad car ventilation is a tale of indifference and half measures. Few prominent car builders took the subject seriously, and those who did seemed to push it aside after a few months of frustrating failures. One of the few railroad engineering figures of any prominence to give ventilation more than passing notice was Charles B. Dudley (1842–1909). For many years he headed the Pennsylvania Railroad research laboratory under the simple title of chemist. Sometime in the mid-1890s Dudley decided to investigate car ventilation, and he was shocked to find that the air was as foul as its reputation.[153] Passenger complaints were actually a reasonable demand for reform. Nevertheless, he said later, during twenty years in railroad engineering he had never faced such a baffling problem. The primary conflict was between heating and the fresh-air supply. It was Dudley who concluded that an acceptable temperature could be maintained if no more than 1,000 cubic feet of air per passenger per hour was brought in. Although not ideal, this com-

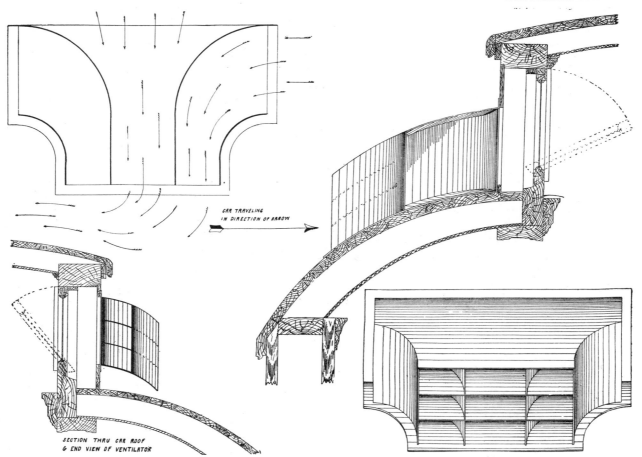

CAR TRAVELING
IN DIRECTION OF ARROW

SECTION THRU CAR ROOF
& END VIEW OF VENTILATOR

Figure 5.42   Garland's ventilator, introduced around 1905, was an extremely efficient aspirator that became standard with Pullman. (Car Builders' Cyclopedia, 1928, figs. 2094–2095)

promise was far better than the old ration of 250 or 300 cubic feet.

Dudley also hoped to salvage what he could of existing heating-ventilating technology. He sought no radical solution but saw merit in the P R R's existing standard, the Spear stove and hot-air register. The main defects of the Spear system arose because the source of heat (the fire) was higher than the outlet registers. This was easily remedied with steam heating, since the pipes could be placed low, near the floor (Figure 5.41). Dudley retained the old air scoop, with modifications. The outside air pulled in by the scoop was directed to an 8- by 14-inch sheet-metal box running under the floor for the full length of the car. Holes cut into the floorboards let the air rise to circulate the heat. The air scoop and its trunk, which were much enlarged compared with the old Spear plan, measured 14 inches square. Scoops were set at diagonal corners of the roof, and a vane inside each scoop could be flipped fore or aft, depending on the direction of travel. A butterfly valve, placed lower in the trunk, could be set to vary the flow of air. It could also be fully closed in tunnels, to prevent the scoop from sucking in smoke, and at terminals, to prevent the heat from escaping through a back draft. Globe-style exhaust vents were installed on top of the clerestory roof. This raised the escape opening 2 feet above the air intake.

Dudley and his associates tinkered with the project for several years, until by the late 1890s they felt that it was just right. Their belief was substantiated when a main-line train was stalled for more than four hours in a snow blockade near Harrisburg. The winds of the blizzard roared against the marooned train, but temperatures in the cars equipped with Dudley's apparatus never fell below 70 degrees. Soon the railroad embarked on a massive conversion, so that Dudley's system became standard on what

was sometimes called the standard railroad of the world. True, there was some grumbling about the necessity of cutting away the cross sills to make room for the ducts below the floor. Iron straps spanned the ruptures. By 1901 over 200 cars had been converted; three years later the total was more than 1,000. Even the B & O was adopting the scheme. Dudley's plan prevailed on the Pennsylvania until the age of air conditioning.

Dudley's efforts substantially increased the comfort of coach passengers, who were normally last in line for any technological improvements. Ironically, it was now first-class passengers who felt that they were being shortchanged. The sleeping car had the reputation of being the worst-ventilated of all railway conveyances, and Pullman was blamed not for its poor system but for the total lack of system.[154] Why pay extra fare to be closeted in an airless box that was upholstered with more suffocating hangings than can be found in a small hotel? For years Pullman smarted under this criticism, weakly contending that the clerestory did the job well enough. Finally in 1905 the company hired a Chicago physician, Thomas R. Crowder, to head its newly established sanitation department.[155] One of his responsibilities was to explore the subject of ventilation. Pullman's action might be explained as the belated recognition of a problem, or a step to quiet the public criticism stirred up by the reform-minded progressive movement. As described earlier, Crowder's investigations led to a dismissal of the old CO_2 bugbear. He also concluded that many complaints about sleeping cars were psychological in nature; claustrophobics inside a confined berth often experienced shortness of breath. Crowder admitted, however, that sleeping-car ventilation was far from perfect.

In the year that Crowder joined the Pullman company, a new device came on the market that promised to revolutionize car ventilating. It was developed by Thomas H. Garland, who had studied the problems of car ventilation while he was superintendent of the Burlington's refrigerator car business. His concern had been fruits and vegetables, not passengers. All previous ventilating systems had placed great emphasis on both the entrance and the exit of air, but Garland concluded that a strong exhaust was the key to good ventilation. Because no car body was perfectly tight, fresh air would find its way through the cracks, voids, holes, and crevices inherent in any man-made structure. Garland's roof-mounted, double-sided air scoop ventilator was based upon the aspirator principle (Figure 5.42):[156]

The theory is sound and well applied. Large openings in the ventilator face toward either end of the car, and these form the ducts through which the air is given most of its application to the purpose of exhaustion. The motion of the car forces the air in quantities into one opening or the other, according to the direction in which the car is moving; and the right angle bend in the duct conveys the air outward and away, leaving the ventilator by the openings at the outer edge of the car roof. In traversing the passage to its exit, the air goes from the intake duct into the transverse duct, and at this point a strong suction is effected which draws the air from the car through the opening provided for that purpose. But this exhausting action is effective also in other ways. Where the air from the intake leaves the ventilator it is drawn backward by the motion of the car and the current of the outside air passing over the mouth of the central duct, making that also effective in ventilating. In fact it is claimed that every passage in the ventilator contributes its share to the work, and the action of the whole is very definite and certain. It makes no difference in which direction the car is moving.

The sheet-metal housing conformed to the arch of the roof, draping over it like an anteater's snout. The suction created was astonishing: at 30 mph each vent pulled out 15,000 cubic feet of air. With ten vents to a car, a complete change of air was effected in less than two minutes.[157]

Pullman was convinced that Garland's was the ventilator it had been seeking. By June 1908 nearly one-half of the Pullman fleet was outfitted with this invention. Clerestory windows were soon a memory, while thousands of Creamer ventilators went on the scrap heap. Garland joined Burton W. Mudge & Company of Chicago, which acquired manufacturing rights to his invention.

Garland's device, though widely successful, had several faults. Like all wind scoop ventilators, it did not work unless the train was underway. Its powerful exhaust tended to waste heat, and in extreme cases where there were too many vents or the train was moving at high speed, it created an interior vacuum strong enough to make door closing difficult and even to cause toilets to overflow when they were flushed.[158]

Since the 1850s various inventors had proposed fans to induce artificial drafts beneficial to ventilation and cooling. Most of them had regarded fans as a means of stirring up the existing inside air for cooling purposes. In 1881 a Philadelphian named Isaac Fridenberg proposed a wind-powered turbine scheme.[159] The roof-mounted turbine was belted to a line shaft running the length of the car within the clerestory. Paddles were fastened to the shaft at intervals. Nearly twenty years after Fridenberg's patent the Reading was reported to be testing a similar plan.[160] In this case, each fan had its own wind turbine. Both systems depended on the train's motion, which meant that when the fans were least required, they were spinning to their utmost. Fans were most needed when the train was stationary or moving slowly. Thus some independent source of power must be supplied for fans not propelled by the train's motion.

The earliest attempt at such a plan for which a record can be found took place on the Milwaukee Road in 1891.[161] In that year George Gibbs, already deeply involved in steam heating and electric lighting experiments for the Milwaukee, placed electric fans aboard several dining cars: one fan in the kitchen, two in the dining compartment. Gibbs proposed that all dining, parlor, and sleeping cars be similarly equipped. The moving air must have seemed like a blessing to the occupants of a sweltering dining car standing in a sun-baked terminal. The idea was gradually adopted by other lines. In 1898 the B & O's Southwestern division was using electric fans to cool sleepers at layover points.[162] Passengers were commonly accepted on board several hours before a late-night departure so that they could bed down at a conventional hour, but the cars were unbearably hot and stuffy after standing in the terminal. Electric fans, plugged into the station power source, provided some relief until departure time.

Within ten years most parlor, buffet, and dining cars had electric fans.[163] As electric lighting was gradually adopted for coaches, electric fans began to become more common on non-first-class cars. By 1920 there were usually small fans at either end of a car, or in better installations a series of ceiling fans. Although they were an improvement, they mostly agitated the stale interior air rather than producing an effective exchange of air. In the mid-1920s Pullman was using powerful clerestory-mounted exhaust fans that created some circulation in stationary cars, provided that the porter was conscientious enough to turn them on. If it had a large enough capacity, an exhaust fan could do much to reduce stratification and equalize temperatures between the ceiling and floor. This was extremely important to the comfort of passengers in both upper and lower berths. A forced-air system, however, was considered too expensive for general use. Blowers of modest size were widely installed where they were most needed—in dining-car kitchens, sleeping cars, and smoking cars—but a large system for a coach was regarded as much too costly.[164] Moreover, a fan large enough to exhaust an entire car would require a substantial power source, whereas most generator-battery systems were barely adequate for electric lighting. It was also contended that a large blower system would be unbearably noisy. Even so, the gradual progress of electric fans was a prelude to the next major advance in car ventilation: air conditioning.

Air conditioning began a new era in railroad travel. It involved more than just lowering the temperature during summer heat.

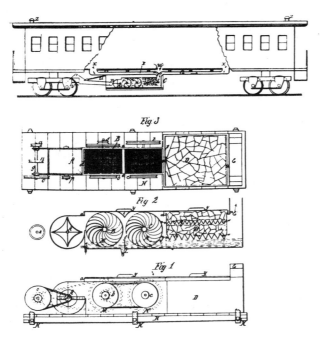

Figure 5.43   *J. R. Barry patented this ice cooling system with axle-powered fans in 1855. (U.S. Patent Office)*

Air conditioning was "manufactured weather," in which mechanized systems controlled the humidity, the purity, and the volume of air and regulated the temperature in all seasons of the year. It eliminated the filth associated with railway travel in the steam age. Because air conditioning was introduced during the same years as streamlining, it was overshadowed by the glamorous polished-aluminum and stainless-steel trains of the 1930s. Yet air conditioning was a major revolution in passenger comfort, a tangible improvement in place of the tin-deep eye appeal of the streamliners.

The origins of air conditioning in railway cars can be traced to a patent (No. 12,851) granted Job R. Barry of Philadelphia on May 15, 1855. Barry envisioned an ice system with axle-driven fans (Figure 5.43). It does not appear that his scheme was given a test, but as mentioned earlier, W. D. Mann employed a similar ice system on many of his sleeping cars. The most persistent effort made during the nineteenth century to perfect some form of car cooling appears to have taken place in India. There an ancient method of cooling was the tatty, a dampened grass mat that was hung in a doorway or window to lower the temperature of the air passing into the dwelling by evaporation of the water. This primitive scheme, improved with a semiautomatic method of wetting the mats, was applied to the Viceroy's private car in 1872–1873.[165] An extended series of articles explaining the system and its use on the Great Indian Peninsular Railway appeared in the British magazine *Engineering* in 1874. Fifteen years later the tatty system was still reported in service, but it was criticized as being ineffective when the train was idle.[166] Ice coolers and better insulation were proposed. In 1914 the G I P tested what is probably the earliest non-ice system used for passenger car cooling.[167] A carbon dioxide refrigerating unit was placed in the baggage car. Cold brine pumped back in pipes through the train was said to lower temperatures by 20 degrees. Because of the humid climate, condensation was a problem. Some passengers also complained that the unnaturally cool temperature was not healthy and petitioned for removal of the system. The outbreak of World War I drew the entire empire to the aid of Britain and was said to offer the Indian railways a convenient excuse for abandoning the experiment.

Meanwhile the Santa Fe hoped to perfect some method of car cooling to make its Southwestern desert route more appealing to transcontinental passengers. Since daytime temperatures often

exceeded 100 degrees, experienced travelers generally chose a more northern route. Late in 1911 the Santa Fe inaugurated a fancy first-class train named the De Luxe. The dining car was outfitted with an ice-cooled ventilating system.[168] The air was cleaned by a jet spray before being directed into the dining compartment. The apparatus was reminiscent of Mann's old system, but with an important difference: a large electric fan pumped 90,000 cubic feet of air through the compartment every hour. As many as thirty-three Santa Fe diners were similarly equipped.[169] As an air cooler, however, the system turned out to be relatively ineffective, and by 1926 it was at last abandoned.

A year or two after the Santa Fe's first ice-cooled diners entered service, the Pullman Company asked Willis H. Carrier, a pioneer in commercial air conditioning, if he could provide a mechanical refrigeration unit for sleeping cars.[170] After some study Carrier concluded that the state of the art was not ready for such an installation, and there the matter rested for more than a decade. In 1927 Pullman decided to reopen the question, and in April of that year the company installed a Servel electromechanical air conditioner aboard the sleeping car *Jacksonville*.[171] The unit, an apparent failure, was removed in August. Pullman next tried an ice system on the *McNair* in September 1929. Other companies were now awakening to the idea of passenger car air conditioning, and experiments began in rapid order on several railroads.

In the summer of 1929 Willis Carrier was asked to test his apparatus aboard a Baltimore and Ohio coach.[172] The experiment was successful enough to warrant a second installation, this time on a B & O dining car, the *Martha Washington*.[173] A 7½-horsepower, 110-volt electric motor worked the ammonia refrigeration compressor. A gear-driven 10-kilowatt generator powered the motor. The refrigeration unit cooled water, which was pumped to coils in roof ducts. A fan blew 2,500 cubic feet of air per minute over the coils into the dining room. (The kitchen was not cooled.) In a test held in April 1930, the steam heat drove the temperature up to 93 degrees, and Carrier's air-conditioning unit lowered it to 73 degrees in just twenty minutes. The car entered regular service a month later and was not retired until 1965. Carrier's original equipment was found unsatisfactory in some way. It is likely that the electrical power setup was defective.

Whatever the reason, the B & O turned almost immediately to the York Ice Machine Corporation for thirty-six gasoline-engine-powered, 4-ton units.[174] Electricity was needed only for the pump that circulated the brine coolant to ceiling coils. In May 1931 the B & O had equipped enough cars to make up the Columbian, which was said to be the first complete air-conditioned train in the United States. By the next spring the road had over 100 air-conditioned cars in service and several air-conditioned name trains.[175]

Air conditioning was introduced in the Far West just a few months after the B & O's first air-conditioned diner entered service. The Santa Fe seized upon the new technology to better its competitive position. In July 1930 it began running a new dining car, Number 1418, that was fitted with a Carrier electromechanical unit.[176] Unlike the apparatus that Carrier produced for the *Martha Washington*, conventional car-lighting current of 32 volts DC was employed. Two direct-drive 7½-kilowatt generators supplied the power at speeds above 15 mph. Batteries took over when the train was running slowly and at station stops. A special coolant was devised by Carrier to replace brine water. With the kitchen at full heat and the desert simmering at 104 degrees, the air conditioning maintained the inside temperature at 72 degrees. Passengers tended to linger at the table; the 1418 was the most popular car on the Chief. Yet the success of Carrier's electromechanical unit won no repeat orders from the Santa Fe; instead the line adopted a steam ejector system devised by Carrier and manufactured by the Safety Car Heating and Lighting Company. Early in 1932 the Santa Fe had ten diners cooled by Carrier's steam apparatus.[177]

Few innovations benefited passengers so directly as air conditioning. It brought immediate and dramatic relief from the worst discomforts of summer railway travel. Lines which did not offer direct routes between two cities found that they were winning business from lines which did, simply because they had installed air conditioning. Patronage on the Chesapeake and Ohio's George Washington jumped 25 percent because of air conditioning.[178] Passengers who had not ridden the train for as much as twenty years decided to take the roundabout way between Washington and Cincinnati just to escape the heat. People were equally delighted by the relief from dust. On May 10, 1934, the B & O's Capitol Limited journeyed through a nationwide dust storm, and during its 17½-hour trip the air conditioning kept the interior pure and dust free.[179] It was reported that the train's intake box had accumulated dust equal to that normally taken in during a two-week period. The spread of such wondrous tales created an air-conditioning bandwagon. All major lines were signing up. A market nonexistent two years before was being served by eight manufacturers in 1932. Minor lines like the M K T pushed in ahead of their slower-moving, big-system competitors. Conservative New England lines such as the Boston and Maine, which could hardly complain of Sahara-like summers, rushed ahead in 1931 with ice-cooled cars.[180]

No one showed more enthusiasm than the Pullman Company. Air conditioning offered a dramatic luxury that might bolster its declining sleeping-car business at a cost far below the cost of new cars. With the Depression, outside orders for new equipment had evaporated. The idle shops and men might as well be put to work, so in 1935 and 1936 Pullman began a massive air-conditioning program. A year beforehand the company had settled on a direct-drive, mechanical system after many experiments. The compressor was driven by a shaft geared to one of the wheel axles (Figure 5.44). An electric motor plugged into station power acted as a standby, but during the journey the cooling plant was dependent on the motion of the train. The blowers were independently powered by electric motors. The shaft drive system became Pullman's standard and was used on most of its heavyweight cars.[181]

Now that satisfactory equipment was on hand, Pullman lost no time in applying it. In 1936 there were 5,800 air-conditioned passenger cars running in this country, about 40 percent of them Pullman-owned.[182] One year later the company had 3,300 air-conditioned cars, on which it had spent 25.5 million dollars.[183] Before World War II Pullman had all but completed its air-conditioning program; only its oldest cars lacked the new equipment. Progress for the industry as a whole is summarized in Table 5.1. (The percentage of passenger cars that were converted to air conditioning is greater than the totals in the table might indicate, since baggage and express cars are included in the total passenger car figures.)

Table 5.1   Air-conditioned Cars in Service Compared with Total Passenger Cars

Year	A-C Cars	Passenger Cars°
1933	2,375	56,100
1935	5,809	50,400
1940	12,204	45,200
1945	12,598	47,200
1950	17,350	43,500
1955	16,084	36,900
1960	12,450	28,300

° Figures include passenger, baggage, and express cars.
Source: Railroad Car Facts, 1961, published by the American Railway Car Institute. Figures given here for air-conditioned cars in the earliest years are considerably greater than those reported elsewhere. The 1936 *Master Car Builders Report* claims that only 658 air-conditioned cars were running in 1933.

Air conditioning had many of the same problems associated with car heating. The passenger car was as difficult to cool as it was to heat because of the large window area, frequently opened doors, short air throws, restricted power supply, and limited space for ducts and machinery. And because cold air will not rise, it had to enter the car through the ceiling, thus necessitating a secondary duct system. The ceiling ducts were used to good advantage in the winter to distribute warm air via an auxiliary heater coil. In summer, railway passengers generated far more heat than that found in most air-conditioned areas because so many warm bodies were crowded into a relatively small structure. As a general rule 60,000 to 80,000 Btu must be absorbed hourly to maintain a comfortable temperature in a passenger car.[184] This required a 7.5-ton unit, and in a mechanical system 10 to 13 horsepower was necessary. Depending on the type of equipment used, weights varied from 4 to 7½ tons. Installation costs in the 1940s ranged from $4,000 to $10,000.[185] In an effort to hold down the power drain and the overall size of the plant, 75 percent of the old air was recycled; the remainder was made up of fresh air.

Three basic systems were adopted for railroad car air conditioning: ice, steam ejector, and mechanical. Ice was the simplest and least expensive in first cost. Its modest power needs were perhaps its greatest attraction. Ice provided the coolant independently of power-hungry compressors. Existing electrical power supplies were often adequate to power the fans and circulating

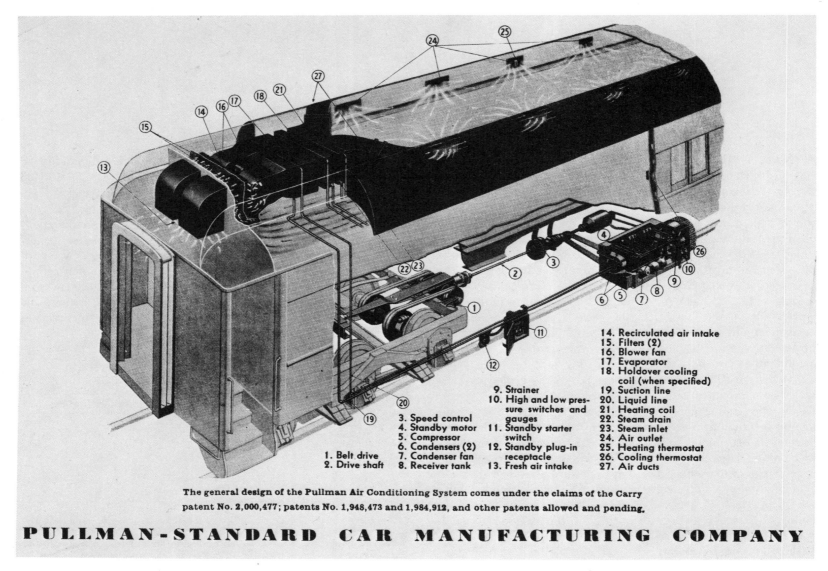

The general design of the Pullman Air Conditioning System comes under the claims of the Carry
patent No. 2,000,477; patents No. 1,948,473 and 1,984,912, and other patents allowed and pending.

PULLMAN-STANDARD CAR MANUFACTURING COMPANY

*Figure 5.44   Pullman's axle-drive air-conditioning system was fully developed by 1932.
A standby electric motor drove the compressor when the train was not moving. (Pullman,
Inc.)*

pumps. If not, a modest auxiliary generator and battery set could handle the extra load. Basic rewiring could thus be avoided. The first cost of an ice system was calculated at 60 percent of a comparable mechanical air-conditioner.[186] The maintenance cost was only one-third that of a mechanical unit. It was all so simple: ice melted in a bunker below the car, the cold water was pumped to a radiator in the ceiling, and a fan blew air over the radiator into the passenger compartment, lowering the temperature as much as 14 degrees. The water returned to the bunker for recooling, and the warm surplus was discharged overboard.

One of the first railroads to adopt the ice system was the Boston and Maine. It was drawn into the experiment through the enlightened self-interest of the Metropolitan Ice Company of Boston, which sponsored the tests.[187] For an average New England summer—short and relatively mild—the ice system made sense. The investment was small, and the trains did not have to haul the dead weight of a mechanical system. In cool months the ice bunkers were empty, thus eliminating the greatest portion of the system's weight.

Yet while ice seemed suitable mostly for the Northeast, there were other large users of this most primitive form of air conditioning: Pullman, the Southern Pacific, and the Pennsylvania. By 1939 the country's railroads had nearly 3,800 ice-cooled cars.

After this time, however, the defects of the system were recognized as overwhelming. Ice was never really effective for sustained hot weather, such as that encountered in the South, Midwest, and Southwest. It was expensive to operate: at $5.29 per 1,000 miles, it cost five times more to use than mechanical air conditioning.[188] This cost was directly related to that seemingly cheap commodity, ice. It was not the price per pound but the quantity consumed and the labor required to handle it at the storage depots en route. A railroad car consumed approximately 500 pounds of ice per hour.[189] Fifteen 300-pound blocks equaled 4,500 pounds of ice—enough for nine hours, or 360 miles. Some roads installed giant 8,000-pound-capacity boxes on their cars, but the basic disadvantages of the ice system remained unsolved. After 1945 almost no new ice-cooled cars entered service, and those already running were gradually converted or retired.

Willis Carrier, who had at first championed electromechanical air conditioning for railway service, decided later that the power problem could best be solved by switching to a system requiring minimal electric input.[190] Steam, after all, was in plentiful supply. Borrowing a technique from elementary physics, where hot water is made to boil at a low temperature in a vacuum chamber and thus to give up much of its heat, Carrier felt that this simplest of all heat exchangers was perfect for car air conditioning.

1. Belt drive
2. Drive shaft
3. Speed control
4. Standby motor
5. Compressor
6. Condensers (2)
7. Condenser fan
8. Receiver tank
9. Strainer
10. High and low pressure switches and gauges
11. Standby starter switch
12. Standby plug-in receptacle
13. Fresh air intake
14. Recirculated air intake
15. Filters (2)
16. Blower fan
17. Evaporator
18. Holdover cooling coil (when specified)
19. Suction line
20. Liquid line
21. Heating coil
22. Steam drain
23. Steam inlet
24. Air outlet
25. Heating thermostat
26. Cooling thermostat
27. Air ducts

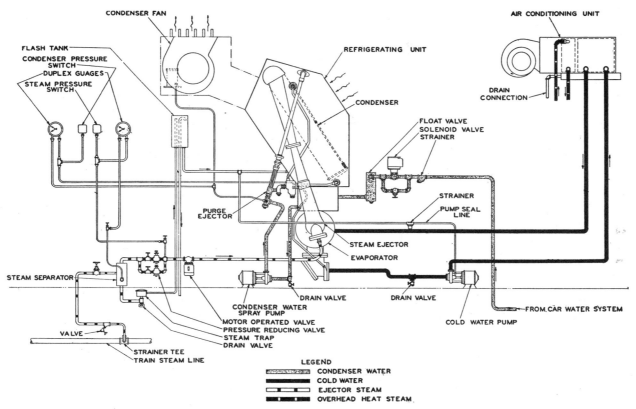

Piping diagram of type RE steam ejector equipment.

The Safety Car Heating and Lighting Company, Inc.

Figure 5.45   Steam air conditioning was popular on some U.S. roads because it required only modest electrical service. But it proved so wasteful of steam that it was replaced by electromechanical units before 1970. (Car Builders' Cyclopedia, 1953, p. 709)

After two years of experimentation, he placed a commercial steam ejector air-conditioning unit rated at 5 tons' capacity aboard a Santa Fe car. Carrier worked jointly with the Safety Car Lighting and Heating Company and the Vapor Heating Corporation in perfecting the details of the 5,000-pound apparatus.[191] Its sponsors claimed that its steam consumption (206 pounds per hour) was less than that required for heating. Standard car-lighting power (32 volts DC) was adequate to run the small auxiliary fans and pumps.

Many railroads accepted the arguments in favor of steam ejectors, and it became a popular method of air conditioning. By 1937 nearly 1,100 cars were being cooled by steam; some lines had a few experimental installations, while others, like the Santa Fe, were enthusiastic converts. However, steam air conditioning had many drawbacks. It was deceptively simple in principle, but like many schemes with "only one moving part," it was bulky, inefficient, and dependent on countless auxiliaries (Figure 5.45). At $8,475 a car, it was the most expensive of the three major types of car air-conditioners.[192] While it scored well on maintenance, it was tricky to operate. And in spite of manufacturers' claims, it turned out to be a steam hog, consuming one-third more steam than that required for heating.[193] As steam locomotives vanished from passenger service during the 1950s, the ready supply of power for ejector air conditioning also disappeared. Some lines converted immediately to mechanical air conditioning; others, like the Seaboard, struggled along with head-end flash boilers until the late 1960s, when they too converted to electromechanical units.[194]

Mechanical air conditioning was recognized as the best method for cooling cars (Figure 5.46). But modest train-board power and inefficient refrigerants inhibited its rise to prominence. Ammonia, successfully employed for generations in artificial ice plants, was the best refrigerant available. However, it was not well suited to compact cooling units because it required a bulky cooling tower and because brine was used as the coolant, since it was not considered safe to introduce compressed ammonia directly into the passenger compartment. A safe artificial coolant called Freon that came on the market around 1932 effectively solved this problem. The power problem was solved by spending money: additional batteries, larger generators, or gasoline-powered auxiliary units were costly but necessary for electromechanical compressors of sufficient capacity (Figures 5.47 and 5.48).

The Waukesha Motor Company produced one of the most popular independent engine-driven AC units. Chunky four-cylinder engines powered 7-ton-capacity ice engines. By 1943 the *Car Builders' Cyclopedia* claimed that twenty-five railroads were using these under-floor units. The Milwaukee Road, the Frisco, the Illinois Central, and the Southern Pacific were all enthusiastic Waukesha customers. Some of these lines also installed a second motor-driven power package to furnish current for lighting and other electrical auxiliaries. After about 1940 Waukesha pioneered the introduction of propane-fueled engines. They were cleaner and easier to maintain than equivalent gasoline or diesel engines, as well as being more convenient to refuel at station platforms. Instead of risking fuel spills in a congested passenger area, it was possible to exchange fuel tanks. But Eastern roads, particularly those entering Manhattan terminals, banned propane-powered units and required removal of the tanks before a car so equipped could enter the approach tunnels. According to W. D. Edson,

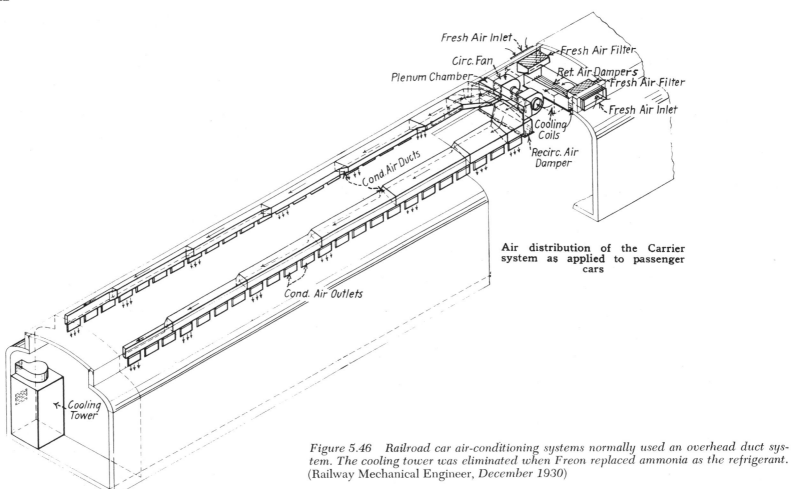

Figure 5.46   Railroad car air-conditioning systems normally used an overhead duct system. The cooling tower was eliminated when Freon replaced ammonia as the refrigerant. (Railway Mechanical Engineer, *December 1930*)

who supplied much of the data in this paragraph, few Eastern lines employed independent under-floor AC or generator units.

Pullman avoided the electric supply problem by devising a direct mechanical drive for the refrigerant compressor.[195] A gear box was fitted to one of the wheel axles, with a shaft to the compressor. A governor protected the compressor from overspeed. The system worked well and was adopted by some railroads for non-Pullman equipment. Although it was not efficient where there were frequent or prolonged stops, for short stops the residual cold air, together with the standby electric-powered blower, acted to maintain the temperature at a comfortable level (Figure 5.44).

Railroad air conditioning entered a final stage following World War II. The trend was toward developing more efficient large-capacity units, for 6- to 8-ton-capacity installations were no longer thought adequate.[196] Engineers now felt that 10-ton units were needed, and compressors of this size could not be economically driven from the 32- or even 64-volt DC service then common. These low-voltage systems were favored because of the train's dependence on batteries. As a general rule, the higher the voltage, the more expensive the battery. But there is a trade-off in favor of 110-volt circuits in terms of the lighter wiring possible; low-voltage systems require very heavy and hence expensive wiring if they are to service power-hungry units such as air conditioners. Thus 32 volts was satisfactory for lighting but not for motor currents. In addition, 110-volt AC systems could make use of conventional and cheaper motors, switches, and other

Figure 5.47   After air conditioning, the size and complexity of the electrical system required a closet in place of the small cabinet shown in Figure 5.48. (Trains *magazine collection*)

Figure 5.48   Before air conditioning, train-board electrical service was essentially a low-voltage lighting system. The modest wire cabinet here was typical of the 1920s. (Pullman Neg. 28894)

components. Most AC cars were equipped to run on 220-volt AC at stations, so that it was possible during layovers to plug standby power into a jack fastened on the underside of the car.

Westinghouse placed a 220-volt AC three-phase unit in service during the spring of 1947. With a more potent power package, overall weight was cut to just over 1 ton—a remarkable improvement over conventional equipment. To remove drag on the locomotive imposed by axle-driven generators, gasoline- and diesel-powered units became more popular in the postwar years. The trend today is toward head-end power.

Lighting

Some form of artificial lighting has always been necessary in the Northern United States, where daylight is short in the winter season. As night travel by railway increased, so did the need for lighting. It was provided at first in niggardly amounts: one or two candles were considered adequate to ensure safe passage in or out of a car. Common road coaches of the day offered nothing better, and in the 1830s even domestic lighting was provisional. In fact travelers from the backwoods had never enjoyed this luxury and were usually content to sleep on board until the journey's end.

In urban areas, however, artificial illumination was more advanced. Parlors, theaters, and stores were well lit; candles and oil lamps were plentiful. By mid-century, central gas systems were common in most large cities, and the affluent city dweller was appalled by the night train's "gloomy cage."[197] It was like stepping into some rude country tavern where thievery flourished in the shadows. Reading was impossible, as was conversation or any other civilized pastime. The educated traveler, who was usually

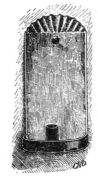

Figure 5.49   The earliest passenger car lighting needs were satisfied by tin-plated candle sconces. (Railway Age Car, *pamphlet, 1883*)

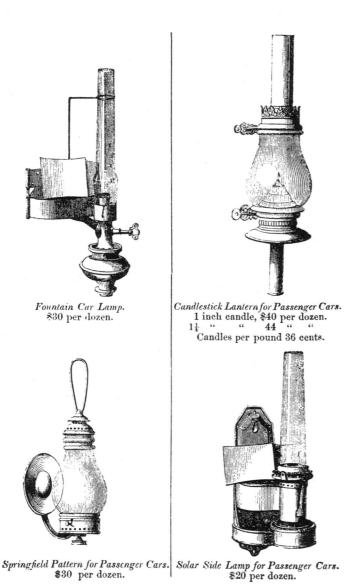

Fountain Car Lamp.
$30 per dozen.

Candlestick Lantern for Passenger Cars.
1 inch candle, $40 per dozen.
1¼ "   "   44 "   "
Candles per pound 36 cents.

Springfield Pattern for Passenger Cars.
$30 per dozen.

Solar Side Lamp for Passenger Cars.
$20 per dozen.

Figure 5.50   By 1860, candle and oil side lamps were used for car lighting. Center lamps came into favor only after the clerestory roof elevated the ceiling. (Bowles Railway Supply Catalog, *1860c., p. 141*)

age. In 1846 a traveler wrote: "A solitary lamp burned at one end of the car and its sickly light fell upon the faces of the sleepers."[198] Almost forty years later another traveler spoke of "the dim religious light" that was sufficient only to delineate the extent of the darkness.[199] Newsboys offered papers and magazines as if in "playful irony." The only diversion was to watch the orange flame of the lamp swimming in its glass globe like a smoky goldfish.

The technology for good lighting was available. Even candles could do an excellent job if enough were provided. Oil lamps were better, but this apparatus was expensive because it required eight burners instead of one or two fixtures. And as long as they could, most railroads kept their purses tightly closed.

Cost was an important consideration to railroad managers. A few cents an hour could sway the decision for or against better lighting. Trivial as such economies might seem, small savings could mount up with a large number of cars. Discussions of lighting in the trade press and the Master Car Builders Reports always emphasized cost. In one case an official discussing the whole matter of passenger comfort agreed that all reasonable efforts should be made to provide better accommodations, such as lighting, but candidly said that he did not feel a responsibility to indulge the public.[200]

It was not until after 1880, when the public outcry had grown loud enough, that the industry was willing to provide good lighting.[201] It is likely that first-class cars had been outfitted with proper lighting a decade earlier, because travelers who paid a premium felt entitled to the small luxury of an extra lamp or candle.

By the time railways became a public carrier, the art of artificial lighting had already passed through many stages and, like an ancient tree, possessed many branches and roots. At one time or another every conceivable burnable material, from yak fat to rape seed oil, had fueled the world's lamps. In the western nations by the early nineteenth century, candles were the most familiar source of lighting. Simple, smokeless, and relatively safe, they were the first source of light aboard railroad cars and enjoyed a long reign—from 1830 to about 1875. Tallow candles were the most common, but they were soft and tended to droop in warm weather and in direct sunlight. They were also eaten by rats and mice. Spermaceti candles, made from a rich wax extracted from the sperm whale, were the finest available, although they were not widely used because of the cost. After about 1860 paraffin wax manufactured from petroleum produced a cheap grade of candle that avoided the worst defects of tallow.

Originally two candles per car was considered adequate. Precise information is unavailable, but the 1836 Camden and Amboy coach may be taken as representative of the period (Figure 1.72) It has a small candle lamp built into the end bulkheads at diagonal ends of the body. At about the same time, tin side lamps were used (Figure 5.49). A flat sheet of tin-plated iron formed the reflector. The lamps were fastened to the interior wall panels between the windows. Lamps of this type were remembered by George Fowler, who said that there were rarely more

not content to bed down at sunset, was obliged to sit and stare out of a dark window because the railroad was too cheap to provide decent light.

Trainmen also cursed the dark cars as they stumbled through the aisles. Their dim lanterns were more than signal lamps; the conductor would raise his to read a ticket or decipher a timetable, and the brakeman needed his to move from one swaying platform to the next or to guide a passenger down the steps at a dark station.

Complaints began early and continued well into the railway

Figure 5.51 *This 1874 advertisement testified to the continuing popularity of candle illumination. The center lamp was mistakenly printed upside down.* (Poor's Manual of Railroads, 1874)

THIS CUT REPRESENTS ONE STYLE OF
CENTRE LAMPS.

THIS CUT REPRESENTS THE SIDE
CAR LAMP FOR CANDLE.

DANE, WESTLAKE & COVERT,

MANUFACTURERS OF

WESTLAKE'S

Loose Globe Car Lamps & Lanterns.

Send for Price List and Circulars.

228 LAKE STREET, CHICAGO, ILL.

than six to a car.[202] Large, 2-inch-diameter candles were used, but the light was too dim for reading. Around the middle of the century, open candle lighting gave way to more elaborate fixtures. Candles were placed in tubular spring-loaded holders. The cap of the tube was rolled over, leaving a small circular opening, and the spring pushed the candle up against the opening as it burned. This kept the flame in one position, making more efficient use of the reflectors. Glass globes or chimneys protected the flame from drafts—a necessity in a vehicle as breezy as the railroad passenger car. Side candle lamps were used in low-ceilinged, arched-roof cars (Figure 5.50). After 1860 when clerestories came into fashion, more elaborate center lamps were used. They provided aisle lighting, and the better class of car also had side lamps for reading.

After 1875 candles were considered obsolete for car lighting, although of course they did not disappear entirely.[203] In Ohio, candles were in fact the only legal form of car lighting after 1869, but the 1877 *Ohio Railroad Report* admits that the law was not actively enforced and that most lines employed oil lighting. Candles remained a secondary form of emergency lighting, much as they do today in most households. As late as 1942 the *Car Builders' Cyclopedia* showed a single candle side lamp for emergency lighting.

Historically, oil lamps predate the candle, but they do not appear to have been much used in railroad cars before the 1850s. They were more expensive and demanded more attention: filling the reservoirs and trimming the wicks was a chore. They gradually gained in popularity, particularly in areas having a good supply of oil, because a good oil burner gave more light at less cost than candles. Before the age of petroleum, however, there was no cheap lamp oil. A great variety of natural vegetable and animal oils—rape seed, whale oil, refined lard, and even Colza, a light oil made from a variety of cabbage—found their way into railroad car lamps. In the 1840s the Boston and Albany was using whale oil, which was said to be smoky.[204] Other New England roads employed the more expensive sperm oil with good success, and one line as late as 1891 thought it was superior to gas lighting.

One of the first artificial fuels introduced was camphene—a flammable mixture of redistilled turpentine and high-proof alcohol. It was first used in 1820 and grew in popularity as the price of whale oil increased. Just how early it was adopted for car lighting is uncertain; the *Illustrated London News* for April 10, 1852, mentions camphene lamps on a New York railroad. An engraving of the car's interior accompanying the article shows end corner lamps, indicating that even though candles were no longer used, the car had only two lamps. Although camphene was cheap and gave good light, it was explosive. It was said that "the friends of temperance will not object to the burning of alcohol," but camphene became notorious as a cause of fires and explosions.[205]

A safer artificial lamp oil was introduced commercially by a Canadian inventor in 1846. The oil was distilled from coal and

was commonly called coal oil, but its inventor preferred the trade name Kerosene. Commercial production began in 1856 at a cost of $1 per gallon. At the time whale oil cost $3 a gallon, lard oil $4, and sperm oil $7. Within four years 30,000 gallons of coal oil were being produced every day.

Coal oil was no sooner established than it was displaced by an even cheaper lamp fuel. Petroleum, known since ancient times as a curiosity or medical curative, was found in great subterranean pools in western Pennsylvania. Oil wells began to pump kerosene-rich crude in 1859. Petroleum was a far better source of kerosene than coal; a gallon of crude would yield nearly 3 quarts of kerosene. By 1865 improved refining methods brought the cost of production down to 5 cents a gallon. Its abundance and cheapness prompted a massive conversion to kerosene lighting. The *Western Railroad Gazette* of July 18, 1863, mentioned the sunlight glow provided by six kerosene center lamps on one of Pullman's new cars. The spread of this form of lighting in later years was attributable not only to its improved illumination but to its economy. A large kerosene fixture produced 24.5 candlepower at a cost just under 1 cent per hour.[206] Five candles could produce an equal amount of light, but at a cost of 12½ cents per hour.

Eighty years before kerosene was available, oil lamp technology had made some major advances. Possibly the most important were achieved by a Swiss, Aime Argand (1750–1803). In 1780 Argand devised an adjustable tubular wick holder with a center

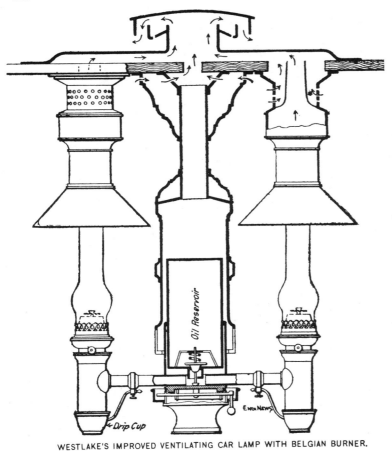

WESTLAKE'S IMPROVED VENTILATING CAR LAMP WITH BELGIAN BURNER.

Figure 5.52   Oil lamps became a major form of car lighting after 1860. The fumes were exhausted through rooftop ventilators. (A. M. Wellington, W. B. D. Penniman, and C. W. Baker, The Comparative Merits of Various Systems of Car Lighting, *1892)*

*Figure 5.53   A large oil-burning side lamp made by Post and Company of Cincinnati in 1882. (*National Car Builder, *1882 advertisement)*

tube for a good air supply. A glass chimney protected the flame from drafts and increased the area of combustion without cutting off the light. The gain in illuminating power was spectacular. All subsequent advances in oil lamp technology were really variations on Argand's original innovations. Beneficial as his design proved, however, oil lamps were less than perfect. The light was always too high or too low, depending on the quality of the oil and the condition of the wick. And air currents inside the car caused the lamps to smoke, although this defect was partially remedied when roof vents were installed above the chimneys (Figure 5.52).

The railroads' stinginess about the quantity of light continued to be a problem. Even at operating costs of a penny an hour, managers were reluctant to install more lamps than were absolutely necessary. In 1878 the Illinois Central began replacing candles with oil lamps on its suburban cars.[207] Three center lamps cost approximately $100 installed and produced approximately 72 candlepower. It was calculated that 120 candlepower was necessary for good illumination, but the railroad was not about to squander an extra lamp per car so that commuters might read their papers.

Actually there appears to have been more concern over lamp safety than over the quantity or quality of illumination. The fire-peril hysteria included oil lights as well as car stoves. Throughout the kerosene lamp era there was a widespread belief that such fixtures added materially to the dangers of railway travel. In reviewing the history of car fires in 1890 after nearly thirty years of oil lighting, *Engineering News* concluded that the hazard was a popular myth.[208] All major disasters involving fires could be

traced to stoves or to the locomotive's firebox. In only one instance could a fatal car fire be blamed on a lighting fixture. This accident occurred on the Great Western Railway of Canada in February 1874. A small lamp in the water closet broke loose and fell to the floor, starting a fire that went undiscovered. Before the train was stopped, eight passengers had been suffocated or burned to death; another thirteen had been injured. The car was totally destroyed. Yet this toll was minor when compared with such horrors as the 1867 Angola tragedy, when forty-two people died. Newspaper accounts blamed the kerosene lamps for the ensuing fire that claimed so many lives, but later reports corrected the error. They said that the wreck had occurred in daytime, and that candles were the only form of lighting on the Lake Shore cars.[209] Nevertheless, the image of flaming oil lamps was fixed in the public's mind; corrections had no effect. A few months earlier a locomotive plowed into the rear of a train on the Cincinnati, Hamilton, and Dayton Railroad. Its kerosene headlamps ruptured, setting the telescoped rear car ablaze. Not long afterward the palace car *City of Chicago* went up in smoke, and

Figure 5.54 An oil-burning center lamp manufactured by Howard and Company of Boston around 1876. (A. A. Merrilees)

again the blame was initially placed on the lamps. Later newspaper stories reported that the stove was the actual source of the fire, but they were lost among the back pages.

When kerosene was first introduced, railroads were foolhardy enough to use a household grade. Some bought from unscrupulous dealers who added naptha, thus lowering the cost but making it a volatile fuel too dangerous for lighting purposes.[210] After a few lamp explosions, railroads adopted less flammable, heavy oils sold under such reassuring trade names as Mineral Sperm Oil. They were free of naptha and were guaranteed to produce no vapor until heated to 300 degrees F.[211] They would not burn at less than 600 degrees. In addition to the built-in safety of high fire test oils, the railroads contended that the sudden movement of a derailment or collision would extinguish the lamp's flame.

Granting that the fire peril was greatly exaggerated, oil lamps did have many disadvantages. For all their ingenious burners and chimneys, they were smoky and evil-smelling. Many travelers objected to the pungent odor of kerosene, and more complained about dripping oil caused by leaking reservoirs, loose caps, or indifferent filling. Oil not only dripped on carpets, seats, and window hangings, but stained the passengers' clothing and luggage, creating a steady flow of work for the claims department.

Gas lighting was the best method of artificial illumination available in the nineteenth century. However, it was generally limited to wealthy homes, public buildings, and commercial establishments, because it was complex and expensive. Attempts had been made in the eighteenth century to use it for domestic lighting, but little real progress was achieved in this field until after 1800. In 1812 a city gasworks was established in London; four years later Baltimore opened a plant; and by 1850 over fifty American cities had gas. The number quadrupled in a decade, so

that gas lighting was common in most U.S. cities by the beginning of the Civil War.

The earliest attempt to use gas for car lighting that can be discovered was reported in 1834.[212] An unnamed British railway was said to be experimenting with a portable gas system. Many years passed before any similar tests were made on this side of the Atlantic. In 1851 the Hudson River Railroad began outfitting its cars with gasometers of sufficient capacity for a trip between New York City and Albany. The apparatus was more than a temporary experiment, since it stayed in service for at least five years.[213] A number of other railroads, including the Galena and Chicago Union, the Troy and Boston, and the Camden and Amboy, also used gas for car lighting during the fifties.[214] Many of these lines adopted a system perfected by Charles Starr of New York (Figure 5.55). Star compressed ordinary city coal gas to a pressure of 200 pounds in tanks with a capacity of 120 cubic feet. The meter was a clockworks device. The entire apparatus cost about 100 dollars per car. Starr's system was also installed with good results on several New York Harbor ferryboats.

The Troy and Boston's system, devised around 1856 by Marcus P. Norton of Vermont, employed a single gasometer in the baggage car for the entire train rather than individual tanks under each car. Norton's plan required some method of flexible coupling between cars, which undoubtedly proved troublesome.

In 1860 the Philadelphia and Reading began what was probably the first large-scale, long-term use of gas for car lighting. This move was encouraged by the success of the New Jersey Railroad and Transportation Company, which had replaced its stearin candles with compressed-coal-gas lamps.[215] Even considering the high cost of city gas—$3 per 1,000 cubic feet—the New Jersey line claimed a cost saving of 30 percent over candle lighting. A railroad could cut the cost of gas by making its own, which both the New Jersey and Reading lines elected to do.

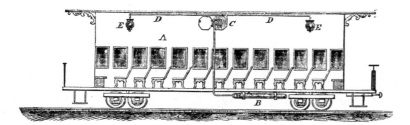

Figure 5.55 *Gas lighting apparatus offered by the New York Car and Steamboat Gas Company in 1860. B marks the gas tank, C the regulator, and E the lamps.*

Within a few years the Reading decided to upgrade its lighting by converting to a better type of gas. In 1874 the road built a new plant in Reading to manufacture gas from petroleum for lighting both its cars and its repair shops.[216] The gas was far richer than the coal gas previously used; one tank of oil gas would last five nights, whereas an equal charge of compressed city gas was consumed in a single evening.

In 1860 or 1861 the Pennsylvania began compressing city coal gas at its Altoona car shops.[217] Tanks measuring 7 feet, 6 inches long by 14 inches in diameter held enough gas for sixteen to eighteen hours of service. The gas was compressed to the extraordinary pressure of 600 psi. Apparently the system was not notably successful; coal gas that was compressed to such a high pressure lost most of its illuminating power. It was also claimed that an accident associated with filling the tanks had started a fire that destroyed several cars.[218] The Pennsylvania temporarily abandoned gas lighting, but by 1871 the road was recommitted to it.[219] Thirty-five cars were already fitted with gas apparatus, and plans were underway to convert the entire fleet. The gas was made from petroleum. Each car contained five burners rated at 17 candlepower each and a tank carrying a sixty-hour supply of gas. By 1876 the line had compressor plants at Altoona and Jersey City.[220] In addition to the gaslights in a large number of its cars, the Pennsylvania's ferryboats were also being lit with compressed gas, though it was stated that city gas was again being used in these installations.

During the decade of the seventies other roads, such as the Chicago and Alton and the Erie, were using gas to some degree. It appeared likely that gas would become a general form of car lighting. Had it not been for the introduction of cheap kerosene a few years earlier, American railroads might have passed swiftly from candles to gas. But oil lamps were so cheap and gas so complex and costly that most roads compromised on kerosene. The Erie, for example, decided against expanding its gas-lighting program when its calculations showed that gas cost three times more than oil.[221] The price of kerosene lamps was half that of gas apparatus, mainly because the lamps did not require tanks or piping.

In addition to cost, early gas lighting was not entirely satisfactory. Coal gas could not stand compression without a substantial loss of illuminating properties, yet it had to be greatly reduced in volume or a reasonable supply could not be carried aboard a railroad car. In addition, it gave off carbon that fogged the lamp globes.

A gas rich in hydrocarbons could undergo considerable compression with no major reduction in its illuminating powers. Many engineers knew that such a gas could be produced from heating petroleum in a retort. As already noted, both the Pennsylvania and the Reading attempted to produce gas in this form during the 1870s. Why it was not successful is unclear. Late in the same decade an independent mechanic, J. M. Foster of Philadelphia, tried to work out a petroleum gas system and had at least moderate success.[222] His gas, manufactured from naptha or crude oil, was compressed to 40 atmospheres. With twenty burn-

ers he could produce 260 candlepower. The lamps were ignited by an electric spark. Foster's system was used both on the New Haven Railroad and in Mann's sleeping cars before an accident in the compressor plant around 1890 put an end to Foster's American Lighting Company.[223]

Just as European railways showed more concern than U.S. lines for improving car heating, so too they led the way in developing better forms of car lighting. While America tinkered with gas, Europeans methodically moved toward its large-scale application. In 1861 a Belgian railway began using enriched gas made from rock shale oil.[224] Among the other Continental inventors drawn to this field was an obscure tinsmith and lamp maker from Berlin, Julius Pintsch (1815–1884).[225] He began to develop a system of gas lighting specifically suited to railway service. Pintsch was not the first to explore the subject or to recognize the defects of coal gas; nor did he invent petroleum gas. But he does seem to have been more single-minded in his effort to perfect every detail both of the gas production and of the lamps and fixtures for use on railway cars. This concentration of effort resulted in the most widely adopted system of gas car lighting in the world. The name Pintsch became synonymous with gas lighting.

Pintsch began his experiments in 1867 and within two years was installing lights in Prussian cars. Between 1871 and 1878 the Baden State, Lower Silesian, and Alsacean railways adopted his system. Several British lines, including the London underground, converted to Pintsch lights at the same time. Steamships and buoys began to use Pintsch gas. By 1880, 6,000 railway cars were lit with lamps manufactured by the Berlin tinsmith.[226] In 1881 Pintsch decided to invade the American market and opened an office in New York.

The Erie appears to have been the first customer, contracting with Pintsch for 100 sets of lights.[227] In July 1882 a special train ran over the Erie with lighting so brilliant that it was actually possible to read in the cars. This was considered miraculous. Each car had a 344-cubic-foot cylindrical gas tank mounted under the floor (Figures 5.56 and 5.57). The gas, compressed to 127 psi., passed through a regulator valve box that was also under the floor and then through ¼-inch-diameter gas pipes to the lamps. The gas was reduced to less than 1-ounce pressure by the regulator before entering each of the thirteen burners. Every "sun burner" was rated at 17 candlepower. The lamps burned 8½ feet of gas an hour at a cost of 1.7 cents.

In 1883 the West Shore contracted for Pintsch installations on eighty-two cars, the New York, Providence, and Boston for installations on five cars, and Pullman for an indeterminate number.[228] A major obstacle to Pintsch in expanding the business was that the company had a lone gas plant in New York. Only lines operating in that locality could be serviced. Progress was slow in the American branch; by 1888 just over 400 cars had Pintsch lighting in the United States. In Europe meanwhile, 27,000 cars carried Pintsch lights, and Count de Rothschild was a major backer.

In 1887 several American capitalists decided that the Pintsch

Figure 5.56 *Pintsch produced the most popular gas lighting system for railroad cars in the world. H is the gas tank, R the regulator, N and M the lamp vents, and V the filler valve. (International Textbook Company, 1912)*

system was a good thing which needed some capital and an aggressive management.[229] George Pullman, Sidney Dillon, and T. C. Platt provided the cash. They hired Arthur W. Soper, a minor but respected railroad manager (and at one time general manager of Wagner), to run the business. The firm was reorganized as the Safety Car Heating and Lighting Company. In 1888 only four railroads were using Pintsch lamps. There were now six compressor plants, all in the Northeast except for a lone outpost in Marion, Ohio.[230] Soper began to push. It was easy to sign Pullman, who had a direct stake in the firm's success. Wagner was forced to follow, because its management could find no superior form of lighting available. By January 1891 twenty-nine railroads with 2,000 cars were using Pintsch lamps. Fourteen compressor plants were in operation in all regions of the United States. From this time on, the firm was a runaway success.

As line after line signed on for Pintsch lights, gas became a major form of car lighting in America. For the first time cars were being well lit, and lit cheaply enough so that gas could be used not just for the luxury sleeper and parlor trade but also for the workaday coach. In 1893 the Lackawanna began gas-lighting its suburban cars. It was reported that passengers were willing to stand in Pintsch-equipped cars rather than sit in kerosene-lit coaches, because at least they could read the evening paper while they were standing. Oil lamps were declared (prematurely, to be sure) "a relic of a past age."[231] The Chicago and North Western found that gas lighting reduced labor costs; the men who had been needed to clean and maintain the oil lamps were let go, and car interiors required less frequent cleaning now that the smoky oil lamps were gone.[232]

The progress of Pintsch lighting was recorded in advertise-

Figure 5.57 *A Pintsch gas center lamp of about 1895. Pintsch lighting was a major system on American railroads between 1885 and 1910.*

ments and news accounts fed to the trade press. By 1897 the Safety Car Heating and Lighting Company claimed that 10,809 cars carried its Pintsch equipment. Two years later its near-monopoly of gas car lighting was strengthened when the Reading Company closed its own gas plant to adopt Pintsch.[233] Foster was already out of business, and only a few other competitors were still feebly stirring. In February 1900 the New York Central announced plans to convert entirely: all 1,500 cars would glow with Pintsch gas, and all oil lamps would be scrapped.[234] Four years later it was reported that on a worldwide basis, 120,000 railway cars used Pintsch lighting. Of this total 20,000 were on U.S. railroads. The number of American compressing plants had grown to seventy. In 1908, probably its peak year, the Safety Company claimed that its system was installed aboard 32,455 cars in the United States and Canada.[235] This represented approximately 60 percent of all passenger cars in service.

Pintsch gas maintained its leadership in the field by offering a good product and by buying or squeezing out rivals. After the Foster company's collapse, Pintsch had few serious competitors. A minor threat was posed by the Frost "dry carburetter" system, which appeared on the market in 1884 (Figure 5.58). It was a gasoline vapor light especially designed for safe use on railway cars. To avoid the hazards of fluid gasoline, the reservoir or tank was packed with wicking, which absorbed the fuel and thus produced the so-called dry carburetor. The gas or vapor was made by passing compressed air, drawn from the air brake system, through the carburetor at a very low pressure.

In the spring of 1885 the Frost system was given a series of tests on the Pennsylvania Railroad.[236] The gas tank was placed inside the car. In later installations, however, individual carburetors were made for each lamp and were placed on top of the roof. This not only removed the gasoline from the passenger compartment but also placed the tank above the lamp chimney in order to heat the reservoir and improve vaporization in cold weather. The new style of carburetor, 30 inches in diameter and 4 inches deep, was like a flat cake pan with a hole in the center to ventilate the lamp. It was made of copper, so that it might crumple but would not burst in the event of a wreck. Each carburetor held 4 gallons of gasoline—enough for forty to fifty hours of lighting.[237] After tests made in 1889, the New York State Board of Underwriters declared it safe "beyond the suspicion of danger."

The Pennsylvania was satisfied not only with its safety but with the brilliance of its light and the economies it made possible. Frost burners were rated at 40 candlepower—more than twice that of Pintsch lights.[238] While they cost 3 cents an hour to operate, which was more expensive than Pintsch, the Frost system required no gas plant. It produced its own from an explosive liquid that distilleries could not sell for lamp oil and usually dumped in the nearest creek.

By 1892 the Pennsylvania, always the main user of Frost lamps, had installed them on 776 cars.[239] The road announced plans to equip all 3,700 cars on its associated lines with the lamps at a cost of 1 million dollars. Although it claimed that all the

Frost system's problems had been solved, some complications apparently developed, because the conversion never took place. Several other railroads, such as the Norfolk and Western, the Reading, the Burlington, and the Wisconsin Central, also used Frost lamps. Estimates made in 1900 were that 1,000 cars had this form of gas lighting—not a significant proportion of the market.[240] Among the drawbacks of the Frost system was that it did not give uniform light. Impure gasoline could foul the apparatus, and in hot weather the mixture tended to be overly rich, causing the flame to glow orange and smoky. In 1900 it was reported that Frost had been absorbed by a rival firm (presumably the Pintsch system's Safety Car Heating and Lighting Company), which was withdrawing the product from the market.[241]

Shortly before Frost was safely out of the way, another gas system was introduced to disturb the Safety Company's comfortable monopoly. Around 1898 the Chicago railway supply firm of Adams and Westlake began to produce acetylene lamps adapted for car lighting.[242] Derived from carbide of calcium and water, the gas had been known to science since 1836 and was one of the most potent illuminants available. The Texas Midland and the Santa Fe railroads tried the acetylene lamp system with some success. At one time the Safety Company even considered mixing 20 percent acetylene with Pintsch gas to double the candlepower. But beyond this mild interest, acetylene made little progress in car lighting. It was considered dangerous, it gave off an offensive odor, and it was highly corrosive.

During the peak years of gas lighting, 1890 to 1910, the only serious rival to Pintsch gas was electricity. But the early electric train-lighting apparatus was handicapped by mechanical defects and high costs. In addition, technical improvements within the field of gas lighting itself prolonged its domination of the market. In 1885 Carl von Welsbach of Vienna patented a mantle made of cotton thread and the oxides of metals called rare earth. The gas flame heated the mantle, making it incandescent and thus vastly increasing the amount of light produced. Mantle fixtures cut gas consumption by two-thirds, while improved burner design boosted efficiency by 300 percent.[243] Thus the cost of gas lighting fell dramatically even as its power of illumination soared. In 1909 a comparison of gas and electric train lighting underlined the superiority of the Pintsch system.[244] With a base of 400 candlepower at six hours, Pintsch cost 12 cents while electricity cost $1.42.

Although Welsbach's mantle was adopted for domestic lighting during the 1890s, it came into favor more slowly in other areas. Just why this was true, considering its remarkable benefits, is not understood. Yet the *Railway Review* of June 8, 1907, noted that the ordinary flat flame jet continued to prevail, that only 2,000 cars were fitted with the mantle-style burners, and that all these had been installed within the past two years.

But the resurgence of gas lighting was a temporary victory; electric lighting was the coming technology, and railroads used it on luxury trains despite the higher cost. The Pintsch system also had its faults, the chief of which was the terrific expense of the

THE FROST CARBURETTER SYSTEM OF CAR LIGHTING

THE GREATEST LIGHT OF THE AGE.

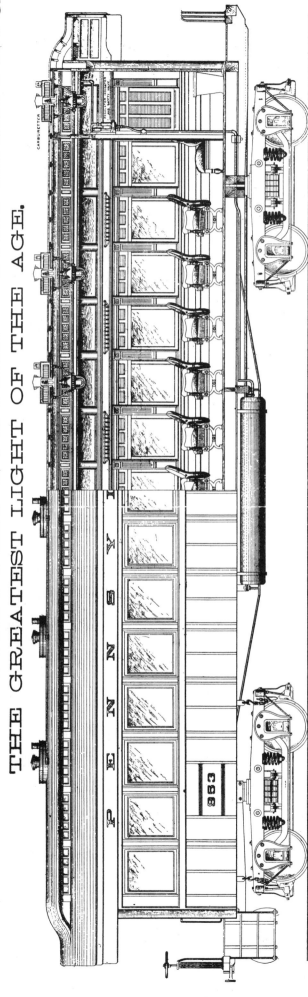

Standard Arrangement of Parts as Adopted by Pennsylvania Railroad.

The The **Frost Dry Carburetter System**, for lighting passenger and other Railroad Cars, is now so thoroughly perfected that it can be presented to Railroad Managers with the utmost confidence in its simplicity, cheapness, safety and vastly enlarged volume of light.

It has been well tested and then adopted by the Railroad management that controls the largest car equipment of the continent, and all that is claimed for it can be so fully and so easily verified, that it needs only unbiased inquiry to assure its prompt adoption.

All that is asked for this wonderful combination of simplicity, safety and economy, with vastly the best lights ever put in a car, is impartial investigation and a test of its merits, and it will speedily command the approval and patronage of the Railroad Systems of the World.

For a general system of car lighting, the **Frost Dry Carburetter System** is the cheapest and best known up to the present time, and it will more than hold its own against every other mode of lighting. The first cost of equipping a car is comparatively small, when the great advantage arising from the ability to obtain a light which will burn at least one hundred and twenty-five (125) hours without attention, is considered; and the economy due to the less frequent moving of cars is most striking, especially at crowded terminals, where every additional train movement causes considerable trouble and embarrassment.

EACH CAR IS ABSOLUTELY INDEPENDENT FOR ITS LIGHT AND REQUIRES NO EXPENSIVE CHARGING PLANTS TO OPERATE IT.

EXTREME DURABILITY AND SIMPLICITY.

BEAUTY AND BRILLIANCY OF LIGHT.

PERFECT SYSTEM OF VENTILATION.

ECONOMICAL BEYOND PRECEDENT.

NOMINAL COST OF MAINTENANCE.

Absolute Safety.

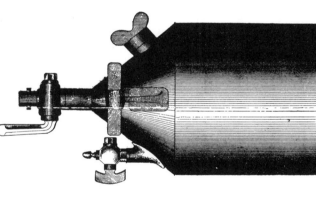

THIS IS OUR ENTIRE PLANT OFF THE CAR

THE

Railway Lighting and Heating Co.,

Office and Factory: 1114 SANSOM STREET, PHILADELPHIA, PA.

Figure 5.58   The Frost lighting system ran on gasoline and was used mainly by the Pennsylvania Railroad and its subsidiary lines. (Railroad Gazette, January 2, 1891)

gas-producing plants. To service its relatively small fleet, the Milwaukee Road had to maintain seven compressor stations. The first cost represented an investment that railroads would have preferred to apply elsewhere. In addition, a minor but continuing annoyance was that the apparatus itself was manufactured in Germany, and the foreign standards, machine threads, and measurements complicated repairs.[245]

The real downfall of the Pintsch system was, of course, the perfection of electric lighting. After 1910 it began to grow in popularity, and its progress inevitably meant Pintsch's decline. The Safety Company now rarely reported hard facts about its business, so that statistics on the number of cars with Pintsch lighting are elusive. C. Malcolm Watkins of the Smithsonian Institution, a frequent commuter on the Boston and Maine in years past, recalls that the line began removing Pintsch lamps in 1928 and completed the process two years later, with the exception of a few head-end cars. He also remembers the faint odor of petroleum when one entered Pintsch-lighted cars, and the intense, slightly blue-green flame of the jets.

In the 1928 edition of the *Car Builders' Cyclopedia* the Safety Company advertised that the Pintsch apparatus was still available. In the 1931 issue there was no such advertisement, although the apparatus was illustrated in the technical portion of the text. By 1937 it was described in the text with a closing statement: "This system is now obsolete on American railroads." Nevertheless, W. D. Edson, identified earlier in the chapter, recalls gas-lights on Milwaukee commuter cars in the 1940s.

Electricity was the first source of flameless lighting. In 1881, only a year or two after its commercial introduction, electric lamps were tried out on railroad cars. Gas had only begun to dominate car lighting, and there was reason to believe that electricity might overtake it and all other established forms of illumination. But this did not happen; instead, electricity waited in the wings for over thirty years while the old stars, somewhat beyond their prime, burned brightly on center stage.

To a limited extent, electric arc lighting had been employed for street and factory lighting since the late 1870s. It was far too intense for domestic service, however, and just how to harness this amazing source of illumination eluded inventors until Thomas Edison and Joseph Swan almost simultaneously perfected a practical incandescent light-bulb. The breakthrough was hailed by some as the greatest wonder since the invention of the locomotive itself. Others were more cautious, suggesting that the new light was dangerous to the human eye and might provoke "a fearful catalogue of nervous diseases."[246] The London, Brighton, and South Coast Railway chose to ignore such gloomy predictions and in October 1881 installed twelve Edison lamps aboard the Pullman parlor car *Beatrice*.[247] Current was furnished by thirty-two batteries. Late the same year the Brighton line had four cars with lights powered by batteries that were recharged at the end of each run.[248]

In the summer of 1882 the Pennsylvania Railroad decided to emulate the British experiment.[249] Batteries were imported from France, and the Edison Company assisted in wiring coach Number 397, the first car in the United States to have electric lights. Seven lamps with two 8-candlepower bulbs each were powered by a set of batteries weighing 1,200 pounds, which were rated sufficient for sixteen hours' service. In August 1882 the car ran between Jersey City and Newark. In 1884 the Pennsy's chemist, Charles Dudley, began a second series of experiments using Brush batteries. During the next year eight parlor cars with electric lights were placed in regular service. In 1887 two of the road's best trains, the Florida Special and the Chicago Limited, were outfitted with electric lights.[250] Steam-driven generators, stationed in the baggage cars, provided the power. From that time on the Pennsylvania persisted in operating at least a few electrically lighted trains, despite the many problems of this technology during its formative years.

Within a few years of the Pennsylvania's initial trial, other American lines began to test electric lighting. The Boston and Albany, the Connecticut River Railroad, and both Wagner and Pullman all had some electrically lit cars by 1886–1887. Although batteries were used in most of their installations, the Connecticut River line sought a practical way to manufacture current independently of accumulators.[251] Since the generator selected for this experiment was too bulky to be mounted on the trucks, it was placed inside the baggage car with a belt passing through an opening in the floor to a pulley fastened on one of the axles. The belt failed; so did the rope and chain drives that the line tried next. A 7-horsepower steam engine that drew steam from the locomotive was then placed in the car and belted to the dynamo. The system worked well enough to light a six-car train, but not well enough to encourage wide adoption of the head-end system.

Batteries, at first regarded as the simplest and best source of power for electric car lighting, proved to be a dismal failure. In November 1889 after a 2½-year trial, the Boston and Albany abandoned battery lighting for Pintsch gas.[252] In England the Midland and the Brighton railways decided to give up electric illumination after equipping nearly 1,000 cars and spending five years in the effort to develop a workable system.[253] The weak link was the wet cell battery, an expensive, short-lived mechanism that was at once heavy and fragile. Rough handling in the process of recharging, and agitation of the acid by the motion of the train, both ruined the delicate pasted grid plates. Batteries with a projected life of 7 hours rarely lasted 5½, and they took 15 to 20 hours to recharge.[254] Few weighed less than 200 pounds each, and with six to a coach, the workmen were hard-pressed to keep a continuous caravan of wet cells moving to and from the recharging shed. A few roads—for example, the C & O and to a lesser degree, the Pennsylvania—remained loyal to the straight battery system for many years, but most lines abandoned electric lighting after a few months of experimentation.

Small steam-driven generators were at first regarded as the best alternative method. Most of these installations, known as head-end plants, involved a single engine-generator set in the baggage car, with steam supplied from the engine or an independent boiler. In 1886 a Philadelphia inventor, H. E. Freese,

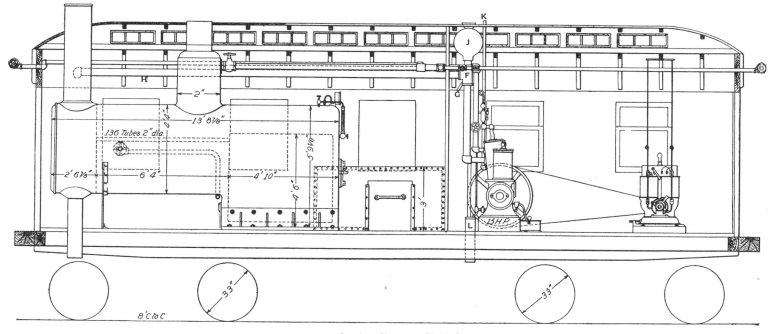

Section Showing Right Side.

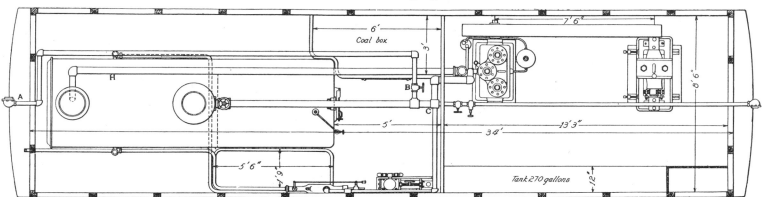

Figure 5.59   The Milwaukee Road's head-end car of 1890 provided steam for heating and electric power for lighting. (Railroad Gazette, *June 13, 1890*)

promoted a steam generator set for each car.[255] Freese proposed using a "package" engine with a kerosene-fired boiler. The 2-horsepower unit could be placed under the car or in a closet. The scheme eliminated batteries and axle belt drives and gave each car an independent power source. But apparently the cost and maintenance expenses of so many individual engine-boiler units discouraged the adoption of the Freese plan.

Inspired by the personal enthusiasm of its electrician, George Gibbs, the Milwaukee Road became a champion of the head-end system (Figures 5.59 and 5.60). During the 1890s, when most of the industry adopted a wait-and-see attitude, the Milwaukee continued to operate electric lights. The road first tested batteries, but quickly discarded them in 1888 for a head-end steam plant.[256] Two years later Gibbs designed special cars (described earlier in the section on steam heating) with boilers, high-speed engines, and generators to replace the makeshift baggage car generator units. When the lighting and heating tenders entered service, the line had forty-five cars wired for electricity. They ran primarily between Chicago and Minneapolis, and during the thirty-six-hour journey there was no need to worry about discharging batteries. In 1892 the Milwaukee had eighty-two cars wired and ran five electrically lit trains a day. The system was expensive both in first cost and in operating costs. It was necessary to have an electrician aboard the tender, and his monthly

salary of $85 was considered extraordinary at the time.[257] Hourly operating costs were figured at 35.5 cents an hour, or ten times the cost of gas. Occasionally in extremely cold weather it was necessary to shut down the generator's engine so that all steam could be used for heating. Despite these drawbacks, the Milwaukee stayed with electric lighting. By 1904 it had electricity on 300 cars—more than any other U.S. railroad.[258] It still favored the head-end steam tenders, although seventeen of the cars in commuter service had battery lights. Not long after this, however, the steam head-end generators were probably replaced by axle dynamos, which had been perfected in the meantime.

Electric lighting remained a novelty on American railroads until about 1910. The railroad trade press continually referred to it as a rarity found only on a few luxury trains. And even in these instances combination gas and electric fixtures were used, because the new system could not be depended on. In 1896 the C & O frankly admitted that it retained electric lights on its first-class cars simply for their advertising value.[259] Electricity was popular with the passengers, but it was not practical or economic. Meanwhile greater progress was being made in Europe: by 1898 an estimated 6,000 cars were electrically lighted.[260] In the United States it is doubtful if more than 150 to 200 were so equipped.

The situation was not unlike the period of transition from

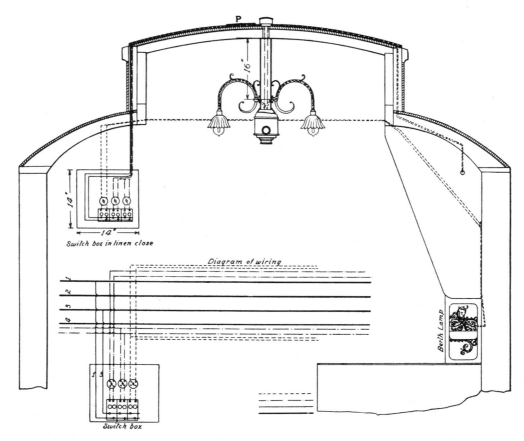

Figure 5.60   *The Milwaukee Road was one of the first American lines to make commercial use of electric lighting for passenger trains. Lamps and wiring details are shown in this drawing.* (Railroad Gazette, June 13, 1890)

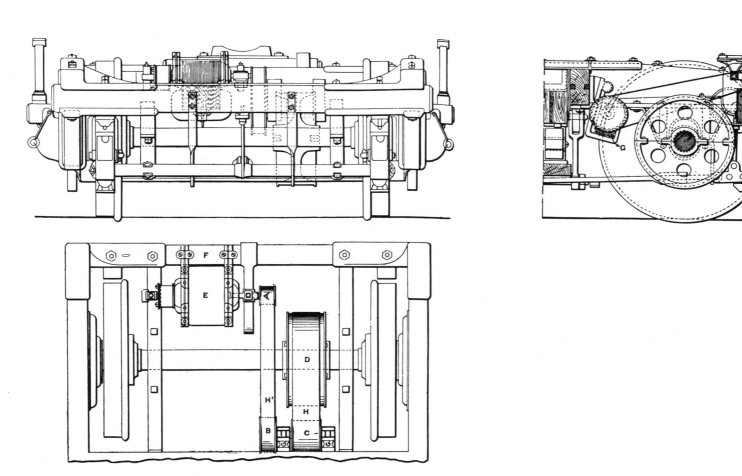

Figure 5.61   *A Moskowitz belt-drive generator of 1896 led to the standard system of passenger car electric-lighting service in this country.* (Railroad Gazette, October 30, 1896)

Figure 5.62   Large flat belts driven by an axle-mounted pulley powered the car-lighting generator. This photograph was made around 1910. (Pullman Neg. 17796)

wooden to steel cars. The advantages of the conversion were obvious and much discussed. The technical problems were solvable, but the industry left the task to outsiders, arguing that the old form was proven, satisfactory, and economical. Cash requirements, inertia, and existing investments in gas and oil lighting reduced the railroads' interest in switching to electricity. Meanwhile, several private firms and individuals worked to produce a more acceptable system of electric lighting.

Morris Moskowitz devoted himself to developing the axle light system, so called because it involved an axle-mounted pulley that was belted to a generator. When the train was in motion, the generator powered the lights. When the car was standing, or when it was moving at very low speeds that did not produce enough current, batteries would take over the power needs. The system seems so simple, yet its perfection baffled mechanical engineers for many years.

The basic problem was how to maintain a constant voltage when the generator's rotation fluctuated with the speed of the train. In the normal range of speeds (zero to 70 mph), the generator's revolutions per minute varied from zero to 800. In addi-

tion, the generator's maximum wattage was reached at intermediate speeds of 22 to 25 mph. A third problem involved reversal of polarity when the train backed up or the car was pulled in the opposite direction. Moskowitz solved the voltage problem by a specially wound field magnet. By passing a current through a secondary large wire coil, the magnetic field was progressively weakened as the armature rotated faster, hence lowering the voltage so as to maintain the desired pressure of 32 volts.

The scheme was tried in Germany as early as 1882, and the first recorded North American test was made two years later.[261] This was almost a decade before Moskowitz became involved in axle lighting. In 1893 he placed a small belt-driven generator inside a car on the Central Railroad of New Jersey.[262] A chain drive from one of the axles powered a countershaft. Within three years Moskowitz had greatly improved the design by mounting the generator on the truck frame (Figure 5.61).[263] The Pennsylvania, Erie, and Santa Fe tested the apparatus, and a Southern Pacific private car traveled across the country demonstrating the reliability of Moskowitz's rig. In the spring of 1897 he was ready

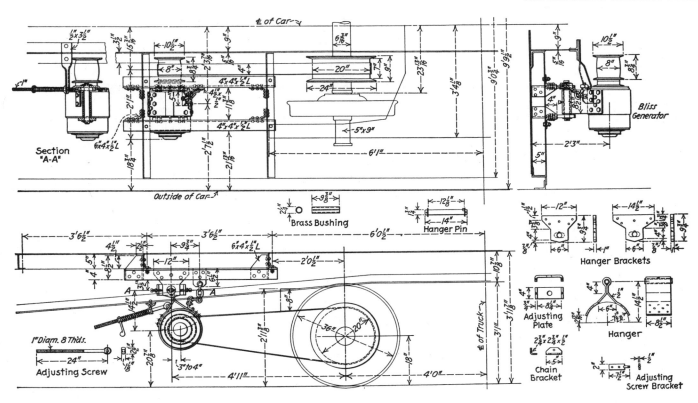

Figure 5.63   Belt drives remained in favor throughout the heavyweight period, but after about 1915 the generator was mounted to the car rather than to the truck's frame.

to make an extended test of the axle light system and installed it on fifty cars.[264] Two years later his apparatus was on one hundred Santa Fe cars. Camel-hair belts were used at first, then friction drives, and finally belts were reinstated. By the summer of 1900 the Santa Fe decided that the Moskowitz system, although promising, was not ready for large-scale application.[265] All the original gear was removed, and a contract was signed for new tests on a dozen cars.

The axle light system thus suffered a temporary setback, but Moskowitz was clearly moving in the right direction. The Franklin Institute gave him an award that year. Other firms and mechanics, notably William L. Bliss, now recognized that axle lighting was the coming method of car illumination (Figure 5.62). The United States Light and Heating Company was formed to take over the Moskowitz patents. In 1902 the Twentieth Century Limited adopted axle lighting, and two years later the Chicago and Alton announced plans to convert all its passenger rolling stock to axle dynamos.[266]

While American railroads had now settled on axle lighting, massive conversion was still in the future. The familiar objections to the high cost and excessive weight of electrical systems were repeated as late as July 1909 in the influential *Railroad Gazette*. According to detractors of electric lighting, it was so expensive that railroad officials suppressed the actual figures to keep shareholders from accusing management of squandering company resources. Equipping a car with axle lighting cost $1,673. It added nearly 2½ tons of dead weight to the car. And annual operating expenses were $631.[267] Pintsch gas fixtures, even figuring in a prorated share of the compressor plant, came to only $550 a car, while operating costs were calculated at $55 to $65 a year.

But other factors began to overwhelm plain economics. When the Pennsylvania and New York Central railroads banned wooden cars from their New York City terminals, they simul-

taneously forbade stoves and open-flame lamps. This ruling suddenly rendered a great deal of serviceable equipment obsolete, and the effect on car lighting was dramatic. In 1902 the United States Light and Heating Company estimated that there were only 1,000 electrically lighted cars on American steam railroads. Within nine years this figure grew to 11,000.[268] In a year another 2,700 were wired for electricity. If cars with combination electric-gas fixtures were counted, nearly 30 percent of the entire passenger fleet now had electric lighting. By 1912 electricity was so clearly coming into its own that the Master Car Builders Association prepared an elaborate series of recommended designs and standards.

Lack of statistics prevents a more complete chronicle of electric car lighting in later years. Its slow progress can be traced for one representative railroad, because the Baltimore and Ohio published an annual booklet, *Summary of Equipment*, that included data on car lighting. The accompanying table provides some figures.

Year	Total B & O Cars	Electrically Lighted B & O Cars
1917	1,207	187
1924	1,400	473
1932	1,636	1,027
1938	1,217	1,049
1946	1,126	1,014

The progress of axle lighting is attributable in large part to the slow maturation which all mechanical devices undergo between the time of their invention to the time that they are finally ready for everyday service. This metamorphosis involves minor and unspectacular alterations in design and material. A slight increase in bearing area or the change from cast iron to malleable iron can make all the difference. This is an unpopular explanation of

Figure 5.64   When power requirements and generator size increased after 1930, shaft-drive generators replaced belt-driven units. (The Budd Company)

mechanical progress, however. And in the case of electric car lighting, there was in fact one dramatic breakthrough that corresponds to the popular notion of progress as something which is achieved in astonishing leaps: in 1907 a method was discovered for producing ductile tungsten lamp filaments. By 1911 bulbs of this type were being manufactured for commercial sale. Tungsten did for electricity what the Welsbach mantle did for gas: it radically reduced the power consumption for a given amount of light. This reduction was critically important to car lighting because it meant smaller generators and batteries—cutting down weight, investment, and operating costs.

In 1912, according to the electrical engineer C. W. T. Stuart, ball bearings replaced plain bronze bearings, making generators more dependable and easier to maintain.[269] A few years later the old standard of truck-mounted generators began to give way to an underbody fastening much on the order of the European plan (Figure 5.63). The underbody mounting had several advantages. It removed that much unsprung weight from the trucks and raised the generator and its belt above snow and flying ballast. Throughout the 1920s the old-fashioned flat belt drives remained in favor. Their defects were well established—rapid

wear, slippage, and complete loss of the belt if the splice failed—yet they were simple and cheap. So long as small 2- to 4-kilowatt generators were common, the flat belt remained secure. By the early twenties, however, there was enough dissatisfaction with the old system to encourage tests with shaft and chain drives. Of the many designs tried, the Spicer drive, developed by the Safety Car Heating and Lighting Company, was recognized as the best (Figure 5.64). The drive gear and case were designed so that the unit could be fitted to a rough-turned axle without dismounting the wheels or performing any machining. A safety clutch would automatically disengage in the event of any undue resistance from the generator. The clutch would also disengage when axle speed fell below the full load speed. It was possible to detach the shaft so that the generator could be manually turned, or "motored," during inspections without fully disassembling the drive mechanism. Spicer drives became increasingly popular during the thirties as the larger generators necessary for air conditioning came into use. By 1942 it was recognized as an industry-wide standard.[270] Flat belt drives continued on non-air-conditioned cars, and some were still in service in recent years.

While axle lighting was the orthodox system in this country,

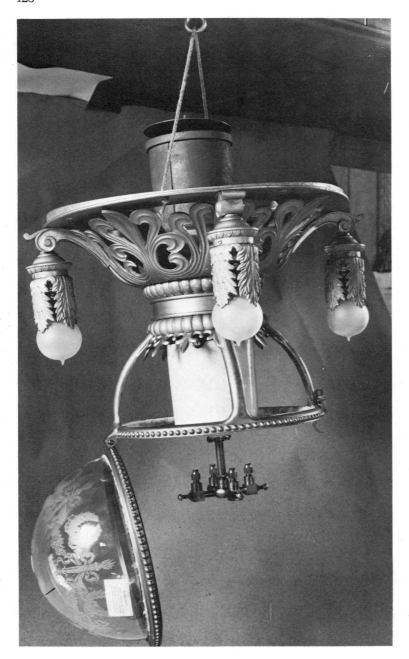

Figure 5.65   A combination gas and electric center lamp of about 1910. (Smithsonian Neg. 49749)

some railroads chose to use head-end power plants. The case of the Milwaukee has already been described. Several other lines were tempted to adopt head-end power in 1907 when General Electric placed a Curtis steam turbine generator set on the market.[271] It was offered in 15-, 20-, and 25-kilowatt sizes and was designed for direct mounting on the locomotive boiler or as an auxiliary in the baggage car. The turbine overcame many of the objections to earlier steam head-end units. The vibration problem was eliminated, and because the sets were self-contained, no attendant was necessary. Head-end power was ideal for trains running point to point and not subject to re-marshaling en route. Through limiteds and commuter trains were then the province of steam turbine lighting. It was widely used on the suburban trains of the Boston and Maine and the Central Railroad of New Jersey. The latter road had 20 locomotives and 100 cars outfitted with turbine lighting by 1944.[272] A year later it had converted all 60 locomotives and 300 cars in suburban service. A 125-volt, 12-kilowatt generator was mounted on the tender. Ironically, these roads gave the very reasons that had once been used for justifying axle lighting to explain why they opposed it; they believed that an independent electrical system for each car was wasteful and unnecessary. A central power source removed 2 tons of dead weight per car. The tremendous "drag" of axle generators was figured at 33 horsepower a car, and on a fifteen-car train that amounted to 500 horsepower—enough to ruin a good schedule and tie up the line.

The motive power drain of axle lighting was also cited by advocates of head-end and under-the-floor gasoline or diesel engine generator units. The early Union Pacific motor trains employed head-end internal combustion plants. By 1941 the U P decided to place self-contained propane gas engine generators of 7½ kilowatts on each car of the new train, City of Los Angeles.[273] The Southern Pacific and the Milwaukee Road also used engine generator sets of this type. Most of the individual sets were 110 volts AC, but on cars not so equipped it was necessary to add small auxiliary 110-volt alternators to power electric shavers and water coolers. Rectifiers were tried out by some lines, but they were found unreliable and prone to burn out.

During the same year that axle lighting was being challenged by other methods of power generation, electric light fixtures were undergoing changes. The earliest lamps, particularly the combination gas-electric fixtures, were traditional in appearance (Figure 5.65). Electric lamp design evolved into a more functional form that was only partially obscured with a decorative overlay in classical, Adam, or mission motifs (Figures 5.66 and 5.67). In 1911 the Santa Fe tried indirect lighting, possibly for the first time on a railway car.[274] Hanging fixtures reflected the light to the ceiling, where it was diffused throughout the car. A few years later the Northern Pacific purchased some cars from Pullman with the lamps built into the ceiling along the line of the clere-

Figure 5.66   An electric side lamp of about 1925. Such lamps were used in parlor and dining cars.

Figure 5.67   This 1928 Kansas City Southern business car had dome center lights and ceiling fans. (Pullman Neg. 32916)

story.[275] They were spaced at 2-foot centers, and no center lamps were used.

The last major change in passenger car lighting was the introduction of fluorescent lamps. Railroads showed a progressive spirit in adopting this form of lighting only about a year after it was placed on the market. In its October 18, 1938, issue *Railway Age* noted that the New York Central had fitted a coach with some of these remarkable tubular lights. The low power consumption of mercury-vapor lamps made them ideally suited for railway car illumination. With only a small increase in wattage, the illumination was nearly quadrupled. A small motor-generator converted the power to AC. In June of the following year the Lehigh Valley began operation of a number of remodeled cars with fluorescent lamps on the John Wilkes.[276] Within a few years the fluorescent system was widely applied to car lighting. Special 60-volt tubes were used where normal 110-volt current was not available.

Toilet and Drinking-Water Facilities

Of all the excruciating discomforts of travel, none can cause greater desperation than the absence of a toilet. To be trapped aboard a crowded public conveyance without hope of relief until the next stop is a torment that seems to have touched even the stony heart of the most unfeeling general manager. And the extra cost of such accommodations was justified by more than compassion, for if each car had its own toilet and drinking water, passengers would not be tempted to try the next car. This would discourage the dangerous passage across the open platforms. It would also discourage through passengers from leaving the train at intermediate stations and either stranding themselves or delaying the train's departure.

In what may have been a coy effort to obscure its existence, the toilet was given such esoteric labels as saloon, lavatory, dressing room, washroom, water closet, and retiring room. In a price quotation for new cars dated May 10, 1855, the Buffalo Car Company delicately referred to it as a "loafing room."[277] Such a proliferation of titles might suggest something grand, but the earliest washrooms were less than sumptuous. Until rather late in the nineteenth century they were airless, cramped closets barely 3 feet square. Although contemporary evidence is sparse, the drawings reproduced in earlier chapters (for example, in Figures 1.81 and 1.82) show that the toilet itself was a primitive dry hopper—really nothing more than a wooden box with a circular hole cut in the top for a seat. It was much like an outhouse, except that the refuse dropped out directly on the tracks rather than into a privy. The first mention of facilities even so elementary as these is found in reference to cars built for the Philadelphia and Columbia Railroad in the middle and late 1830s.[278] It is possible that the special car *Victory* of 1835 had a toilet. The surviving contemporary model pictured in Figure 1.68 shows a small anteroom with two commodes side by side. One is quite low, indicating that it was for children.

How quickly such facilities spread is difficult to document; probably they were rather common by the early years of the following decade and became increasingly so as the length of rail travel increased. Passengers could no longer be expected to tend to their needs at the beginning or end of the journey. In 1842 the luxury of the washbasin was introduced.[279] Perhaps earlier cases could be found, but the pictorial evidence available indicates that few coach passengers could perform their ablutions on board the train much before the 1870s or 1880s. Such niceties were available to first-class passengers at an earlier date, of course, but even this favored group was obliged to wait in line during the morning and evening rush for a turn at the sink. Once inside the sanctuary, a passenger was obliged to make his toilet with as many companions as the tiny 3- by 6-foot compartment could hold. The commode itself was in a separate closet, often on the opposite side of the aisle. To draw water, one worked a plunger pump that pulled water from a tank (Figure 5.68). A public hairbrush and comb, tethered to a cord, were available for all who cared to use them. The towel was equally democratic. This practice persisted as late as 1914.[280]

Separate washrooms for men and women were relatively common on the first generation of sleeping cars, but apparently not on coaches until the 1880s. By then some luxury cars had bathtubs. One inventor envisioned a three-compartment bathing car modeled on a fully equipped Turkish bath.[281] The average traveler, however, felt well satisfied if water and a clean towel were available.

Air-activated water systems helped to bring the railway car bathroom out of the dark ages. The idea was patented in 1879 (No. 221,680) by David H. Jones of Belleview, Pennsylvania. A similar scheme was tried on the Fitchburg Railroad three years later.[282] Now that the water was under pressure, the old-fashioned sink pump could be retired. Even more important was the introduction of flush toilets and urinals. The aesthetic as well as sanitary benefits were substantial. Pullman began to use the air water system in 1888 and within five years was moving rapidly to install it on all cars.[283] At that point, however, serious technical problems were encountered, because the air water apparatus tapped into the brake system for the air supply.[284] A toilet that

Figure 5.68   A spacious men's washroom on a Pullman sleeping car of about 1885. Note the hand pumps and massive marble-top sink. (Car Builder's Dictionary, 1895, fig. 3468)

was flushed at the wrong moment could upset the pneumatic balance of the Westinghouse system enough to cause an unexpected release or application of the brakes. The leakage of water into the brake system was another troublesome side effect. The complexity of the system, if not its actual workings, is indicated by Figure 5.69.

Washroom furnishings had also improved substantially by the 1880s. The rude box commode was replaced by elegant ceramic bowls. Handsome paneled washstands were capped with white marble tops and nickel-plated fixtures. The disreputable bare wooden floors were now often covered with zinc or copper. The Michigan Central found slate an even more durable and sanitary flooring.[285] Linoleum and plastic tile came into favor in more recent times. The abandonment of marble-top sinks was caused by more than changing fashion, for these heavy slabs became lethal flying platters during a wreck. There is one reported incident of a Pullman porter being decapitated by a loose sink top.[286] In and around 1910, lightweight sink bowls of nickel silver were introduced (Figure 5.70). Vitreous ware or porcelain enamel also came into use (Figure 5.71). Stainless steel fixtures typified the modern period.

Despite all these improvements, the disposal of human excreta is no more advanced now than it was in 1835. Raw human waste continues to be dumped out of railroad cars onto the tracks. In 1947 the Association of American Railroads voluntarily investigated the problem, but no reforms were instituted. Around 1970 the consumer advocate Ralph Nader took an interest in the problem, and a few press statements appeared. Greater issues soon overtook this quixotic crusade, however, and the railroads were allowed to continue as before.

The American mania for ice water was endlessly recorded by British visitors during the nineteenth century. This passion was based on more than a local fancy; baking heat and billowing dust created a thirst best satisfied by cold water. As the duration of journeys increased with the expanding rail network, so did the need for drinking water aboard the cars.

In 1847 a Virginia railroad specified 4-gallon water coolers for its cars.[287] It was more common, however, for a boy to dispense water to thirsty passengers. Throughout the journey he walked the train from end to end, carrying glass tumblers and a brass or tin can made much like a tea kettle (Figure 5.72). On the Boston and Worcester, according to an 1853 account, the water boy used a silver urn and goblet.[288] Water boys were reported on trains as late as 1898.[289] No matter what system each road used, the industry as a whole was credited with liberal arrangements for providing drinking water.[290]

Ice coolers placed in each car allowed passengers to help themselves rather than await the water boy's next passage. But

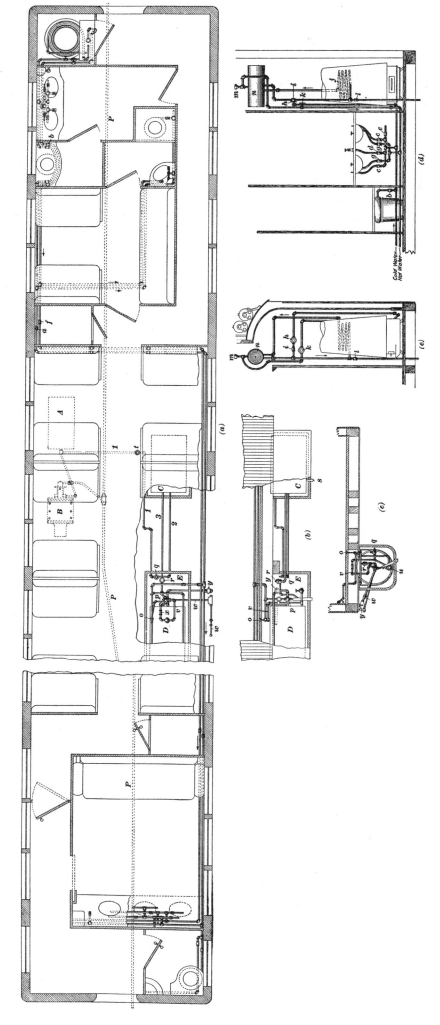

Figure 5.69 Air-activated water systems eliminated hand pumps. The air was bled from the braking system. (International Textbook Company, 1912)

Figure 5.70   *The men's washroom aboard a 1928 sleeping car was furnished with leather sofas and nickel-silver sinks. (Baltimore and Ohio Railroad)*

the democratic cup remained, and everyone was expected to use it. When new, it was a polished, inviting vessel, but time reduced its bright surface to a grimy patina. A repulsive green crust formed around the lip, making it as appealing as a soiled napkin.

Public sanitation was a burning cause among the reformers of the Progressive era. Elimination of the public drinking cup was viewed as a necessity. The evils of the common glass were explored in the liberal press, and state legislatures were persuaded to outlaw its use. There was merit in this reform, but as in the case of Prohibition, not all of the public was pleased by the efforts of others to protect its well-being. *Railway and Locomotive Engineering* of October 1912 summarized the opposition's sentiments:

The cranks whose senseless agitation has eliminated the public drinking cup, even in Pullman cars, have inflicted much discomfort upon ordinary people and have largely increased the business of saloon keepers. The victims that deserve the greatest pity are the poor immigrant children who have need to make long journeys in hot dusty cars. When this pernicious movement against free water drinking has passed, people will wonder at the blindness which enabled cranks and saloon keepers to spread so much discomfort among honest people.

At last disposable paper cups eliminated the health hazard of the public glass (Figure 5.73). Just when they were first placed aboard railway cars has not been determined, but an early user was surely the Lackawanna, which began installing paper cups in 1909.[291] The scheme spread as local and state statutes began to outlaw the common cup. By 1917 it was reported that even where no such laws existed, the public water glass had vanished from railroad cars.[292]

To further assure the sanitation of its drinking water, Pullman made a double tank for ice and drinking water so that the ice which cooled the water did not come into direct contact with it. In the older style of tank where ice and water were mixed together, employees had simply been admonished not to handle the ice with their bare hands. With the advent of air conditioning, drinking water was cooled by small refrigerating units. After about 1940 Pullman provided almost all room-type cars with circulating ice-water systems that delivered drinking water to each compartment. The Pennsylvania continued the more traditional method of placing individual thermos bottles in each room. These heavy, silver hotelware bottles were kept filled by the sleeping-car porters. They presented an unexpected health hazard, because some patrons unaccountably mistook the water carafes for miniature urinals.

Decoration and Painting

The passenger car is the traveler's home. If it is neat and attractive, the pleasant surroundings can do much to ameliorate the tedium of the journey. Passengers have traditionally expected more than a weatherproof box with adequate lighting and heating; the interior must also reflect the taste of the times. Car builders responded to this demand, often with an exuberance that drew more censure than praise. While the industry could point to noted architects and interior designers who served as consultants, much of the planning was actually left to the chief draftsman. These solid, practical men turned to existing examples of the decorative arts for inspiration. Some translated successfully from classical forms, but the less talented car designers inadvertently transformed coach interiors into perambulating

Figure 5.71   A washroom and toilet of about 1950. (B & O)

museums of ancient decorative styles, from the Tuscan to the Hindu.

The more ostentatious embellishments were reserved for first-class cars. Car builders answered the popular demand, which seemed insatiable by the last decade of the nineteenth century, for ever more adornment. Their migratory products were boastfully described as "an active messenger of fame for the place of its fabrication."[293] The best minds of the day saw little to admire in the pretentious finery of the palace cars. Less discriminating critics rode the whirlwind of changing fashion, one season praising the design that they declared banal in the next.

To domesticate the new technology of steam railroads for the traveling public, measures were taken to introduce the familiar and traditional. The adoption of old-fashioned road coaches may not have been solely for the car builder's convenience; their appearance was also reassuring to the first railway travelers. So too the orthodox decorating motifs—the reproductions of old masters and scenes along the line—disguised the raw mechanism of the cars. It was reassuring to look at "The Wounded Tiger," "Napoleon Crossing the Alps," or "Washington at Trenton" rather than at a barren wooden box with a smoking metal steed at its head. The interior of the railroad car was also brightly colored, creating such a cheerful appearance that a German visitor wrote in 1859:[294]

The equipment of the cars must be called elegant in all cases, and sometimes even luxurious. Floor and ceiling are covered with oil-cloth or carpets, the walls are decorated with paintings and gold mouldings; in some cases, the pillars between the windows are upholstered. Good woods are used for the window frames, Venetian blinds, and seats; they are neatly polished; the upholstery is covered with red or blue plush. The outside of the cars is painted in a light color and varnished; above the windows, they show the names of the railroad to which they belong. All this, together with the spaciousness and good lighting of the interiors, gives them a friendly, inviting look.

The very highest style of exterior painting was evident by 1835, when John Baker of Lancaster, Pennsylvania, produced a car whose ends were embellished with scenes based on Byron's poem *Mazeppa* (appropriately, the verse revolved about a tale of fast running).[295] On the sides were finely executed paintings representing the seasons. During the same period a car built for the Tonawanda Railroad was pleasingly decorated in the Grecian idiom. Stephenson produced a car for the Auburn and Syracuse "gorgeously flowered and varnished." Heroic battle scenes, landscapes, panoramas, all painted in rainbow hues and as festive as a new shop window, were studied with tireless interest by city and country folk alike. In New England, scenes of everyday life were featured on the New London and Northern cars of the 1850s.[296] An old seaman would be pleased to step aboard a coach featuring an exterior painting of "Homeward Bound," "Cutting in a Whale," or "Storm on the Whaling Grounds." The most grandiloquent of the exterior paint jobs, and one for which a complete description has survived, concerns some cars produced by Eaton and Gilbert in 1835 for the Rensselaer and Saratoga Railroad:[297]

The outside of the cars is painted of a beautiful fawn colour, with buff shading, painted in 'picture panels,' with rose, pink, and gold borders, and deep lake shading, the small mouldings of delicate stripes of vermilion and opaque black. Within the panels are 'transferred' some of the most splendid productions of the ancient and modern masters, among which are copies from 'Leonardo da Vinci,' 'Horace Vernet,' 'David,' (the celebrated painter to Napoleon,) 'Stuart,' and many more of the modern school. The whole number of the subjects of the twenty-four cars, cannot fall far short of two hundred, as each car averages from six to ten subjects; among which may be enumerated, several copies from the antique, Napoleon crossing the Alps, the two splendid scenes in Byron's Mazeppa, the Hospital Mount St. Bernard, portraits of most of the distinguished men of our own country, among whom Washington (from Stuart's original) stands conspicuous, the wounded tiger, the avalanche, portraits of distinguished women, views of several of our popular steamboats, the railroad bridge near Philadelphia, and several views in the south. The 'tout ensemble,' is more like a moveable gallery of the fine arts, than like a train of railroad cars.

As late as 1848 this tradition of railroad car art remained very much alive. The South Boston *Gazette* of December 30, 1848, praised the work of local craftsmen:[298]

For some time past the painters employed by the Old Colony Railroad have been engaged in painting a passenger car. We had the

Figure 5.72   Water boys once patrolled American trains to dispense ice water to thirsty passengers. This scene dates from about 1880. (Dayton Rubber Company)

pleasure of visiting the shop a few days ago and examining this car, and pronounce it the most splendid specimen of car painting ever produced. It is of a vermilion hue, shaded with brown and black, while either corner is ornamented with gold scrolls done in a superior manner. But by far the greater attractions are the pictures upon the sides. These were executed by Mr. Kelley, a young artist of very promising talents, and are pronounced by excellent judges worthy of a place in any drawing room. Four in number, they represent scenes in such a manner as to make it appear as if the events portrayed were transpiring before us. A lion springing upon a gazelle as it is drinking from a brook is the subject of one of the pictures. A lady riding upon a powerful charger and accompanied by a dog is also portrayed, while on the other side is a fine view of a buffalo hunt. The fourth, and by far the most beautiful, is a painting descriptive of the inhuman mother and the wolves. The almost fiendish look of the mother, the agonizing features of the children, the appearance of the hungry wolves as they surround the sleigh, and the noble bearing of the horse as he exerts himself to the utmost to escape the ferocious animals are all portrayed in a most beautiful manner. The pictures are surrounded by gilt frames, and as if to make the delusion more perfect, these frames appear to be hung by small cords upon the sides of the car. The painter of the vehicle is Mr. James Hazeltine, who has acquired an enviable reputation as a car painter.

The talent required to execute such paintings did not come cheaply. John Stephenson paid his artists $30 a week in the 1840s. One of these men, James Walker, made his way into more respectable art circles in later years and is remembered for the heroic battle scenes in the National Capitol building's ro-

tunda.[299] The most highly paid artisans at Wason's car shop drew $8,000 a year—an extraordinary wage for a workingman in the 1850s.[300] William R. Tyler, Eaton and Gilbert's top-ranking artist since 1833, retired in 1858 when individual-portraiture styles of fancy paint jobs were replaced by simpler methods of ornament.[301] He opened a studio and lived to see his works exhibited in many public galleries. Possibly a more typical career was that of Jacob Scheller.[302] A native of Zurich, Scheller painted china, carriages, signs, and frescoes before leaving his native land during the 1848 revolution. Two years later he was working in the repair shops of the P W & B. At first his talents were well employed, but as simpler forms of car painting gained favor, Scheller's work became routine and he indulged his artistic bent as a Sunday painter.

From 1830 almost to 1860, car ornamentation was generally limited to what could be laid on by the brush (Figure 5.74). The bodies were plain boxy structures, largely devoid of any architectural ornament. The work was solid carpentry rather than fancy cabinetry. Stripped of their elaborate paint jobs, American cars of this period were as unimposing as a shoe box. The few surviving photographs of early cars almost all date from the 1880s or later, after the original ornate styles had been discarded for very plain painting. The plainness of the unadorned arched-roof car is clearly evident from Figures 1.24 and 1.26 in Chapter 1.

Probably it was very early in the history of car design that the costly mantle of exterior scenes gave way to a more serviceable exterior finish. The surviving Harlan and Hollingsworth draw-

Figure 5.73   Paper cups began to replace the common glass or cup around 1910. A Dixie cup dispenser is visible on the right side of the car. Note the water tank mounted in the ceiling over the washroom. (B & O)

ings, several of which are reproduced in Figures 1.88, 1.89, and 1.92, show that at least a few roads continued to honor the old ways into the mid-1860s by painting vignettes in at least a few panels. By this time most roads seemed content with the effect created by striping, lettering, and scrolls on the corner and door posts. It was a very rich finish: the lettering was double-shaded, and every visible nut and washer, even on the trucks, was high-lighted by striping.[303] All surfaces were varnished and rubbed out to a mirror-like finish.

The basic body color before 1870 was yellow—described as a very pale, creamy yellow, much like linen or straw.[304] Other colors that were occasionally used included light blue, pea green, gray green, drab, and Indian red. In the first decades of the American passenger car, bodies were sometimes painted dark

Figure 5.74   This panel from a head cloth of 1892 is representative of the decorative painting style common in American passenger cars forty years earlier. (Hall of Records, Dover, Delaware)

colors, but again pale yellow seems to have been the general favorite. It served as an effective ground color for accent decorations in vermilion, ultramarine, fine lakes, and gold and silver leaf. Yellow was also basic camouflage for the clay dust churned up by the train; a yellow car coated with yellow dust presented a tolerable appearance.

Interiors were done in much the same fashion and colors. Artwork there made more sense because it was inside, protected from the weather and placed where passengers could study and enjoy such primitive masterpieces. There was little opportunity to admire the artist's product while standing in the rush and confusion of a station platform.

The artist's main canvases were placed between the window panels and end bulkheads, but his greatest opportunities were in the large uncluttered ceiling. The roof ribs were covered with a head cloth. An extraordinary variety of materials was used for this purpose. Before 1880, colorful oilcloth or painted cotton duck were the most popular fabrics. Silk and German damask were more elegant, but they were difficult to clean and easily soiled by minor roof leaks until enameled cloth was developed. Common cotton duck was filled with sizing and then painted in bright colors, often in geometric patterns, that were hand-printed with wooden blocks. Head cloths for the best class of cars were decorated freehand at a cost of $125 each.[305] In 1850 Charles Stodder of Boston was said to have manufactured painted head linings for three years that were now extensively used by Northern railroads.[306] An Englishman, George Holt, settled in Worcester, Massachusetts, and became the major producer of "tasty and elegant" painted head cloths.[307] His business grew to such proportions by 1867 that he had to open a new factory. Within a decade, however, the chromo head cloth was out of fashion.

The brightly painted head cloth was in fact the last token of

the old order; a new and more somber mode of decoration came into fashion at the end of the Civil War. The cheerful painted gaiety of the thirties, forties, and fifties gave way to what Lewis Mumford has named the brown decades. He explains this sudden change in taste as a gloomy reflection of a tragic war. In England, Victoria began a period of mourning for her late consort that was to continue for generations. Yet funereal as these times might appear, there was such richness in the dark walnut and heavily draped windows that the era was also called the gilded age.

Even before 1865, which Mumford designates as the beginning of the brown decades, there was evidence of discontent with garish car painting. In 1856, for example, the *Railroad Advocate* pleaded for simpler paint schemes within the capacity of an ordinary house painter.[308] It pointed out that once-beautiful cars became eyesores when their elaborate festoons were redone by the amateur decorators employed by most repair shops. And the best cars, according to the *Advocate*, were used on night trains, where fancy paint schemes go unnoticed. By 1870 strong condemnations of intemperate color schemes were a standard theme in the railroad press.[309] Cars decked out in showy garments were ridiculed as artistically and morally wrong. The "toy shop mixture" of colors was discordant; it offended the eye of people with taste. The *Manufacturer and Builder* stated that we judge cars, as we do men, by their dress, and an outfit wanting in taste shows that "some of the moral chords" are out of tune. Railroads were advised that the $500 wasted on showy paint jobs could be put to better use.

Pale yellow was replaced by olive green, chocolate brown, plum, and Munich lake. Car painters now argued that the darker hues were more practical.[310] Only three coats were needed for a well-finished job, compared with four to five coats of light color.

Figure 5.75   A car painter at work during the 1880s in Pullman's Detroit shop. Note the stencil in the extreme right foreground. (Chicago Historical Society)

Dark shades were more durable, and they helped to hide the black soot belched out by the coal-burning locomotives which rapidly came into favor after 1860.

Since polychrome decoration was not appropriate for the deep exterior colors, highlighting was accomplished with gilt and silver ornament, often magnificently shaded and outlined. Side and end panels, corner and door posts were embellished with luxurious scrolls and line work. The ebullience of the bright-color era was replaced by the opulent contrast of gold leaf on a dark base. The design was laid out by a stencil, but much hand work was

required.[311] Six men might work on a car at one time, each devoting fifty to sixty hours to the job (Figures 5.75 and 5.76). In the 1880s the cost of ornamenting a car was calculated at $120 (less the lettering). Of this figure $50 went for the seventy-five or more "books" of gold leaf required. A cheaper job could be done with transfers or decals, and in 1878 Wason was using this method with a notable saving in time.[312] The idea was credited to the sewing machine industry, which had eliminated much hand work through the employment of decals.

The darkness of the exterior was repeated in the car's interior.

Figure 5.76   An exterior decorative scheme for a car produced by Gilbert in 1885.
(National Car Builder, *October 1885*)

Rather than somber painted surfaces, however, interiors were being finished in varnished hardwood panels: mahogany, walnut, rosewood, ebony, and many other rare cabinet woods. Ornament consisted of moldings, raised panels, carvings, inlays, and veneers. Contrasting wood tones and grains added to the effect— and to the cost of joinery and finishing. Cabinetmakers and wood-carvers replaced painters as the most skilled men in the shop. Often this elite corps was composed of Germans who sang songs of the old country or exchanged studio jokes with comrades from a Berlin academy while carving an antique head for a door-way.[313] Such artisans would spend up to 140 hours executing an end door panel.[314] Much carving and all moldings, traceries, and beadings were machine-made, yet the fitting of these elements required expert woodworkers.

Marquetry offered another opportunity for the decorative dis-play of varnished wood. Simple straight-line inlays created at-tractive borders in contrasting colors. Rosewood, tiger maple, tulip, amboyna, and teak offered gorgeous hues and variegated patterns. Where nature failed, craftsmen would dye veneers to the tone desired. To imitate the crumple of a rose leaf, a minute piece of satinwood was lightly scorched in a bed of hot sand.[315] This and other tricks created the special effects of flowers, flying dragons, and allegorical representations of "Music" or "Night and Morning" that were emblazoned on berth fronts, bulkheads, doors, and transoms (Figure 5.77). No blank space was left to suggest that something might be lacking.

The beauty of natural wood was displayed on the outside of the car by naturally finished window sashes and doors, often of cherry wood. A very few U.S. railroads sheathed the entire ex-terior in naturally finished mahogany, and this scheme was enthusiastically followed by Canadian lines.

There were other minor revolts against the uniform solemnity of dark olive green. Surely during the Centennial year, it was argued, something more cheerful should be adopted, even if it were red, white, and blue.[316] A train of darkly finished cars was too much like a funeral cortege or a black-coated, black-hatted Hibernian procession on St. Patrick's Day. Some years later the Reading decided to lift the gloom by switching to brighter col-ors.[317] At an earlier period its cars had been red; then it had changed over to Pullman green. In 1888 the road adopted lemon yellow, with the letter board in a light orange. But this sunburst was short-lived, for within four years the Reading returned to the standard dark green.[318] The Pennsylvania was one of the few mavericks strong enough to follow an independent course, using Tuscan red rather than the standard green.

Varnished wooden interiors remained in high fashion until the advent of steel cars, but within that forty-year period—roughly 1865 to 1905—at least one major change occurred. In the early 1880s the fancy veneers, with their thin moldings painted in black, cream, and gilt, gave way to solid mahogany and oak paneling.[319] No contrasting moldings were used. Considering the times, the carving and other decorations were severe. Al-though to us the coach interiors of 1875 and 1885 may look equally elaborate, at the time the shift to honest, plain, solid paneling was regarded as a commendable return to basics. Pull-man, however, did not adopt this reform, and first-class equip-

Figure 5.77   Rare woods highlighted by inlays typified the first-class cars of the 1870s to 1900. The berth fronts of this sleeper were decorated with the popular pineapple pattern of inlay. (Pullman Neg. 2137)

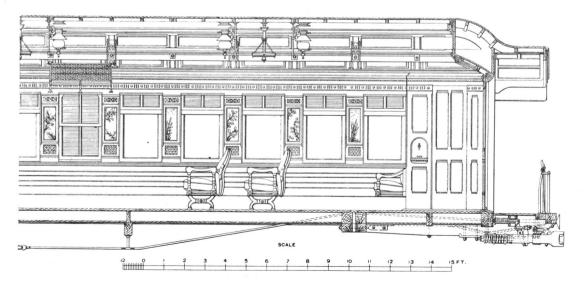

Figure 5.78   The Pennsylvania Railroad championed what were extremely plain interiors for the times. The example here, dating from about 1880, was one of the Pennsylvania's more elaborate designs which featured carved panels between the windows.

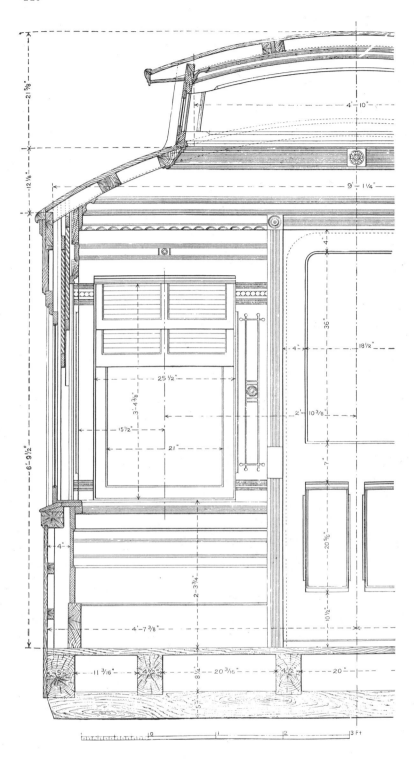

Figure 5.79   A cross-section drawing of a Pennsylvania coach, published in 1881, shows the shallow beading used to decorate the oak interior. (Railroad Gazette, *August 19, 1881*)

taste-maker Charles L. Eastlake (1836–1906). In 1868 Eastlake had published a modest work, *Hints on Household Taste in Furniture, Upholstery and Other Details*, which called for more directness in design. He took up Ruskin's precept to ornament the construction and not construct the ornament. Eastlake especially disliked the French Renaissance furnishings that were popular at the time, with their rounded lines and deep carving. He suggested simple, straight lines to show clearly how the parts of the furniture were put together. For carving, he recommended shallow grooves or beadings that followed the lines of each part. Eastlake wanted to eliminate the pretentiousness of machine-made furniture. Somehow his little book sparked a passionate response, and the so-called Eastlake style soon raged like an epidemic throughout England. It crossed the Atlantic and took hold of American decorators, architects, and furniture makers. Car builders became equally intoxicated with the new fashion. But Eastlake's notion of purity in design was soon perverted as manufacturers tooled up for mass production, and it became a cliché by 1884.[321] Car interiors decorated in this style were criticized for being as angular as a carpenter's square. The Reading's round-end cars, with their angular, beaded, chamfered, and notched interiors, were compared to "a Swiss cottage on wheels."[322]

Despite such wholesale disparagement, elements that can be classified as Eastlake continued to appear in such praiseworthy efforts as the oak-interior cars of the Pennsylvania Railroad. Production of this series of coaches was begun in 1878 under the classification PD. The interior decoration was a conscious effort on the part of the car department's management, according to the *Railroad Gazette*, to "abandon the old styles of barbaric finish."[323] There was to be no jugglery—no wooden puzzles, no veneers, no fancy imported woods, and no superfluous molding. American oak or ash was the basic paneling material used. There was a clean, republican honesty about the selection of a native wood. It was common, yet cool and elegant. Its strong grain and light color made it right for a coach interior. Mahogany was considered handsomer, but it was a relatively soft wood not well suited to the rough-and-tumble service of coach traffic. Mahogany darkened with age, a characteristic disliked by some Victorian car builders, such as William Voss, yet even its critics admitted that it remained the most popular wood for parlor and sleeping-car interiors.

The Queen Anne style, an outgrowth of Eastlake's teachings, was the basic architectural form followed in class PD cars. Shallow, machine-cut beads, rosettes, and diapered work called for little hand finishing except in the window panels, where floral patterns were modeled on delicate reproductions of oak leaves, bulrushes, and mountain laurel (Figures 5.78 and 5.79). The patterns varied from one group of cars to the next, and patterns were alternated from panel to panel to relieve the repetition. Yet each piece of cabinetry was standard in size—a necessity because of the hundreds and perhaps thousands of coaches (classes PD, PE, and PF) built on this general plan by the Pennsylvania between 1878 and 1891.

ment in general became ever more baroque until the wooden car itself passed away.

The revolt against flashy burls and overdecorated details condemmed the painted head cloths in particular calling them "common and vulgar in their gaudy paint and large figures to spoil the simplicity of the wood finish."[320] Plain, thin, wooden head linings were praised as chaste and harmonious. Beautiful as the new form was considered, some difficulties were experienced. Wooden head linings were built up in thin veneers to prevent splitting, but a leaking roof would loosen the glue and destroy the ceiling.

Ironically, one victim of the reform movement was the English

Figure 5.80 Luggage racks fastened above the seats were generally made of brass or bronze. This example dates from 1870. (Lyles Railway Manual, 1870–1871)

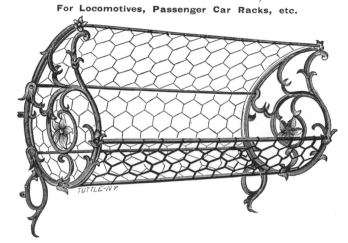

The discipline of the designers was relaxed enough to allow the mixture of two other contrasting woods in class PD coaches. The window sash and blind frame were made of cherry, and the slats of the blinds were of selected bird's-eye maple. The head cloths were painted in unobtrusive patterns of deep red, drab, or grayish brown. The seats were covered in golden brown or cherry red. The hardware was polished bronze. For the times, these cars were considered models of restraint and good taste.

Much if not all of the actual interior design for coaches was left to the anonymous draftsmen or cabinetmakers. The decoration of first-class cars, however, was thought too important to be trusted entirely to car shop employees and their uninspired layouts, which were copied from a standard handbook of ornament or from architectural style manuals. By the 1880s the best palace cars had interiors created by the top-ranking decorators and architects. Consultants with famous names were undoubtedly costly, but the publicity received was considered a fair return on the money. In 1881 the Wagner Sleeping Car Company was having its interiors styled by the prestigious New York decorating firm of Pottier and Stymus.[324] A few years later Wagner hired Louis C. Tiffany, "an artist unsurpassed in this line," to decorate its new Chicago Limited; "proof that the Company is determined to turn out a train unequaled in appointments by any in the world."[325] Pullman's consultant was the architect Solon S. Beman, whose monthly retainer amounted to $5,000 by 1892.[326]

Railroads and car builders also sought expert advice. The Boston and Albany turned to the architect H. H. Richardson, the master of Romanesque revival, not only for stations but also for suburban car designs. His contemporary Bruce Price (1845–1903), who was famous for his churches and hotels, also devoted a great deal of time to car architecture. In 1884 he developed the bay-window parlor car for the Pennsylvania.[327] In addition, he worked with the Boston and Albany and with such car builders as Wason and Barney and Smith.[328] His dark, Byzantine interiors, while considered refined and restfully elegant (rather like a brown study), were regarded as too melancholy for the average American traveler. The Canadian Pacific obtained a more cheerful design from an imported French decorator, who installed flamboyant interiors in the Louis XV style.[329]

The educated taste of professional architects and decorators was meaningless without the encouragement of industry leaders. During the nineteenth century such a patron was found in George M. Pullman. In his own words:[330]

I have always held that people are greatly influenced by their physical surroundings. Take the roughest man, a man whose lines have brought him into the coarsest and poorest surroundings, and bring him into a room elegantly carpeted and finished, and the effect upon his bearing is immediate. The more artistic and refined the external surroundings, the better and more refined the man.

Pullman's interpretation of elegance resulted in a superabundance of ornament, for he began with the assumption that more was better. As a former cabinetmaker he had a love of beautiful woods and elaborate marquetry. While others pointed the way to chastened oak interiors, Pullman remained loyal to the styles of the early 1870s. In fact he favored still more elaboration, and his influence was considerable because he oversaw the construction of the great majority of first-class cars in America.

That Pullman correctly assessed the public's liking for ostentatious finery seems undeniable. Many years later, H. L. Mencken said that no one ever lost money by underestimating American taste. And for generations leaders of the American literary and art worlds heaped ridicule on the Pullman palace car. It symbolized all that they despised about American life in the gilded age. The novelist William Dean Howells gave it a heavy broadside in his *Letters of An Altrurian Traveler* (1893–1894). Musing over moral and physical ugliness, he wrote:

But, in fact, do not the two kinds of ugliness go together? I asked myself the question as I looked about me in the ridiculous sleeping-car I had taken passage in from Chicago. Money had been lavished upon its appointments, as if it had been designed for the state progress of some barbarous prince through his dominions, instead of the conveyance of simple republican citizens from one place to another, on business. It was as expensively upholstered as the bad taste of its designer could contrive, and a rich carpet under foot caught and kept whatever disease-germs were thrown off by the slumbering occupants in their long journey; on the floor, at every seat, a silver-plated spittoon ministered to the filthy national habit. The interior was of costly foreign wood, which was everywhere covered with a foolish and meaningless carving; mirrors framed into the panels reflected the spendthrift absurdity through the whole length of the saloon.

Edward Bok, editor of the *Ladies' Home Journal*, used the Pullman car as an example of how not to decorate the home.[331] He crusaded to cleanse America of its penchant for filigree, turrets, figured carpets, and all other machine-made gewgaws (Figures 5.80 and 5.81). He wanted to raise the level of taste above frippery and the carnival arts, and he believed that railroad cars should be no exception. A train ride, he said, is about the only time that the average person finds free for meditation, and it should not be ruined by the absurdly busy trappings of the average parlor car.

By 1900 the relentless criticism of Howells, Bok, and other American Brahmins was having an effect. Popular humorists like George Ade described palace cars as a "chambermaid's dream of Heaven." They were part steamboat, part vestry, and part cuckoo clock (Figure 5.82). Even the trade press joined the attack. In its obituary notice of George Pullman, the *Railroad Gazette* praised his life and work but deplored the misapplica-

No. 5.

No. 1, Back .. . .5 inches high, 5 inches wide.
No. 3,   "   .... 3½ "   "   7   "   "
No. 5,   "   5   "   " 7⅞ "   "

Figure 5.81   Late-nineteenth-century hardware was generally highly decorated cast bronze. The device shown here is a comb and brush rack. (Dayton Manufacturing Company catalog, 1888c.)

tion of money and labor in car decoration and Pullman's bad influence on public taste for so many years.[332]

The reaction against Pullman's excesses was in full swing by 1905, when the reformers who set out to cleanse the spirit of American decorating were poetically advocating the mission style (Figure 5.83). The railway industry was swept along with this popular movement. An uncompromising austerity banished all unnecessary detail. Plain, unadorned furnishings and bland colors were combined to create an understated elegance. Fusty damask curtains and ancient wooden slat blinds were replaced by cloth roller blinds.[333]

Very plain wooden paneling came into vogue at the same time. Steel car interiors were painted in muted shades of grey, green, and cream. Pullman favored wood graining to soften the stark metal interiors. The wood patterns reproduced—a process that required fifteen days—were subtle and without fanciful highlights or imitation inlays.

For the next twenty years car decoration remained spartan. By the middle twenties, however, car designers and the traveling public seemed a trifle bored with the monotonous appearance of the steel cars, inside and out. Perhaps there had been an overreaction to the excesses of the Victorian era, and the railroads should devise some more imaginative scheme of decoration (Figure 5.84). Early in 1924 the Baltimore and Ohio introduced colonial interiors for some new dining cars.[334] Hepplewhite chairs, fanlights over the windows, and Georgian pilasters added a comfortable element of pleasure to railway travel. Woodcarvers were again in demand as the Chesapeake and Ohio outfitted an entire train in the same fashion. Pullman happily returned to its traditional ways, though with considerably less abandon than in former times.[335]

The final era in railroad car decoration began in 1933 with the two aluminum cars shown by Pullman at the Century of Progress Exposition. They looked like no other cars seen before by the American public, and their interiors were designed in a startling new fashion that was called "modern Empire" for lack of a better name.[336] The seat frames were of square, polished aluminum tubes; the upholstery was jade green. Indirect lighting came from a gutter-like trough above the windows. Polished strips of aluminum were used for trim, and the floor was covered with marbleized rubber tile. In a complete break with the classical models which had been the basis of all decorating styles, streamlining thus came to the railroad car.

Streamlining as a decorative style was an outgrowth of art deco, which itself was introduced at a 1925 Paris exposition on the decorative arts. The intent was to unite art with industry. Hard lines, geometric patterns, and flowing curves characterized the radical new style (Figure 5.85). It was meant to be fresh, pure, and metallic. Futuristic materials like plastic, aluminum, and stainless steel emphasized its break with the past.

The railroads' enthusiastic acceptance of streamlining has been described in Chapter 2. Their decision to hire the best talent available indicates the depth of this enthusiasm, particularly during the formative years of the streamlined style.[337] The leading industrial designers—Henry Dreyfuss, Raymond Loewy, Walter D. Teague, and Paul P. Cret—were engaged to work out the shape, color, and decorative treatment of passenger cars. Some worked directly for the railroads, while others were engaged by car builders. The stylish trains created by these men produced revenues that confirmed the importance of appearance: good car decoration was clearly not a frivolous indulgence.

In the decades of the 1940s and 1950s streamlining became the convention. As more modern trains entered service, the streamlined effect became less dramatic and the railroads' interest in devising new designs seemed to dwindle. Except for top-ranking name trains, design was entrusted to lesser talents already on the payroll of the major car builders. The cessation of large-scale new passenger car orders in and around 1955 ended the incentive for more work in this area. In fact, it would be difficult to name any striking advance made in passenger car decoration during the past thirty years. A coach of 1940 is not materially different in appearance from coaches commonly seen in service today.

Passenger car painting during most of the nineteenth century required an extraordinary amount of labor and time to produce a durable, high-gloss finish. In the late 1860s a full ninety days, some of them devoted to the fancy decorative work, were considered necessary for a first-class job.[338] By the early 1880s sixty days were required for quality results.[339] Some slipshod painters turned out cars in as little as fifteen days, but most car builders agreed that good work could not be done in this time. Experienced men knew that haste guaranteed poor results; there were no shortcuts. A proper painting schedule included the following steps:[340]

Prime bare wood with raw linseed oil—allow to dry for at least 1 week.

Putty holes.
Begin with one light coat of color—dry for 12 to 24 hours.
Put on two coats of "rough stuff" (filler).
Apply one light coat of color—dry for 12 to 24 hours.
Add three coats of thin color—dry for 12 to 24 hours between each coat.
Stripe and letter.
Apply one coat of rubbing varnish.
Conclude with two coats of finishing varnish—drying time, 96 hours.

Figure 5.82   Pullman interiors were noted for their superfluity of decoration. In 1898 one critic spoke of their "Barbaric splendor and gloomy magnificence." (Pullman Neg. 4690)

It was necessary to sand and rub down (with pumice or rotten stone) between each coat. The drying time between the successive coats listed above was figured at 288 hours.[341] This horrendous labor was justified by the slick, waterproof finish produced. Beauty and elegance was required in a passenger car's paint job. In addition, a hard, smooth surface sheds dirt, water, and cinders and is easier to wipe clean than a rough or porous surface.

After the car had left the shop, it was returned once a year for a fresh coat of varnish. Faded or scratched areas were "cut in" with new color, and the entire surface was rubbed out and varnished. Every seven to ten years it was necessary to burn off the old finish to the bare wood and then repeat the steps listed above.[342]

Car painters argued about the life of dark- and light-colored finishes. Some claimed that dark color attracted the sun's heat, causing the varnish to crack and alligator, and that it came off with the varnish during the annual rubdown.[343] Light color could be made from lead-base pigments, the most durable paint then available. But the enemies of light color said that it must be

Figure 5.83   Plain mission-style interiors came into fashion after 1905 in reaction to the elaborate palace car decor of former years. (Pullman Neg. 9254)

applied thickly and tended to crack. They also argued that lead paints "suck up" the varnish.[344] Fashion rather than durability, however, was the main factor in color selection.

The painter's workbench was the conservative stronghold in every car shop. No one seemed more resistant to new methods than the master car painter. He looked on the products of commercial suppliers with contempt; such materials were possibly fit for the household, but had no place in his domain. He boiled his own linseed oil and japan, ground his own pigments on a screeching dry mill, and hand powdered the fancy colors for lettering on a stone slab. The paint shop "was a Tartarus of hideous smokes and smells, and possessed many dark secrets over which the boss painter was the presiding alchemist."[345]

Slowly the old prejudices broke down. Until the end of the Civil War few car painters would accept anything but English varnish, but they grudgingly began to use American varnish and by 1881 it had become the favored top coating.[346] A few years earlier commercial primers began to find acceptance, because they saved time and much of the heavy sanding necessary with "rough stuff."[347] The old guard insisted, however, that this shortcut produced a less durable finish. During the 1890s experiments were conducted with colored varnish and enamel. The Burlington switched to a four-coat enamel paint job without the usual varnish topcoat, a method which was heatedly criticized in the car painters' column of the *Railroad Car Journal* in April 1896.

A far greater heresy was announced the next year: William P. Appleyard, master car builder for the New Haven, proposed a plan for an outer sheathing of sheet copper that would entirely eliminate the need for exterior painting.[348] Thin copper (no. 30

Figure 5.84   The severity of the mission era gave way to more sprightly painted interiors during the 1920s. This car is decorated in the Grecian manner. (Pullman Neg. 28991)

gauge) was wrapped around each window post, sheathing plank, letter board, window frame, and every other exterior stick of wood. The sheets were carefully rolled around the pieces so that no edge was left exposed. It took only ten days to sheathe a car—a great saving in time compared with the traditional hand-painting method. The new system required 1,000 pounds of copper, which represented no serious increase in dead weight, since 275 to 300 pounds of paint and varnish were needed to do the same job. (Thirteen years earlier the Southeastern Railway in England had anticipated Appleyard's plan by paneling a car in sheet copper.[349] Its silver-plated moulding and lettering made it the only true "silver palace" car in existence. But apparently the

British experiment did not spawn a large family, nor did it employ Appleyard's wraparound technique.)

Appleyard claimed other advantages for his scheme. Copper was a durable, nearly permanent material proved by centuries of use for gutters, flashing, and so forth. It was maintenance-free. No undercoating or preparation was needed. Cheaper grades of wood could be used. It had a high scrap value, recoverable when the car was retired. A handsome, glossy, brown-green surface could be produced by skillful anodizing. Because annual varnishing and periodic repainting were eliminated, cars would spend more time in revenue service.

Early in 1897 the first copper-sheathed New Haven coach,

Figure 5.85   Streamlining introduced polished aluminum, chrome, and bright colors in the 1930s. (Pullman Neg. 39615)

Number 365, entered service.[350] After a year's test, two more New Haven cars were sheathed. The neighboring Fitchburg Railroad encased several of its cars in Appleyard's copper mantle. Predictions that it would wrinkle like a rhinoceros hide as wood expanded and contracted did not discourage the Boston and Maine, the Erie, and even the Southern Pacific from trying Appleyard's plan. New York's Interborough subway purchased 500 wooden cars with Appleyard's sheathing, but diverted them to elevated service after an accident in 1905 proved that they were not fireproof.[351] The New Haven itself had 800 to 900 copper-covered cars, but with Appleyard's death in 1905 the scheme lost its chief promoter. Some of this equipment remained in service until about 1940.[352]

After the initial excitement had subsided, old-line painters

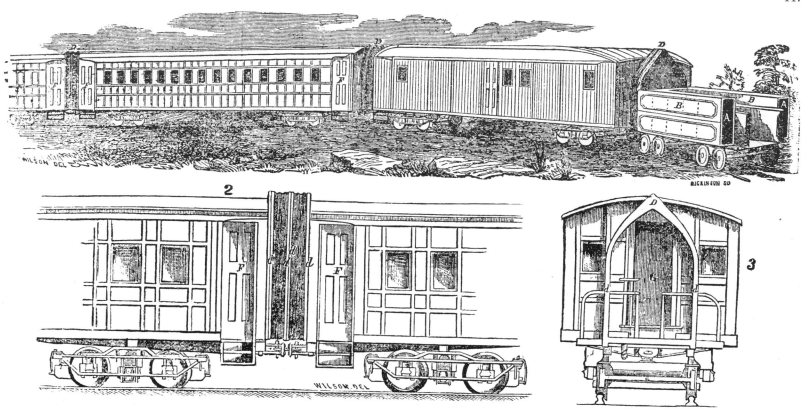

Figure 5.86   Waterbury and Atwood patented a vestibule in 1854 that was actually used on several Eastern railroads. (American Railroad Journal, *July 1, 1854*)

were pleased to discount copper plating as a "red herring"—not at all the important improvement that had been claimed. However, the traditionalists soon faced a competing method that was not so easily put to rest. Compressed air was finding many uses around the repair shops. Already it powered riveting hammers, chipping guns, and wrenches; next some genius would think of a way to paint pneumatically. The earliest case of spray painting applied to railway work has been described by the prominent railway journalist Angus Sinclair, who visited a Southern Pacific engine house near San Francisco late in the 1880s that was in the process of spray painting.[353] He was told that the scheme would next be applied to freight cars. The use of air guns for painting some of the Columbian Exposition buildings in 1893 gave the technique even greater exposure in the mechanical trades. Many railroads adopted it for freight car finishing. But although spray painting came to be considered suitable for rough equipment, maybe even for passenger car trucks, it was certainly not regarded as a technique to use on the superstructure. The spray gun was only slowly accepted for passenger equipment; the brush and varnish pail were to remain the principal implements of the painter for many years.

Other innovations were adopted more readily. The most important was baked enamel finishing, which was first tried in January 1913 by the Pennsylvania Railroad.[354] A normal varnish job required attention every eighteen months, but the baked car still looked good after forty months of service. Painting time was also reduced from fourteen to six days. The baking time was three hours at 150 to 160 degrees F. The following year the B & O began testing baked finishes, but the road insisted that a good job could not be done in less than twenty-three days and that two coats of varnish were a necessity.[355] All agreed that the faster-drying oven jobs were not only more durable but smoother, because dust had less opportunity to settle on the surface.

Spray painting for large objects was facilitated by the introduction of the canopy exhaust hood. Powerful fans removed overspray and dust, improving the quality of the paint job and at the same time lessening health and fire hazards. The Milwaukee Road installed such a unit in 1931 and achieved estimated savings of $12,000 to $15,000 yearly through reduced labor costs.[356] Spraying consumed nearly 10 percent more paint than brushing, but the work went five times faster. However, the Milwaukee remained loyal to oil paint and varnish finishes. Pullman, after a fleeting enthusiasm for enamel and lacquer, also stood by oleoresinous paint topped with varnish.[357] For hard outdoor service no better finish could be found, in the opinion of the Pullman maintenance force. With a solid, red-lead base, seven finish coats, and fourteen days' labor, a Pullman finish would last for two years, or about 360,000 miles of running through sun, rain, snow, salt air, and alkali dust.

Synthetic lacquers were soon to overthrow old-fashioned varnish finishes. The automative industry had been using the tough, fast-drying synthetics with great success since the 1920s, but it took the railroads almost twenty more years to fully accept the new technique.

Vestibules

Vestibules form an enclosed hallway between the cars. The accordian-like appendages, called diaphragms, at each end of the vestibules unite the individual cars of the train into a single unit, just as the couplers and drawbars hold it together physically. Passengers and crewmen alike can pass from one end of the train to the other, safeguarded against falling under the wheels and fully protected from the elements and the train's own smoky trail.

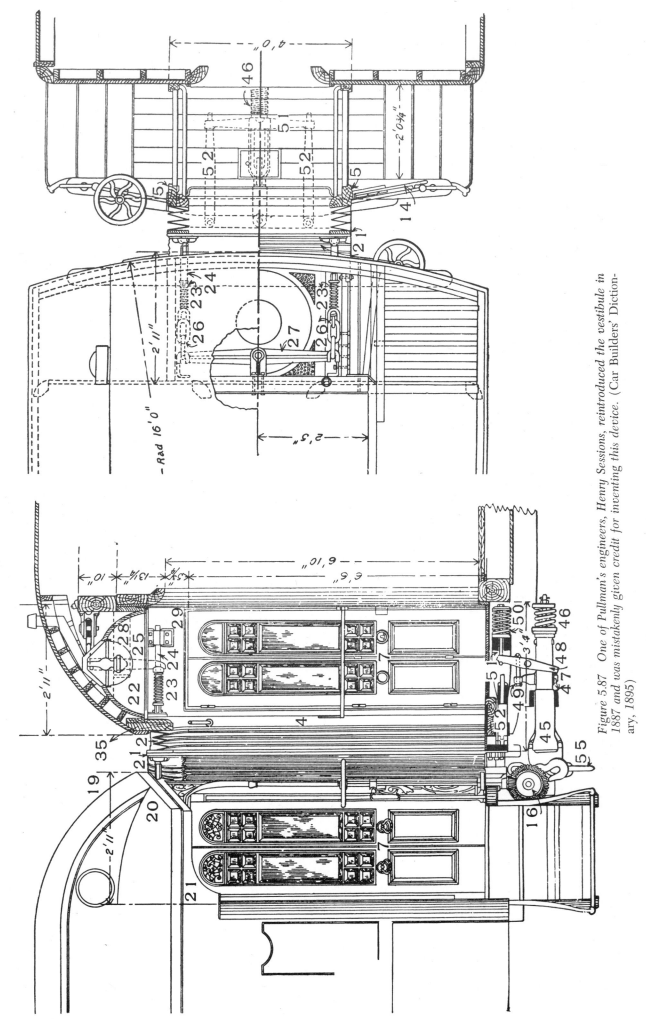

Figure 5.87   One of Pullman's engineers, Henry Sessions, reintroduced the vestibule in 1887 and was mistakenly given credit for inventing this device. (Car Builders' Dictionary, 1895)

Figure 5.88   Narrow vestibules prevailed until 1893. Most featured decorative end hand-rails. (Hall of Records, Dover, Delaware)

Pullman is popularly credited with inventing the vestibule, but like so many widely held beliefs about inventors, the facts do not bear this out. The experiments and applications that were made before Pullman form a long list. In 1836 W. Bridges Adams, an English inventor with many schemes for both highway and railway vehicles, patented a connecting passageway between cars.[358] Eleven years later Sir Henry Bessemer developed a vestibule in connection with experiments on high-speed trains. The device was actually intended to reduce wind resistance by filling in the space between the cars and making the train a reasonably uniform tube. S. R. Calthrop obtained a U.S. patent in 1865 with a similar purpose in mind, but his vestibule was also designed to accommodate human traffic between the pleated folds of the enclosure.

Ideas and patent papers, however, can be partly discounted as speculative rather than practical; the actual testing of vestibules on full-sized trains seems to have been the work of two New England railway employees, Charles Atwood and Charles Waterbury. Waterbury, the superintendent of the Naugatuck Railroad, obtained a patent in 1852 for platform canopies. The main purpose of the invention, particularly as explained in a second patent of 1855 taken out in conjunction with Atwood, was actually ventilation rather than passage between cars. The vestibules made the train into something of a wind tunnel, the air being scooped in at the open front of the baggage car (Figure 5.86). However, one of the patent claims specifically mentions safe transit between cars. The device was tested on the Naugatuck Railroad in the summer of 1853.[359] Over the next year or two it was tried on the Housatonic, the Erie, and the New Jersey Railroad and Transportation Company.[360] For all its originality, the plan did not work out well because of unequal platform heights and slack couplers. Also, the cars were said to become insuffer-

The Wagner Company, which could not afford to be outclassed by Pullman, inaugurated a new vestibule train for the Lake Shore route with much fanfare late in 1887.[366] A vestibule craze swept the industry, stimulated by the enthusiastic public response. By June 1887 the Pennsylvania had three vestibule trains in service, with more on order. The Santa Fe, the Chicago and Alton, and the Rock Island were quick to respond. In August 1888 the B & O announced that all its best cars would be vestibuled within sixty days.[367] Apparently, however, industry-wide conversion of name-train equipment was not completed for another decade.[368]

Vestibules were primarily installed in order to improve accommodations for first-class passengers. Supporters of the vestibule also claimed that the friction of the diaphragm face plates steadied the ride by retarding the swaying of the car. The B & O considered this a very good reason to press ahead rapidly with vestibules because of its serpentine line through the Alleghenies. The vestibule was also said to be a safeguard against telescopes. But the claim that it was effective up to speeds of 40 mph was undoubtedly an exaggeration.[369]

The vestibule had critics as well. Some roads found it troublesome and expensive: cost estimates ranged from $400 to $1,800 per car, depending on the materials, ornament, and patent fees involved.[370] While end platform enclosures were useful in keeping heat in and cold air and dust out, they also blocked the free flow of air when it was most needed during the summertime. Some managers considered the stabilizing effect questionable; they did not believe that the friction created by springs and relatively small face plates had any substantial effect on the sway of a 30-ton car.

During the vestibule's period of introduction a legal battle developed between Pullman and Wagner. Only days after the first Wagner train entered service, Pullman began suit for patent infringement.[371] After hearing the evidence, the U.S. District Court in Chicago ruled that Pullman could claim rights only to the equalized tension spring covered in Sessions's patent of April 29, 1887 (No. 373,098)—not to the vestibule concept in general. Wagner was enjoined to remove the spring apparatus and did so willingly, noting that such devices were not really necessary on the smooth "Water Level Route," because the vestibule ends stayed together of their own accord. Unsatisfied, Pullman instituted two more suits in 1889 and 1890 and in the latter case succeeded in gaining a broader reading of its other patents. Wagner was given forty days to remove the vestibules from the 205 cars it had already fitted.[372]

To counteract this monopoly by preparing alternate designs, Wagner's chief mechanic, T. A. Bissell, devised a vestibule diaphragm with built-in springs that created the necessary end pressure. The steel face plates were replaced with soft, horseshoe-shaped pads. In 1892 he obtained a third vestibule patent that used a lever or pendulum face-plate pressure device.[373] This design was manufactured by the Gould Coupler Company of Depew, New York, and was widely used on all New York Central properties, including the Wagner line of cars. Meanwhile

ably hot when the train was stationary. After a few years the experiment was abandoned.[361]

The vestibule reappeared occasionally over the succeeding years. Between 1868 and 1877 the Burlington is reported to have used some form of car-to-car passageway.[362] The Michigan Central stretched canvas roof canopies between its cars, but the sides remained open. In 1875 the Lake Shore's fast mail train had a primitive sort of vestibule arrangement to enable the clerk to go from car to car.[363] These wooden partitions also discouraged tramps and robbers from riding the platforms. In 1882 Vanderbilt's twin private cars were joined by enclosed platforms. During the same year Pullman was reported to be fabricating a train for the Nickel Plate Road with enclosed platforms, "thus making it practically one continuous car."[364]

In about 1885 Henry H. Sessions, superintendent of Pullman's Chicago plant, began to investigate the vestibule problem. The most casual search would have revealed that the basic idea was an old one; thirty-four U.S. and British patents existed, in addition to the actual tests already described. But when the first cars to be fitted with Sessions's vestibules left the Pullman shops in April 1887, the scheme was proclaimed an innovation in railway travel (Figures 5.87 and 5.88).[365] The cars were operated by the Pennsylvania Railroad on a New York–Chicago express after a trial run on the Illinois Central. And while the claims for originality were unfounded, the vestibule train created the desired sensation. It became the talk of the traveling public, and for a time the name "vestibule" was used to indicate an extra-deluxe train.

Figure 5.90   At stops, it is necessary to raise the trap door (visible at right) to gain access to the steps. (B & O)

Figure 5.91   Full-width diaphragms improved the appearance of the train and had some effect on wind resistance, but they were expensive to install and maintain. (B & O)

Wagner appealed the 1890 decision, and early in 1892 a Chicago judge set aside Pullman's short-lived vestibule monopoly.

A year later Pullman produced an entirely new vestibule design.[374] Instead of the narrow aisle enclosed by double-hinged doors, the new wide vestibule extended the width of the platform. Trap doors covered the stairwells. Large side doors offered easier access, and the additional floor space made room for extra luggage and a place for a polite trainman to step aside for passengers. The vestibule became less of an appendage and more of an integral part of the car body. The narrow diaphragm passageway, however, remained unchanged. Essentially the wide vestibule patented on June 5, 1893, by Pullman is the same as that used on modern cars (Figures 5.89 and 5.90).

In later vestibules parts that were formerly wooden became

steel, but the rubberized fabric of the first Sessions diaphragm continued to be used for generations. On some classes of cars, notably dining, baggage, and mail cars, the vestibule and steps were eliminated from one or both ends. Diaphragms and end doors were retained so that passage between the cars was not interrupted (Figure 5.91). By the mid-1930s designers recognized the practicality of saving space by eliminating one vestibule in coaches, and Colonel E. J. W. Ragsdale of the Budd Company estimated that it saved costs of $4,000 a car.[375] He also opposed the outer diaphragm covers that were popular with some designers because they hid the unsightly end space of the exposed diaphragm. While they did unite the train of cars as a smooth tube, the covers were such a maintenance headache that Ragsdale started an "anti–outer diaphragm society."[376]

CHAPTER SIX

Head-End Cars

BAGGAGE EXPRESS, COMBINATION, EMIGRANT, AND MAIL CARS

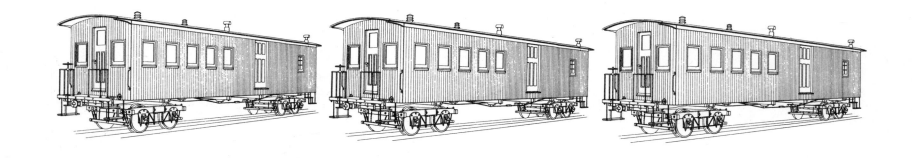

BECAUSE BAGGAGE, MAIL, AND EXPRESS CARS are commonly placed at the front of the train, they are called head-end cars. They are the least elegant members of the passenger car family, so humble that in some way they resemble boxcars. Their proletarian connections are confirmed by their freight car–like side framing: since they have no continuous row of windows, they are built with full trusses and, like stocky peasants, tend to be shorter than the more aristocratic coaches, sleepers, and parlors.

Consistent with their utilitarian purpose, head-end cars are often plainer in finish, less influenced by changing fashion, and longer in service life than other cars in the passenger train. The trade press has traditionally regarded them as unnewsworthy and rarely refers to them, except for an occasional drawing or photograph accompanied by a caption that gives the basic dimensions. Because these cars are so poorly documented, it is necessary to depend almost entirely on surviving illustrations to trace their history. The evidence is especially weak in the early period, and since photographs are variously identified, hard-and-fast dates are not always available. All these pictures cannot be reproduced here, but most of them are at least cited in the notes to this chapter.

Baggage Express Cars

Baggage is integral to the whole business of passenger transport. Few travelers go on a journey without a bag or at least a package; even the commuter has his briefcase. Since the beginning of public transportation, carriers have been expected to make some provision for baggage. The earliest stagecoach lines set a limit of 14 pounds per passenger and charged a fee for baggage over that weight.[1] Steam railroads, however, could afford to carry extra baggage, and to win passengers from the highway, they offered to carry it free. Here was a service unique to America, and one that drew praise from European travelers. Passengers took their small grips on board with them, but larger pieces were placed in the baggage car. The traveler was given a brass check in receipt, and to the astonishment of most foreigners, the railroad assumed responsibility for the luggage in addition to transporting it without charge. To hold this traffic within some bounds, most lines agreed only to the free carriage of clothing; otherwise their lucrative express trade might disappear.[2] It was also necessary to place a top limit on weight—usually 100 pounds per passenger. In 1855 the General Ticket Agents Association began to suggest that 100 pounds be made the uniform free poundage limit, which apparently most railroads were willing to accept as a national standard.[3] By 1870 the limit was raised to 150 pounds and remained at that level for many years.

Generations of railroad managers must have looked upon baggage service as a costly nuisance that was unfortunately necessary in a competitive business. Large quantities of baggage during seasonal rush periods disrupted the smooth flow of train operations, whereas mail and express brought in extra revenue.

In the palmy days of the passenger train this additional income may have seemed incidental, but as passenger traffic itself declined after 1945, head-end earnings became a major source of revenue. Before a Federal subsidy was granted in 1971, mail and express business was vital to American passenger train operations.

Perhaps the earliest published reference to an American baggage car appeared in the January 19, 1833, *American Railroad Journal*. On the front page a paragraph reported a fire, presumably started by a spark from the locomotive, that consumed a Newcastle and Frenchtown baggage car and its contents, including $60,000 in greenbacks. The paragraph did not describe the car itself. Later that year the Baltimore and Ohio Railroad's annual report carried the following statement:

The conveyance of baggage on the tops of passenger-cars has been attended with so much labor and inconvenience in loading, unloading and shifting the baggage at so great a height, added to the manifest injury done the cars in the hurry and bustle of these operations, that the necessity of some improvement on the present plan is obviously indispensable. It is proposed to construct some cars and tenders to the passenger-cars for the exclusive purpose of carrying baggage and in constructing them to make two distinct apartments, one for the baggage going to the ultimate point of destination, which will remain undisturbed for the whole trip, the other for the accommodation of way passengers.

The Philadelphia and Columbia tried to overcome the problem of storing baggage by building possum-belly compartments under the floors of cars.[4] These compartments had side doors and, being close to the ground, were convenient to load and unload. In 1838 it was reported that the Morris and Essex Railroad had a similar car in service.[5] Most railroads, however, concluded that a separate car for baggage was more practical; it had greater capacity and was undoubtedly cheaper than fitting luggage compartments to every coach.

Little is known about the earliest baggage cars. Railroad annual reports list them but rarely give particulars beyond the fact that they had four or eight wheels. The Austrian engineer F. A. Ritter Von Gerstner named a number of lines that had eight-wheel baggage cars when he traveled in America from 1839 to 1840, but again with few details. Possibly the best picture of a first-generation head-end car is the engraving in Figure 6.1. It might be dismissed as just another fanciful printer's engraving, except that the locomotive is a faithful drawing of a Locks and Canals locomotive delivered to the Boston and Worcester Railroad in 1839. Since the entire engraving seems to be the work of the same artist, the rest of the train can be accepted as reasonably accurate. Another printer's train cut showing a larger four-wheel baggage car with outside bracing is reproduced on page 132 of August Mencken's book *The Railroad Passenger Car*. Judging by its mechanical details, the locomotive shown in the engraving would date from 1850, certainly no later than 1855. Yet the engraving is printed on a broadside dated November, 1858, and cuts of the same vintage were reproduced much later

Figure 6.1   The earliest style of four-wheel American baggage car is illustrated in an engraving of a Boston and Worcester train of about 1839. (Smithsonian Neg. 60600)

Figure 6.2   The first eight-wheel baggage cars were little more than boxcars with end platforms. This Old Colony Railroad train was photographed about 1855. (Chaney Neg. 5911)

in the nineteenth century. Caution must therefore be used when citing such illustrations as evidence of dates.

One of the oldest actual photographs of a baggage car is shown in Figure 6.2, which pictures a train on the Old Colony Railroad around 1855. The locomotive *Randolph* was built in 1846 by Hinkley, and although the age of the baggage car cannot be determined, it may have been no newer than the locomotive. An 1842 engraving of a Western Railroad train reproduced back in Figure 1.19 was made from a daguerreotype. The arched-roof, eight-wheel baggage car is similar to the Fall River car except for the lack of end platforms and doors. Another double-truck mid-century baggage car, shown in a lithograph of a scene on the Georgia Railroad, has end platforms, a central side door, four side windows, and the lettering on either side of the center door: "U.S. Mail" and "Baggage."[6]

Figures 1.21 and 1.22 in Chapter 1 provide two other pictures of baggage cars from the 1850s. And Figure 6.3, reproduced from A. Bendel's 1862 volume on American railroads, furnishes a scale drawing of a New York and Harlem car. It measured only 46 feet

over the platform ends; notice the extremely short 4-foot wheel-base of the trucks. During his visit of 1859 Bendel, an official sent by the German government to study American trains, was impressed by the popularity of combination express and baggage cars. He saw few full baggage cars in service, reporting that most were divided into at least two compartments.[7]

While conventional baggage cars served the needs of most railroads, lines that had steamship or ferry connections sought a better system. The transfer of luggage at such terminals was laborious and time-consuming. As early as 1839 (and perhaps even earlier) several railroads began to pack baggage in containers to expedite the transfer. Trunks, parcels, hatboxes, and bags were placed in crates measuring roughly 7 feet square by 5 feet high. The crates or luggage boxes were mounted on small wheels. As many as five could be carried on a flatcar. A few lines labeled individual crates for major stations and picked them up or dropped them off en route.[8] There was no waiting for the head-end crews to find and unload each bag; the crate was quickly deposited on the platform, the local agent distributed the

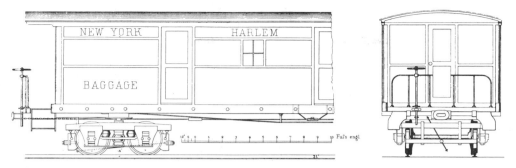

Figure 6.3 *A New York and Harlem baggage car drawing made by a visiting German engineer in 1859.* (A. Bendel, Aufsätze Eisenbahnwesen in Nord-Amerika, *Berlin, 1862, plate 15*)

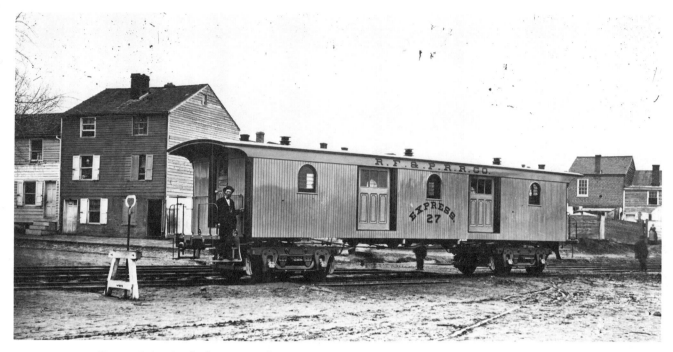

Figure 6.4 *By the late 1860s, baggage cars took on a more finished appearance. The car here was produced by Jackson and Sharp around 1867 for the Richmond, Fredericksburg, and Potomac Railroad.* (Hall of Records, Delaware Archives)

individual pieces, and the empty crate was ready to be filled for the next train. The system was even more useful for through passengers on lines that had a rail-water portage. To ride the Erie, New York passengers were obliged to take a steamer up-river to Piermont. Evidence that the crate system was employed is available in a contract made with Davenport and Bridges in September 1840 for the necessary equipment.[9] The Camden and Amboy used baggage crates on its land-sea New York–Philadelphia route, as did the Philadelphia, Wilmington, and Baltimore Railroad at its Susquehanna River ferry crossing. The Stonington Line undoubtedly carried baggage in crates longer than any other American railroad. A line of fast steamers took passengers from New York to Stonington, Connecticut, where they transferred to the New York, Providence, and Boston Railroad. During his visit of 1839, Von Gerstner was impressed by the efficiency of the baggage crate system at Stonington. The boxes, fitted with 15-inch-diameter wheels, were easily managed by the deckhands. The competing Fall River Line also used the container system. In time both the Fall River and the Stonington lines were taken over by the Old Colony Railroad, whose management thought so highly of the scheme that a baggage crate car was included in the engravings on its stock certificate.[10] Reference was also made to the system in the railroad's annual re-port of 1853. By 1896 the scheme was abandoned because of a change in operations.[11]

While the crate system was prophetic of what would one day revolutionize merchandise transport, it was not to become the prevailing system of handling baggage in America. Open-compartment baggage cars were found more suitable for this work, and their development closely parallels that of other passenger equipment. By the late 1860s the pictorial evidence shows more elaborate head-end cars, with clerestory roofs, arched windows, and doors (Figures 6.4 and 6.5). They seemed less alien to passenger service than their roughly finished ancestors, for the rustic outside body framing had been abandoned. Architecturally the head-end cars blended with the rest of the train, forming a suitable introduction to the elegant palace cars trailing behind. They were shorter and had fewer windows, but in elegance of form they duplicated passenger cars. In 1880 the Santa Fe brought out a new standard baggage car of graceful pattern.[12] The body measured 50 feet, 1⅝ inches by 9 feet, 9⅝ inches. The floor frame was composed of six 5- by 8-inch sills. The fine-looking Tudor arched doors offered a 6-foot, 2-inch by 4-foot, 4-inch opening (Figure 6.6). The normally utilitarian function of such rolling stock was occasionally upgraded for a special occasion, as when the Milwaukee Road outfitted a baggage car

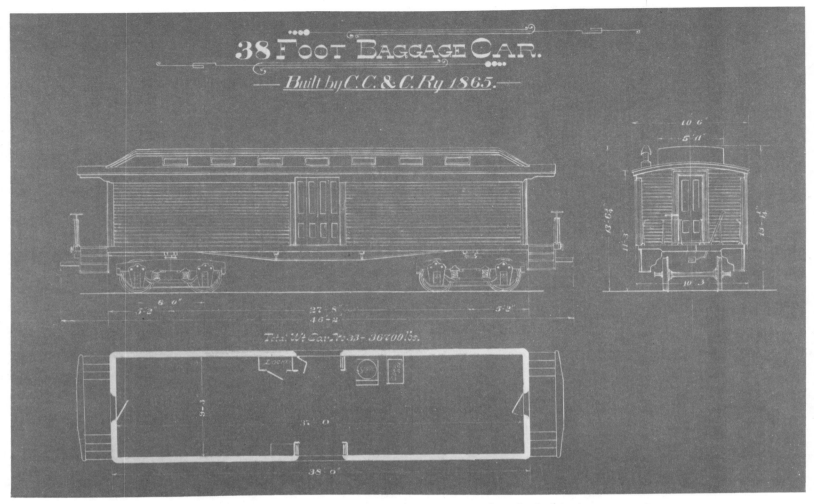

Figure 6.5   The Cleveland, Columbus, and Cincinnati Railway built this clerestory-roof car in its shops in 1865. (Railway and Locomotive Historical Society)

as a ballroom car to enliven an excursion.[13] Carpets were laid and folding chairs were stationed around the side walls, which were hung with pictures and garlands of evergreens and roses. A grand piano and an eight piece orchestra played in one corner for the dancing excursion party.

A modern counterpart of the ballroom car can be found in the ski-train bar cars operated by such lines as the Denver and Rio Grande Western. In 1953 this road refitted an aging baggage car with ice lockers, stand-up counters, and other basic equipment for a primitive saloon. Holiday spirits were in plentiful supply.

In the final decades of the nineteenth century a material growth was evident in the baggage car's size and weight (Figures 6.7 and 6.8). The Pennsylvania's standard class BA baggage car, adopted in 1881, measured 46 feet long and weighed just over 17 tons. Eleven years later, the new class BC design measured 60 feet and weighed 33½ tons. After the turn of the century 60-foot cars remained popular, but heavier framing and six-wheel trucks jumped the weights to as much as 41 tons.[14] Eliminating the end platforms gained space and saved weight; in 1899 the B & O was reportedly removing them from all its baggage, express, and mail cars.[15] In general 60-foot cars were regarded as the optimum size, although 50-foot cars were acceptable for less heavily patronized trains, and 70-footers were considered necessary on runs involving regular movements of theater scenery.[16] Four doors, two near each end, were also favored because they made it possible to separate baggage and express handling, while both commodities could be piled in the large central space.

The shift from wooden to metal baggage cars occurred at the same time as the conversion of other passenger equipment. In the early years of the twentieth century, steel and steel-under-frame baggage, mail, and combine cars began to appear on American railways. This change had been anticipated in earlier decades by various patented schemes for burglar-proof express and mail cars. Except for the Green and Murison car of 1889 (Figure 2.16), none appear to have reached the full-size proto-type stage. Safety was the most potent argument for fireproof, collision-proof passenger cars ("safety" meaning the safety of paying customers, not of employees). The loss of mail and express shipments, however, entailed major claims—as did lawsuits initiated by widows of railway clerks. Furthermore, no train was fireproof if the head-end cars were not. The Pennsylvania Railroad did not exclude head-end equipment when it banned wooden rolling stock from the tunnels leading to Penn Station. A burning baggage car might not ignite a steel coach, but it could suffocate everyone on the train if the fire occurred inside one of the tubes. The railroads had not only a public duty but also a financial interest in upgrading every type of car. The mixture of wood and steel cars was a lethal blend. The temptation was to put "perfectly serviceable" wooden cars ahead of the coaches because they carried only nonrevenue passengers, but in practice it proved so dangerous that the railway mail clerks eventually mustered Congressional action against it.

As it happened, at the beginning of the steel car era when the first steel coaches were only inventors' dreams, a full-sized steel baggage car was already in service. In 1904 the Standard Steel

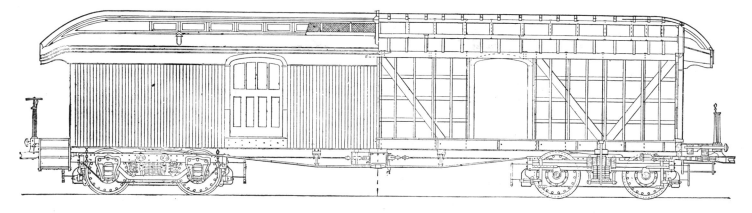

Side Elevation and Section.

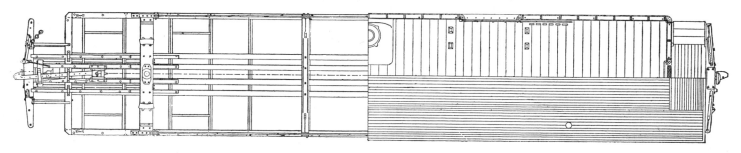

Bottom of Floor Frame.   Floor and Roof.

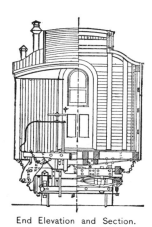

End Elevation and Section.

Figure 6.6   A graceful four-door baggage car designed by the Santa Fe's mechanical department in 1880. The body measured 50 feet, 1⅝ inches long by 9 feet, 9⅝ inches wide. Miller couplers and paper wheels were used. (National Car Builder, January 1881)

Car Company delivered an express-baggage car, Number 599, to the Erie.[17] The twelve-wheel, 60-foot car weighed 53 tons. The side wall sheets measured ³/₁₆ inch thick, and the roof plates were ⅛ inch thick. The car was built with a wooden lining, which was later removed to make it truly fireproof. However, crews complained so much about the lack of insulation that a fire-resistant composite board was installed.

Two years passed before another steel baggage car appeared, this time operated by the Pennsylvania Railroad. But in 1908, the same line had over twenty steel baggage cars in service and was publicly committed to restock its main-line fleet with them as rapidly as possible. In the Middle West the St. Louis and San Francisco Railroad decided to compromise by purchasing composite steel and wooden cars. In 1908 Pullman produced twenty-five baggage cars with distinctive fishbelly underframes (Figure 6.9). All four sills, including the visible outside members, were

deep (28-inch) plate girders. The effect was that of an old wooden body dropped onto a gigantic twelve-wheel steel flat-car.[18] Actually the body was steel-framed, and the wooden sheathing was laminated with sheet metal. The composite cars complemented existing wooden equipment, cost less than all-metal fabrication, and saved 5 tons in dead weight. The Santa Fe purchased a quantity of nearly identical cars from American Car and Foundry Company, but few other lines chose to buy composite cars. New all-metal or steel underframes for existing wooden cars were the most common direction for head-end equipment in the standard era.

Like coaches and sleepers, the new metal head-end stock simply perpetuated orthodox designs in a different material. Four doors and 60- to 70-foot lengths prevailed as before, though weight, strength, durability, and cost all increased. Individual designs continued to flourish after the introduction of steel con-

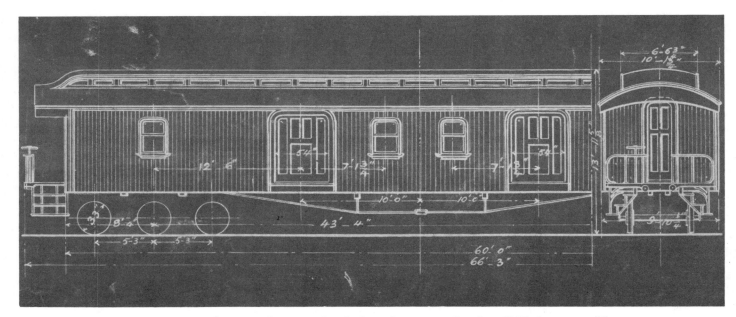

Figure 6.7   The Pennsylvania Railroad class Bb was introduced in 1887. Large cars like the Bb were needed for well-patronized through trains.

Figure 6.8   The K C S baggage-express car is representative of the final all-wooden cars. Pullman built it in 1901. (Pullman Neg. 5834)

struction. The variations were modest, but each railroad had its own plan. For a brief time the Harriman Lines agreed upon a common standard, and in the years just before World War I it was possible to see new baggage cars of a single design on the Union Pacific, the Illinois Central, and other Harriman roads. The U.S. Railroad Administration developed a series of standard

passenger car designs, but only its plan for a 64-foot baggage car seems to have been fully developed or produced.

The lightweight era represents the final phase in the development of head-end equipment. Low, arched-roof baggage cars were built as part of the dazzling motor trains of the mid-1930s. Full-sized lightweights of aluminum, Cor-Ten, and stainless steel

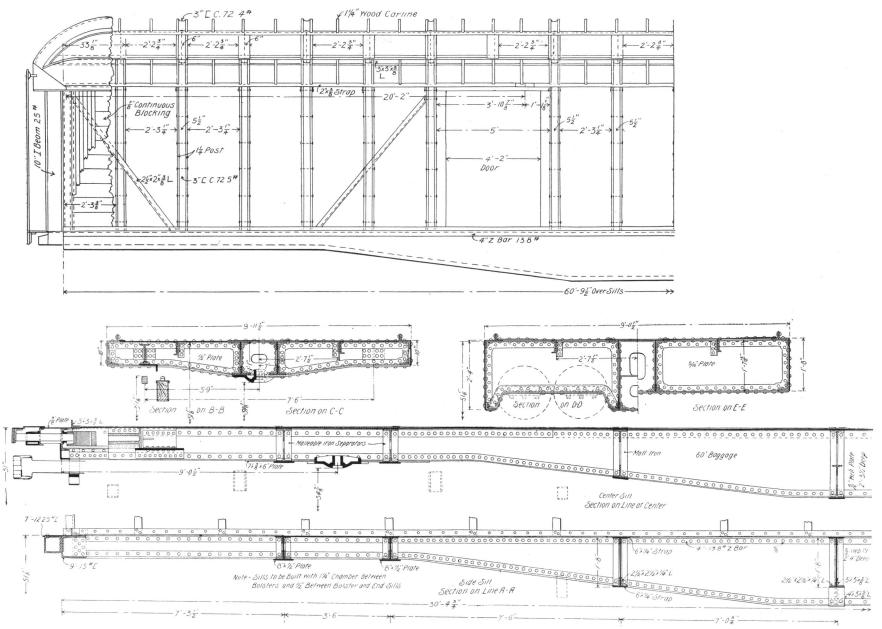

Figure 6.9   *The Frisco bought twenty-five steel-frame, wooden-sheathed baggage cars from Pullman in 1908. They measured 60 feet, 9 inches long and weighed 48.2 tons.* (Railroad Gazette, June 18, 1908)

Figure 6.10   *The Southern Pacific was a leader in adopting all-steel cars. The Number 6400 was produced by Pullman in May 1911.*

Figure 6.11   The massiveness of standard-era equipment is illustrated by this Central Railroad of New Jersey baggage car. It was built by A.C.F. in 1923.

Figure 6.12   Standard Steel Car Company built the New York Central's Number 3361 around 1925.

Figure 6.13 *Baggage car interiors were open and spartan. The Atlantic Coast Line's 668 was a Pullman product of 1917. (Pullman Neg. 21096)*

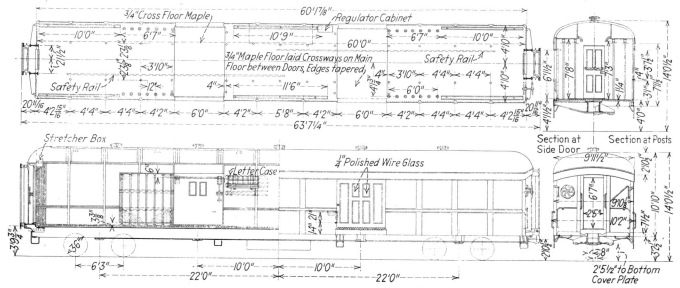

Figure 6.14 *A Pennsylvania Railroad all-steel baggage-express car from about 1925. Cars of this design weighed 43.8 tons and carried 20 tons. (Car Builders' Cyclopedia, 1928)*

Figure 6.15   A lightweight baggage-express car built in September 1950 by A.C.F.'s St. Charles, Missouri, plant. (Santa Fe Railway)

soon followed, as discussed in Chapter 2. Style dictated the need for new baggage cars, because nothing ruined the appearance of a streamliner more than hulking, clerestory-roof heavyweights at the head end. In later years some railroads reduced first costs by economizing on the materials in new head-end construction. By using conventional rather than specialty steels, heavier but cheaper cars could be produced. In 1956 the Santa Fe began to acquire "semilight" express-baggage cars.[19] To cut corners, Pullman fabricated the basic body while the Santa Fe finished the cars for service. Several hundred were produced between 1956 and 1964. Other Western lines continued to buy new baggage cars into the 1960s, and in surprisingly large numbers, considering the downward plunge of passenger traffic. Now much of this relatively new rolling stock has been converted to maintenance service, sold, or scrapped.

Combination Cars

No other passenger cars are more difficult to classify than combines. While relatively few in number, they were highly dissimilar in floor plan. Their taxonomy is a frustrating scramble of variations, exceptions, and special cases. The industry was satisfied to define them as passenger cars divided into two or more compartments to accommodate different classes of traffic. Officially they were termed as combined, combination, or occasionally composite cars. Colloquially they were called combines.

Cars that might technically be designated as combines include coach-baggage-mail, lounge-dormitory, café-coach, coach-buffet-parlor, baggage-dormitory-boiler, and coach-baggage cars. It could even be claimed that Pullman was largely an operator of combination cars, for its fleet was laden with buffet-parlors, bedroom-observations, coach-sleepers, and parlor-sleepers. In theory, all the variations of composite mail, baggage, and express cars were also combines. Yet what was commonly considered a combine was a simple coach-baggage car. A small baggage compartment occupied one end, with the larger part of the interior space given over to the passenger compartment. For the most part, this style of car was found on branch lines or accommodation trains. It was a serviceable maid-of-all-work vehicle suited to short hauls where few passengers or little baggage could be expected. On such routes combines were the last car on the mixed trains, which typically consisted of a dozen or more freight cars and a single combine. Combines also facilitated branch-line–main-line interchanges. A composite car could be switched directly onto a connecting main-line train, thus avoiding the transfer of passengers, baggage, and express. And on the main line itself, a combine was often part of the best name trains; high-class combines were found on such specials as the Royal Blue and the Empire State Express.

The passenger compartment of such a combine was normally set aside for smokers. It became a male retreat—a hideaway from wife and family where a man could sit down to enjoy a good cigar in solitude or in conversation with other men. In 1859 a German traveler wrote: "In these cars, smoking, which is considered bad manners in America just as it is in England, and

Figure 6.16   A combination car on a Providence and Worcester train around 1870. The car itself dates from about 1855 to 1860. (Thomas Norrell Collection)

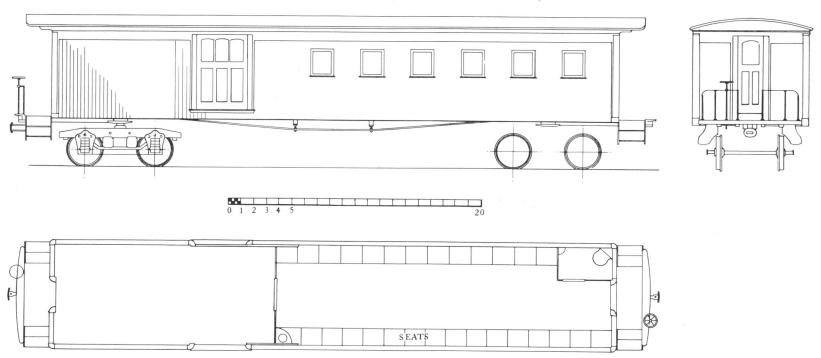

Figure 6.17   The Lake Superior and Mississippi combine Number 2 was built by Harlan and Hollingsworth in 1869. (Traced by John H. White, Jr.)

which is always restricted to rooms specially designated for it, is permitted. This room contains plain wooden seats and is similar to our third class; it is easy to do without a smoke after looking into that room. Wherever such a room is not available, Negroes and smokers have to be satisfied to stay in the baggage room."[20]

The origins of the combine may indeed be found in a desire to separate Negroes from the regular coach passengers. Von Gerstner observed this form of segregation on several American lines during his journey of 1839–1840. He mentions baggage car passenger compartments that were set aside for Negroes on the Petersburg, Baltimore and Susquehanna, and Richmond, Fredericksburg, and Potomac railroads.[21] On the Petersburg Railroad Negroes might ride in the baggage car for half fare but could travel in the coaches if they paid full fare—a surprisingly liberal attitude compared with the policy of Northern railroads, where no blacks were permitted in the coaches at that time. Ironically, the antislavery sentiments so widespread in the North just before the Civil War did nothing to end segregation on board trains.

No illustrations of the first combines have been found; the earliest are a few photographs showing arched-roof cars of the

1850s. In the photograph in Figure 6.16, which appears to date from 1860 to 1870, the rear section was clearly for express, the center may have been for baggage or mail, and the forward end was for second-class passengers. A slightly later but remarkably similar combine is shown in the Harlan and Hollingsworth drawing of the late 1860s in Figure 6.17. This plain car was built for the Lake Superior and Mississippi Railroad. Unfortunately the complete date is missing from the drawing. The L S & M (which eventually became part of the Northern Pacific) started construction of its track in 1868 and opened the first section in the following year. Because the car is numbered 2, it must date from 1868 to 1869.

The most interesting style of combine—the open-side, corridor, or gallery car—also comes from this period. It was a clever scheme to create a separate room on one side of the car and yet allow free passage, so that trainmen and passengers could walk through the car without entering the room itself. Such security was desirable for express, mail, and even ordinary baggage. Photographs show that gallery cars were used on the New Haven, the Central Pacific, and the California Pacific railroads.[22]

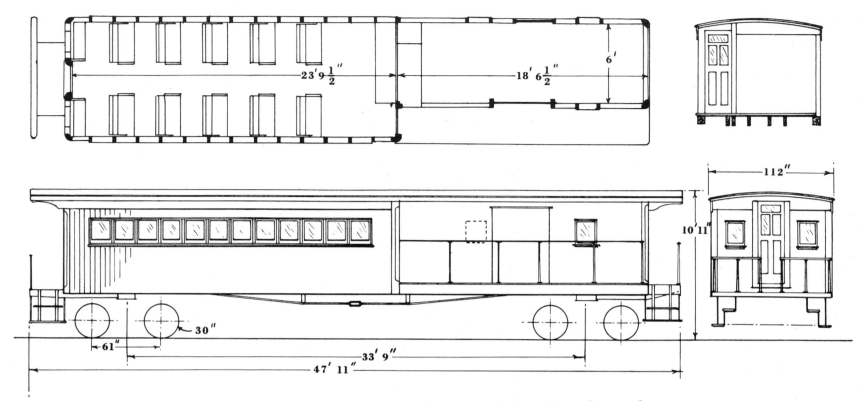

Figure 6.18 This Cumberland Valley combine car was built in the road's own shops around 1855. The car is now exhibited by the Pennsylvania State Railroad Museum, Strasburg, Pennsylvania. (Traced by John H. White, Jr.)

Figure 6.19 Combines, like baggage cars, tended to be more elaborately finished after about 1865. The example here was completed in 1872 by Jackson and Sharp. (Hall of Records)

The Cumberland Valley Railroad also had a gallery car, which is now in the Pennsylvania State Railroad Museum in Strasburg, Pennsylvania. It was reportedly built by the Cumberland Valley in its repair shops around 1855. The passenger compartment seats 24, and the total weight is given as 31,500 pounds. The principal dimensions are shown in Figure 6.18.

Conventional cars could provide equal security without the space lost to the gallery simply by facing the baggage end toward the locomotive. Passengers had no need to pass beyond this point, and trainmen communicated with the locomotive by signals. Thus conventional body styles became the standard for combine cars. In general appearance there was little to distinguish them from other passenger cars, except for the side door. Unlike baggage and mail cars, combines were usually built full-length. In finish they matched any coach, and for special service they were decorated grandly in the palace car style.

An example of a richly scrolled and outlined job is the Sodus Point and Southern's Number 1, shown in Figure 6.19. Two West

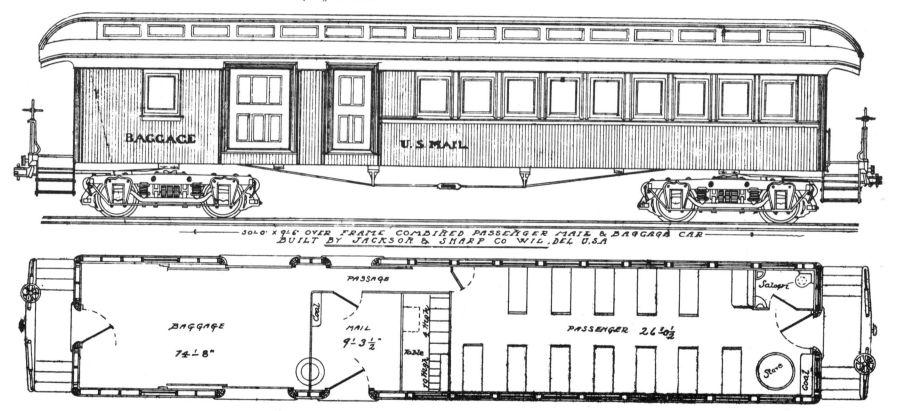

Figure 6.20   *Virginia and Truckee combine cars Numbers 15 and 16 were built by the Detroit Car Company in 1874.*

Figure 6.21   *Combines often did more than carry passengers and baggage; some had mail or buffet compartments as well. The example here dates from about 1885.*

Coast contemporaries now devoid of their original ornamentation have survived to document the combine of the 1870s: Numbers 15 and 16 of the Virginia and Truckee Railroad. The first is held by the Railway and Locomotive Historical Society's Pacific Coast Chapter awaiting the opening of the railway museum in Sacramento. The second car was sold to Paramount Pictures in 1938, and its future is uncertain. Both were built in 1874 by the Detroit Car Company. A diagram made from measurements supplied by Gerald M. Best is shown in Figure 6.20.

What might be labeled a typical combine of the next decade is the 50-foot car shown in Figure 6.21. It was built by Jackson and Sharp for an unspecified customer. The small mail apartment and the common stove straddling the partition between the mail and baggage rooms are of particular interest. The car was probably intended for a branch line; such roads were partial to small, multipurpose combines. One of the most compact combines, and

something of a parody on main-line practice, was a coach-baggage-mail car produced in 1917 by A.C.F.'s Jackson and Sharp plant for the East Tennessee and Western North Carolina Railroad. All three working areas were crammed into the ridiculously short 40-foot body (Figure 6.22). The 3-foot track gauge further restricted the interior space. The car, antique in appearance even for 1917, had steel floor and body framing. The roof, sides, and interior were wooden. It should also be noted that the car is posed in the photograph on standard-gauge shop trucks for shifting in and around the maker's yard trackage.

A remarkable contrast to the stumpy, narrow-gauge triplex can be found in the long, elegant combines built for main-line service. A 60-passenger Erie Railway combine constructed in 1898 for through service is shown in Figure 6.23.[23] Only 14 feet, 6 inches of the 60-foot body was given over to baggage space, because the car was used on first-class trains carrying little lug-

Figure 6.22 The East Tennessee and Western North Carolina's diminutive narrow-gauge combine offered space for passengers, mail, and baggage. The Number 15, produced in 1917 by A.C.F.'s Jackson and Sharp plant, is shown on standard-gauge shop dollies. (Hall of Records)

gage. The conventional coach style of body framing and the freight car–like full trussing at the baggage end are clearly shown in the engraving. An even more elegant example is the New Haven stateroom car Number 2263 (Figure 6.24). Obviously this grand conveyance, manufactured by Pullman in 1900, was no country cousin intended for branch-line runs. (More on this level of combine is given in Chapter 4.)

The combine began as a lowly accommodation for emigrants, smokers, and Negroes. By 1890 it was occasionally adopted for most elite classes of service, and like its fancy relations toward the rear of the train, it evolved in the twentieth century from heavyweight steel to streamlined sleekness (Figures 6.25 and 6.26).

Emigrant Cars

Although emigrants were usually carried in broken-down coaches, some lines built special cars for emigrant traffic. Consequently it is impossible to characterize emigrant rolling stock exactly, but it shows more similarity to head-end cars than to any other general passenger car grouping.

As early as 1836 the Utica and Schenectady Railroad was running half-fare second-class cars which were said to be so inferior that only emigrants would ride in them.[24] After three profitless years they were withdrawn from service. During the next decade occasional mention was made of emigrant cars; the B & O's 1847 annual report, for example, listed two in the road's possession. As the tempo of Western settlement increased, the need for equipment that was adapted to the lowest possible fares became evident. A happy solution was found in specially outfitted boxcars.

The Illinois Central ordered twenty in 1855 at just $200 above the normal boxcar unit price.[25] Windows were placed on either side of the center doors, end doors were cut in, and seats were installed. In the same year the Michigan Central reported that it had fifty cars on the same general plan.[26] End platforms were added and the seats were removable, so that emigrants could be carried west and freight east. Here was passenger transport at the most basic level: it was not so much what the traffic would bear, but what the traffic could afford.

On the Grand Trunk Railroad emigrant accommodations reached a new low, according to a report published by *Railway World* in July 1867. An estimated 800 to 900 Germans were crammed into ten boxcars that had not been adapted to passenger service except for boards nailed partway up at the side door openings and a small hole cut into the upper rear corner for ventilation. No seats, windows, stoves, water coolers, or toilets were provided. The wretched assembly had endured these grim quarters for three days when they passed through Guelph, Ontario, and were seen by a local reporter. They had many more miles to travel before reaching their destination in western Canada. As the train stopped at Guelph, several pails of water were passed up into the cars, but the train pulled out before more could be provided. The reporter called it the most inhumane and shameful scene he had ever witnessed. A *Railroad Magazine* article written in the 1950s claimed that some emigrant families were content to load themselves, their livestock, and their baggage and tools into a boxcar for the journey west, and that such ragtag assemblies were called "Zulus" by contemptuous trainmen.[27] No descriptions of Zulu cars written by eyewitnesses have been uncovered so far, but the literature on the history of emigration has yet to be thoroughly researched.

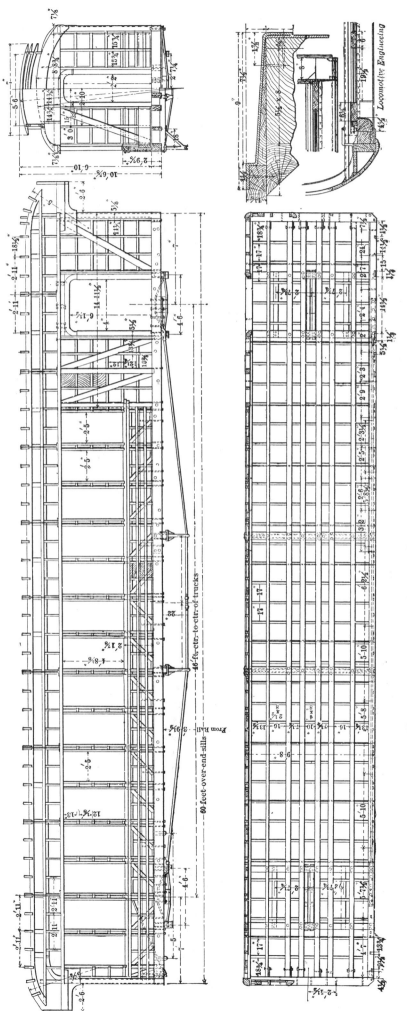

Figure 6.23   An Erie Railway combine car, constructed in 1898, for first-class trains. The car was carried on six-wheel trucks.   (Locomotive Engineering, July 1898)

Figure 6.24   Combines are usually associated with branch-line or secondary trains, but some, such as this New Haven stateroom car, were built for deluxe runs that did not service large quantities of baggage. The Number 2263 was completed by Pullman in 1900. (Pullman Neg. 5208)

Figure 6.25   The Standard Steel Car Company produced this steel combine for Chicago suburban service in 1929.

Figure 6.26   Combines survived into the lightweight era. The Number 281, which seated 48 passengers, was manufactured by A.C.F. in 1947. (New York Central)

The essential meanness of the emigrant coach is shown in Figure 6.27's startlingly clear photograph of the Central Pacific emigrant car Number 319. This is hard evidence that such equipment was stark and comfortless, in an age that furnished opulence and plush ease for the traveler who could pay. Compared with the Central Pacific, the Pennsylvania Railroad pampered its steerage customers with padded seats, a clerestory, a stove, and a water cooler (Figure 6.28). But the severe interior and the presence of only two lamps indicate that there were no frivolous indulgences. A further sign of the Pennsylvania's emigrant policy was the reassignment of its class PC coaches (for example, see

Figure 1.44), built as cheap excursion cars for the 1876 Centennial Exposition, to emigrant service after the fair had closed. (Some of these cars were also used to carry miners in the anthracite coalfields.) Of course, the poorest grade of car could be justified because most emigrants could afford only the cheapest accommodations. Many of them were accustomed to such poverty that the coaches seemed no worse than their homes in the Old Country, or the stinking vessel that had landed them at Castle Garden on Manhattan.

The need for better emigrant cars became more obvious after 1870. For the next quarter century immigration reached its peak,

Figure 6.27   This emigrant car was completed around 1869 for service on the newly opened Transcontinental Railroad. Note the Creamer spring brake on the end platform railings. (Hall of Records)

Figure 6.28   A rare engraving of an emigrant car interior from a Pennsylvania timetable of about 1870.

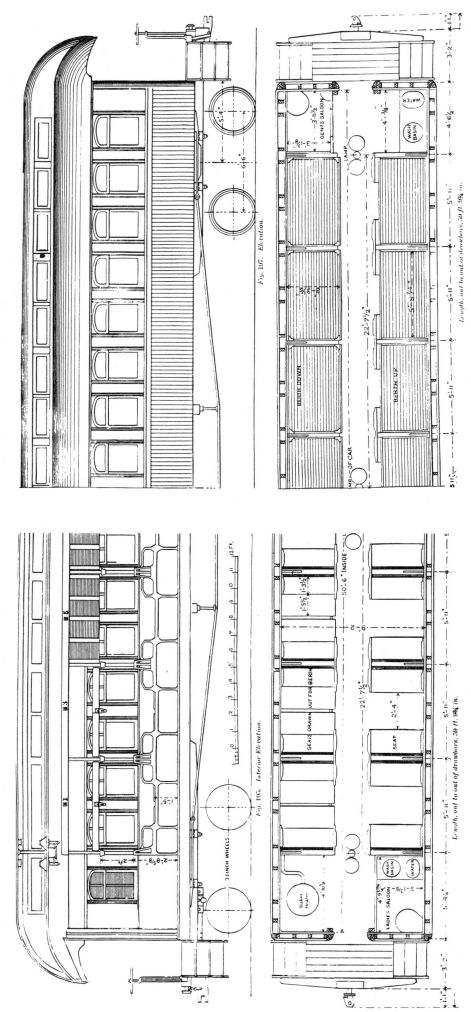

Figure 6.29 *The Western railroads began to furnish plainly finished emigrant sleeping cars in 1879. The example here was designed around 1885 by the Union Pacific.* (Car Builders' Dictionary, 1888)

Figure 6.30   The Pullman tourist sleeper Number 260, from a photograph made in 1897. The car was very similar to the emigrant sleepers in the preceding drawings. (Pullman Neg. 3835)

with over 10 million persons landing on our shores. Those who chose to leave the Atlantic seaports and go west faced a wearisome journey. At least one Western railway official, Alban N. Towne (1829–1895), was sympathetic to their plight. Perhaps he remembered his poor New England boyhood and the discomforts that humble people face daily. As general manager of the Central Pacific Railroad he was in a position to reform passenger service on all levels. He chose to improve the emigrants' lot by providing supereconomy sleeping cars. In its basic design the emigrant sleeper was a duplicate of an open-section Pullman, but it was wholly devoid of ornament and frills (Figures 6.29 to 6.31). Plain oak-plank interiors and board seats with cast-iron ends made up the interior furnishings. A common cooking stove was provided, along with toilet facilities and water. Passengers were expected to furnish their own bedding, including mattresses. In April 1879 the road's Sacramento shops were busy fitting up twenty-five emigrant sleepers.[28] The Central Pacific's annual report for the year noted that sixty-eight second-class coaches had been converted into rudimentary sleepers.

Naturally the other Western lines had to match the improvements provided by Towne. The Santa Fe, the Rio Grande, the Union Pacific, and the Northern Pacific railroads all had emi-

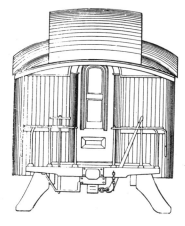
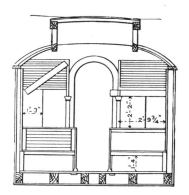
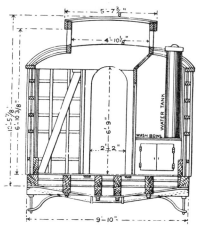

Figure 6.31   Details of the Union Pacific emigrant sleeper shown in Figure 6.29 (Car Builders' Dictionary, 1888)

grant sleepers running before 1885. The desire to surpass the Central Pacific's standard led to the development of the tourist sleeper, in which the ideal of ultimate economy was abandoned for plate-glass windows, shutters, fancy lamps, paper wheels, and ice chests. The Denver and Rio Grande built six fancy tourist sleepers at its Burnham shops in 1883.[29] Within a few years the Santa Fe added steam heat, electric lights, and bedding. In 1889 Pullman succeeded in taking over tourist car operations on most major Western lines.[30] By this time they were distressingly similar to the palace sleepers, even including the services of a porter, and Pullman viewed them as a serious rival. Although tourist fares of 50 cents a night for the four- to five-day trip seemed wonderfully cheap to middle-class travelers, they were beyond the means of the typical emigrant, who was content with boarding cars and welcomed the largesse of such roads as the Santa Fe, which offered a mattress for the entire journey for an extra 75 cents.[31]

The primitive emigrant sleepers appear to have disappeared rapidly after 1900, for by 1915 the I.C.C. listed only 50 in service. The tourist cars remained, however; in fact Pullman had nearly 300. The Santa Fe promoted economical family travel via the tourist car in an elaborate color brochure.[32] Not only was the fare low, but meals could be prepared in the car's diminutive kitchen, thus avoiding the costly dining car. During the Depression, the Great Northern reintroduced the tourist sleeper in an effort to compete with the automobile and the bus.[33] The fare, including the berth, was only 2 cents a mile. By October 1933 such cars were running on eight trains.

The tourist sleeper fleet was expanded temporarily in 1943–1944 to handle troop movements. Even Pullman's large reserve of heavyweight open-section cars could not accommodate the traffic. The Federal government agreed to purchase 1,200 cars if Pullman would operate them. These boxcar-like carriers were extremely utilitarian in appearance and appointments. They measured 51 feet long and weighed 37½ tons. They had no vestibules; a center door was provided, as were twelve small side windows. Diaphragms and end doors provided a passageway to other cars in the train. The open, three-tiered berths were set across the car. Beds for 29 soldiers were provided. In addition to the sleepers, 400 kitchen cars built on the same general lines were produced to operate in conjunction with the troop trains. Both groups of cars were carried on a short-wheelbase, express-car style of truck that offered a safe (if not notably smooth) ride. After the war these cars were sold to various railroads for conversion to mail, express, or mail storage service. Others were retained as built and served admirably as bunk and kitchen cars on work trains.

None of the former troop-train sleepers were used for regular passengers, because the tourist sleeper went into a decline in the postwar period. During the war, rates had been boosted to discourage travel. Before 1941 tourist fare was normally one-half that of first class, but by 1942 it had jumped to 76.6 percent of Pullman fare.[34] In 1947 the Union Pacific and many other Western lines decided to eliminate tourist car service (such cars had

rarely, if ever, been run in the East at any period).[35] The roads were losing money on the tourist car, and they claimed that it was no longer necessary because new coaches with reclining seats offered the same comfort and carried larger revenue loads. On the other hand, the Great Northern decided to continue operation of its old-fashioned tourist cars, while the Milwaukee built new streamline equipment of the same class. Since both of these lines were weak contenders in the competition for Western rail travel, they found it necessary to offer special concessions. For all practical purposes, however, the tourist sleeper died in 1947.

Mail Cars

Because mail cars are the most complicated and interesting members of the head-end family, and because their link with government ensures documentation, their history has been very well recorded. Although their purpose was to carry goods and not people, they were intimately involved with passenger train history. Since mail cars were normally placed at the head of passenger trains, they were designed to resemble passenger cars in size, shape, and construction. Postal traffic represented a sizable volume of business and revenue.[36] During its peak years mail was carried on 9,000 trains over some 200,000 route miles; 30,000 clerks were employed; and gross annual revenues topped 50 million dollars. Some trains made up exclusively of mail cars carried 300 tons of mail daily between major cities such as Philadelphia and Pittsburgh. The Fast Mail was the most important train on the railroad, with precedence over everything on the line, including deluxe expresses. Every trunk line hoped to become known one day as the route of the Fast Mail.

For over a century the Post Office shared the industry's enthusiasm for moving mail by rail. Even after the coming of the airplane, the railroad remained unbeatable for speed, cost, and reliability. The mail service literally followed the railway; the two were interdependent, until eventually one of the partners faltered and was replaced by a more efficient rival.

The beginnings of the partnership were unglamorous. The railroad was viewed by the Post Office as another carrier to move letters in bags, even as it transported crates of shoes and baskets of vegetables in one corner of the baggage car. Mail was shipped as a bulk commodity from one large central post office to another for redistribution to smaller hamlets and substations within a hundred-mile radius. It is claimed that mail was first carried by rail in this country on the South Carolina Railroad in 1831. According to one source, the first official contract for moving the mail was granted to the Baltimore and Ohio in November 1834.[37] Another source claims that no contract was signed until August of the following year.[38] There does seem to be general agreement on the legal starting date of July 7, 1838, when Congress designated all railroads as official postal routes. Within seven years 3,900 of the 4,600 miles of American railroads then in operation were carrying mail.

Nevertheless, although mail was being transported on the majority of lines, the new technology of moving the mail by steam transit was poorly exploited in America. It was in Britain that the best coordination was achieved between railway and postal service. Early in 1838 the Grand Junction Railway tested an imaginative scheme for sorting the mail en route.[39] Clerks, stationed inside a car fitted with tables and pigeonhole boxes, prepared it for delivery as the Flying Mail traveled between Birmingham and Liverpool. Thus the mail was not being simply hauled but actually processed during the journey, so that it arrived ready for immediate distribution. At the same time a net permitted mailbags to be picked up at stations along the line without stopping the train, and pouches were dropped off through a side hatch. The premiere "traveling post office" thus had all the features of the modern mail car. The idea was adopted throughout Great Britain and in many parts of the Empire. A British journal's description in 1849 of clerks dexterously filing letters into thirty-five pigeonholes of a London–Glasgow mail car reads like an account written a century later.[40]

According to British authorities, the American Post Office waited twenty-five years before introducing this practical system, and even then viewed it with skepticism. At least, that is the story told by the published histories of the railway mail service, including those written by former officials of that agency. But it seems to be something of a half-truth, since there is evidence that the British system was being practiced in this country, at least in an elementary form, in the very year that the first Grand Junction car entered service. The following paragraphs from the *Niles Register* of May 18, 1838, make it clear that the mail was being sorted, picked up, and dropped off en route.

New mail arrangement. Mail cars, constructed under the directions of the post office department, are now running on the railroads between Washington and Philadelphia. They contain two apartments; one appropriated to the accommodation of the great mails, and the other to the way mails, and a post office agent. The latter apartment is fitted up with boxes, labelled with the names of all the small offices on or near the railroad lines. It has also a letter-box in front, into which letters may be put up to the moment of starting the cars, and any where on the road.

The agent of the post office department attends the mail from the post offices at the ends of the route, and sees it safely deposited in his car. As soon as the cars start, he opens the letter-box and takes out all the letters, marking them so as to designate the place where they are put in. He then opens the way mail bag, and distributes its contents into the several boxes. As the cars approach a post office, the agent takes out the contents of the proper box and puts them into a pouch. The engineer slackens the speed of the train, and the agent hands the pouch to the postmaster, or carrier, who stands beside the track to take it, receiving from him, at the same time, another pouch, with the matter to be sent from that office. This the agent immediately opens, and distributes its contents into the proper boxes. Having supplied thus all the way offices, the agent, when arrived at the end of the route, sees the mail safely delivered into the post office.

The mail leaves Washington at 6 o'clock, A.M. All letter-writers should, if possible, put their letters into the post office the preceding evening. The letter box at the cars wll not be open until after 5 o'clock in the morning. It is not *a post office*; and letters will not be received at it while the post office is open. Its object is to allow all persons, until the last moment before departure, to send letters by mail, even although the mail has been closed at the post office, and even placed in the car.

The letter box will be open after the mail closes in the post office at Baltimore, and all the intermediate points. Philadelphia cannot at present enjoy its advantages fully, because the cars do not run into the city.

In this arrangement, the post office department has sought to give the greatest security to the mails, and to afford the community the best possible accommodation. Well executed, the plan must be almost the perfection of mail arrangements. It is intended, when it can be conveniently done, to extend a similar arrangement through to New York.

A year or so after the *Niles* article, Von Gerstner mentions seeing four mail cars with heated offices that collected, sorted, and delivered letters during the journey.[41] There were other reports of such service on the Boston and Albany in 1852 and the Richmond and Terre Haute in 1855.[42] Although their documentation is not conclusive, certainly the *Niles* article and Von Gerstner's account are indisputable. Furthermore, the Post Master General's report for 1840 speaks enthusiastically of the British system and provides a paper on the subject prepared by agents sent overseas to make a study of European methods.[43] Thus the Post Office was not unaware of the system. In addition, Canada began a traveling post office service in 1854, and in his report for 1859 the Post Master General praised its effects on cost and speed of delivery.

The facts seem to be that mail was sorted on some routes in this country, but that the system was not vigorously developed by the Post Office Department. Many routes were served by apartment cars such as the mail-baggage combine shown in Figures 6.32 and 6.33. The illustrations date from the 1860s and 1870s, but they might equally well represent cars of the previous decades, when even smaller, more poorly furnished compartments were the rule. In 1855 the Post Office began to press for better facilities so that its route agents could work more efficiently.[44] Their jobs encompassed what is now regarded as the function of a railway post office car. Letters were received, stamped, sorted, and dropped off (though exactly how much sorting was done during the trip is not clear). Pouches were picked up and dispatched en route. The Post Office asked for two rooms. The agents' room was to be no less than 7 by 12 feet, with a counter, boxes with sliding lids, a window and a door on each side, and locks on all doors. The second room, to be no less than 7 by 13 feet, was for through mail in locked pouches or canvas bags. In response to these requirements it is known that at least one railroad, the Lake Shore line, fitted three baggage cars in 1856 with a 7½- by 17-foot compartment for the Toledo-Chicago run.[45] By 1860 traffic had developed to the point where the road furnished cars with even larger compartments. In 1866 it was necessary to provide 35-foot-long cars devoted exclusively to mail.

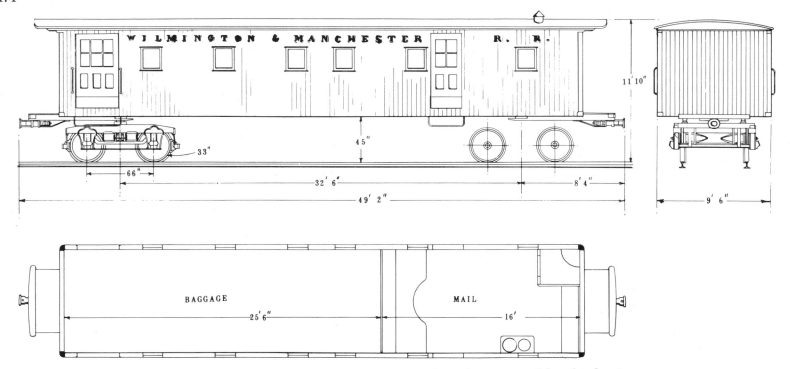

Figure 6.32   No illustrations of first-generation mail cars have survived, but this drawing is a good representation of them. The antique-looking combine was built in 1866 by Harlan and Hollingsworth. (Traced by John H. White, Jr.)

Figure 6.33   A Denver Pacific arched-roof combine mail-express-baggage car, constructed around 1870. (Hall of Records)

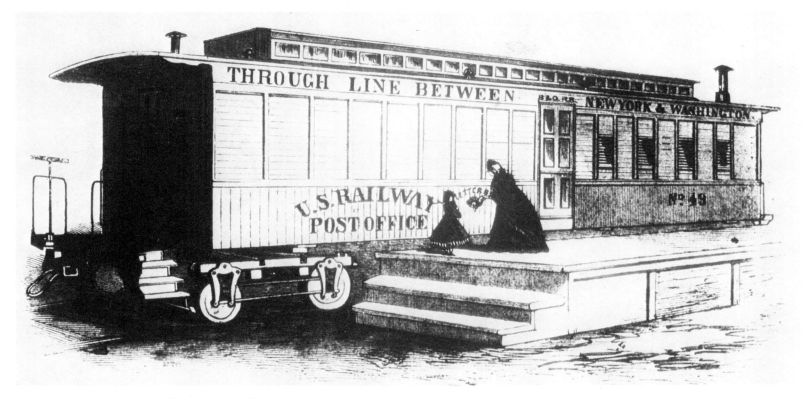

Figure 6.34   The B & O's Mt. Clare shops built this elaborate mail car in 1864. It was one of the finest produced up to that time. (Leslie's Illustrated Magazine, *October 8, 1864*)

The Civil War created an unexpected volume of mail that overwhelmed the postal service. Official correspondence was only part of it. Letters between soldiers and their families and sweethearts created an avalanche of paper that the existing system could not handle. As letters stacked up, the public grew more impatient, and its complaints were forcefully passed on by Congress. Radical reform was needed.

Postal officials were ready to try any scheme to expedite the mails, so that William A. Davis, assistant postmaster in St. Joseph, Missouri, had little trouble in obtaining permission to introduce train-board classification of the Western mails passing through his home town in such volume. By sorting aboard the cars en route to the California Pony Express connection, considerable time would be saved. The Hannibal and St. Joseph Railroad agreed to fit two baggage cars with tables and pigeonhole filing boxes. The service began in late July 1862, but was discontinued after a year or two because of a shift in the California mail route. Many years later some of Davis's friends and relations —a son-in-law was particularly active—insisted that he had originated the railway post office scheme.[46] It was a foolish claim in light of the much earlier British and American examples already noted, but Davis's supporters went so far as to fake a photograph by grafting the image of a circa 1900 mail car to a picture of a wood-burning locomotive from the 1860s and labeling the fraud as the first railway post office car. Davis himself, it should be noted, never claimed the honor.[47] He was a lifelong postal employee and knew better.

It was George B. Armstrong (1822–1871) who might better be called the father of the railway mail service in America. He envisioned, and around 1864 proposed in writing, a plan for a national system of railway mail service which would fully utilize the British concept of traveling post offices. A supervisor in the Chicago post office, Armstrong was appalled by the backlog of

mails created by the war. In cooperation with Eastern officials of the Post Office Department, he experimented with mail cars in August 1864. The first car, worked by two men, was run between Chicago and Clinton, Iowa. It was a 40-foot-long route agent's car with blind ends. It had been remodeled with a bank of seventy-seven pigeonholes and a cluster of newspaper boxes.[48]

On almost the same day that the Armstrong car began its first trip, a similar experiment was begun in the East. Actually, A. N. Zevely, third assistant post master general, had conducted tests since the spring of 1864. In a remodeled baggage car his crews had been able to sort 5,000 letters on the New York–Washington run. Zevely wanted a first-class Railway Post Office built to order and persuaded the B & O to fabricate such a unit at its Mt. Clare shops. The result was a magnificent eight-wheel car with a clerestory roof and bold lettering which read "Through Line between New York & Washington."[49] The Number 49 was ready for service by August 31, 1864. The inside body measurements were 9 feet, 4 inches by 45 feet, 5 inches. The interior furnishings included a 400-module pigeonhole box, tables, a stove, and space for newspapers and packages. A second car was under construction at the Camden and Amboy shops. The B & O car attracted so much attention that *Leslie's Magazine* published engravings in its October 8, 1864, issue (Figures 6.34 and 6.35). The car represented a remarkable advance, but the schedule of 12½ hours left something to be desired in the way of speedy service.

The successes of Armstrong and Zevely encouraged the mass application of the Railway Post Office idea. In 1865 such cars began to appear on the Burlington, the Alton, the Illinois Central, and the Lake Shore railroads.[50] Fewer R.P.O. cars were found in the East, where service was limited to the original New York–Washington route and two longer hauls between Albany and Buffalo and New York and Erie, Pennsylvania.[51] During that final year of the Civil War, sixty-four clerks were working

Figure 6.35   The B & O's 1864 mail car had 400 pigeonholes so that letters might be sorted en route. (Leslie's Illustrated Magazine, *October 8, 1864*)

R.P.O.s over 1,041 miles of track.[52] Little progress was made during the next few years, but the plan was firmly established in spite of heated criticism from the district postal supervisors, who feared that it would eliminate their patronage if not their very jobs. In 1869 the Railway Mail Service was officially inaugurated, with Armstrong as its head.[53]

After Armstrong's death in 1871, George S. Bangs took charge of the agency and set out to expand it. At the end of his first fiscal year he was in charge of 57 lines (covering 14,117 miles) and 649 clerks.[54] Traffic was very heavy on some routes, such as the Buffalo-Chicago line, which daily carried 50,000 to 60,000 letters each way and 3 to 10 tons of newspapers, magazines, and parcels.

More and better cars were now being run. Full mail cars were used on all major lines; the once standard apartment car (the mail-baggage combine) had been downgraded to lesser runs. A crook-arm mail catcher patented by L. F. Ward of Elyria, Ohio, on January 29, 1867 (No. 61,584), was now widely used.[55] In 1870 the Michigan Central turned out several finely finished cars with 48-foot bodies, 450 letter boxes, and 60 newspaper cases placed in a semicircular arrangement.[56] In addition to providing sorting tables and unusually good lighting, the Michigan Central

added two velvet-covered sofas so that the clerks could take quick naps. One-third of the space was open for pouches. The exterior was painted a buff color. Two years later the Central Pacific shops turned out a palace mail car that featured a parlor-dining-kitchen compartment in the center of the body.[57] It had a cooking stove, a table, chairs, a washstand, and several beds. More typical of the 1870s, however, is the utilitarian Lake Shore car depicted in Figure 6.36. In 1877 the Milwaukee produced a car after the design of J. E. White, later named head of the Railway Mail Service.[58] The 50-foot body was carried on six-wheel trucks and cost about $5,000. White's interior plan was designed to speed distribution. He dispensed with the semicircular box case and used a long case on one side of the car. Bags were held by iron racks resting on the floor. White claimed that his floor plan boosted capacity by 50 percent, but ultimately the three-sided end case prevailed.

These technical refinements were pale compared with the most sensational achievement of the R.M.S. in its early years: the Fast Mail. Rapid sorting aboard the cars meant little when the fastest train between New York and Chicago lumbered along at around 25 mph.[59] The 979-mile journey required just four minutes under thirty-seven hours. No wonder that beds were placed aboard the

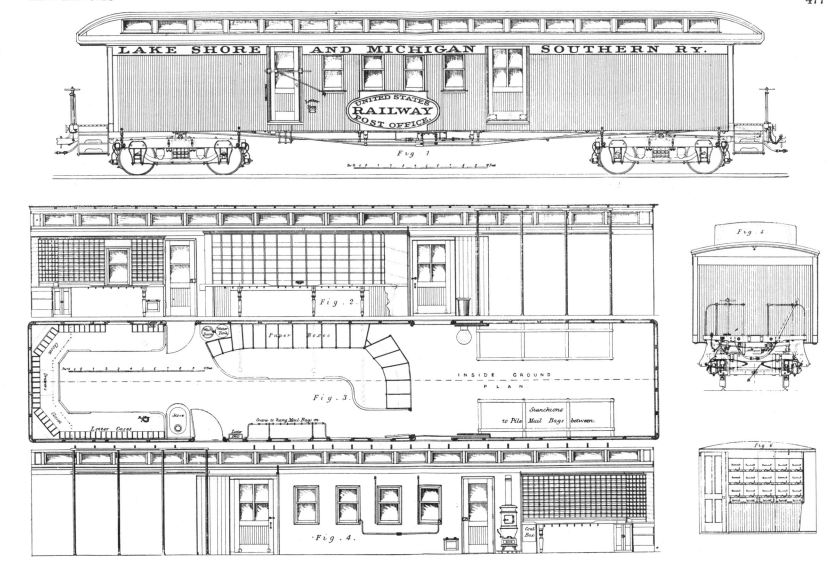

Figure 6.36   The British trade magazine Engineering *published this detailed drawing in its July 28, 1876, issue. The brief description did not mention the designer or builder.*

Figure 6.37   The Lake Shore and Michigan Southern formed an important link in the New York Central's New York–Chicago fast mail train of 1875. (Chaney Neg. 13666)

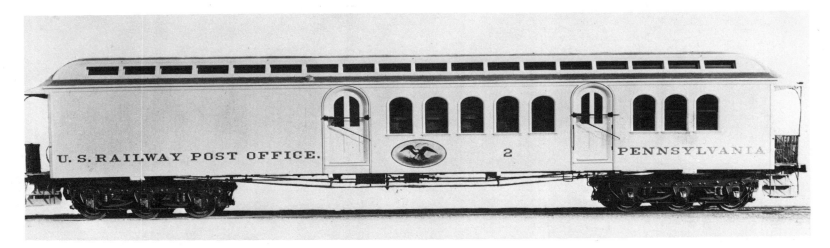

Figure 6.38 The Number 2 is believed to be one of the Philadelphia–St. Louis fast mail cars produced in 1875. (Smithsonian Neg. 71262)

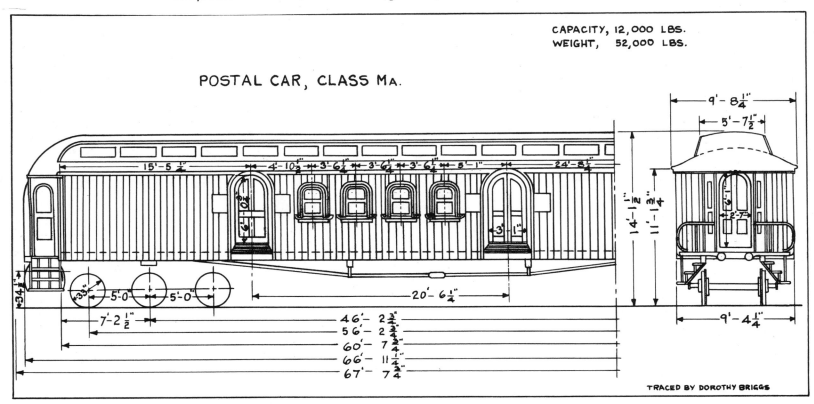

Figure 6.39 The Pennsylvania's class Ma mail car was introduced in 1876.

R. P. O.s. The average speed of passenger trains was not the only problem, for mail movements were tied to existing schedules. Departures were usually not geared to the next morning's deliveries, and connections were poorly scheduled for mail transfers. Bangs recognized that no material improvement in deliveries could be made until extra-fast mail trains ran independently of regular passenger service. He dreamed of a twenty-four-hour train between Manhattan and Chicago, and railroad men agreed that the existing equipment was fully capable of such a performance. But special train service and the twenty new cars that Bangs envisioned would be expensive. The R.M.S. had no funds and no prospect of obtaining any from Congress, which was even then quarreling with the railroads over a fair rate of compensation for carrying the mails.

In 1874 William H. Vanderbilt heard of Bangs's idea and became enthralled with it, refusing to be dissuaded by his father's caustic opinions on dealing with those rascals in Washing-ton.[60] Support for the Fast Mail was explained as a public-spirited gesture on the part of the younger Vanderbilt. Privately, he was convinced that the publicity value of the project would be invaluable and that if it did improve service, public opinion would force Congress to fund it properly. Bangs was apparently willing to accept the circus aspects of Vanderbilt's promotion to see life breathed into his pet, and Vanderbilt for his part cut no corners. The New York Central and the Lake Shore shops at Albany, Cleveland, and Adrian, Michigan, were ordered to build a fleet of the finest mail cars ever produced and to do it quickly.[61] Three styles were devised: R.P.O. letter cars, newspaper cars, and bulk storage cars for through mail that was not to be sorted enroute (Figure 6.37). Other valuable details are given in a notice from a *New York Graphic* dated Sept. 15, 1875:

The letter-distributing cars are 50 feet in length, while those designed for the newspaper mail are 10 feet longer. All are uniform in width, 9 feet 8 inches, and 6 feet 9 inches high in the clear. The finish of the exterior does not differ, all of them being painted white, with cream-colored borderings and gilt ornamentation, highly varnished within and without. Midway on the outside and below the windows of each car is a large oval gilt-finished frame, within which is painted the name of the car, with the words "United States Post-Office" above and below. Along the upper edge and center are painted the words, in large gilt letters, "The Fast Mail," while on a line with these words, at either end, in a square, are the words, in like lettering, "New York Central" and "Lake Shore." The frieze and minute trimmings around the windows are also of gilt finish. At the lower sides and ends of the cars are ovals corresponding to those on which the names are painted, and inclosed at one end a painted landscape scene background, and in the relief an all-seeing eye, beneath which is a pyramid inscribed with gilt Roman figures, "MDCCCLXXV," and the motto "Novus ordo seclorum." At the opposite end, in the same colors, on a blue background, is the United States coat-of-arms.

In running appointments, as well as all others, the most important patents have been selected. The platforms are inclosed on either side by swinging doors, which can be fastened open at will to prevent disturbances by wind when the mail matter is being trucked from the tenders to the distributing cars, and also as a guard against any danger of its falling off. The cars are supplied with patent air-brakes and patent spring-brake, and the end doors of the cars are protected by an inclosed platform. On either side of each car are two doors supplied with patent mail-bag catchers for securing the mail along the route. The 60-foot cars are mounted on 6-wheel trucks, giving in all twelve wheels to each car, while the 50-foot cars are on 4-wheel trucks.

The article reported that the cars were named for state governors, such as Todd, Dix, Allen, and Hayes. The letter cars had a room for the chief clerk outfitted with a folding bunk, desk, and closet. The distributing room was furnished with the usual counters and pigeonhole boxes. An iron safe was provided for valuables. Washstands, water coolers, and hot-water heating were also available. The letter cars cost $4,200 each; the newspaper cars $3,300.

Every effort was made to draw public attention to the inaugural run. One hundred railroad officials, politicians, members of the press, and dignitaries such as the U.S. Vice President were aboard for the first trip on September 16, 1875. An equally impressive load of mail was placed on the cars—so much, in fact, that some of it overflowed into the V.I.P.s' drawing-room car attached to the rear of the four mail cars. The contents totalled 33 tons, a burden increased by another 150 letter pouches picked up at Albany. The first lap of the trip to Buffalo was made at the astonishing average rate of 40.9 mph (including stops), which equaled express-train speeds in England.[62] Once on the Lake Shore portion of the run, the average speed dropped to 33 mph. Even so, the train completed its run in just twenty-seven hours and fifteen minutes, a saving of nearly ten hours over the fastest regularly scheduled express train. The lightning run was to be more than a onetime stunt; this schedule would be followed on a daily basis.

The president of the Pennsylvania did not want to be outdone by the Vanderbilts, and just weeks before the first run of the Fast Mail, Tom Scott initiated a crash program to produce a competing train.[63] The cars were built in ten days (Figures 6.38 and 6.39). The Limited Mail was ready on September 13, 1875, but its inaugural trip from Jersey City to St. Louis was less well publicized than the New York Central's and its remarkable record of thirty-three hours received relatively little attention. Even so, Bangs's contention that faster train speeds were practical was again demonstrated.

What should have been the beginning of faster mail service, however, proved to be something of a false start. Congress refused to support such a luxury, just as the old Commodore had predicted. In fact, the two Houses reduced the overall rate of compensation to railroads by 10 percent in July 1876. Vanderbilt was infuriated and withdrew the Fast Mail service in a matter of days.[64] The Pennsylvania followed suit. The panic of 1873 was bottoming out, and the volume of mail declined as business slowed down. Congress reacted again in June 1878 with another 5 percent cut. This action reopened the long-standing battle between Congress and the railroads over a fair rate compensation for carrying the mails. In the early 1870s the industry protested against the niggardly rates paid, claiming that they were less than second-class freight rates.[65] Yet the cars required by the R.M.S. were expensive, highly specialized equipment, which spent as much time in the station as on the road. At one point the railroads threatened to withdraw all postal cars unless relief was granted. Refusing to yield, Congress hinted that it would take equally drastic action if the railroads interfered with the transport of mail. The battle continued, in various degrees of Sturm und Drang, throughout the history of the Railway Mail Service.

As the economy recovered, the volume of mail began to surpass old records. Congress relented somewhat, and the Fast Mail concept was restored to life. In 1881 the New York–Chicago train was revived, and a fast train from New York to Washington, Atlanta, and New Orleans brought the wonders of rapid mail delivery to the South.[66] In the same year a total of twenty-four hours was slashed from the transcontinental mail train schedules. In 1882 the B & O produced five elegant white cars, trimmed in ultramarine, for the Baltimore–St. Louis Fast Mail.[67] After the wonder of the new service had faded slightly, humorists satirized the Fast Mails as moving so rapidly that cows along the line were given timecards so that they could withdraw from the tracks in good time.[68] It was also declared necessary to tie down light stations, anchor handcar shanties, and use extra-large nails in fences because of the trains' terrific suction.

The R.M.S. was entering a period of enormous growth. In 1881 nearly 3,200 clerks and 1,892 cars ran over 90,200 route miles.[69] In 1888 another 1,900 clerks were traveling over an additional 53,000 route miles. By 1894, 3,059 mail cars were in service.[70] This growth in personnel, equipment, and mileage paralleled a similar expansion in the railroad industry. And even as the railroad business reached its peak in 1915, so the R.M.S. extended to its widest range in that year, when some 20,000 clerks and 4,000 to 6,000 cars (depending on which figures one cares to believe) moved the mails over 216,000 route miles.[71]

Figure 6.40   Jackson and Sharp built the Georgia *in 1879. It ran between Richmond and New Orleans over several Southern lines. (Hall of Records)*

Figure 6.41   Interior of the Georgia *of 1879. (Hall of Records)*

During these decades of growth the mail car followed the general development in size and materials of other passenger equipment. Full R.P.O.s, like baggage cars, remained on the short side, but combination mail and passenger cars were usually made full length. Until sometime in the early 1890s exteriors were painted white, with buff, red, or blue trim (Figures 6.40 and

6.41). The color scheme set mail cars apart from the rest of the train, but in later years colors became drabber and more standardized. A special mail car shown at the 1893 Columbian Exposition was one of the last known to have been painted in the old tradition. A more important development was the elimination of the end platforms. Many of the earliest R.P.O.s were built with blind ends; platform steps and handrails were retained for the convenience of the brakeman, but there were no end doors (Figures 6.42 and 6.43). However, the platforms were an inviting place for thieves and hoboes, some of whom would actually break through the blank end wall in search of plunder.[72] Around 1875 the Louisville and Nashville began to build head-end cars without end platforms. A drawing of such a car was reproduced in a French report on American railroads published in 1880 and 1882 (Figure 6.44).[73] Some of the data were gathered as early as 1876. By the mid-1880s nonvestibule mail cars were the general rule.

Representative postal cars for the period include the Michigan Central's *Indiana*, built in 1880, which measured 55 feet in length and weighed 26 tons.[74] Later in the decade the Milwaukee Road produced a 69-foot car weighing 33.5 tons.[75] At the 1893 Exposition two fine mail cars were shown, both featuring 60-foot bodies, which the Post Office fancied for many years as the optimum length for a full apartment mail car. A drawing of the 45-ton Pullman-built entry is shown in Figure 6.45.

Mail car furniture was also standardized during the final decades of the century. The most notable single development in this area was iron rack frames. In the mid-1870s wooden frames or racks with hooks replaced the circular cases formerly used to hold full,unsorted letter pouches.[76] The pouches hung on the

Figure 6.42   An early Railway Post Office car stands behind the locomotive's tender in this 1867 scene. It was photographed at Rockford, Illinois, on the Chicago and North Western Railway. (Smithsonian Neg. 73-403)

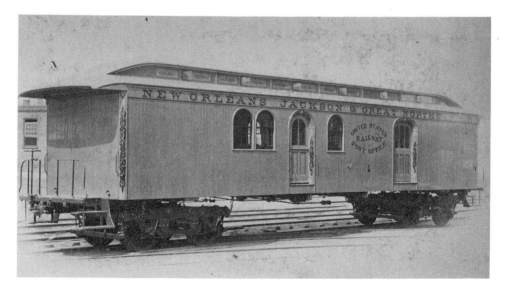

Figure 6.43   The N O J & G N's first Railway Post Office car was built by Jackson and Sharp around 1875. Note the blind ends.

racks were more accessible to the clerks. Gas-pipe racks were tested as a possible improvement over the wooden frames. On seeing the pipe racks, Charles R. Harrison, a R.M.S. employee, was inspired to devise a cheaper, more flexible design. Around 1879 Harrison worked out a plan for hinged, cast-iron mail car furniture (Figures 6.46 to 6.48). Pouches, racks, and sorting table could be set up, moved, or folded away as necessary, giving the individual car great flexibility.[77] It was even possible to clear the car for baggage or express without removing the R.P.O. fixtures. Costs were reduced, since the standard fittings were cheaper to make and replace. Harrison also claimed that his equipment was safer, because cast iron would not splinter and

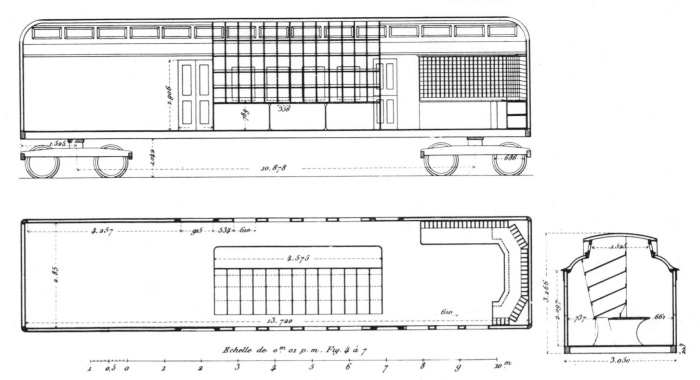

Figure 6.44   The drawings for a Louisville and Nashville mail car of about 1875 were published in a French work on American railroads. The center table and rack was probably for newspapers. (E. Lavoinne and E. Pontzen, Les Chemins de Fer en Amerique, Paris, 1882, plate 6)

burn. Harrison opened a factory in Fond Du Lac, Wisconsin, in 1881.[78] Within a few years a number of railroads were using his furniture and by 1910 it was declared to be almost universally employed in American mail cars. In later years iron furniture manufactured under the trade name of Otis superseded the Harrison fixtures.

Although most car design was the exclusive province of railroad and supply trade engineers, mail cars were a minor exception to this normally inviolate rule. Since the beginning of the R.M.S., Post Office managers took a direct hand in the interior arrangement of the cars. It is true that frame, body, and running gear designs were left to the railroads, but even this fundamental arrangement was to change. Eventually the government was insisting that its standards and general specifications be carefully followed. In 1885 came what appears to be the first effort to achieve a system-wide set of standard floor plans.[79] Charles W. Vickery, head of the R.M.S., approved detailed drawings for a 40- and a 50-foot mail car. A U-shaped letter box at one end was to have 450 openings, 4 inches by 4 inches by 7½ inches deep. A sixteen-box letter case was also specified, as were particular door openings and inside dimensions. A number of railroads were said to be constructing new cars in accordance with Vickery's plan.

The Post Office also sought to improve productivity by requiring better facilities for clerks. Few jobs involved conditions of such stress. A man was expected to have unusual powers of concentration, dexterity, and precision, as well as the ability to work rapidly while standing for hours in a smoky, swaying railway car (Figures 6.49 and 6.50). To qualify he had to have the eyes of a hawk, the nervous system of a surgeon, and the balance of a sailor. The combination of exacting yet physically demanding labor was more than the average man could long endure. Hence the demands for better lighting, ventilation, and heating were not unreasonable.

Lighting was one reform that received early attention. Excel-

lent lights were necessary to read handwritten envelopes, and by the early 1880s special mail car lamps with fantastic umbrella-like reflectors had been developed (Figure 6.51). By the mid-1890s the Post Office was calling for an end to oil lamps, more for reasons of safety than of improved illumination.[80] It was contended that thirty fires aboard mail cars in a single year had been caused by kerosene lamps—a claim that industry called patently untrue. Even so, by 1897 one-third of the R.P.O.s had gas lighting, and the Post Office, looking beyond gas to electricity as a final solution, sponsored a test of electrical lighting in 1912 that ran for several weeks.[81] It was found that the style of reflector used could affect lighting efficiency by as much as 55 percent. A published report and new lighting specifications followed the tests.

Owing to pressure from top management, the clerks' union, and interested members of Congress, the Post Office exhibited increasing concern for the safety of its employees. In 1877 statistics on deaths and injuries began to appear in the annual reports of the Post Master General. During that year only 2 clerks were killed in accidents; 10 were seriously injured.[82] By 1884 7 were killed and 28 were seriously wounded. The total accidents jumped from 27 to 154. Increased train speeds and the growing number of mail cars caused the toll to rise. Between 1890 and 1900 mail cars were involved in 5,000 accidents that resulted in 75 deaths.[83] In 1897 alone, 14 clerks lost their lives. By 1907 a total of 210 clerks had been killed since 1875. During the next eight years an average of 15 names were added annually to the death rolls, but after 1915 fatalities fell off to a uniform average of 2 a year.[84] The improvement can be credited to more rigorous safety measures, the growth of automatic signaling, and the introduction of steel mail cars.

In the 1870s and 1880s the Post Office established standard floor plans and pressed for better lighting and heating. Yet the actual car structure was left to the judgment of the individual

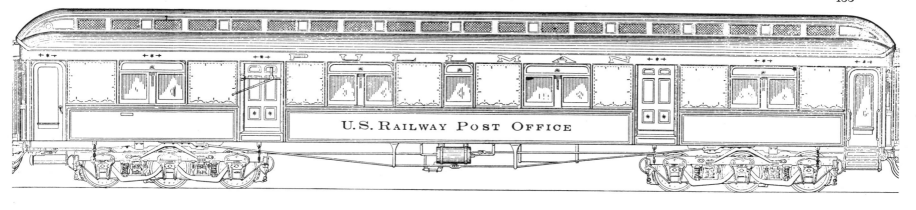

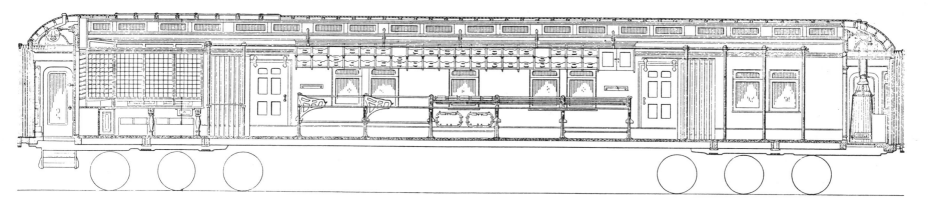

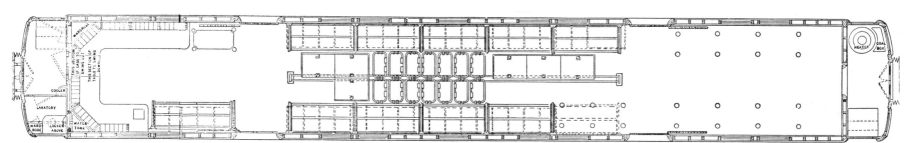

Figure 6.45   Pullman built a mail car as part of its 1893 Columbian Exposition train. The car was fitted with the popular Harrison folding tables and fixtures. (Pullman's Souvenir Booklet, Columbian Exposition, 1893)

master car builders, most of whom were satisfied with a stock baggage car design.[85] Thus mail cars were generally the most lightly built of all passenger cars and, being placed at the head of the train, were the most likely to be crushed in an accident (Figures 6.52 and 6.53). Finally in September 1891, the R.M.S. formulated its own minimum standards for floor framing.[86] The specifications included more and larger sills as well as metal plate reinforcements and iron bolsters. In 1904 the R.M.S. requested even stronger framing, with end platform frames made from steel I beams, and larger-diameter axles.[87] Car builders contended that these changes would increase car weight by 14 tons. Even so, the R.M.S. pointed to a wreck of a Fast Mail on the Southern Railway the year before in which five clerks were killed. The existing framing was clearly not adequate, but the R.M.S. specifications were only recommended practice and no railroad could be forced to adopt them.

With the introduction of the steel passenger car, better wooden framing became a forgotten issue. Agitation began for the universal adoption of steel construction for postal cars. It was

argued that because a Federal agency paid the fare and its employees were carried upon equipment furnished by the railroads, the government was justified in insisting upon the best and safest equipment available. In March 1911 a law requiring steel or steel-framed mail cars was passed by Congress (Figure 6.54).[88] The original bill had called for immediate suspension of payment to any line running wooden mail cars, but the final version was much weakened and allowed the use of wooden cars so long as they were not adjacent to steel cars. After July 1, 1916, the Post Master was to withhold payment to any line running wooden cars in trains having a majority of steel equipment in their consist. Railroads were also encouraged to adopt all-steel R.P.O.s as soon as possible. The statute was ineffectual compared with the strong legislation desired by the mail clerks, but it was a beginning.

In the fall of 1911 the Post Office sought the advice of the best talent in the car-building industry in preparing exact specifications.[89] Barney and Smith, A.C.F., Pressed Steel, and Pullman sent their chief engineers to Washington. Several leading rail-

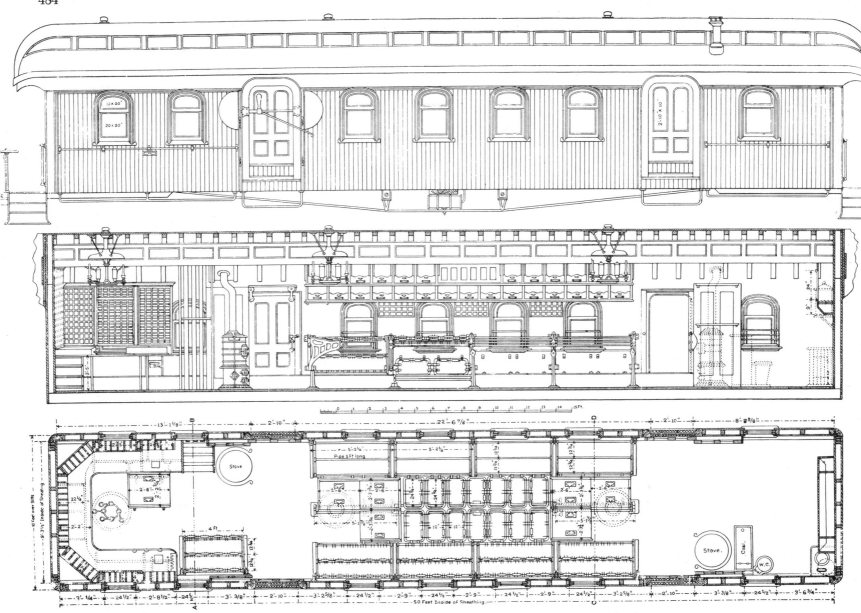

Figure 6.46   A mail car of 1886 equipped with interior furnishings supplied by the Harrison Postal Bag Rack Company, Fond du Lac, Wisconsin. (Railroad Gazette, January 29, 1886)

roads also lent key men. Charles A. Seley, mechanical chief of the Rock Island, was named chairman of the committee. The engineers decided to formulate guidelines that would outline desired results rather than to compose ironclad specifications tied to particular designs. The committee continued to meet for nearly three years. Its most important action was the establishment of the 400,000-pound impact load as the minimum standard for floor and end framing. Until all-steel cars could be acquired, the committee recommended several short-term safety reforms. Only all-steel, steel-underframe, or steel-plate-reinforced wooden mail cars should be in main-line service. Wooden cars with weak ends (only six to eight posts) should be limited to branch-line service on trains of no more than four cars and at speeds no greater than 21 mph. A cheap end post reinforcement was suggested as well: old steel rails, four to an end, could be bolted on vertically to provide a very effective end bumper. The cost was figured at only $100 to $160 per car.

In 1912 the law was modified so that beginning in June 1913, one-quarter of the wooden mail car fleet would be replaced annually by steel.[90] Thus the conversion would be accomplished in

four years. It seemed that the steel postal car law at last had some real teeth and that the changeover was assured. The industry was cooperating, after all; long before either bill had been passed, some railroads had voluntarily acquired all-steel mail cars. The Erie was well in the lead. In 1905 its president, F. D. Underwood, ordered an all-steel, 68-foot R.P.O. from the Pressed Steel Car Company (Figure 6.55). The Number 699 weighed 70 tons and was soon running with two sister cars. During the same year the Santa Fe purchased thirty-nine steel-underframe R.P.O.s from A.C.F. The 60-footers had unusual four-girder fish-belly underframes. The outside sills were visible, making the cars look as though they were mounted on a long steel flatcar. But all-steel construction was obviously superior. Within a year after the Santa Fe composite cars entered service, the Pennsylvania and the Southern Pacific developed long-range plans to retire all wooden equipment. In 1906–1907 two of the Harriman Lines, the U P and the S P, produced sample all-steel R.P.O.s[91] By June 1907 they were ready to order the cars for regular use (Figures 6.56 and 6.57). Within five years they had twenty-eight in service and another sixty-nine on order. The Pennsylvania's first steel

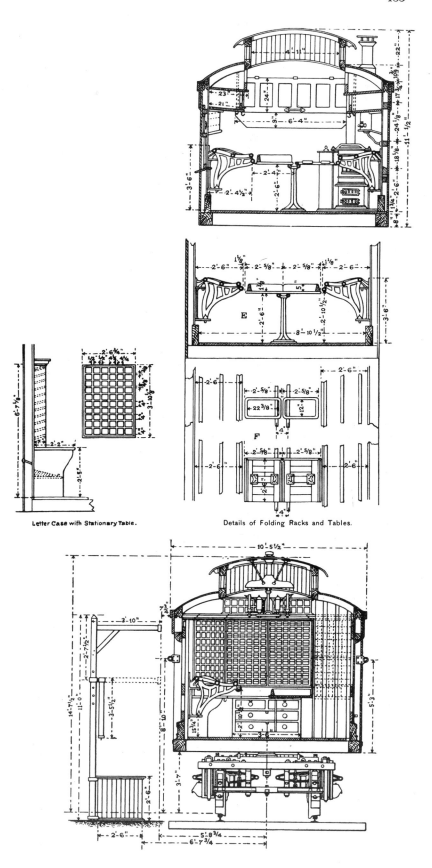

Figure 6.47   *Details of the 1886 Harrison-equipped mail car. Note the trackside mail crane in the lower end-elevation drawing.* (Railroad Gazette, *January 29, 1886*)

Letter Case with Stationary Table.

Details of Folding Racks and Tables.

R.P.O., Number 6546, built at a cost of $15,200, was ready for service in March 1907.[92] Nineteen more were running by 1909.

Other roads, however, showed less enthusiasm. Of the 5,866 mail cars in service by December 1914, 1,587 were all-steel and 677 were steel-frame, wooden-bodied cars.[93] Thus nearly two-thirds of the postal car fleet was still wooden. The railroads looked better when the statistics were limited to full R.P.O.s: of the 1,566 nonapartment mail cars, only 461 were wooden. Even so, no rush was being made to comply with the law. In 1922 the number of full R.P.O.s was down to 1,087, yet 67 steel-reinforced wooden-frame and 4 all-wooden cars were still running.[94] The ratio was much worse for the 4,074 apartment cars then in service: 1,104 were steel, 641 were steel-framed, 1,947 had reinforced wooden frames, and 382 were all-wood. It was not just the impoverished short lines that were reluctant to convert to steel; some wealthy trunk lines were remarkably indifferent to the Congressional mandate. The Santa Fe, for example, purchased no steel R.P.O.s until 1924.[95] The road was convinced that the best mail cars were still the steel-frame, wooden-body equipment purchased in 1905.

In the spring of 1926 the House Post Office Committee opened hearings on the situation.[96] Congressmen expressed considerable frustration over the industry's languid reaction to past legislation and introduced a tough bill calling for an immediate substitution of steel cars. Curiously, the Post Office Department now sided with the railroads in opposing the measure. Both parties denied the need for any legislation. They said that an instant conversion would amount to confiscation of valuable private property, because the railroads would be forced to retire serviceable cars. According to their argument, the cost involved would be unreasonable even if the cars could be produced immediately, which was impossible. Also, mail service would be disrupted in the interim. Post Office officials contended that they had not been negligent in carrying out the wishes of Congress, that the 1911 and 1912 bills applied exclusively to full R.P.O.s and not to apartment cars, or so the Post Office had interpreted their wording. After all, most full R.P.O.s were steel or steel-underframe, and all-steel cars were used on the heavy, fast trains likely to be most dangerous to life and limb. Thus 14,000 of the 18,000 postal clerks were riding in safe cars—that is, steel or steel-framed cars. Most wooden cars were in slow-speed, branch-line service. The new floor plans that had been introduced around 1915 did much to safeguard the clerks. The letter case was now placed toward the center of the car rather than at one end as before. This rearrangement, according to the Post Office, removed the clerks from the fatal end position.

The railroad spokesmen dwelled on the immense cost involved. The price of a new mail car was $24,000, and enough cars to satisfy the pending law would cost over $67,000,000. This investment in the face of Post Office cutbacks (for more mail was being presorted at central mail depots) made the investment in new cars unwarranted. Many were already in storage. And steel cars in themselves did not end the perils of railway travel. The great decline in fatalities was credited to safer operations that resulted in fewer wrecks. In 1906, mail cars were involved in 2,588 accidents; by 1922 the number fell to 757. The vulnerability of steel cars was underscored by the fact that of the 8 clerks killed between 1923 and 1925, 7 died in all-steel equipment.

The counterarguments centered on the railroads' unwillingness to abide by the law and the dangers that thousands of clerks faced daily aboard unsafe cars. Supporters of the tougher bill pointed to the suffering of the families of clerks who had been

Figure 6.48   Interior of a New York Central Line's mail car with Harrison racks. The car was built by Pullman in 1909. (Pullman Neg. 11926)

Figure 6.49   In 1873 an artist depicted the frenzied actions of clerks sorting mail en route. (Scribner's Magazine, June 1873)

Figure 6.50   This posed photograph made aboard a B & O mail car in 1938 fails to capture the rapid work of the clerks. (Baltimore and Ohio Railroad)

killed or maimed in the service. They also emphasized the fact that for each death, the government paid claims averaging $20,000—enough to buy a new steel car. However, they could not muster the votes to pass the bill.

Nevertheless, the hearings had some effect. The railroads added a large number of steel mail cars, including over 250 apartment cars, during the next two years.[97] In 1927 the Post Master General prohibited the operation of wooden or steel-frame cars in trains with a majority of all-steel cars. Steel-frame cars could no longer be operated between cars of all-steel con-

struction. Railway Mail Service specifications now contained sixty-one headings and filled thirteen pages of fine print.[98] The Post Master's authority was recognized as covering both full and apartment cars. In new Congressional hearings in 1928 the railroads repeated the arguments they had made two years earlier, but they could not claim that they had shown any better faith in carrying out the wishes of Congress. Again nothing was done; Congress decided that the old laws were strict enough. And then came the Depression, when Capitol Hill was swamped by problems greater than the existence of a few antique mail cars.

Figure 6.52   Mail cars, often built without end platforms to discourage thieves, tended to look unusually short. The Number 19 was produced by Jackson and Sharp in 1898. (Hall of Records)

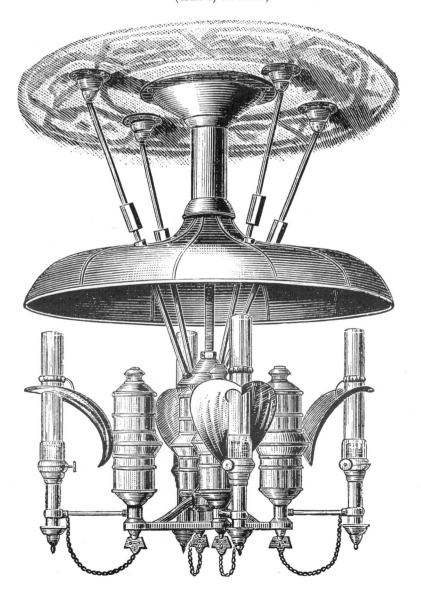

Wooden cars were left to disappear by attrition at an exceedingly slow pace. As late as 1950, thirty-seven steel-frame and eight wooden R.P.O.s were reported in service, although none were on main lines.[99]

The interior arrangement of mail cars changed very little during the standard heavyweight period. Size was fixed by the Post Office decision in 1916 to pay for service by the floor space of the compartment. Units of 15, 30, and 60 feet became standard. Full-sized cars were set at 60 feet and were occasionally placed within an 80-foot car, the remaining space being used for baggage. Technically such a unit could be classified as an apartment car, yet the Post Office would insist that it was a full R.P.O. Mail cars followed the improvements introduced in the lightweight era (Figures 6.60 and 6.61). An 85-foot apartment car of the postwar period registered only 62 tons while a 74-footer of the standard era weighed 72 tons.[100] In 1940 mail clerks were first given the benefits of air conditioning.[101] A few years later the Pennsylvania produced a so-called dream car, with such luxuries as an enclosed lavatory, an electric coffee maker, and a refrigerator (Figure 6.62).

In a way, the superficial improvements just mentioned foretold the coming disappearance of the Railway Mail car. It had reached its apogee generations before, and had become an increasingly expensive and obsolete method for mail distribution in the postwar period. The basic problem was its operating cost: it was labor-intensive. A single car would have a crew of six to fourteen men. The upward spiral of wages, coupled with the forty-hour work week, soon made the mail car an economic impossibility. Machine sorting of letters eliminated much expensive personnel. This could best be done at a central post office, where

Figure 6.51   Before the time of electric lighting, mail cars were lighted with massive oil or gas lamps. The engraving here dates from 1883. (Engineering, August 31, 1883)

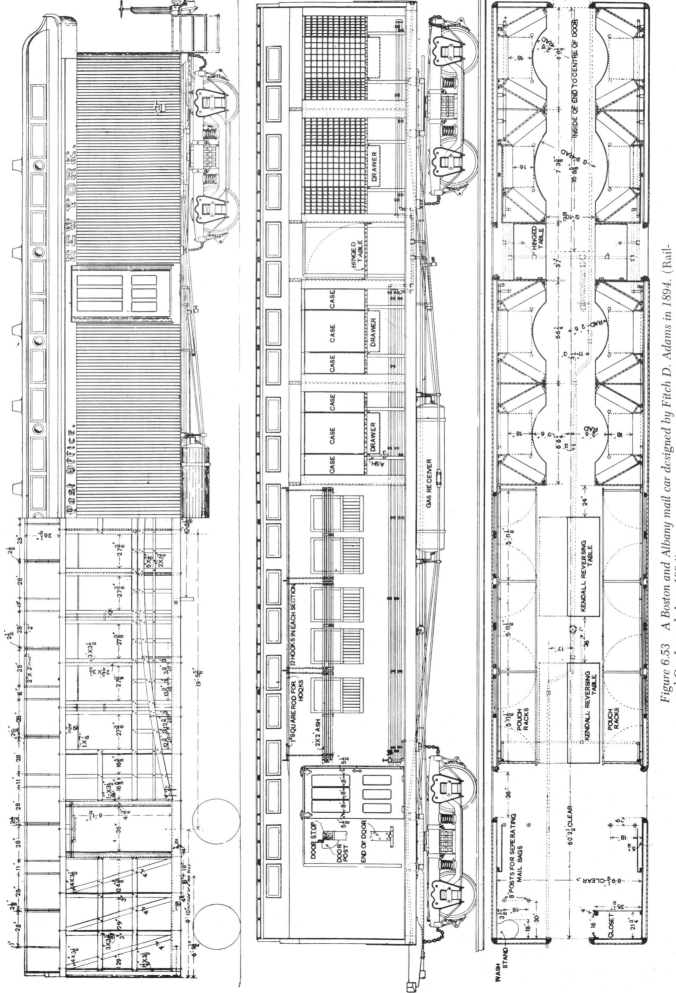

Figure 6.53 A Boston and Albany mail car designed by Fitch D. Adams in 1894. (Railroad Car Journal, June 1894)

Figure 6.54   *The Railway Mail Service pressed the railroads to provide safe cars to protect its clerks. Mail cars were among the first to be built with steel frames. The Number 20 was constructed in 1912. (Hall of Records)*

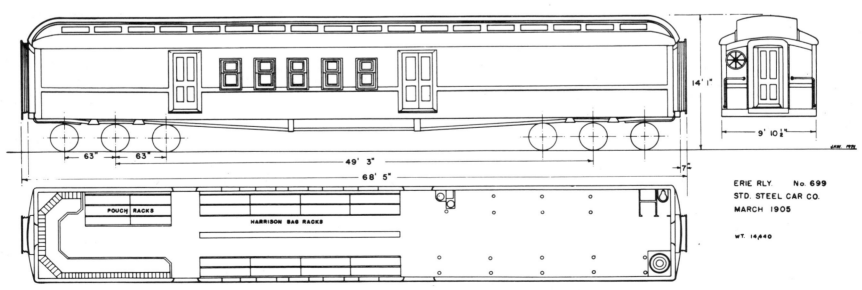

Figure 6.55   *The Erie Railway's mail car Number 699 was among the very first all-steel cars in America. (Traced by John H. White, Jr.)*

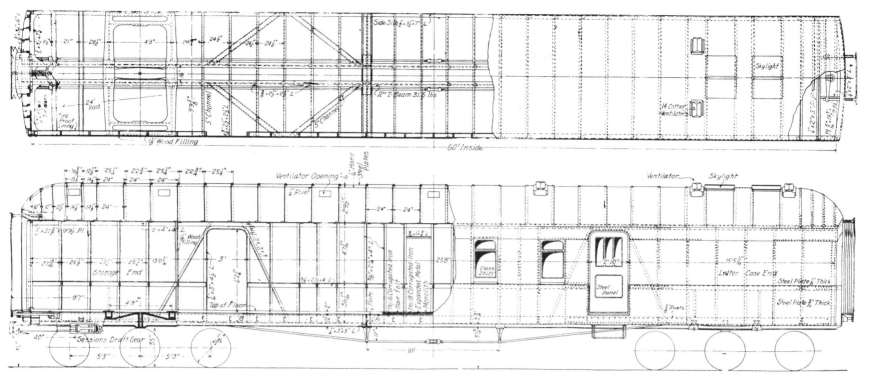

Figure 6.56   *The Harriman Lines' all-steel R.P.O. car, designed in 1906. The first two sample cars, illustrated by this drawing, were produced in the following year.* (Railway Review, *June 15, 1907*)

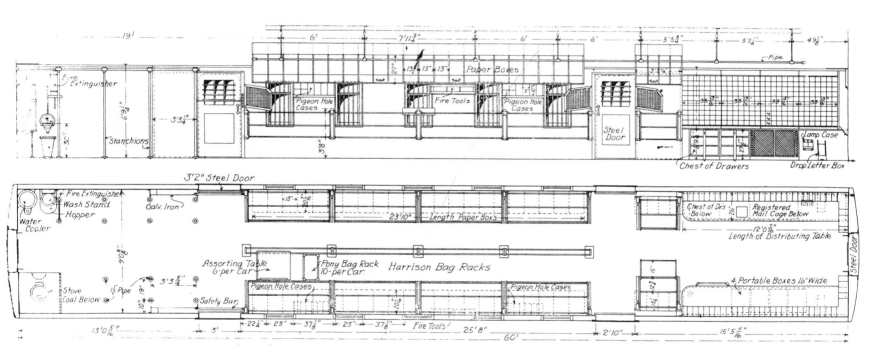

Figure 6.57   *Interior arrangement of the Harriman Lines steel R.P.O.* (Railway Review, *June 15, 1907*)

sorted mail was pouched and sent as a bulk commodity. The system of mail distribution had reverted to methods used before Armstrong introduced the R.P.O.

Actually, the decline of the R.P.O. began well before 1950. Twenty years earlier the Post Office had begun to turn to motor trucks, in which clerks sorted the mail just as they did in railroad cars.[102] Between 1929 and 1932 the railroads cut out 650 trains, forcing the Post Office to seek another medium of transit, which in turn forced the road to cut more trains; the trend was unending.

By 1951 long-distance truck transport had begun. A few years later first-class mail began to move by air on a space-available basis. By 1955 the volume of mail that was carried by rail was one-half of the peak wartime volume.[103] The railroads were hauling bulk mail, which produced the smallest margin of profit. Passenger train abandonments accelerated the decline of the R.M.S. In 1951, 700 routes were in service; within ten years only 262 remained.[104] As the Post Office shifted more of its traffic, the railroads were compelled to take off more trains. The Pennsylvania claimed that mail revenues amounted to 11½ percent of all

Figure 6.58   Because mail service was such an important source of revenue, a few lines named cars for Post Office officials. The W. Irving Glover *dates from about 1925.*

Figure 6.59   Pullman built this combine mail-baggage car in 1927. (Pullman Neg. 31681)

passenger train–related income, forming a vital subsidy to the continued operation of many trains.[105] In 1960 the Northern Pacific pulled off two trains when it was deprived of a $145,000 mail contract. The curtailment of such service had a devastating effect on overall railroad revenue figures. In 1961, the railroads' gross mail income was 341 million dollars; by 1973 this total had dwindled to 83 million dollars.[106] Following the introduction of the ZIP code in July 1963 came the mass abandonment of the surviving R.P.O.s. The last R.P.O. route—the New York–Washington run—was abandoned on July 1, 1977.

A few railroads that were slow to recognize how soon the R.P.O. would disappear unwisely bought new cars. In 1963 the

Union Pacific purchased nine new cars from Budd, and a year later the Santa Fe bought a dozen similar units.[107] Within three years the Santa Fe had thirty lightweight mail cars that were useless for any other service. Few were more than twenty years old—modern for an American passenger car. Yet because of the small side doors and window configuration, it was not practical to rebuild them for any other service. All were either scrapped or sold to Mexico.

Some other lines, such as the Missouri Pacific, avoided the mistake of investing in new R.P.O.s and instead bought some extremely long storage cars (Figure 6.63). Fifty cars were produced for the M P by the St. Louis Car Company in 1964. Here

Figure 6.60 *The interiors of lightweight mail cars did not differ materially from the standard steel cars of twenty years before. This Great Northern car was built by A.C.F. in 1950. (Great Northern Railway)*

Figure 6.61 *The lightweight R.P.O. Number 4907 was part of the New York Central's reequipment program after World War II. (New York Central)*

Figure 6.62   Before the lightweight era, mail clerks enjoyed few luxuries. Enclosed toilets and electric lighting came after the Number 21 entered service in 1907. (Pullman Neg. 9489)

is a good example of a car design influenced by the method of payment: the ceilings were kept low because the price was based on linear footage rather than weight or volume. An ordinary high-ceiling, 60-foot baggage car would probably carry the same load as these specially contrived 80-footers.

Railway mail service today is a high-speed freight operation with few ties to the passenger side of the business. In the early 1920s the New York Central experimented with containerized mail cars. The idea was revived in 1958 and has been expanded since that time. Some mail trains run with as many as forty cars, usually in a mixed consist of piggybacks and standard baggage car–like mail storage units. Attracted by these economical methods, some first-class mail has returned to the rails in recent years.

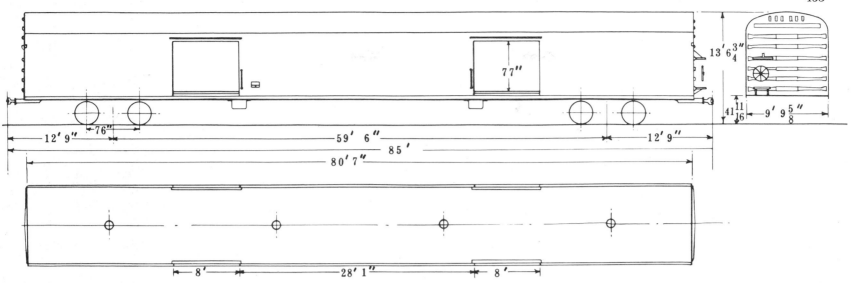

Figure 6.63 Because the Post Office paid by the linear foot rather than by cubic area, some railroads bought long, low-ceiling cars to improve revenues. The St. Louis Car Company built this bulk-mail car for the Missouri Pacific in 1964. (William D. Edson Collection)

CHAPTER SEVEN

Running Gears

PREVIOUS CHAPTERS have examined the obvious, eye-level portions of the passenger car that are most visible to the traveler; this chapter describes everything below the floor. In the shadows of the undercarriage lie many mechanisms which, although they arouse little interest in most passengers, are vital to their comfort and safety.

Trucks

Next to the body, the largest and most important subassembly of a railroad car is the trucks. Each truck is in effect a small independent car, consisting of a frame, wheels, axles, journal bearings, springs, and brake rigging. Not only do the trucks support the body, but because they can move laterally and turn on a center pin, they also materially reduce the car's resistance when entering a curve. The truck is expected to cushion the ride, and since no track is in perfect level or alignment, lateral and transverse irregularities must be absorbed by the springs and swing bolsters.

Spiral springs capable of a large deflection and hence a low frequency of vibration were favored throughout the history of the railroad passenger car. To dampen the bounciness of the "soft" coil springs, stiffer leaf or elliptical springs were arranged, usually through equalizing levers, to act in concert with the spirals. In modern times hydraulic, automotive-style shock absorbers were used as well. Side bearings were employed in yet another effort to control the car's roll, especially when it entered a steeply banked curve. Rubbing plates or roller-type bearings served this function. Side bearings could also be expected to avert derailments. If a wheel were to break, the side bearing would support the car and allow it to run on three wheels until the train could be stopped. In the days of cast-iron wheels this was an important duty of the side bearings.

With so many functions to perform, truck design presented the car builder with a major challenge. Advances in car weight and train speed compounded the difficulties. Faster schedules, particularly those requiring speeds above 85 miles per hour, dramatically increased noise and oscillation. Yet the car builder was expected to produce a safe, smooth-riding, lightweight truck for a reasonable cost. Ingenious plans were devised to satisfy this need, some of which will be outlined in the next pages. But no truck, no matter how well designed, could perform satisfactorily if not properly maintained. Hard-riding cars were usually the result of poorly maintained trucks. The most common ailments are listed below.[1]

Improper side bearing clearance—puts dead weight on bolster springs, which transmit vibrations to the body most noticeably when transversing a curve.

Weak elliptical springs—normally attributable to age, but also to improper tempering by the manufacturer.

Loose center plate—causes a rotating sensation.

Coil spring clearance—2½ inches are required. If clearance is not allowed, the equalizer will strike the truck frame a sharp blow and cause the car to vibrate for miles after the point of contact.

Worn equalizer ends—cause journal boxes to tip, creating a rocking motion.

Flat or out-of-round wheels.

Throughout their history American passenger car trucks have remained true to form: the four-wheel truck has always been the standard. Less consistency is found in the details of design, and less agreement on precisely what features produce a smooth-riding truck. This is normal, of course, in any active area of technology. Truck size has grown since the early days of the railroad. In a single century the wheelbase length increased from a scant 3 feet to 9 feet, the truck weight from less than 1 ton to nearly 10 tons.

The earliest trucks were indeed primitive affairs. Wooden frames built up from 4- by 6-inch timbers were tied together by a few light iron rods. Slender axles and wheels that would be considered barely sufficient for a modern handcar were expected to safely carry a carload of passengers. The wheels were set as close together as possible to minimize resistance in traversing curves. Leaf springs over each journal box provided the only suspension; no equalizers were used. Drawings of these pioneers are given in Figures 1.18, 1.19, 1.74, and other illustrations in Chapter 1.

In the 1840s the basic wooden-beam truck underwent gradual refinement. The so-called square truck came into favor during this decade.[2] It was thought that if the wheelbase equaled the track gauge, the truck could not fail to tram precisely because the geometry was so beautifully balanced. This myth prevailed for many years, and even after the steady-riding qualities of the spread- or long-wheelbase truck were established by locomotive designers, some conservative car builders held fast to the chattering square truck.

Two fundamental advances in truck design were introduced during the 1840s. The first and most important was the swing bolster. In an ordinary truck the bolster, the central transverse beam, is rigidly bolted to the "wheel pieces," the truck's side-frame members. The truck has no lateral flexibility, and each sideways shock the wheels receive from a poorly aligned truck or from their passage through switches or crossovers is transmitted directly to the body. The springs are effective only in absorbing vertical shocks. Charles Davenport of the Cambridge, Massachusetts, car-building firm Davenport and Bridges devised a simple but effective scheme to overcome this problem (Figure 7.1). The bolster was not fastened directly to the wheel pieces but tied to the truck frame with swing links connected to a secondary bolster, called the spring plank. The spring plank in turn was the lower support for springs fastened to the wheel pieces. The bolster was thus free to swing as much as an inch. This free motion, together with the semi-independent spring mounting, absorbed the lateral shocks encountered.

Davenport patented the swing bolster on May 4, 1841 (No. 2071). Some years later during the eight-wheel car case (covered in Chapter 1), Henry Waterman claimed that he had used swing-motion trucks as early as 1839.[3] Twenty years later a patent

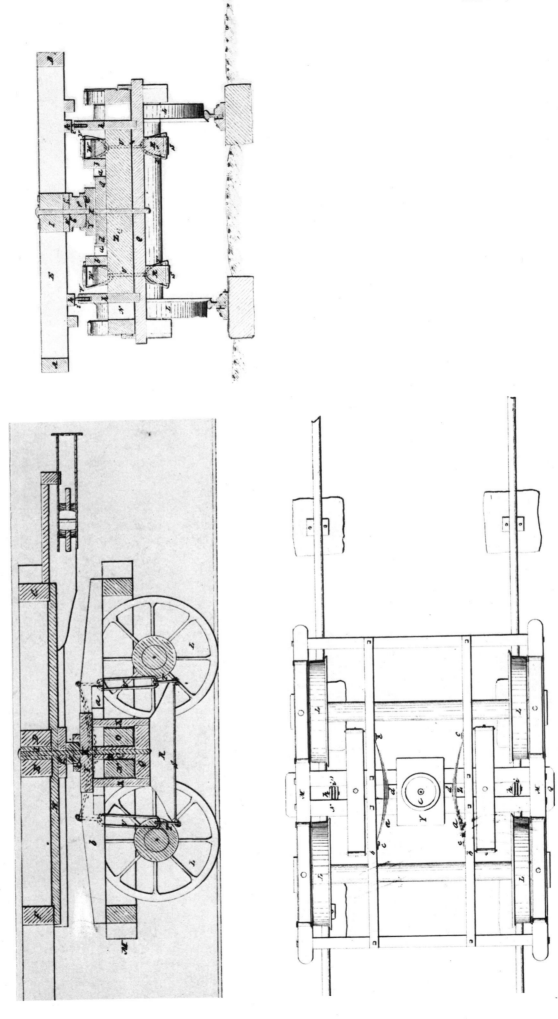

Figure 7.1   Davenport and Bridges patented the swing bolster truck in 1841. It provided a fundamental improvement in the riding quality of railroad cars by dissipating lateral motion. (U.S. Patent Office)

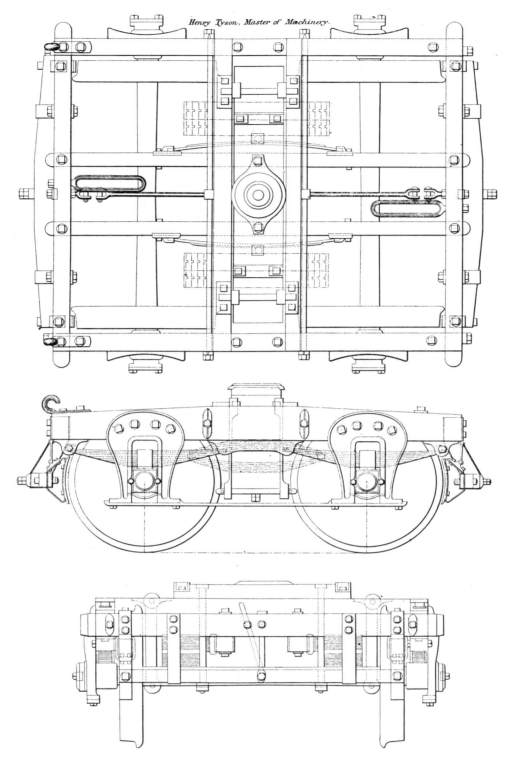

Figure 7.2 Wooden-beam, four-wheel passenger car trucks changed little during the nineteenth century except for an elongation in wheelbase. This truck was designed in 1856. (Douglas Galton, Report on U.S. Railways, London, *1857–1858)*

examiner overlooked both Waterman and Davenport's inventions and awarded a patent for the same idea to Henry Kipple and Jacob Bullock of Philadelphia (No. 26,502, December 20, 1859). The landmark invention of the swing bolster has been mistakenly credited to Kipple and Bullock in at least one modern account.[4]

While car builders came to accept the swing bolster as a necessary feature of passenger car trucks, it was less widely used for freight cars because of the additional cost. A few car builders even complained that it was not really effective for passenger service unless long links were employed.[5] The trend to short links actually worsened the ride, imparting a series of uneasy

jerks when the wheels struck a curve. Rapid flange and journal wear was a secondary result, because these heavy side thrusts pushed the truck out of square.

The second reform in truck design to appear in the 1840s was equalization. A lever or equalizer connected the axles, so that a bump or shock received by one wheel set was transferred to and partially absorbed by the other. The scheme had been used by American locomotive designers since 1838, but apparently it was not applied to cars much before the mid-forties. Early in 1845 a British engineering journal published a drawing of an eight-wheel American car with an elementary form of equalized trucks.

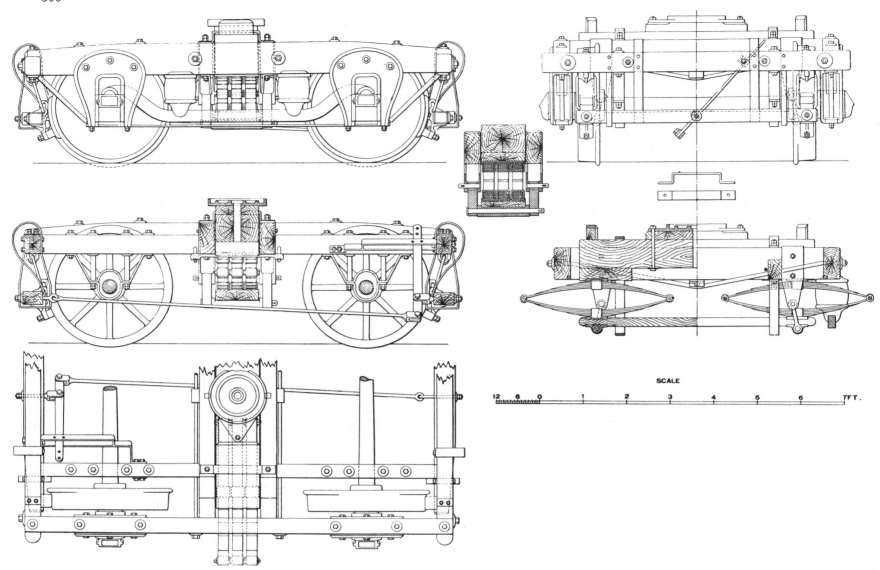

Figure 7.3   A Pennsylvania passenger car truck of 1866 incorporated such improvements as equalizing levers and swing bolsters. The rubber springs shown were obsolete by that date, however.

The drawing together with the citation is reproduced in Figure 1.78. Two leaf springs over each axle were linked to a common central spring mounted below the wheel piece. Surviving drawings indicate that the equalizer was rarely applied before the middle or even late 1860s; just why remains a mystery, since it was universally installed on American locomotives generations earlier. Perhaps car builders felt that it was not necessary for very light cars traveling at low speeds. After the Civil War cars rapidly became heavier and speeds faster, which may account for the late adoption of equalizers. Once accepted they became universal, and they remain a basic part of modern truck design.

However, the merits of the equalizers have been challenged by critics who point to its ineffectiveness at very high speeds. They claim that it is actually counterproductive because the lever occasionally acts against, rather than in concert with, the springs and causes the truck to jump.[6] In the course of braking, the equalizer makes the truck tilt. Moreover, equalizers are heavy, adding to the car's dead weight; they break frequently, and they block access to the brake rigging. Yet most car designers seem to feel that the equalizer's merits outweigh its shortcomings and that it is necessary for a smooth ride.

Few new ideas gained acceptance in basic truck design after the introduction of the equalizer and the swing bolster. From this

time forward, the truck's development was a process of gradual growth in size within the rigid confines of a single pattern. By the late 1860s an archetypical design had evolved that was to remain in favor until the present day. Even the conversion from wood to steel did little to affect the basic plan. This universal truck is well represented by Figures 7.2 to 7.4. Its general features included open-jaw pedestals fastened to a rectangular frame and cross timbers with a floating swing bolster between them. A shallow U-shaped equalizer rests on each journal box, with a pair of coil springs mounted between it and the truck's main frame. The bolster is cushioned with elliptic springs independent of the equalizer. In appearance, wooden-frame four-wheel trucks twenty years apart in age look like duplicates. It is only when the actual dimensions are compared that the increasing size becomes obvious.

Six-wheel trucks are the second most common style found on American railways (Figures 7.5 and 7.6). They appeared as early as 1845 and, as mentioned in Chapter 1, a number were adopted by several railroads, including the Michigan Central. They were not widely used by the industry as a whole until after the introduction of heavyweight palace cars in the late 1860s. As sleeping, parlor, and dining cars became more numerous, six-wheel trucks grew in popularity. The extra axle at either end of the car was

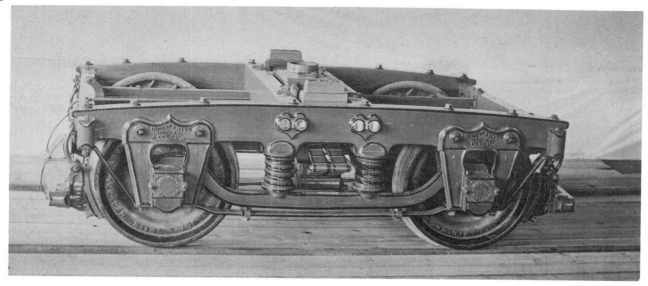

Figure 7.4 *Barney and Smith, car manufacturers of Dayton, Ohio, produced this wooden-frame truck around 1880.*

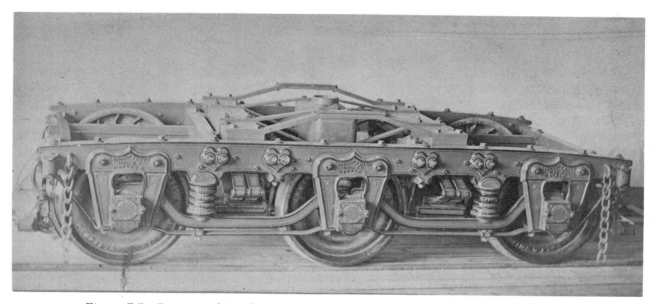

Figure 7.5 *Barney and Smith constructed this typical six-wheel truck around 1880. Note the decorative striping.*

Table 7.1   Growth in Truck Size, 1867–1887

	1867	1887
Wheelbase	6 feet	7 feet
Wheel piece (frame top rail)	9 × 4½ inches	9 × 4½ inches
Equalizer springs	Rubber cylinders	Steel coil
Bolster springs	Leaf steel	Leaf steel
Brake shoes	4	8
Wheels	33-inch cast iron	36-inch cast iron
Axle (diameter at center)	3 $^{15}\!/_{16}$ inches	4 $^{1}\!/_{8}$ inches

Source: Railroad Gazette, December 9, 1887, p. 792.

justified chiefly because of the car's great weight. Concentrating the load on eight bearings was considered undesirable for several reasons. It promoted hotboxes and axle and wheel failures.[7] In a time of cast-iron wheels, small journals, and faggoted wrought-iron axles, there was good reason for concern. The track was generally not up to supporting such a load safely; light ballast, slender rails, and widely spaced ties provided a shaky foundation. It was necessary to spread the load out as far as possible and divide it over not eight but twelve bearings. The longer wheelbase had the added advantage of imparting less motion to the car and thus in theory providing a smoother ride. The center axle had a generous amount of end play to help it ease around curves. Because brake shoes could be mounted to twelve rather than eight wheels, six-wheel-truck cars were considered safer. It was also claimed that brake shoe life was extended because the wear was more widely spread. In the wooden age, any car weighing over 25 tons was given six-wheel trucks.[8] After steel cars came into favor, cars weighing 42.5 tons and over were normally fitted with twelve wheels.[9]

At an early date critics gathered in opposition to the twelve-wheel truck. Their arguments were convincing enough: they cited the greater first cost and the increased weight, both of which easily exceeded the cost and weight of the conventional

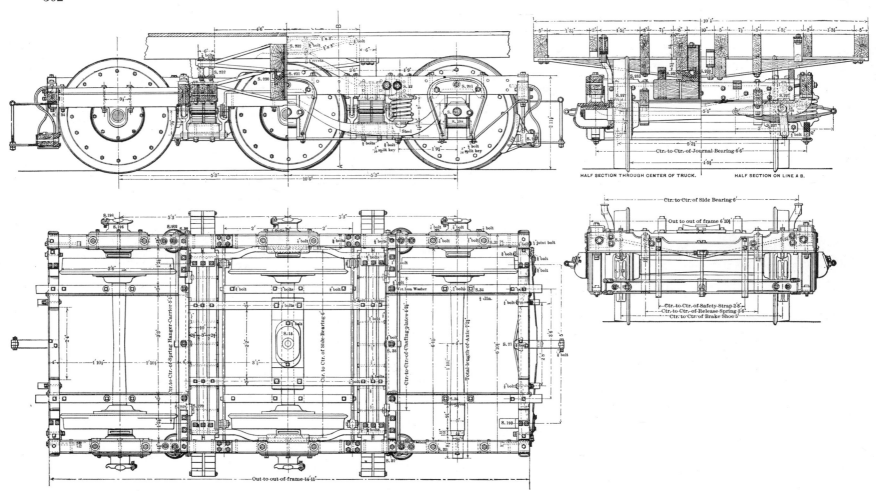

Figure 7.6   An iron-plated, wooden-frame, six-wheel truck with paper wheels made by the Chicago, Burlington, and Quincy in 1886. (Railroad Gazette, April 2, 1886)

four-wheel variety by one-third. In 1890 a four-wheel truck weighed 5 tons, a six-wheeler 7 to 8 tons. Six-wheel framing was much more complicated, because the center or kingpin plate was directly over the middle axle. It was necessary to bridge over this wheel set and hang the truck on two auxiliary transoms or bolster beams. The bridge bolster was costly, complicated, and weak. The extra wheel bearings amounted to that much more drag on the train. The 10- to 11-foot wheelbase impeded the passage of curves, increasing friction and damaging the track. The supposed steadier ride resulting from the long wheelbase was a myth, according to tests made on the Milwaukee Road.[10] In this test the center wheel set was removed. The truck rode no better than a 9-foot-wheelbase truck and was considerably noisier. The critics claimed that long-wheelbase, four-wheel trucks with large journals could match any six-wheeler. But the majority of railroad men seemed to regard the six-wheel truck as the best form for passenger cars. Such an old-line executive as Daniel Willard of the B & O was seen walking along station platforms with eyes intent on the undercarriage of the cars. He would pick out a twelve-wheeler and climb aboard. A car man explained: "He won't ride in an eight wheeled job, no matter how many fancy stripes they paint on the sides."[11]

During the heavyweight era, twelve-wheel cars reached their peak of popularity. Six-wheel trucks were largely rejected for lightweight cars. They were rarely used after 1935 and then only for very special cars, such as the Milwaukee Road's superdomes (1952). An exception to this general rule was the New Haven, which continued to order twelve-wheelers and bought some twelve-wheel dining cars as late as 1949.

There was a short-lived interest in eight-wheel trucks, and a number of centipede-like sixteen-wheel cars were constructed in this country. Mechanics are occasionally inspired to indulge in excesses—reasoning, for example, that if twelve wheels are better than eight, then sixteen must be better yet. But in fact the belief that more wheels would ensure a smoother ride was ill-founded. They ground through curves and occasionally upset, and with so many axles, springs, and levers, they seemed to exaggerate rather than dampen the defects of a poor track. Lincoln's sixteen-wheel private car was abandoned by its second owner because of its galloping ride. The journalist Zerah Colburn said that traveling in a sixteen-wheeler "is more like sailing than rolling."[12]

It has been reported that such cars first appeared in 1839. In 1858 Wason built some sixteen-wheel sleeping cars for the Michigan Southern. Two years later the same firm produced an elaborate private car similarly mounted for the Viceroy of Egypt.[13] The 20-ton, 67-foot-long car had an open center section covered by a fanciful canopy and surrounded by scrolled iron railings. The special iron-frame trucks are illustrated in Figure 7.7. The car, minus most of its trimmings, is still occasionally used by the Egyptian government.

Pullman approved of the sixteen-wheel car and had twenty-three in his fleet.[14] His acceptance was doubtless inspired by his mechanical chief, C. F. Allen, who received a patent (No. 65,788) for a sixteen-wheel design on June 18, 1867 (Figures 7.8 and 7.9). Allen had built such cars for Pullman since 1863 and apparently had also been successful in converting his former employer, the Chicago, Burlington, and Quincy. Long after 1869, when Pullman had abandoned them, the Burlington continued to

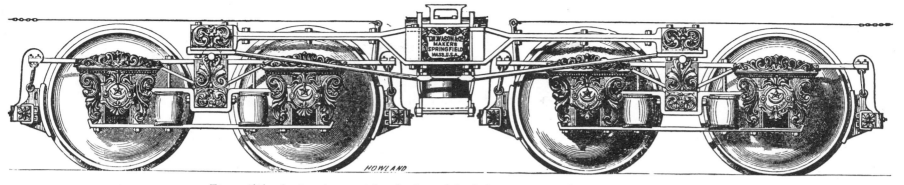

Figure 7.7   An iron-frame eight-wheel truck built by Wason for the Viceroy of Egypt's private car, 1859–1860. (American Railway Review, *January 19, 1860*)

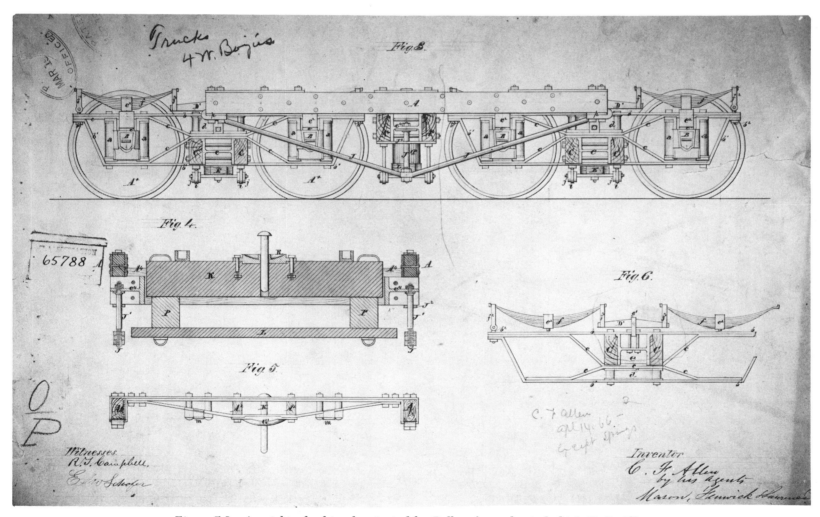

Figure 7.8   An eight-wheel truck patented by Pullman's mechanical chief, C. F. Allen, in 1867. (National Archives)

produce the sixteen-wheel centipedes. In 1876, for example, the road's Aurora shops were reported to have completed a new sixteen-wheel diner.[15] Travel brochures spoke of the splendid sixteen-wheel giants available to patrons traveling west. And then unexpectedly, the railroad reversed its policy in 1879 and condemned sixteen-wheel cars as a foolish mistake.[16] It rebuilt all its sixteen-wheelers with conventional trucks, ending this curious misadventure in passenger car suspension.

Late in the last century, cars grew to such proportions that wooden-beam trucks were no longer practical. To build a truck with a wooden frame of the requisite strength would call for such massive timbers that the structure would be absurd. The same compromise that was found so effective in body construction was

also employed for the undercarriage parts. Iron plates and rods stiffened the wheel pieces and transom members. Flitch plates covering the backs and fronts of the major truck-frame timbers became common by the late 1880s. The outer plate was normally ½-inch thick, while the inner plate measured ⅜ inch. Button-headed, ½-inch bolts secured the plates. By the 1890s composite construction was standard, except for the very lightest passenger cars.

Composite construction naturally suggests the possibility of all-metal trucks, but even before the coming of the commercial steel car, many attempts were made to introduce trucks of this type. Probably the earliest builder to try it was Ross Winans. Around 1835 he produced an all-metal truck with a large leaf spring that

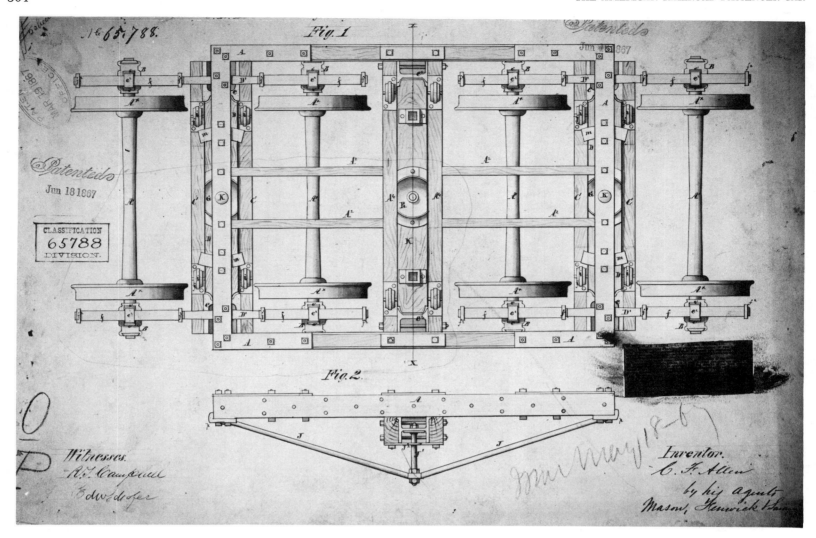

Figure 7.9   A plan view of Allen's eight-wheel truck patented in 1867. (National Archives)

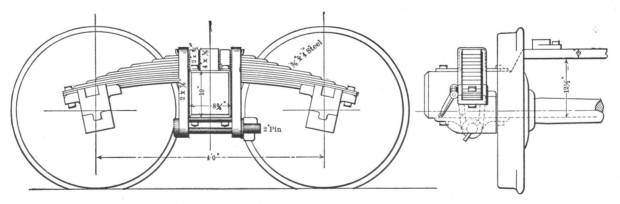

Figure 7.10   Winans's live spring truck, similar to the type that was used on B & O passenger cars in the 1830s and 1840s. The example here is a much later model that was in service on the Georgia Railroad's freight cars. (National Car Builder, January 1891)

served as both the side frame and the suspension. A stirrup bracket at the spring's center held the leaves together and formed a connection with the forged wrought-iron bolster. The cylindrical ends of the bolster were free to swivel in the bracket, giving the truck great flexibility. A Winans spring or live truck is illustrated in Figure 7.10. The B & O continued to use the live truck for its passenger cars until at least 1840. The truck's serious failing was its tendency to run out of square because there was no cross framing. This caused excessive flange wear and sus-

ceptibility to derailments. The constant flexing of the spring-side member led to metal fatigue, resulting in broken leaves and ultimately in total failure. It was practical only for very short wheelbases, a characteristic originally thought advantageous, but one that was soon recognized as a liability. The fact that it was impossible to mount brake shoes on trucks of this design was a defect so serious that it is a wonder live trucks were ever built at all. Yet they were simple, cheap, and well suited to light, slow-moving cars. In 1860 several Southern lines were still using

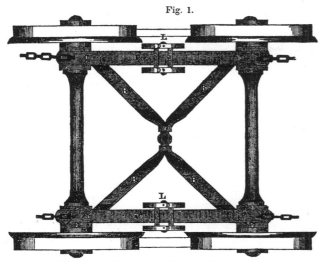

Fig. 1.

Fig. 2.

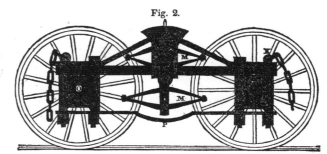

DAVENPORT & BRIDGES' IMPROVED PATENT IRON TRUCK FOR RAILROAD CARS, is presented above, and the attention of Railroad Companies is respectfully invited to the following description of their justly-celebrated invention:

Fig. 1 of the drawing above represents a top view or plan of our Improved Railroad Truck. Fig. 2 is a central, longitudinal, and vertical section. C, Fig. 1 and 2, represents the arched bars of the side trusses: they consist of two long bars of plate iron (about three inches wide by seven-eighths of an inch thick,) bent into the shape as seen in drawing 2. Each of them is placed directly over a flat and straight tie bar, A, which extends from one end to the other, as seen in Fig. 2. These parts, so arranged, receive between their ends the ends of diagonal cross bars or braces, B, which are united at their centres y being clasped and welded, as seen in Fig. 1. The bars so composing what may be considered as side trusses and diagonal cross braces, rest at their ends upon four pedestals, F, F, which receive the bearings or boxes for the axles to run on. Another flat tie bar, P, extends from the under side of one of the pedestals to that of the other, on the same side of the frame, and the whole is secured together by eight bolts, J, J, passing down through the ends of the several ars, A, B, C, and the pedestals, and on each side of the journals of the axles, O, O, in the positions represented in the drawings. From the above it will be seen that there are two bolts to each pedestal, and that this number is all that is requisite for the full security of the bars and pedestals together. The body rests and moves upon two sectional supports, D D, arranged on the sides of the truss frames, as seen in Fig. 2; they extend somewhat, or a sufficient distance above the truss frames, and are jointed at their lower ends by means of a bolt, L, which rests upon the top of the lower spring, M, which spring rests upon a bolt passing through the lower part of the inverted strap, E, which strap passes over and rests upon the top part of the upper spring, M, which is placed within the truss frame, and rests upon the top of the bar, A.

Two bands, N, N, are passed entirely around the central part of each truss frame, the object of the same being to transfer the strain, or a portion thereof, of the spring, from the tie bar, A, to the arched bar, C.

These Trucks are adapted as well for eight-wheeled passenger cars as for baggage and freight cars, giving to each a more agreeable and easy motion than any other Truck heretofore constructed or in use. They are simple in their construction, combining strength and great durability, although weighing at least twelve hundred pounds less than the common Trucks. Besides these excellences, by reason of the elasticity of the braces, B, B, B, as seen in the drawing, and the other peculiarities of construction, the weight is *equalized* upon all the wheels, and yet any one may be raised so as to pass any inequality on the rails without lifting either of the other wheels from the track, thus rendering it almost impossible to run a car off. Being bound, and having as it were but four joinings, they are protected from injury by lateral strains, and in case of damage are easily repaired.

These excellences have been fully tested by use, for a long time, on the Eastern, the Fitchburg and Long Island railroads; and for proof of the above stated superiority of these Trucks over all others, we refer to the experience of those who have used and run them.
CAMBRIDGEPORT, April 1, 1845.

DAVENPORT & BRIDGES.

Figure 7.11 An inside-bearing, all-metal truck patented by Davenport and Bridges in 1844. (American Railroad Journal, July 24, 1845)

Winans's trucks on their freight cars.[17] The side-spring weight was given as 250 pounds each. More than thirty years later, the Georgia Railroad still had about 200 cars running on the old-fashioned live truck.[18]

In the 1840s other builders were promoting the cause of the iron truck. Charles Davenport patented such a carriage fabricated from iron straps on August 10, 1844 (No. 3697). Four double elliptic springs cushioned the ride (Figure 7.11). The inventor claimed savings of $100 in cost and 1,200 pounds in weight over a comparable wooden truck.[19] The device was promoted by Davenport's huge car plant—at the time probably the largest in the country—and received tests upon the Eastern, Fitchburg, and Long Island railroads. A similar design was patented by Fowler M. Ray on March 21, 1845 (No. 3962). Ray claimed that his limber truck could pass over obstacles such as logs or large stones without derailing.[20] He said that each truck weighed a mere 1,000 pounds, yet produced "delightful riding cars." Advertisements running in the *American Railroad Journal* throughout 1847 carried testimonials from several prominent lines—including the Philadelphia and Reading, which stated that it was making extensive use of Ray's miraculous invention. At the time that Davenport and Ray were promoting iron trucks, Eaton and Gilbert placed a competing design on the market. It was more conventional in appearance and, because of its relatively long wheelbase, must have offered a superior ride. How extensively it was used is not known, but a fine drawing of it is available in the plates of the American-style car built for Germany by Eaton and Gilbert (see Figure 1.79).

The initial flurry of interest in iron trucks produced few lasting results. Wood remained in favor, although mechanics did continue to experiment. In addition to the Wason and Allen eight-

wheel iron-frame trucks of the 1860s., Allen patented four- and six-wheel trucks with strap iron side frames. He continued to use wooden transoms, however. Many years later, in 1884, the Philadelphia and Reading produced an all-iron six-wheel truck with a 10-foot wheelbase.[21] The trucks, with their 6-inch channel wheel pieces, were used under one or more parlor cars. A few years later the Lehigh Valley Railroad adapted a six-wheel tender truck design of 1884 for passenger car service. The frame was forged very much like a locomotive bar frame. The top rail measured 1¾ by 4 inches. Unlike a locomotive's, however, the frame was placed outside of the wheels. The success or failure of these two designs is unrecorded, but apparently they were not widely copied. Even so, all-metal trucks were coming into broader use at this very time, at least in the area of freight cars. Pressed-steel trucks were being introduced with some success in the late eighties. In 1889 Sampson Fox, a leading British manufacturer of such trucks, made a set of 7-foot-wheelbase bogies for passenger car service.[22] Praised by the railroad press as simple and handsome, they were surprisingly modern in appearance. The Pennsylvania and perhaps other American lines ran tests with Fox's experimental truck.

A far more durable and solid truck frame could be made from cast steel. A one-piece casting assured rigidity, guaranteeing that the frame would be square and that the axles and wheels would always run true. As early as 1896 Chicago's Lake Street Elevated was purchasing cast, malleable iron truck side frames.[23] Two years later the same line switched to cast steel, with bolsters made from 10-inch channels.[24] Steam railroads adopted cast-steel trucks somewhat later for freight cars. The earliest case that can be found of a cast-steel frame for a passenger car was in 1904 on the Big Four, a subsidiary of the New York Central. Cast-steel

Figure 7.12   A six-wheel, cast-steel truck with bolt-on pedestals, made in 1918. (Pullman Neg. 23662)

Figure 7.13   This cast-steel truck of 1929 had integral pedestals. Note its similarity to the wooden-frame, six-wheel trucks shown in Figures 7.5 and 7.6. (Pullman Neg. 32230)

four-wheel trucks weighing 13,000 pounds each were placed under a baggage car.[25] They proved so successful that within two years sixteen baggage cars had similar trucks. They replaced ordinary six-wheel trucks, with a weight saving of 9,500 pounds per car.[26] With the coming of steel cars, steel trucks were mandatory. The adjectives strong, massive, and heavy describe these ponderous subassemblies. In the heavyweight era four-wheel trucks weighed 7½ tons each, while a six-wheeler weighed 10½ tons (Figures 7.12 and 7.13). Designers would continue to agonize over the ratio between truck weight and total car weight. It was discouraging to find that even in the lightweight era a set

of trucks weighed nearly 10 tons each and equaled one-third of the car's total weight.

The age of steel revolutionized the materials of construction but did little to alter basic design. The steel truck was a literal facsimile in metal of what formerly had been made of wood. There were, of course, some exceptions, but only one variant design had any widespread use. Ironically, this radical truck style was chosen by the standard railroad of the world: the Pennsylvania. At the outset of the steel car age the Pennsylvania abandoned the conventional pattern for both four- and six-wheel trucks. The four-wheel-truck design that it adopted had several

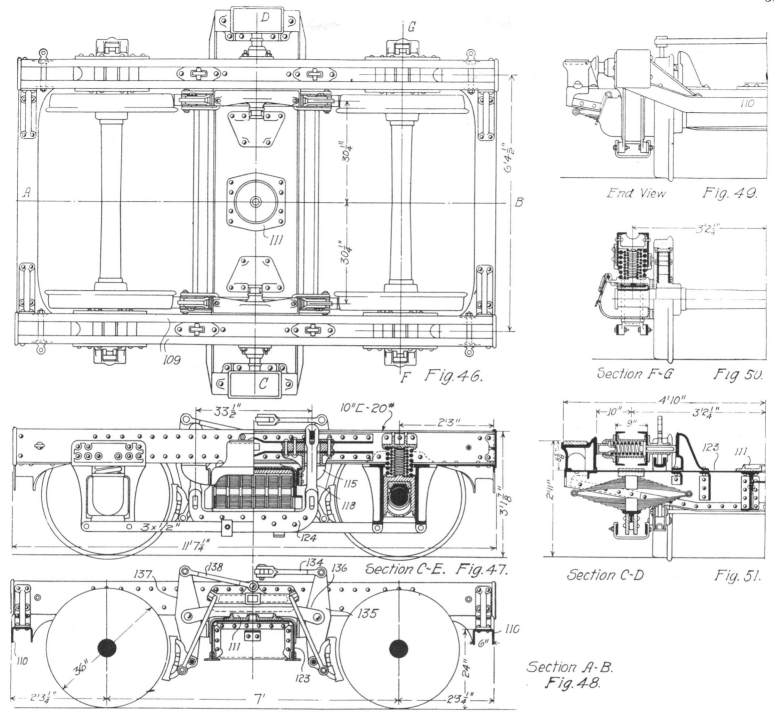

Figure 7.14 *The Pennsylvania's distinctive style of fabricated steel truck used on the first series of P-70 coaches.* (Railway Review, June 8, 1907)

unorthodox features. No equalizer was used; a triple nest of helical springs was mounted over each journal box (Figure 7.14). Unlike traditional wooden-frame construction, the bolster and center plate were placed very low to provide clearance for the center sill. The brake dead levers were anchored to the car frame rather than to the trucks, eliminating the tilt and resulting jerky motion produced when the brake levers use the truck as a fulcrum. The Pennsylvania's six-wheel truck exhibited similar peculiarities (Figure 7.15). The first group of trucks was fabricated from pressed shapes or mill stock, but by 1915 one-piece cast-steel pedestals and end brackets were substituted.[27] Within the next decade one-piece cast-steel frames were adopted, and the equalizer was readopted.[28]

It is not surprising that efforts were made to reform truck

design during the early stages of the lightweight era. Critics of the railroad engineering establishment contended that no new ideas had been seriously considered for generations. One of the most vocal opponents of conventional truck design was the Milwaukee Road's chief mechanical officer, Karl F. Nystrom. Nystrom tried every reasonable alternative to the conventional patterns. He tested very long wheelbases, experimented with placing springs in the immediate vicinity of the boxes (in theory the ideal location), and studied such European innovations as the triple bolster. Yet after years of testing, he was forced to admit that the conventional pattern rode about as well under normal operating conditions and was cheaper than the exotic schemes.[29] It was one thing to produce a good riding truck for speeds up to 85 mph, but for travel above this rate the answers

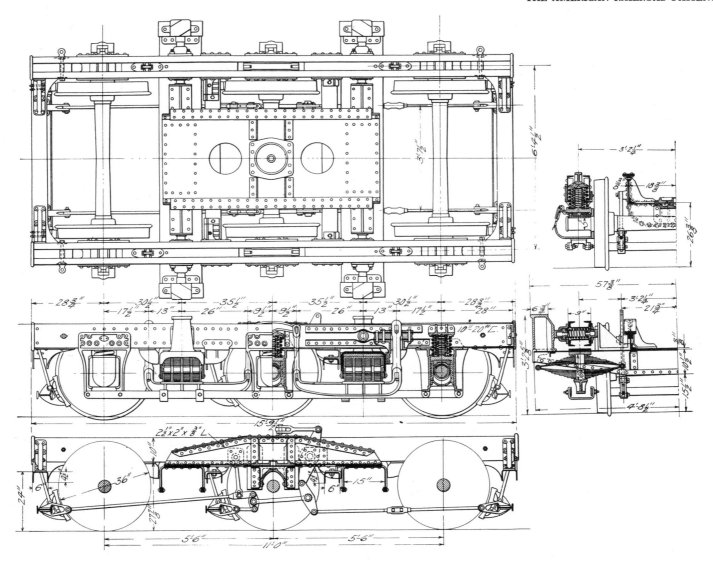

Figure 7.15   This Pennsylvania six-wheel, all-steel truck of 1907 featured a massive plate-steel bridge bolster. (Railway Age, *June 7, 1907*)

came very hard. One of Nystrom's assistants summarized the lessons learned after ten years of testing:[30]

(1) High-speed passenger car operation begins above 85 m.p.h. Below this speed, ground wheels and good track maintenance on conventional four-wheel truck will give satisfactory results.

(2) No high-speed truck of any design will function properly with wheels that are eccentric more than .020 in.

(3) The vertical riding qualities of a truck are a function of spring deflection. The bolster spring of a high-speed truck must have a sensitivity ratio (spring rate divided by the load carried) of not more than .15, preferably .10.

(4) The truck must have two spring systems, namely, primary and secondary systems. The equalizer springs system must not have a sensitivity ratio lower than .25, or spring surging will result. Springs of high deflection cannot be set down on completely unsprung parts without spring surging. A rule of thumb is that the equalizer spring deflection should be not more than ½ that of the bolster.

(5) The truck must have equalizers, or truck frame gallop will result if operated above 100 m.p.h.

(6) Friction surfaces should be reduced to a minimum.

(7) Car-body-roll stability is a function of the square of the lateral spread of the bolster spring.

(8) The value of swing hangers, especially long ones, is badly overrated in producing a good lateral ride.

(9) Vibration dampers and insulators are necessary to give a quiet car body.

(10) Bolster springs should be placed as high as possible to reduce instability.

(11) Wheel base is not a function of good riding either vertical or lateral.

(12) Wheel shimmy is caused by the taper of the worn wheel metal and is not a function of the shape of the tread when new. Wheel shimmy cannot be controlled by truck design, snubber or other devices. What causes a wheel to wear a critical shimmy tread is not known. The Milwaukee has one truck design that is not plagued with this phenomenon while all the others at times will develop wheel shimmy. There are so many differences between the truck designs that it would be impossible at this time to state the cause.

(13) High-speed passenger car trucks have excessive vertical amplitudes on rough branch line tracks when operated at low speeds. When designing branch line or suburban cars, high speed truck design should not be followed.

The triple-bolster idea that Nystrom studied was modified for American requirements by the Pullman Company.[31] These trucks were distinguished by wing-like pedestals with coil springs at each end, quite near the axle boxes (Figures 7.16 and 7.17). The heavy equalizer was eliminated, and with it a large part of the objectionable unsprung weight of the conventional truck. The bolster, although carried by full elliptic springs, was not pivoted to the truck frame but rather supported by intermediate bolsters hung from the frame on short hangers and coil springs.

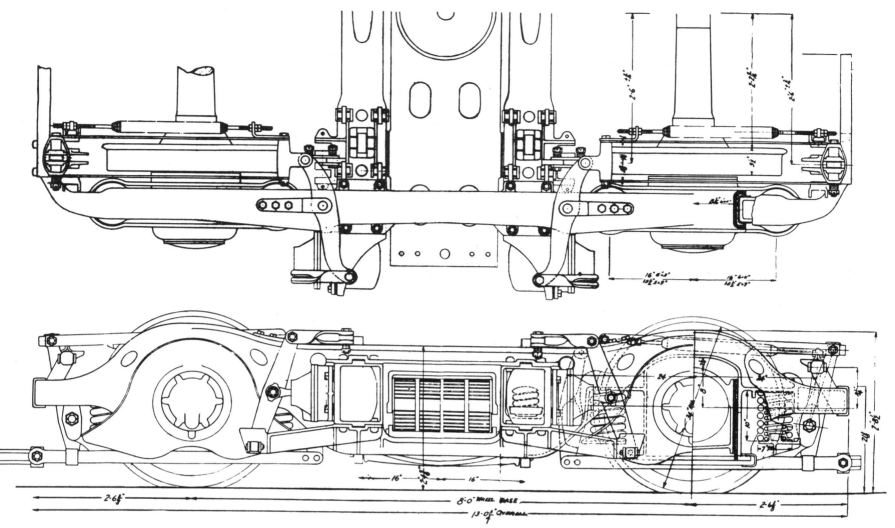

Figure 7.16   A cast-steel, triple-bolster truck with roller bearings, dating from about
1937. (Car Builders' Cyclopedia, 1937)

Figure 7.17   A triple-bolster, cast-steel truck of 1936 that was used on the Illinois Central's motor train, the Green Diamond. Note the truck-mounted air brake cylinders. (Pullman Neg. 39521)

The triple bolsters offered a soft ride; however, they were also prone to weave or teeter, and around 1940 they fell from favor.

The railroad supply industry played an important part in modern truck design, and in fact began to dominate this area after the passing of the wooden car. Many railroads and car builders were content to accept the stock models, especially for freight cars. Even in the passenger car field, a large proportion of the American fleet rode for many years on trucks manufactured by the Commonwealth Steel Company of Granite City, Illinois (Figure 7.18). This firm, later part of the General Steel Casting Corporation, continued to play a leading role in passenger truck design until the market for such equipment expired. Around 1940 Commonwealth introduced the floating bolster, a scheme that freed the transom from the restrictions and wear of guide pedestals.[32] The bolster was held in place by a heavy tie rod, one end of which was fastened to the truck side frame and the other to a

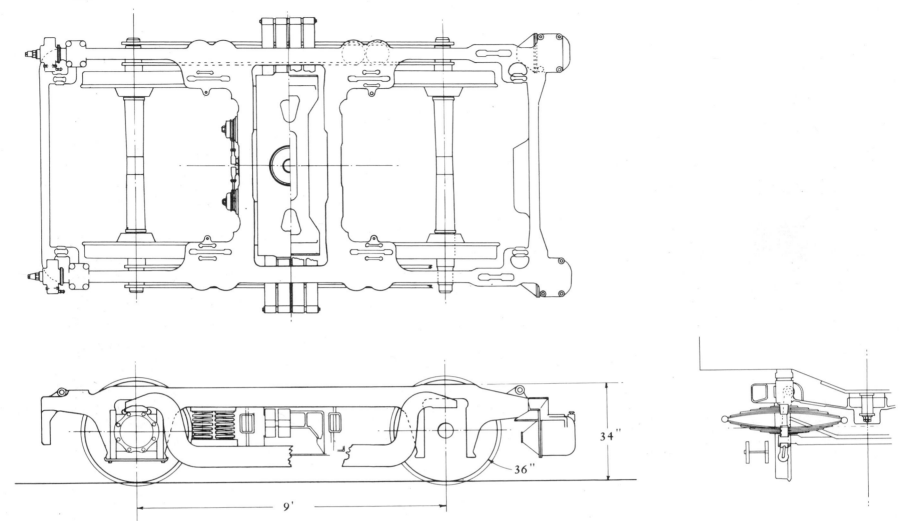

*Figure 7.18   A double-equalizer truck with inside swing hangers and roll stabilizers,
used on a 1939 Budd-built Seaboard Air Line coach. The cantilevered air brake cylinder
proved a poor arrangement, and side-mounted cylinders became the favored position in
later years. (Seaboard Air Line)*

bracket on one end of the bolster (Figures 7.19 and 7.20). Rubber discs at both unions allowed the necessary play. This became a common feature on four-wheel trucks. Another less widely used but promising innovation by Commonwealth was the outside swing hanger introduced in 1948 on a group of cars produced for the New Haven.[33] The bolster extended well beyond the side frames, so that the swing links were outside rather than inside (Figure 7.21). Because they were so far from the center pin, these trucks provided great stability and did much to prevent unwanted rolling. Only coil springs were used; the manufacturer claimed that they offered high static deflection. The trucks produced a soft ride, but their full potential was never realized because it was necessary to restrict the free swing of links to stay within existing side clearances.

Springs

As indicated in the previous section, various styles of coil and leaf springs were the basic forms used for railway car trucks. Since about 1870 they have been combined with great success, for the elliptics serve as an effective damper against the harmonic vibrations characteristic of the coil springs. Tempered steel was the only material used for metal railway car springs.

It was assumed by some early engineers that railways would furnish such a perfect pathway that no springs would be necessary. Smooth wheels running over an equally smooth permanent way should result in a faultless ride. But the motion of heavy trains, combined with the action of rain, sun, and frost, soon added unwanted kinks and buckles to what was once straight and true. In its annual report for 1833 the Baltimore and Ohio confessed:

Experience has demonstrated that the use of springs on all the carriages, whether for the transportation of passengers, or of merchandize, would contribute greatly to the preservation, both of the cars and of the road. Such an arrangement would assist in regulating the motion, and serve in some measure to alleviate the sharp jar inseparable from the sudden and violent collision of two hard bodies and no mitigating medium, and thus save both the cars and the road from that racking concussion which no perfection of construction, nor any reasonable care and foresight to prevent every accident, can entirely obviate.

In like fashion the Philadelphia and Columbia was rueful over the miserable performance of its initial lot of unsprung cars, most of which had pounded to pieces after little more than a year.[34] It was agreed at a very early date that springs were no idle luxury.

Figure 7.19   General Steel Casting Corporation made this floating-bolster truck for the New York Central in 1942. The horizontal stabilizer or drag link, visible above the right set of coil springs, held the bolster in line but allowed it to move vertically. (Pullman Neg. 45955)

Figure 7.20   A truck very similar to the cast-steel, all-coil-spring truck shown in Figure 7.19. Elimination of elliptical springs reduced maintenance costs, because the coil springs did not wear. (Pullman Neg. 48061)

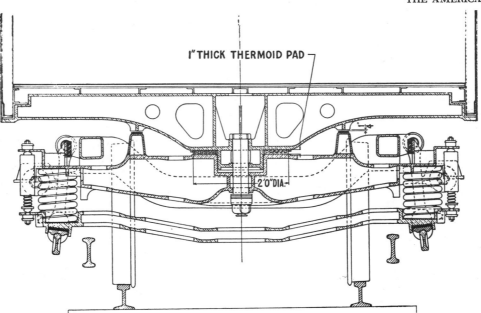

Figure 7.21   Cross section of an outside swing hanger truck manufactured and introduced in 1948 by the General Steel Casting Corporation. (Car Builders' Cyclopedia, 1953)

In the pre-Bessemer period, steel was a scarce commodity. It was difficult and expensive to manufacture, and it could be exceedingly unreliable unless it was made by a warranted master. Since the American iron industry was generally unable to produce such rarified metals as spring steel, most orders went to overseas suppliers. There was a great need for some alternative— some new, resilient material that was cheap and within the capabilities of domestic manufacturers. Both practical and foolish schemes were proposed. Like alchemists of an earlier age, Yankee mechanics seemed ready to try any combination of materials or methods in the hope that one would give the desired results. They tinkered with rubber, felt, wood, and compressed-air cylinders.

Rubber was the most promising and widely used nonmetallic spring adopted for railway work. It is said to have been suggested as early as 1825 by Thomas Tredgold. Six years later Conduce Gatch claims to have used india-rubber blocks under the bolster of the eight-wheel car *Columbus*.[35] In 1835 a British patent was granted covering the application of such springs to vehicles.[36] The idea was obvious enough to have occurred to many men with an inventive turn of mind, but the difficulty lay in the material itself. Natural rubber was suitable to only a few uses because of its extreme reaction to changes in the temperature. It melted in the summer and froze in the winter. In 1839 Charles Goodyear perfected a practical method of stabilizing it through the so-called vulcanizing process. The method was not patented until 1844, but news of it appears to have spread quickly to those interested in developing a rubber car spring.

Fowler M. Ray of New York began experimenting with india-rubber springs in 1843.[37] In the following year his tests were transferred to the Osgood Bradley car plant in Worcester, Massachusetts. At that time Ray adopted vulcanized rubber, with results so good that actual road tests were made by the New Jersey Railroad and Transportation Company in the winter of 1845–1846. Rubber springs worked well in the light cars of the day. Ray convinced other railroads of the utility of his "invention" and secured a patent (No. 5696) on August 9, 1848. He claimed that rubber springs saved 444 pounds and $59.06 per truck compared with steel springs.[38] Ray's New England Car

Spring Company announced that it had sold 1½ million pounds of springs worth over 1 million dollars between 1847 and 1852.[39] During these years of expansion Ray quarreled with an English inventor, W. C. Fuller, who claimed exclusive rights to the manufacture of rubber springs. In 1850 the U.S. Patent Office decided in Ray's favor, but Fuller continued to protest the decision through the courts. Within a few years, however, the controversy became an empty battle, because the rubber spring fell from favor.

The declining fortunes of the gutta-percha spring were announced as early as 1857, when the *American Railroad Journal* said that "vegetable gum" was not the thing for car springs.[40] When it was new it was too lively and made the car dance, yet after little more than a year it crushed down into an inelastic mass. And in the winter when the tracks were rigid with frost and good springs were most needed, the gum block had frozen solid. In the same year the Illinois Central annual report announced its intention of reconverting to steel springs, and within two years all the rubber springs had been removed from its passenger cars. Some other roads remained loyal to rubber springs, and the fact that the New England Car Spring Company continued to advertise in the *American Railroad Journal* into the 1860s indicates that the market had not entirely disappeared (Figure 7.22). Rubber also continued to be used in place of coil springs on some lines into the 1870s, but the bolster (and hence the main weight of the car) was supported by leaf springs. An example of this form of spring was shown in 1879 in the first *Car Builders' Dictionary*.[41]

Air and wooden springs were tried briefly at the time that rubber was being accepted into railway service, but they never achieved any wide application. The attempts to use them indicate the extent and the intensity of the search for a substitute for steel springs. The first serious effort to introduce air springs was made by Levi Bissell of New York City, an inventor best remembered for his locomotive safety truck. Bissell secured a patent (No. 2307) on October 11, 1841. Each spring consisted of a cylinder 6 inches in diameter by 10 inches long. Leather rings and a 2-inch reservoir of oil and white lead formed the seal.[42] In the following year the Schenectady and Troy Railroad was re-

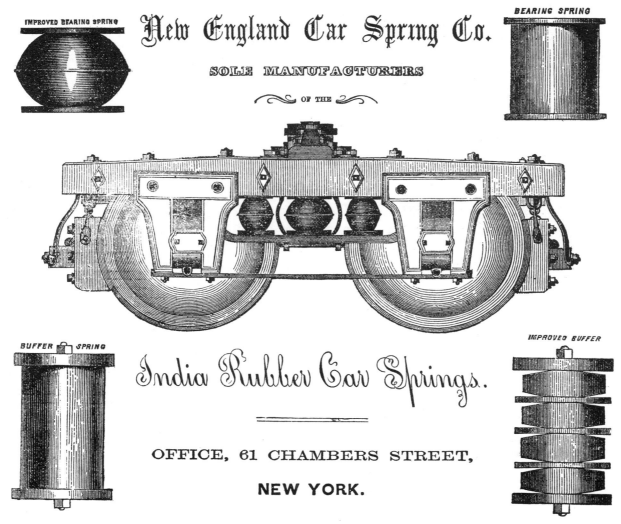

Figure 7.22   *Rubber springs were widely used by American car builders between 1845 and 1860, though they did not replace steel springs.* (American Railroad Journal, *February 8, 1862*)

ported testing "atmospheric springs" on some of its coaches.

Baldwin also experimented with the Bissell air cushion. Several years later the Boston and Maine was said to be applying air springs to new cars under construction at the Andover shops.[43] In 1861 the Philadelphia, Wilmington, and Baltimore claimed that it was still getting good results from such springs after four years of service.[44] Although the air spring never really challenged tempered steel, it remained a favorite with inventive minds and has been periodically reintroduced as the solution to the problems of truck suspension. As recently as 1969 General Steel Industries was promoting the idea for rapid-transit cars, and since that time air springs have been used more widely.[45]

Alert designers are always hopeful of employing some common material to solve a persistent mechanical problem. An example was the effort of James Millholland, at that time mechanical superintendent of the Baltimore and Susquehanna, to perfect a wooden car spring.[46] He reasoned that ash, being tough and resilient, would make a cheap, serviceable spring. Two pieces 8 feet long and 2 by 6 inches in cross section formed the spring for the six-wheel cars used on the B & S line. At least one passenger car was similarly equipped. The scheme was never widely used for passenger cars, but thousands of four-wheel coal jimmies were built on Millholland's scheme. The idea was patented on September 23, 1843 (No. 3276).

In the mainstream of car spring development, the most impor-

tant event during the second half of the nineteenth century was the general introduction of coil springs. They were rarely applied to railway cars much before 1870; the veteran car builder John Kirby stated that elliptic steel springs were about the only type used earlier.[47] Coil springs were too limited in capacity or were too high in price. Coil or "worm" springs were tried in carriage suspensions at least as early as the 1790s, but they exhibited a fatal weakness not found in leaf springs: a single break rendered the coil spring useless. In an elliptic, on the other hand, one or more leaves might fracture and the suspension would still be operative, although reduced in effectiveness.

Spring making, like all metal goods production in the early period, was an art. The design and manufacture was entrusted to an experienced blacksmith, "who looks wise and mysterious as he draws hieroglyphics on the dirt floor of his shop, or as he plunges the heated plates into a vessel of inscrutable liquid to give them the requisite temper."[48] The need for more rational analysis and less guesswork in this important area was felt by many railroad managers. Dependable springs were essential. European steelmakers were making advances in spring manufacture at the time of the U.S. Centennial. And a year or two before the Centennial, the Watertown Arsenal conducted a series of spring tests for the Pennsylvania Railroad. The trust in wise blacksmiths was disappearing; spring making became an increasingly specialized process carried on by large commercial producers.

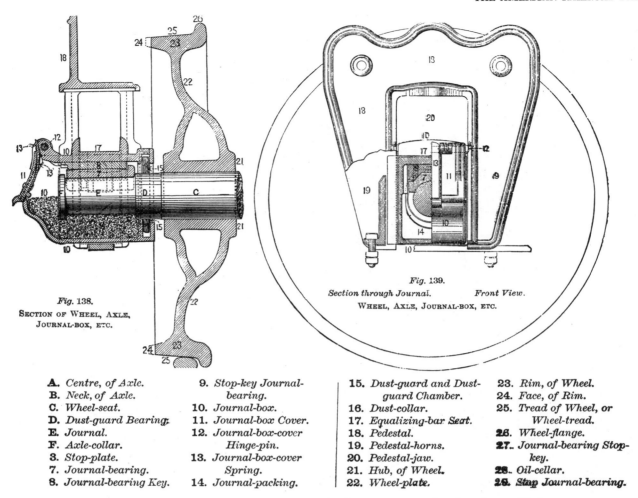

Fig. 138.
SECTION OF WHEEL, AXLE,
JOURNAL-BOX, ETC.

Fig. 139.
Section through Journal.   *Front View.*
WHEEL, AXLE, JOURNAL-BOX, ETC.

A. *Centre, of Axle.*	**9.** *Stop-key Journal-bearing.*	**15.** *Dust-guard and Dust-guard Chamber.*	**23.** *Rim, of Wheel.*
B. *Neck, of Axle.*	**10.** *Journal-box.*	**16.** *Dust-collar.*	**24.** *Face, of Rim.*
C. *Wheel-seat.*	**11.** *Journal-box Cover.*	**17.** *Equalizing-bar Seat.*	**25.** *Tread of Wheel, or Wheel-tread.*
D. *Dust-guard Bearing.*	**12.** *Journal-box-cover Hinge-pin.*	**18.** *Pedestal.*	**26.** *Wheel-flange.*
E. *Journal.*	**13.** *Journal-box-cover Spring.*	**19.** *Pedestal-horns.*	**27.** *Journal-bearing Stop-key.*
F. *Axle-collar.*	**14.** *Journal-packing.*	**20.** *Pedestal-jaw.*	**28.** *Oil-cellar.*
3. *Stop-plate.*		**21.** *Hub, of Wheel.*	**29.** *Stop Journal-bearing.*
7. *Journal-bearing.*		**22.** *Wheel-plate.*	
8. *Journal-bearing Key.*			

Figure 7.23   A typical journal box, bearing, and axle arrangement that has changed little in its general plan since about 1850. (Car Builders' Dictionary, 1879)

Journal Bearings and Lubrication

In the family of vehicles, the railroad car is a simple mechanism. It has few moving parts, unlike the automobile's and the locomotive's hundreds of bearings and wear points, all potentially troublesome and all requiring regular attention to ensure satisfactory operation. A car has only eight main bearings. If the axle boxes are properly serviced, the car should run trouble-free, always excepting the auxiliary equipment such as heating or lighting systems. Yet the simplicity of the car's basic running gear is deceptive. The possession of few moving parts does not necessarily guarantee trouble-free operation. To the contrary, the axle bearings were a notorious source of trouble to both railroad workers and the traveling public, particularly in the early period. Overheated journals caused delays and in some instances frightful disasters. The term "hotbox" became part of the railroad vernacular. The perfection of a reasonably dependable axle bearing has occupied the car builder's attention since the industry's infancy. Enormous progress has been made over the years, but the search continues for a perfect style of bearing.

Through almost the entire history of the American railroad car, bearings have followed one basic design (Figure 7.23). The housing, or box, was outside of the wheels. The axle ends protruded through the wheel and into the box. The smoothly finished ends were called journals. The single half bearing, a crescent-shaped bronze casting, rested on top of the journal. Properly called journal bearings, these pieces were frequently referred to as brasses, even though the metal was a bronze alloy. The lower portion of the box was a cavity used for an oil reser-

voir. It was packed with loose strands of heavy thread, called waste, which acted as a wick to wipe oil on the underside of the journal. It was a simple, self-lubricating bearing that in general proved very satisfactory.

The precise form of the earliest journal bearing is not known, but the oil reservoir, or oil cellar, was used in this country before 1832.[49] In some instances it had a floating cork in place of waste, though wicking was probably also used during the same period. Brass bearings were the standard in Britain at the time; the oil cellar, however, appears to be of native origins. Unfortunately, no detailed drawings can be found for the pioneer period of the American axle box. The general drawings that are available only hint at the nature of the arrangement. It is apparent from the illustrations in Chapter 1 (for example, Figures 1.2, 1.10, and 1.76) that the earliest boxes were very small and that some were undoubtedly more like grease boxes than waste-filled oil reservoirs.

Plain bearings offer too much resistance to the free rolling of the train, according to some pioneer mechanics, who sought to introduce frictionless running gears. When the American railroad era opened, the most promising effort in this direction was a device patented by Ross Winans on October 11, 1828. At the time Winans was an unknown farmer from Vernon, New Jersey. The spectacular performance of his so-called friction wheel bearing made him an instant celebrity in American and British railway circles. Although the device ultimately proved a failure, it did launch its originator on a distinguished and occasionally stormy engineering career. Winans proposed to substitute rolling for rubbing friction and to reduce the surface area at the contact

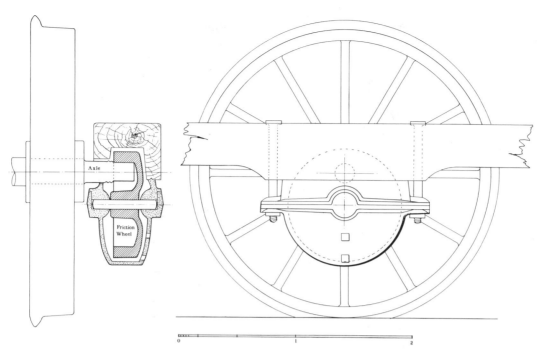

Figure 7.24 Winans's short-lived "friction wheel," or roller bearing, was patented in 1828. (B & O Annual Report, 1831)

point of the secondary axle. He accomplished it by providing an outside, cup-shaped bearing wheel. The axle bore against the inside rim of the friction wheel, which itself turned on a small-diameter shaft (Figure 7.24). Thus the main bearing was a rolling one. Actually, the axle was floating. Winans said that this was an advantage because there would be less flange wear in traversing curves, since the axles, and hence the wheels, were free to depart slightly from their normally fixed parallelism. He did not discuss what effect it might have on the car's steadiness—nor did he explain how brakes might be applied.

Not many months after the patent was approved, Winans exhibited a large-scale model of his friction wheel in Baltimore. The tests astonished all who saw them: the 125-pound car, carrying a load of 860 pounds, was propelled by a piece of twine running over a pulley with a ½-pound weight at its end.[50] Friction was reduced by 40 percent. A larger car was built, and it performed in an equally astonishing manner. In February 1829 four full-sized cars were at work, hauling away spoils from a deep cut being made for the railroad near Baltimore.[51] A single horse could move 15 tons over the rough temporary track. News of the great technical breakthrough reached other fledgling railroads. The South Carolina, the New Castle and Frenchtown, and the Ponchartrain were all eager to try Winans's scheme. Excitement over the friction wheel peaked when Winans demonstrated a car in England at the famous Rainhill Trials. He was now in the company of such masters as Stephenson and Hackworth. The prestige of American engineering was advanced by the remarkably free-wheeling handcar that Winans propelled back and forth over the tracks of the Liverpool and Manchester Railway.[52] His triumph was short-lived, however. The directors of the B & O had adopted the invention, and in service the friction wheel exhibited a fatal defect: the journal and inner periphery of the friction wheel wore very rapidly.[53] White-metal liners were tried, but to no avail. Its extra weight and high cost were against the Winans bearing as well. By 1831 the B & O was replacing the friction wheels with plain bearings.[54]

The B & O, still willing to experiment in spite of the friction wheel's failure, adopted the invention of an associate of Winans, John Elgar (1780–1854), who used new materials to build a better plain bearing. In a patent specification dated October 1, 1830, he proposed to retain the outside-bearing scheme introduced by Winans, but he envisioned a conventional bearing made from more durable materials. The axle ends would be "steeled," that is, a wrapper of steel would be forge-welded around the journal portion of the axle and would form a bearing shaft 2 inches in diameter by 4½ inches long. He would use chill-hardened cast-iron bearings rather than brass. An oil box with a capacity of ½ gill and a dust seal were other features of Elgar's patent. Many years afterwards Benjamin H. Latrobe recalled that Elgar's bearing was used on the B & O for many years with good results.[55] The B & O conducted a test in 1844 which demonstrated the superior antifriction qualities of soft, white-metal-lined brass journals, but the Elgar bearing was so durable and so cheap that it continued as the standard.[56] Other roads also used chilled-iron journal bearings. Von Gerstner reported seeing them on cars of the Boston and Worcester. The South Carolina tested Elgar's iron boxes and steeled journals in 1838 with such good results that within two years all the eight-wheel cars on the line were fitted with them.[57] The materials were at first purchased from the New Castle Manufacturing Company, but after a few years they were made at the railroad's own shops. The boxes showed very little wear and flourished on cheap tallow lubricants. Elgar's style of bearing performed well so long as very slow speeds and extremely light cars were in fashion. Just when it was abandoned is not recorded, but it was undoubtedly unsuitable for passenger train service by the Civil War period, when operating conditions began to change dramatically.

Meanwhile the conventional journal box was being improved. New features outlined in a patent (No. 1390) granted to John H. Tims of Newark, New Jersey, on October 31, 1839, may even be the source of what soon became the basic American railway box. It had such modern components as an inverted half bearing and an oil cellar with wicking brushing the underside of the axle (Figure 7.25). The inventor admitted that underside lubrication of journals was already known, but he claimed that the use of a sponge (his preference) or other absorbent materials to convey the oil to the journal was an original idea. If this was true, Tims's patent contained a basic innovation that came into universal use.

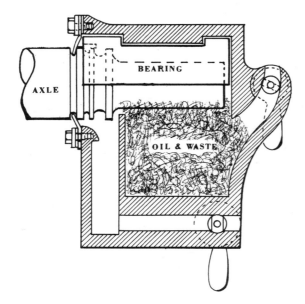

Figure 7.25   John H. Tims's patent of 1838 may be the origin of the conventional American journal box. (U.S. Patent Office)

Other features of his design were not retained, however. The oil was introduced through a petcock. The front of the box could not be opened, so that it was necessary to remove the rear dust seal and pull the bearing out through the rear opening. A more awkward and troublesome maneuver is difficult to imagine. An opening in the front of the box with a hinged or bolt-on lid made the bearing more accessible for inspection or replacement, but extracting the bearing was still difficult. It was necessary to remove the truck, disassemble the box, or take other extreme measures.

John Lightner, master car builder for the Boston and Providence, suggested a simple solution of the problem. He placed a thin wedge or spacer block above the bearing (Figure 7.26). A jack under one corner of the car or the bolster would remove the weight so that the wedge could be slid out, leaving a free overhead space. The bearing could be tipped up and over the end lip or collar of the axle and thus removed after a few minutes' work, with no displacement of the car or the truck. Lightner received a patent (No. 5935) on November 21, 1848. The merit of the scheme was quickly recognized, and by 1855, 150 railroads were using it.[58] Within a very few years the wedge became a standard feature of American railway boxes—and one that remains in use to the present day. Yet apparently Lightner received a rather small reward for his contribution to railway science. In 1850 he sold the rights to the patent to other parties, who decided to demand a license fee of $5 a year for each car fitted with the axle box wedge.[59] The revenues from such a fee would have been enormous, and the users went to court rather than pay it. They claimed that the idea had actually been developed a decade before Lightner's patent by Seth Boyden, an ingenious mechanic in Newark. Boyden, best known for such innovations as malleable iron and patent leather, also found time to manufacture locomotives and cars. Some of the cars he produced for the Morris and Essex Railroad in 1838 were said to have axle boxes with bearing wedges.[60] The scheme was not patented, but one of the original boxes was recovered from an old car in 1860 and introduced as evidence to support the claim.

Perhaps even more important than the general arrangements for car boxes perfected by Tims, Lightner, and Boyden were the improvements being made in bearing metals. All agreed that bronze was the best material for bearings. The precise formula followed was a matter of individual opinion. The Syracuse and Utica preferred a mix of 8 parts copper to 1 part tin, while the Union Pacific held that a more exotic combination of copper, tin, lead, and antimony was necessary for a proper car brass.[61] Other

roads began to favor phosphor bronze, an alloy that first received attention in the late 1850s.[62] Phosphorous proved to be a deoxidizing agent that produced a tough, homogeneous bronze, well suited to the hard service imposed on railway car bearings. After 1870 its commercial potential was recognized and it became something of an industry standard.

Years before the coming of phosphor bronze, railroads recognized the benefits of soft lead- and tin-based liners for locomotive and car brasses. A new journal was smooth and rode snugly in its polished brass bearing, but in time it became scored and rough. Wrought iron was the only material available for axles in the early period, and being fabricated, it was full of flaws and tiny seams which opened up in time, forming an irregular surface. Turning the journal was troublesome and expensive. A soft-metal liner, however, was self-fitting, even on a rough journal surface. It would swage or mold itself in and around the ridges and grooves of the worn journal. The basic idea is generally credited to Isaac Babbitt of Taunton, Massachusetts, who introduced the scheme in 1839. Babbitt advocated a high-tin alloy that proved very effective but was too expensive for such large-scale applications as car brasses. The industry soon substituted a cheaper alloy made of 80 percent lead and 20 percent antimony, although the term "Babbitt metal" persisted. Car builders, while willing to compromise on the ingredients, were not slow to adopt Babbitt's idea. In 1841 the Philadelphia, Wilmington, and Baltimore's annual report credited the soft-metal-lined boxes with reducing friction and saving oil. The Philadelphia and Reading was employing Babbitt bearings, according to a note in the Directors' Minute Books for April 2, 1842. By the mid-1870s, ½-inch-thick soft-metal liners were widely adopted. The Master Car Builders Association made them recommended practice in 1893 and standard practice in 1915, and so they remain to this day.

The consistency over the years in the general design and materials of journals did not hold true for box size. This grew steadily decade by decade until about 1920, when passenger car size reached maturity. The earliest car builders believed that they could reduce friction by employing the smallest possible journal, and some shops constructed journals as small as 1¾ inches in diameter by 3¼ inches long.[63] The size was grudgingly increased as larger cars appeared, but the belief in the small bearing remained in vogue during the pre–Civil War years, when 3- by 5-inch bearings were considered adequate. Around 1860 practical engineers began to realize that small bearings ran hot because too much of a load was concentrated on the bearing surface.[64] The pressure was so great that the lubricant was forced out and friction was increased, not reduced. Then large bearings came into favor. In 1869 the Master Car Builders recommended a standard bearing size of 3¾ inches by 7 inches. Some builders thought that it was overly generous when the standard was first introduced, but the rapid growth of American cars rendered it too small within a period of twenty years. In 1889 the 4¼- by 8-inch journal became standard. Within seven years it gave way to the 5- by 9-inch journal. Standard sizes in 1920 are given in Table 7.2.

The M.C.B. action in setting journal standards was a futile

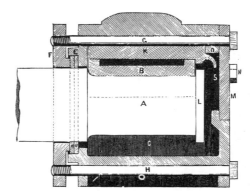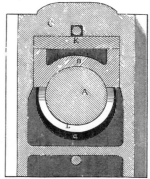

Figure 7.26   John Lightner's 1848 patent added a space plate, K, which permitted easy removal of the bearing, B. This modest invention has remained a basic feature of car journals.

effort to effect a rapid conversion to a single car bearing. In reality the conversion was slow and erratic. Some roads did not like the M.C.B. design. Among them was the North Western, which continued to install a peculiar style of brass that it claimed ran cooler than any other.[65] Some railroads agreed on the sizes specified by the M.C.B. but not on the other critical dimensions, so that brasses could not be freely exchanged. The situation remained chaotic as late as 1900, when the *American Railroad Journal* reported:[66]

One hundred and eight patterns of car brasses make quite a formidable array. It is a silent commentary upon the meaning of the "M.C.B. standard," when such a lot of miscellaneous and grotesque metal shapes must be carried in stock on a single railroad to meet the combined requirements of home and foreign cars. Upon investigating a reason for carrying one of these patterns, we found that a certain prominent road is responsible for putting out brasses and wedges bearing the symbol "M.C.B.," of which the wedges are too narrow to work in standard M.C.B. journal boxes without allowing the brass to tilt over on its side and with side lips so deep that the lugs bear upon the lugs of the brass and prevent the top of the brass from touching the bottom of the body of the wedge. The brass will not work with an M.C.B. wedge and the wedge will not work with an M.C.B. brass. In justice to the railroads the name of this road ought to be made public.

Just when or how completely this problem was resolved has not been uncovered; presumably it was gradually rectified by more rigid enforcement of the interchange rules.

A lubricated bearing requires five times less power to turn than a dry one: the friction is reduced that much by a simple slick of oil or grease. Engineers have continually searched for more effective lubricants and better ways to keep the journal well coated with oil. During the first decades of railroading only natural oils were available. There are few animal and vegetable derivatives that were not tried at some time or other in a car journal box. The Boston and Worcester ran some road tests in October 1838 to determine the best lubricant available locally.[67] The line was using sperm oil at $1 a gallon and wanted to find a cheaper substitute. Tallow, which cost only 10 cents a pound, worked admirably in the summer but froze nearly solid in winter. Common whale oil, priced at 35 cents a gallon, won the contest. Many years later the Michigan Central became a convert to whale oil after testing it against mineral oil.[68] An express train ran eight trips for a total of 1,584 miles. Boxes on one side were lubricated with whale oil and wood shavings, while those on the other side were filled with mineral oil and ordinary waste packing. The

whale oil boxes ran cool, whereas there were five hotboxes on the mineral-oil side. Whale oil cost 60 cents a gallon, mineral oil $1.34 a gallon—a clear victory for the fishing fleets of New Bedford.

Palm oil was another widely used car lubricant. The Newcastle and Frenchtown reduced costs by diluting it with tallow.[69] Boxes would run without attention for eight months. The B & O managed nicely on plain tallow, as did many Southern lines. The Housatonic Railroad recommended pork fat, noting that it cost only 30 cents a year per axle box.[70] Animal fat had its hazards, however. A Southern traveler in the 1840s was astonished by the numbers of hogs that gathered at every station when the train pulled in. Then at one stop he noticed that "gaunt and hungry hogs sauntered up to cars, and with their snouts raised the lids of the axle-boxes, and greedily drained the tallow."[71]

Even after the introduction of petroleum, some railroads stayed with natural lubricants. In the late 1860s the Camden and Amboy concluded that "earth oils" were unsuitable for journal bearings and reverted to its old standards of lard and sperm oil.[72] In 1875 the Master Car Builders report continued to exhibit some skepticism about petroleum, although admitting that because of its cheapness, most lines were employing mineral or "West Virginia oil." It was thought safer to mix the "unnatural" petroleum derivatives with something dependable like tallow or lard. Some lines also believed in additives like "polar grease." The Texas and Pacific claimed that this cooling compound improved the oil so that a quart would last for 2,600 miles.[73] Others argued that adding expensive cooling compounds to cheap oil cost more than buying good oil in the first place.[74] The purchasing agent was singled out as the villain here for his false economy in seeking out the cheapest lubricant. His accounts looked good, but the operating department was left to contend with an epidemic of hotboxes. Adding to the confusion was a blind dependence on brand names such as Black Oil, Well Oil, Stock Oil, and Cleveland Cup.

Laboratory testing was the only sure way to secure a reliable product. In 1875 the Pennsylvania established a chemical laboratory for this purpose.[75] New supplies were tested on a monthly basis, and inferior or adulterated oils were rejected. In some instances the lab discovered that 50 percent cottonseed oil had been added, making the lubricant nearly useless. By the late 1880s many major lines employed their own chemists.[76] The freewheeling days of the shady oil dealer were over, at least so far as journal bearing lubricants were concerned.

The process of journal oiling was to go through another important reform: winter- and summer-weight oils were adopted. Because lubricants suitable for one season were not right for the other, railroads began to use thin oil in winter and heavy oil in summer. On the New York Central the changeover began every November 15.[77] The old waste and oil were removed and set aside for the spring, and a tin tag was attached to the journal box lid noting the date and place of the oil exchange. Regular inspec-

Table 7.2   Standard Journal Sizes Adopted in 1920 by the Master Car Builders

Journal Size, in.	Axle Capacity, lbs.	Car Weight, tons
A. $3\frac{3}{4} \times 7$	12,500	33
B. $4\frac{1}{4} \times 8$	20,500	$56\frac{1}{2}$
C. 5×9	27,000	68
D. $5\frac{1}{2} \times 10$	34,000	$84\frac{1}{2}$
E. 6×11	42,500	105
F. $6\frac{1}{2} \times 12$	51,000	$125\frac{1}{2}$

*Figure 7.27   A typical car journal with the front lid or cover removed. The space
plate, bearing, axle end, and waste are clearly shown.*

tion was another feature of the New York Central rules. Through
cars were oiled at New York City and Buffalo. Inspectors touch-
tested the boxes at intermittent stops but had orders not to open
them or add oil unless they were running hot. Local cars traveling shorter distances at lower speeds were oiled on a mileage
basis—once every 3,000 miles.

The quality and the handling of waste were as important to
journal lubrication as the oil itself. Waste is a tangled mass of
threads, usually salvaged from short ends or remnants of cops, a
by-product of textile mills (Figure 7.27). They were a haphazard
mixture of cotton, wool, and hemp. Waste was clean, absorbent,
and of course, very cheap—just the thing for wiping up around
the machine shop or stuffing into a journal oil box for wicking.
Pure-wool carpet yard was considered the best waste, whereas
mixed waste, which was often nothing more than floor sweep-
ings, was regarded as undesirable.[78] It was important to have
long strands, so that the waste would intertwine to form a pad
with few free loose ends that could work up between the brass
and journal. While wool and cotton were popular, some lines
mixed in cured hair or other fillers.[79]

A few lines experimented with the felt or cotton pad as early

as 1875, but they were criticized for extravagance. Some mechanics also claimed that pads threw so much oil around the inside of
the box that it found even more places than usual to leak out.[80]
Enough oil was already lost through the notoriously inefficient
rear dust seal. Waste remained the standard packing for many
generations, but interest in developing a practical lubricating pad
continued. Advertisements for the latest device regularly appeared in the trade press, but in general the railroads were not
buying. Following World War II the industry at last conducted a
serious investigation of the subject, with results both favorable
and unfavorable to the pad. Because of better oil distribution,
pads proved effective in holding bearing temperatures as much
as 40 degrees below those possible with waste.[81] Yet none of the
pads tested would stand up in regular service. Synthetic materials did much to improve the situation, however, and within a few
years long-lasting pads were available. The most revolutionary
conversion was made in the freight car fleet, because by that
time a large part of the passenger car fleet had roller bearings. In
1957 pad lubricators were required on all new freight cars.[82] By
1961 loose waste was a thing of the past, and during the following year no cars so equipped would be accepted for interchange

service, according to a rule established by the Association of American Railroads.

Improved lubrication techniques all but eliminated bearing failures. In 1950 99.95 percent of railroad cars were running trouble free on this account, yet the hotbox remained "the scourge of railroading."[83] Since the beginning of the steam railroad, operating men had lived with this bogeyman that confounded schedules and demolished equipment. One passenger described the everyday scene in nineteenth-century America:[84]

The advent of a "Hot Box" was heralded by an "aroma" modest at first but growing more insistent as the train sped along and causing travelers to sniff until an outspoken one called across the aisle to a friend—"Hot Box?" and received a nod in reply. Then came rattling of bell cord and quick answer from whistle. The train drew to a halt. Many male passengers swung to the ground and clustered about the fuming journal. Into the throng strode our brakeman with his surgical equipment—a pail of water in one hand—in the other a pail slopping with a dreadful looking mess. His short-hooked poker scalped the journal lid and exposed the heated parts belching smoke. He hooked out the smouldering waste and dashed water inside which forthwith rushed away in sizzling bursts of steam. Then he poulticed the feverish interior with that dreadful looking dressing poked in by rod and hand. Back he slapped the lid and back to their seats scrambled the passengers. The whistle blew. The train resumed its journey while the brakeman busied himself in cleaning his filthy hands.

The great majority of hotboxes ended on this cheerful note—a mere diversion to break the tedium of travel. But some boxes were not so easily cured, as illustrated by the halting passage of the Pacific Express in 1870.[85] A half hour was lost in Harrisburg while two journals on a front-end car were doused and repacked. The train was forced to stop again at Lancaster to quench a blazing hotbox on another car. Three more stops were necessary before reaching the final terminal in Philadelphia. Hours were lost, tempers frayed, other trains sidetracked, and the whole main line put into a furor because of a simple bearing failure. And a hotbox posed an even greater threat than the possibility of alienating passengers or disrupting schedules. First, any breakdown on the road was a setup for a rear-end collision. Second and still more dangerous, a hotbox occurring at high speed could actually burn the journal off, causing a derailment. Steam service had hardly begun before such an accident took place. In November 1833 a train on the Camden and Amboy was running fast to recover lost time. An axle on one of the cars overheated and broke off, causing a wreck that killed two people and injured many others. Among the passengers were former President John Quincy Adams and Cornelius Vanderbilt. By far the gravest disaster attributable to a hotbox occurred 110 years later, near Philadelphia.[86] In 1943 the Congressional Limited was whipping along at nearly 60 miles an hour when a journal on the seventh car in the sixteen-car train burned off. A colossal derailment followed, resulting in 79 deaths and 129 injuries. It was one of the worst rail disasters in American history. An examination of the defective journal showed that its surface reached 1400 degrees before the failure occurred. This disaster cost the Pennsylvania Railroad millions of dollars. And less sensational hotbox failures also caused a heavy drain on the line's treasury; for example, in the early 1950s the P R R claimed annual losses of 4 million dollars due to hotboxes.[87] The vast majority, of course, occurred on freight cars, which received less care than the passenger fleet and were subject to overloading. Unfortunately, no separate cost figures seem to have been maintained for freight and passenger trains, but the 4-million-dollar total noted above for a single major trunk line gives some idea of the staggering cost to the entire industry.

A long, sad list of causes of the hotbox problem can be assembled. Inferior oil was an obvious villain. Cheap brasses made from an indiscriminate mixture of scrap bronze were blamed by some authorities.[88] Loose bearing liners were another cause—a problem that was ultimately solved by a change in the white-metal alloy and more careful fluxing. Imperfect truck design and badly maintained, out-of-square trucks contributed to hot bearings. Overheated journals also resulted from too small a bearing for the normal load or, in the case of freight cars, serious overloading. Ineffective dust seals at the front and rear of the box permitted grit to enter and mix with the oil. Because the boxes were at track level, dust swirled around the trucks in a furious storm, yet it was not uncommon to see freight cars with missing journal box lids.[89] The rear seal was a wooden board with a hole cut in it so that the axle might pass through into the box. Sometimes it split and fell out, and its absence was not noticed for months because of its inaccessible position.

Even the weather was blamed for hotboxes. According to some mechanics, spring and summer were the worst seasons, because it was then that the ballast was renewed and the tracks were at their dirtiest. Other mechanics dreaded the winter, when the oil thickened and moved sluggishly through the waste.[90] Snow could blow through a partially cracked journal lid and emulsify the oil. In the extreme climates of New England and the Northwestern states, the oil and waste would freeze into a solid rock-like mass, ending all possibilities of lubrication.

Waste grabs were long considered a major cause of hotboxes. Loose threads tended to travel up under the edge of the bearing on the rising side of the journal.[91] They would work their way between the bearing and the journal, overheat because of the intense friction and pressure, and burn, setting the box ablaze. A freshly packed box was often the worst risk, because in new waste there are so many lint ends that can be easily picked up by the journal. Buffering impacts encountered in ordinary switching were often sharp enough to cause the bearing to jump up off the journal. A strand or two of waste might then flop up and be caught between bearing and journal when the bearing reseated itself. Waste grabs were not eliminated until lubricating pads were finally adopted.

There was a belief in the railroad industry that all these causes of the hotbox problem were secondary compared with the human element. According to many managers, the careless workman was the major cause of hotboxes. Carmen, oilers, dopers, car tenders—whatever they were called, they were all the same: a lazy, indifferent, dollar-a-day breed of laborer.[92] Even

Figure 7.28   A Timken roller bearing assembly of about 1950.

with the commercial beginnings of roller bearings and represent a late effort to perpetuate the old-fashioned friction bearing for passenger car service. In 1933 the New York Central's West Albany testing laboratory began work on a smoke-odor capsule that would ignite, calling attention to a hot bearing before a serious rise in temperature occurred.[98] Two small capsules with soft plugs which melted at 320 degrees were placed inside the box. One was a stink bomb, the other a smoke bomb. The device was used in some cars during the next year, and in 1936 the bombs were credited with uncovering sixteen hotboxes. The Central began to install the alarms on all its passenger equipment. By 1950, 3,500 cars were carrying them.[99] Despite their effectiveness, the Southern Pacific was the only other line to show much interest. Later the industry was drawn to an even more promising hotbox detector: in the late 1950s heat-sensitive electronic units could be set up at the side of the track to detect a warm journal on a passing train. By 1969 1,500 units had been installed, and some roads were placing them every 25 miles along the track.[100]

Centuries before the introduction of steam railways, mechanics sought to perfect something superior to the plain bearing. The use of rollers and wheels were a daily reminder of the great friction savings possible in rolling rather than sliding a load. The ancient Romans employed roller bearings in the first century A.D.[101] Fourteen hundred years later, Leonardo sketched roller bearings of strikingly modern appearance. Large weather vanes, such as the one atop Independence Hall, were mounted on roller bearings during the eighteenth century. The Patent Office began to receive applications for many ingenious types of antifriction bearings late in that century. In 1802 the tapered roller bearing was patented. The notion that ball and roller bearings are a recent invention is clearly false, but it is true that they did not come into commercial use until relatively modern times. The railroad industry was slow to adopt them, and was perhaps even more reluctant than other segments of American business to accept the antifriction bearing. It was not until about 1940 that the railroads seemed ready to utilize the free-rolling products of Timken, S.K.F., and Hyatt (Figures 7.28 and 7.29).

Part of this reluctance can be explained by the railway itself: the metal wheel running on a metal rail was in effect an antifriction bearing. The rolling friction of a railway car is very small compared with a highway vehicle. The small gain possible with antifriction bearings on the wheel axles would be offset by their extra cost. Furthermore, the first generations of ball and roller bearings did not prove very dependable for heavy machinery. They may have been adequate for a sewing machine or a bicycle, but they could not handle the crushing tonnage of a railroad car. Railroad men were content with solid bronze and Babbitt metal journals and saw little reason to change.

Among those eager to wipe out this prejudice was Ross Winans, whose friction-wheel invention brought him fame but ultimately proved impractical. For nearly fifty years the railroads apparently ignored any other inventor so foolhardy as to propose a similar experiment. Street railway men were more receptive,

when they tried, few oilers could pack a box properly. In their zeal they would overstuff the box so that it wiped the oil off rather than brushed it on, or they would pack it so tightly in the back of the box that the oil could not penetrate and only the front half of the bearing received lubrication. Managers also charged that oilers did not even look inside the boxes because before about 1870, it was necessary to remove two bolts to gain access to most journals.[93] After this time the so-called Fletcher lid came into use. A small coil spring on one bolt allowed the cover to be snapped open for easy inspection.[94] This simple improvement probably saved many a bearing. In time hinged lids came into favor.

Careful management, combined with good materials, could reduce the incidence of hot journals. Why, it was asked, did the Lake Shore have 1,100 hotboxes a month, the Boston and Albany only 2? Indifference was not a fault restricted to simple car tenders; it was the superintendent's responsibility to see that his men knew their jobs and to discipline them if they did not follow orders. On a casually run line cars might go four and five years without repacking; on a well-run road they were repacked every month.[95] The Delaware and Hudson concentrated on hotbox reduction and showed what regular maintenance, periodic checks, and better employee training could do.[96] In reviewing its passenger car fleet alone, the line reported 215 hotboxes in 1915. By 1920 the number had dropped to 45. Ten years later only 10 hotboxes occurred. And the entire industry was making progress. In 1928 a passenger car that ran 100,000 miles between hotboxes was considered a record setter.[97] By 1934 some cars ran as many as 1,000,000 miles between hotbox failures. The problem was at last all but eliminated by the roller bearing.

Hotbox alarms and detectors were introduced coincidentally

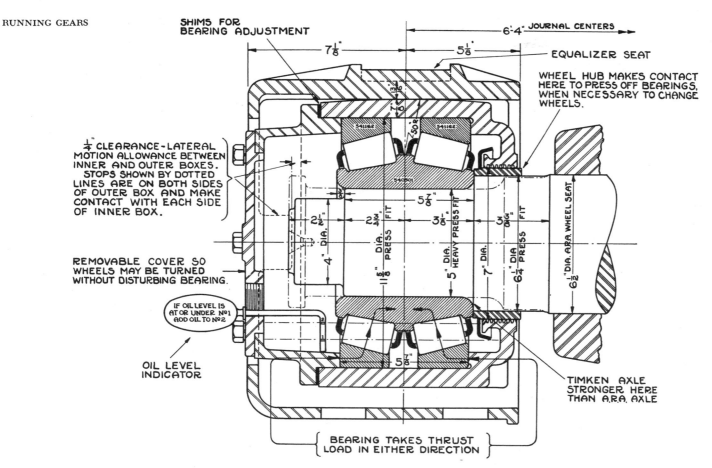

Application of Timken Bearings for 5″x9″ size axles. Wheels can be turned without removing Bearing from axle. Dot and dash line shows contour of A. R. A. standard axle. Note how oil level is checked.

Figure 7.29   A Timken roller bearing assembly drawing of about 1950.

and in the 1870s and 1880s a number of antifriction bearings were tested on horsecars.[102] Results were encouraging on these very light but underpowered vehicles and naturally suggested applications in the larger arena of the steam railroad. In the late 1880s several hopeful inventors began to seek permission from various roads to test their wares. In 1886 Charles D. Meneely of Troy, New York, tried some of his tubular roller bearings on the Delaware and Hudson Railroad.[103] The first sets were a failure because of alignment problems. By the early 1890s Meneely had a new design that proved more serviceable, and the D & H applied it to four cars.[104] The inventor claimed such startling advantages as a 90 percent saving in starting power, a 25 percent saving in fuel, an 80 percent saving in lubrication, and an "absolute" elimination of hotboxes. Meanwhile, the Rock Island was trying the Sharpneck frictionless roller bearing. Tests began in May 1888, and again gratifying results were at first reported.[105] After 38,000 miles on the experimental car in suburban service out of Chicago, several more suburban cars were fitted with the bearings. It was said that a single man could push one of these cars on the level. In the East, the Tripp Manufacturing Company of Boston prevailed upon the New York and New England to try a set of its bearings, which had been used with great success on streetcars.[106] Tripp was more moderate in its claims of friction savings over plain bearings, but the company did maintain that its journal could run for two years or more on the original lubrication.

The optimistic statements of the antifriction advocates were soon contradicted by railway mechanical men. One reported that a patented bearing ran trouble-free for five months and then seized up without warning, never to turn again. The inventor had

meanwhile disappeared. M. N. Forney, probably the most respected man in railway mechanical affairs in nineteenth-century America, spoke out against the roller bearings. He noted that the friction and fuel savings claimed were nonsense, because journal friction represented the smallest portion of the train's resistance. This resistance could actually be measured, chiefly by calculating the weight of the train, the gradients, the curves, and the air resistance.[107] Others objected to the size of the journals (Sharpneck's, for example, required a box nearly 12 inches square) and to the fact that they could not be fitted into a standard-sized journal box. Their first cost—$50 each in the late 1880s—was far more than that of a plain bearing.[108] And of course, a basic argument against roller or ball bearings is that they are no more friction-free than plain bearings except at very slow speeds. They were beneficial in starting the train—an important advantage, to be sure; but once the train was rolling their usefulness diminished.

Roller-bearing manufacturers found other markets outside the railroad industry that led to the perfection of their products for commercial use. With the early roller bearing, design and basic form were not the problems; what the makers needed to discover were the correct materials and manufacturing methods. At an early date they found out that the balls or rollers must not be allowed to rub against one another or a "grindstone" effect was created; they had to be separated and held parallel by a cage or ring. Ordinary mild steel was not up to the demands of the design, for very hard, precise surfaces were needed for a durable bearing. Thus bearings required better tooling, like centerless grinders, and more specialized heat-treated steels, together with refined production methods to ensure accuracy. After these elements were combined, antifriction bearings were successfully manufactured on a commercial basis. This appears to have taken place in the 1890s, although more research is necessary to establish the course of development.

The roller bearing was once again being considered for railway work during the early years of the twentieth century. In 1906 the Strang gas-electric car was equipped with antifriction bearings. William McKeen, mechanical chief of the Union Pacific, adopted roller bearings for his gasoline rail cars. In 1911 the Bangor and Aroostook began testing the bearings on a medium-weight coach.[109] A few years later the Boston and Albany tried them on three passenger cars. More applications were contemplated, but the bearing was slightly oversized, requiring new truck pedestals. These experiments only suggested what was to take place in the next two decades.

Its large-scale application in automobiles had helped to establish the roller bearing as an economically feasible device. The antifriction bearing maker was now marketing a thoroughly practical product, and the railroads represented a lush, untapped field for new sales. In 1921 the Swedish-based firm S.K.F. outfitted six passenger cars on the Pennsylvania.[110] The success of this installation led the road to purchase bearings for 166 cars five years later. These were rugged fixtures that did not burn up after a few months of running. Some of the original sets, in fact, traveled 2.6 million miles before being retired late in 1946.

After a few years the Milwaukee Road followed the Pennsylvania in trying roller bearings but surpassed it in enthusiastic acceptance of the antifriction idea. In 1924 the road began testing.[111] Bearing resistance was calculated at 54.4 pounds with plain bearings but only 7.5 pounds with rollers. Lubrication was managed by an annual greasing. In 600,000 miles of running, no failures occurred. An unexpected benefit was discovered during brake tests: roller-bearing-equipped cars showed less tendency to slide during emergency applications. The reason, mechanical men theorized, was that the action of the bearings eliminated the transfer of weight characteristic of plain bearings during fast stops. There was ample reason to convert, and late in 1926 the Milwaukee began to equip 127 cars with Timken bearings. The New Haven and others began to follow the lead of the Pennsylvania and the Milwaukee, so that by 1929 there were 850 roller-bearing-equipped passenger cars operating in the United States. In 1930 nearly 500 passenger cars were fitted with the new style of journal bearing.

During the next few years, however, the pace of conversion fell off sharply. Strict economy was necessary because of the Great Depression. And some mechanical men continued to question the benefits of roller bearings. A bronze bearing complete with wedge and box cost only $14.40; an equivalent roller bearing cost $260 to $400, depending on the manufacturer.[112] Skepticism about the value of roller bearings was illustrated by the reluctance of some lines to use them even for new lightweight cars. The first stainless-steel streamline equipment purchased by the Santa Fe in 1936 and 1937, for example, had conventional bronze friction bearings.

But in a few more years there was a massive shift of opinion on the subject. Between 1935 and 1940, despite the bad times, an average of 300 roller-bearing-equipped cars joined the passenger fleet each year through the purchase of new cars or the rebuilding of existing stock.[113] Another national emergency, World War II, again interrupted the switch from plain to roller bearings. After the war's end, however, almost no new passenger cars were ordered with plain bearings, according to the *Car Builders' Cyclopedia* of 1961. The higher cost of roller bearings, which had once been a convincing argument against them was now demonstrably offset by the wonderful economies realized after several years of service. Cars ran 30,000 miles a month, trouble-free.[114] It was not unusual to receive ten years' service before replacement was necessary. Long, trouble-free operation is probably what really sold roller bearings to practical railroaders. Free rolling was apparently seen as something of a fringe benefit. And while passenger cars were converted to antifriction bearings rapidly in the early postwar years, work was just beginning on a similar substitution for the larger fleet of freight cars.

Axles

Axles were once the weakest of the components forming the railway car's undercarriage. The previous section discussed axle journal end failures as the hotbox problem, but experienced car masters knew that the entire iron shaft was a source of unremitting trouble. Bent and broken axles littering the yards and shops of every major line in the country gave silent testimony to countless accidents, delayed trains, and sometimes major disasters. Yet small axles were favored because they reduced dead weight and cost. During most of the nineteenth century few car axles were made much larger than 4 inches in diameter. These slender, 350-pound shafts were expected to safely support high-speed express trains or heavily burdened freight cars. Most were fabricated from scrap iron for the sake of economy. The soundness of their manufacture was left largely to mechanics whose knowledge of metallurgy was based on tradition and the superficial appearance of things. If the finished product looked good, it was sent out on the road. If its core was rotten, this could only be discovered when it broke. Even though statistics available for the early years are sketchy, the record of axle failures was a dismal chronicle.

In 1834 the South Carolina Railroad stated in its annual report that broken axles were the principal cause of accidents on its line. A few years later the record books of the Boston and Worcester list a series of wrecks due to breaking axles.[115] In one car, three journals broke off. In June 1837 a broken axle on one car caused a wreck that damaged five other cars. In 1843 a Long Island Railroad train was delayed two hours for the same reason.[116] Such mishaps were apparently so common that they were reported only incidentally, and no one organization or publication made a complete record of them. In 1856 the 150-mile-long Philadelphia and Reading Railroad was breaking 300 axles a year, and other roads were reportedly doing no better.[117]

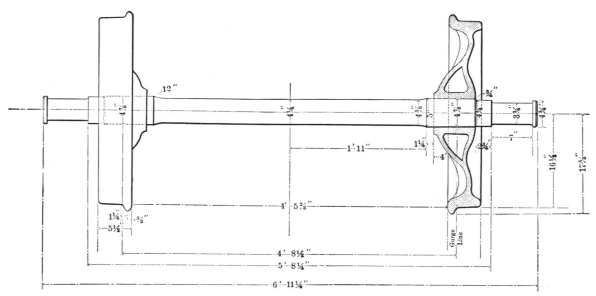

Figure 7.30   A typical axle and wheel set of the 1880–1890 period. Standardization of this assembly was a major accomplishment of the Master Car Builders Association. (William Voss, Railway Car Construction, *New York, 1892)*

Between July 1848 and November 1852 the Reading Railroad had 1,061 accidents because of broken axles.[118] In later years the *Railroad Gazette* served as the industry's semiofficial compiler of accident statistics, but because it depended on facts casually collected from scattered newspapers, its totals can only be viewed as a sampling. Even so, between 1873 and 1882 the *Gazette* reported an average of roughly 35 broken axles a year.[119]

It can be argued that most of these failures did not end in tragedy; yet a few did. In 1850 three passengers were killed on the Western Railroad because of a broken axle.[120] Three years later an accident on the Boston and Maine resulted in the death of President Franklin Pierce's son. A broken axle derailed the train. The shaft measured only 2½ inches in diameter and showed evidence of an old crack. A grimmer wreck at New Hamburg, New York, in 1871 demonstrated the dangers of axle failures even on freight cars.[121] An axle on a tank car broke, derailing it and several other cars on a bridge just as the Pacific Express was passing on the opposing track. The pileup resulted in twenty-two deaths.

Clearly there was an urgent need to improve the strength and dependability of car axles. Increasing their size was an obvious precaution. In its report for 1832 the South Carolina noted that it was enlarging its axle diameter size from 2¼ to 2⅞ inches. In 1839 Von Gerstner reported that the lines running into Boston were using 3- to 3½-inch axles. In the early 1850s several New York railroads noted that they were employing 4-inch-diameter axles.[122] By the turn of the century the Master Car Builders Association was specifying axles for passenger cars with a 5⅜-inch diameter at their centers and 6⅜ inches at the hubs.

Car axles were made smallest in diameter at their centers for several reasons (Figure 7.30). The taper saved iron where the full diameter was least needed. It also allowed the axle to flex or spring slightly, thus helping to dissipate the damaging lateral shocks received when the flanges struck the rail on entering a curve or a switch. The taper was subtle, often no more than 3/16 inch. Axles were made thickest at the hub; during the nineteenth century they were a full inch larger in diameter at this point than at the center. This was necessary because axles tended to break at the hub of the wheel. The hub acted as a fulcrum. Since the operation of the wheel results in a flexing of the axle (with the

axle acting as a long lever), a bending movement occurs just inside the rear of the wheel's hub. Here is the point of fracture and the commonest point of breakage.

The direct path to safer axles was better standards of manufacture and stronger materials. These improvements did come about gradually, but early mechanics could not wait for the industrial millennium; they required safer axles in their time. Since complete reform was not immediately attainable, an ingenious compromise was worked out. That axles were liable to break was accepted as a fact of everyday railroading. The worst consequences could be averted by offering the broken axle some fall-back position—some auxiliary means of support until the train could be stopped. This idea was developed by Joseph S. Kite, superintendent of the Philadelphia, Germantown, and Norristown Railroad. A pair of beams was fastened to the truck frame inside of the wheels. Oblong iron brackets loosely encircled the axles. In normal service the axles could gyrate freely inside the safety beam brackets, but should they break, the brackets offered a secondary point of bearing. The simple and effective plan was patented by Kite on July 14, 1834. Good reports on the invention began to appear in the trade press. A car with a broken axle made four trips on the Philadelphia and Columbia line without incident.[123] Only then was it discovered that the axle was being held in line by the safety beam stirrups. A similar event was reported on the P W & B, prompting the *American Railroad Journal* to congratulate Kite on the effectiveness of his patent in "these days of accidents and disasters."[124] The safety beam arrangement remained a part of American car design until the end of the wooden truck.

Since time in service was also found to have an effect on safety, axle-wheel sets were retired from passenger service after one or two years, or 50,000 miles of running.[125] They could then be used on freight cars. These wheel sets might continue in freight service for as much as twenty years, during which time the journals might wear down from 3¾ to 2⅝ inches.[126] By the late eighties, however, such long service was not considered economic or safe; a journal showing ⅛ inch of wear was considered unfit for use.[127] Most axles could be expected to run for four or five years and deliver 150,000 miles of travel. A few even doubled that mileage.

Inferior material was the chief cause of unsound axles, yet for generations wrought iron was the best metal available. Really high-quality wrought or puddled iron made a good axle, but too often cheap grades were accepted. There was a great variation in price and quality. In the mid-1850s a car axle might cost as little as $40 or as much as $100.[128] The cheap axle contained a high proportion of cinders and was imperfectly forged. Hot-blast anthracite iron was regarded as decidedly inferior to charcoal iron. Even when good iron was used, the fabrication was not always well performed. Few furnaces could produce blooms large enough to make a car axle. It was necessary to stack several muck bars (a term for raw wrought iron after its first rolling) and reroll it to form a blank large enough for forging. The bloom was too often drawn at two heats when it should have had six. It was then hammered into its final form, ready for machining. Frequently the product was a coarse, loose-grained iron that was subject to crystallization, "cold shuts," breaks, intermixtures of raw iron, and other flaws.

Poor as muck bar axles might appear, many railroads purchased cheaper ones made from scrap. Some authorities condemned scrap axles, while others felt that they were superior to axles made from new iron.[129] Imperfect welding was the major drawback of scrap axles, for bits of steel or cast iron could ruin a weld. The long fibers necessary for a strong shaft were possible only with carefully selected scrap and the use of only a few short pieces in the pile. Much sounder scrap axles could be produced by upsetting the piles and devoting enough time to the furnace and the hammer. The process is described here as it was practiced around 1900:[130]

The scrap is made up into packages or small piles, and placed in a furnace and heated to a welding heat, and then hammered by a steam hammer into a slab. This slab is cut nearly in two, then doubled over and returned to the furnace for a second heat; it is then hammered into slabs about five inches wide by two inches thick. Three of these slabs are again heated to a welding heat and hammered into a single slab, one-half of which is finished into axle shape and the other half (on account of not being hot enough to finish) is again reheated and hammered. After being allowed to cool, the metal is made the required length by a cutting-off machine.

Buyers in search of safe axles in the first decades of railroading relied largely on the reputation of the maker. Good materials and workmanship were his trademark. Car men later turned to sample testing as a more positive method of securing the best-quality product. For every lot of axles purchased, one or two were picked out at random for a physical test. At first the only practical method of assessment was the drop hammer test. Such methods were the subject of a report in the transactions of the British-based Institution of Civil Engineers in March 1850. Several years later a series of similar tests were conducted here in Detroit, where a 150-pound, 12-foot drop hammer was set up.[131] Some axles broke after only 11 blows. The winner of the series was E. Corning & Company of New York, which produced a 4½-inch-diameter axle that withstood 193 blows. Similar tests were

conducted by railroads or axle makers, as often as not to the standards established by the individuals involved. The Master Car Builders Association considered the question of a standard testing procedure for nearly thirty years. Finally in 1899 the body agreed on an industry-wide testing formula. The drop hammer remained the single test, and any lot of axles was rejected if the sample ruptured or fractured in any way. But the lot could also be condemned if the sample registered a deflection greater than 8⅛ inches for a 4¼- by 8-inch journal-sized axle. (The permissible deflection varied, of course, for each axle size.) The full test involved five blows, a 1,640-pound hammer, a fall of 16 feet for an iron axle, and a 23-foot fall for a steel axle.[132]

In the twentieth century axle testing and design moved out of the forge shop into the laboratory. More accurate testing machines led to the production of better metals and a more scientific design. Simulated road tests conducted in 1938, for example, inspired new designs that proved 60 to 80 percent more resistant to metal fatigue than conventional patterns.[133] Magnaflux, and in more recent years ultrasonic, inspection has ensured quality control for the quantity production of railway car axles.

The efforts to produce safer car axles through better materials, design, and testing did not satisfy a radical element within and without the industry. This faction contended that the conventional design was all wrong, that the entire concept of wheels solidly mounted on the axle should be discarded. In theory this school had some convincing arguments. Conventional wheel sets are suitable only for absolutely straight railroads. On a curve the inside wheel must drag or slip, because it moves a shorter distance than the outside wheel. This causes unnecessary wear of the wheel and rails and creates an extra drag of 30 percent on the train as it passes through curves.[134] The axle is twisted because of the torsion produced by one wheel attempting to revolve faster than the other. To eliminate these disastrous conditions, various mechanics proposed that one or both wheels be loosely mounted on the axle. Railroad men rejected their recommendation because of the danger of a wobbling wheel, which would be inevitable once the bearing had worn. A more attractive alternative was to split the axle so that each wheel could turn independently of the other. Axles of this type became known as divided, joined, or compound axles.

As early as 1834 James Stimpson patented a divided railway car axle. Little more attention was given to the idea until the 1850s. Josiah Copley of Kittanning, Pennsylvania, devised a hollow divided axle that was tried on the Cleveland and Pittsburgh Railroad. In 1854 Samuel L. Denney patented a divided axle, which he advertised in the *American Railroad Journal*. But like the inventions of his predecessors, Denney's axle did not seem to stir enthusiasm among railroad men, who probably viewed it as a gadget that traded one set of problems for another. A bearing box at the axle's center was just another new trouble spot. Located under the car, surrounded by the truck, it would be a miserable thing to inspect, lubricate, and repair. In 1872 another inventor, G. W. Miltimore, took up the idea. His divided axle

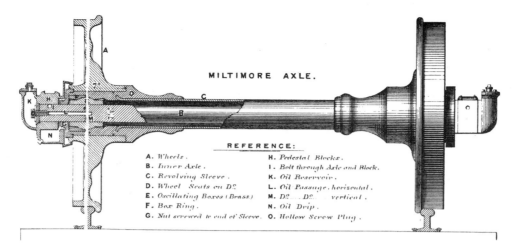

Figure 7.31 Miltimore's divided axle of about 1875 was devised to eliminate axle torque. (Institution of Civil Engineers, vol. 53, 1878)

was tried on the Central Vermont and the Burlington and was quickly pronounced a failure.[135] In four years Miltimore was back with a new design far more elaborate—and costly—than any compound axle yet devised (Figure 7.31). Yet its creator felt certain of success, and indeed it was given a good reception at the U.S. Centennial Exhibit, in company with the other novelties that are an indigenous part of world's fairs. It was even tried on the New York Elevated and the Boston and Albany Railroad. The Master Car Builders, however, continued to show distrust, and eventually the road tests indicated that Miltimore's device became noisy and friction-bound after a few months' service. Railroad history demonstrates that old ideas, no matter how foolish, have a way of being periodically reintroduced: early in 1970 an advertisement appeared in *Railway Age* for a compound axle, with a text that could well have been written by Miltimore himself.

A definitive solution to the car axle problem seemed to have arrived with the introduction of cheap steel in the late 1850s. It was a dense, homogeneous, and elastic metal with a greater tensile strength than wrought iron. With the advent of the Bessemer process, steel became available in large quantities at prices the railroads could afford. Yet they did not make a rapid conversion to the new metal. The resistance to steel axles that lasted for nearly fifty years after their introduction can be explained by more than thick-headed conservatism, for the much advertised superiority of steel over iron axles did not prove itself in actual service. Steelmakers found a friendly reception for their products among track superintendents, a slower acceptance among locomotive builders, and positive rejection from car builders. Steel rails were widely used by the 1870s.[136] Steel made only moderate progress with locomotive men, except for driving wheel tires, which became widely used by 1870.[137] All-steel boilers did not generally appear for another generation. But the persistent holdouts were the car builders, who showed no enthusiasm for steel axles until after 1900.

This is not to say that steel axles did not receive an early trial. Even in 1856 the Buffalo and Erie and the Michigan Southern railroads were testing them, but the experiment seems to have attracted few converts.[138] In 1861 it was reported that steel axles were being used in Europe but not here.[139] Apparently only the Pennsylvania continued to test them on this side of the Atlantic. The road's annual report for 1866 noted that 1,907 steel car axles had been placed in service beginning the previous July. In 1867 the mechanical department was reported to be convinced that

steel axles were superior to iron. Steel had performed well in actual service, and all passenger cars would be equipped with it. In time the road hoped to convert freight cars also.

Practical car builders elsewhere, however, emphasized the defects of steel axles. The early Bessemer steels had too much carbon, and the quality was highly variable.[140] Not enough care was taken in their manufacture, so that some were brittle and others were roughly finished or had sharp corners that invited cracks. They broke easily and unpredictably. As late as 1894 a car superintendent delivered a paper before the Northwestern Railroad Club declaring that steel axles were a menace to railway travel.[141]

Other factors contributed to car builders' preference for iron. Steel journals wore out more rapidly than iron. In one test a Bessemer axle gave only 84,000 miles of service, compared with 120,000 for iron.[142] Steel journals were said to be more "liable to heat."[143] Cost was a major consideration as well. Although according to one report, the price of steel between 1865 and 1878 fell dramatically from $140 to $55 a ton, steel car axles were still twice as expensive as iron axles.[144] Mass production of steel lowered its price, but iron held the cost advantage insofar as car axles were concerned. It remained cheaper then steel at least until the late 1880s, and perhaps even later.[145]

Steelmakers persisted, however, and in time a cheaper, more dependable axle was manufactured. Milder and more uniform steel alloys made a great difference. Improved steel had an elastic limit more than double that of iron, and while wrought iron was more ductile, it deteriorated with the constant flexing characteristic of railroad service. More time under the hammer was necessary for a sound axle.[146] Heat treatment normalized the steel-forged axles. But the improvement in their quality was not enough to break down the long-standing prejudice against steel axles. In 1904 builders were still arguing in favor of iron.[147] Ironically, it was the scarcity of cheap, good-quality wrought iron that at last persuaded car builders to accept steel. In 1908 Marshall Kirkman observed that steel axles were now the most common type, not because they were believed superior but only because they were more readily available.[148] Steel won at last, but only by default.

After the reluctant acceptance of the steel axle, there were few important developments except for the growth in axle size, which was treated in the previous section on roller bearings. One event late in the history of car axles was the invention of a hollow or tubular axle. It was considerably lighter than the conventional solid variety, offering not only less total dead weight, but more important, less unsprung dead weight. Hollow axles were tried early—around 1835—but did not receive much serious attention until the 1940s.[149] The idea was revived by H. C. Urschel, whose design was promoted and manufactured by the Pittsburgh Steel Company.[150] The 9¼-inch-diameter main body of the axle was a thin-wall tube. The walls were considerably thicker at the wheel seat and the journals. The hollow construction saved 275 pounds in a standard 5½- by 10-inch journal-sized axle, which amounted to a weight saving per car of more than ½ ton. The manufacturer claimed a 50 percent improvement in fatigue strength. Urschel's axle was no one-time experiment; it was seriously considered by the trade despite its greater cost. Thousands were manufactured for main-line service. Despite its advantages, however, the Urschel axle fell rapidly from favor when its propensity for cracking was discovered.[151] It was eventually outlawed for interchange service.

Wheels

Wheels are fundamental to land vehicles, for to move is to roll. If railway mechanical men seemed preoccupied with wheels, it was because they understood that good wheels were fundamental to the railway's success. The terrible demand imposed on car wheels makes them the most highly stressed single element connected with railway service. They support the car and its cargo; they steer the train along the track; they are subjected to terrific torsional stress when rounding curves, and they must withstand the heating stresses caused by the brake shoes.

The first railroad car wheels were patterned after common wooden-spoked wagon wheels. They had a flanged tire fastened over the wooden felloe or rim. Europeans remained loyal to the tired wheel, although they soon devised stronger centers in place of the wooden spokes. American mechanics, however, discarded the tired wheel for one-piece cast-iron wheels, which became the standard by about 1840. Around 1875 tired wheels again came into favor here for passenger cars, although cast-iron wheels were retained for freight cars until relatively recent years. After the coming of the steel car around 1910, one-piece wrought-steel units replaced the fabricated-tire style of wheels. They were all but universally adopted within a decade and remain the standard railway passenger car wheel today.

Since the beginning of the modern steam railway, wheels have been rigidly mounted on the axles. This practice naturally created difficulties in rounding curves, because the outer wheel was traveling further than the inner one. If the wheels were loose on the axles they could adjust their rate of travel without difficulty, as highway vehicles do, but the solid axle fastening forces both to travel at a uniform rate. To negotiate curves one wheel must slip, and part of the resulting torsion must be absorbed by the axle. To relieve the undesirable twisting action as much as possible, a coned or tapered wheel tread was introduced. Being sloped, the

wheel tread now had many diameters, although the difference was only about ¼ inch from one extreme to the other. Advocates of the cone claimed that it was effective in allowing the wheel set to seek a more equal rate of travel on curves, because the outer wheel rode on its largest diameter while the inner rode on a smaller diameter. The benefits of this technique may sound dubious to the novice, but tapered wheel treads have stayed in fashion to the present day. Such eminent railway men as M. N. Forney have argued against the coned tread, insisting that flat or cylindrical treads are as good or better.[152] Other authorities have challenged the coned wheel by noting that it does nothing to relieve flange wear on curves, because in actual tests the outer flange always bears hard against the rail no matter what degree of taper is used.[153] An even more serious defect of the cone tread is that it promotes "hunting" on straight tracks as the wheels wander back and forth on their unequal diameters. This characteristic does reduce flange wear, but at the expense of a steady ride. So far, however, no one has managed to overthrow the coned wheel tread.

James Wright of Baltimore is said to have invented the coned wheel in 1829. This assumption is based on a patent issued to Wright in September of that year.[154] Actually the idea can be traced to the tramways of Shropshire nearly a century earlier.[155] Horatio Allen reported on coned car wheels in England during his visit of 1828.[156] If its origins are in dispute, so is the exact degree of taper that a coned wheel should have. The B & O favored a steep taper of 1 in 14, while the nearly level New York Central employed a gentle taper of only 1 in 38.[157] In 1906 the Master Car Builders Association adopted the taper of 1 in 20, which has continued as the industry standard to the present time.

The overall size of railroad car wheels has remained remarkably constant over almost a century and a half. With few exceptions, freight cars have ridden on 33-inch-diameter wheels for the entire period. (The exceptions today are the thousands of "low-level" TTX cars for TOFC and auto-rack service, which use 28-inch wheels, and all 100-ton-capacity cars, which require 36-inch wheels. The biggest eight-wheel cars, with 125-ton capacity, use 38-inch wheels.) Many passenger cars have employed 33-inch wheels, although larger wheels were generally used. In the 1830s, 36-inch wheels became popular and, except for a brief period late in the nineteenth century when 42-inch wheels were in vogue, remain the standard of the present day.

Big wheels offer several advantages. They make fewer revolutions than a small-diameter wheel, resulting in less tire wear and cooler running brasses. They ride better and are said to pass through frogs and crossovers more easily.[158] Large wheels deflect or flatten out less on the rail under a given load and hence require less power to draw. Their larger tread and flange surface provide more braking area, and their greater mass absorbs and dissipates heat from the brake shoes more readily.[159] Of course, many arguments were made over the years against large wheels: they weigh and cost more; they tend to bend axles; and when they are made in the largest size of 42 inches, they interfere with air and steam pipes and brake rigging. But the only really con-

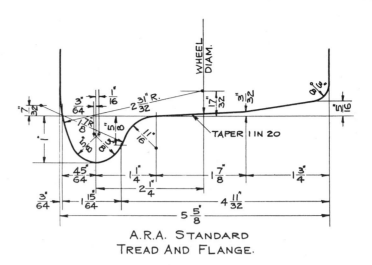

Figure 7.32 The cone or taper of a modern car wheel is shown in this drawing. The idea dates from the beginning of American railroading.

A.R.A. STANDARD
TREAD AND FLANGE.

vincing argument is the last one, against the 42-inch size. Around 1900 very large wheels gave way to the compromise size of 36 inches, which is still the standard for American passenger cars.

The earliest wheels measured 4½ inches wide overall at the tread and flange. They seem to have quickly grown another inch and to have remained at the 5½- to 5⅝-inch girth to the present. The profile of the tread and flange has also shown remarkable constancy, at least since the 1880s.[160] Except for a reduction in flange depth by ⅛ inch for better frog clearance, the dimensions of a modern tire are nearly identical with those of its ancestors (Figure 7.32). An exception to this pattern of consistency was the broad-tread wheel that enjoyed a decade or so of popularity following the close of the Civil War. Much of the Midwestern railway system was built to 4-foot, 9-inch and 4-foot, 10-inch gauges and variables in between. The gauges were just enough at variance to prohibit the interchange of cars. A Pittsburgh businessman, William Thaw, reasoned that broad-tread wheels would allow free intercourse between the Northern lines, allowing shipments east and west in one car.[161] The cars might track poorly on these compromise wheels, being too loose on one and too tight on the other, but at the very slow speeds of the time, it should work well enough. Thaw's scheme was first tried in July 1864, and it was not many years before 175,000 cars—all, or mostly all, freight—were wobbling across the country on treads 5 inches wide.[162] By the late 1870s, however, most of the 4-foot, 10-inch lines had regauged. The cars with broad-tread wheels gradually disappeared, much to the relief of the railroads that had had to use them.

The first wheels in American passenger service were either imported from Britain or made after similar designs. It was natural that the pioneer car builders borrowed from existing technology and modeled the first car wheels on those used by highway vehicles. A flanged tire was substituted for the smooth roadway tire, and solider cast-iron hubs took the place of the traditional wooden ones. One surviving first-generation wheel is today in the collection of the Smithsonian Institution. The iron parts of this 36-inch-diameter wheel were uncovered during some excavation at the old shops of the New Jersey Railroad and Transportation Company in Jersey City. The missing wooden parts were restored, and the wheel, which is believed to date from about 1834, was given to the Smithsonian in 1890. It is currently on loan to the B & O Transportation Museum in Baltimore. The cast-iron hub has six U-shaped arms, which are bolted to every other spoke (Figure 7.33). The axle opening is bored out to a diameter of 3 inches; note the keyway. The flange is worn thin and sharp. The tire measures 4½ inches wide. It is probable that a secondary plain tire was fitted to the wheel to facilitate remounting of the flanged tire, but it is not present on the restored wheel.

Wooden spoked wheels worked satisfactorily with the light, horsedrawn passenger cars operated by many pioneer railroads, but once steam operations began, higher speeds and heavier cars rendered them hopelessly inadequate. Finding a dependable substitute was not easy. A great variety of designs and materials

were tried. Isaac Dripps, onetime master mechanic of the Camden and Amboy, described how this was accomplished on his line:[163]

In 1834 commenced to make a wrought iron wheel for the driving wheels for locomotives and wheels for passenger cars. These wheels had round wrought iron spokes riveted to flanges on a cast iron hub, with outer end of spokes riveted to two iron rings forming the periphery of wheel center; these rings are kept separate the proper distance by blocks and studs riveted between them. On these rings are shrunk a wide wrought iron band, on which is shrunk and bolted to it the wrought iron tire as shown by accompanying drawing marked A.

These iron wheels proved in practice but very little better than the wooden wheels as the rivets would jar loose, and found it difficult to keep them tight. So that in 1835 the cast iron center with wrought iron tire was tried and finally adapted for wheels for locomotives. The wooden and wrought iron wheels were still continued in use under passenger cars. About the year 1838 we could not procure any wrought iron tire, and were compelled to try the experiment of using chilled cast iron tire for both locomotives and car wheels. At this time we also used chilled wheels for freight locomotives, but found a great difficulty in getting them properly chilled.

The first chilled tire we tried had counter sunk holes cast in them for receiving the bolts to fasten tire to the body of the wheel, but with the very light chilled tire we then used they broke through the bolt holes. To remedy this and make the tire secure from further damage, should it break, I cast a flange or rim projecting inwards 1½ inch below the metal of tire on outside of tire.

This flange also added material strength to the tire, and was also used to bolt it to the body of wheels, the bolts passing through this flange and wheel with the nuts screwing up on inside of wheels, so that if the chilled tire did break no portion could leave the wheel. I consider this plan the best way of securing any kind of tire to the wheel center. In time we found both the wooden and wrought wheels very troublesome under passenger cars, and next tried a new wrought iron wheel with flat wrought iron spokes, radiating from a cast iron hub with flanges cast on to receive these spokes. With this wheel was used a tubular axle made of ³⁄₁₆ iron, 10 inch diam., secured to the hubs by rivets. On the outside of these arms or spokes was shrunk on and bolted the wrought iron tire. The arms or spokes were also rivetted to the cast flanges on hub. This wheel is shown by accompanying drawing.

This wheel stood well, proved a strong and efficient wheel, but proved too expensive and was soon abandoned.

The next experiment in wheels for passenger coaches was a solid wooden center with eight sectional pieces fitted into a cast iron hub, extending about half the diameter of solid center, with screw bolts passing through a wrought iron ring in front through the wooden pieces and securely bolted to this flange. On the outside of wooden center is a wrought iron band which held the wooden center firmly together so that the tire could not be removed at any time without injuring the center pieces. On some of these wheels were shrunk an

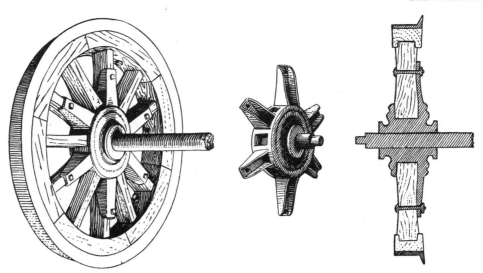

Figure 7.33   A wooden, cast-iron, and wrought-iron 36-inch-diameter wheel used on the New Jersey Railroad and Transportation Company around 1834.

ordinary wrought iron tire, but on others were used cast iron chilled tires, having on the outside of tire a flange projecting inward toward center of wheel with bolts passing through this flanged wooden center, and a wrought iron on back of wooden pieces, thereby fastening the chilled tire very securely to the wooden center.

By this arrangement of fastening there could be no risk or damage should the chilled tire break, as no part of it could possibly get out of place, being held there by 8 ⅝ bolts. After a fair trial this plan of wheel was adapted as the standard wheel for passenger coaches. These wheels were used on an ordinary wrought shaft, the tubular shaft being found too expensive. The back part of hub of these wheels was cast a sleeve extending over shaft 4 inch., so that in case of an axle breaking this sleeve preventing the parts from separating; and in practice has prevented many serious accidents to trains. The ordinary chilled cast iron wheel has been used on the road for freight cars from the commencement of the road, and in practice were found to stand well and give satisfaction, and were finally adapted and used under passenger coaches.

Philadelphia, September 6, 1885   Isaac Dripps

Other railroads sought other solutions. The Petersburg Railroad, after a poor experience with wooden-center and cast-iron wheels, turned to British makers for wrought-iron spoked wheels on the pattern used by the Liverpool and Manchester Railway.[164] They were more expensive than those available locally, but their excellent safety record (in 1835 not one broke during the entire year) made them a wise investment. In the following year the Rensselaer and Saratoga Railroad obtained some creditable domestic wheels which were described in a newspaper account of the day: "The wheels were cast in Troy by Mr. Starbuck, with patent rolled-iron rimmed tire, well annealed and wrought, being put on the cast wheel [center] while red hot, the cooling of the tire and the contraction of the iron, renders it impossible to be ever deviated from its place, the whole wheel is then turned in a steam lathe by machine tools, rendering the circle of the wheel, perfect from its centre, which is a great desideratum."[165] In February 1837 M. W. Baldwin corresponded with one of his customers on the subject of railroad cars.[166] He mentioned that iron-tired wheels with wooden centers—he did not say whether they were spoked or solid—were available in place of the cheaper cast-iron wheels. They would be guaranteed for two years' service. Though not directly stated, the implication was that the tired wheels were superior to

cast-iron wheels. But whatever Baldwin's true opinion, the tired form of wheel was about to disappear from American practice.

One-piece cast-iron wheels were not new, nor were they first made in the United States. British tramways had been using them since the mid-eighteenth century, but when the steam railway was developed, British mechanics felt that cast wheels were not only antique but inadequate for high-speed service. Americans did not harbor the same prejudice, and they worked to perfect a better cast wheel. Cast-iron wheels were admittedly primitive and in some ways inferior to fabricated wrought-iron wheels, but they were very cheap. Here again, capital-poor American lines were forced to devise less expensive supplies if they were to survive and expand. It was also true that American roads were not attempting any fast running in the early years, and a poor grade of wheel might serve their needs adequately.

It was found that cast wheels had several surprising advantages in addition to their low first cost. The very hard chilled tread and flange provided a long-wearing running surface, and one that could withstand the grinding friction of brake shoes. The center or plate of the wheel was not chilled and remained tough, fibrous, and able to absorb normal road shocks without cracking. The hub stayed soft for easy machining. A cast wheel weighed only five-eighths as much as an equivalent tired wheel. A large part of the original cost could be recovered through the high scrap value, for worn-out wheels could easily be broken up with a hammer and remelted. Recovering the metal from a wrought-iron wheel was difficult before the acetylene torch was perfected.

Cast wheels had some disadvantages, of course. They were more likely to fracture than wrought wheels because of the essentially brittle nature of cast iron. Hidden flaws are possible in any casting. Furthermore, the entire wheel was ruined once the chilled tread wore through; it was a "one wear" wheel. And all cast wheels are slightly out of round. Before large grinding lathes were commonly available, there was no practical way to machine the hard running surface of the casting. These defects were serious but not catastrophic, and American railway men came to champion the chilled cast-iron wheel even while they admitted its failings.

What made the cast wheel so attractive was the incredibly durable running surface offered by the chilled tread. Anyone who has worked with a raw iron casting is familiar with its tough outer skin, which always makes the first cuts difficult when the

36 inch 475 lb. Wheel

Used on South Carolina & other Southern Roads

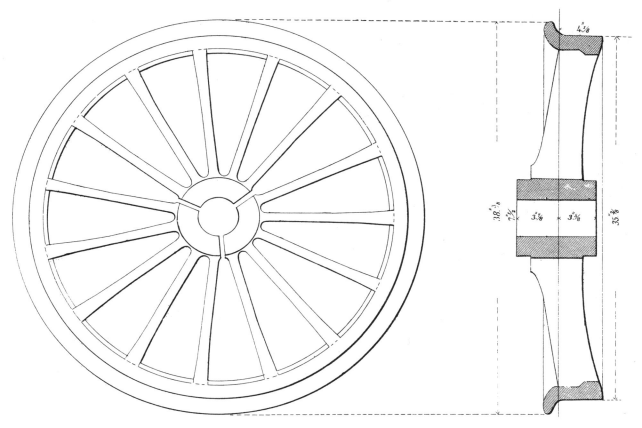

Figure 7.34 Cast-iron car wheel produced by Ross Winans with a chilled tread and split hub. The design dates from about 1835. (American Railroad Journal, June 26, 1847)

piece is turned or filed. This outer surface is caused by the molten metal's quick cooling upon touching the cold surface of the mold. The phenomenon was well known to all foundrymen. The wheel maker attempted to capitalize on the normally undesirable chilled surface by intensifying it for the purpose of strengthening the tread and flange of the wheel. By making the rim of the mold from cast iron rather than sand, an extremely deep chill could be obtained: anywhere from $\frac{1}{2}$- to 1-inch deep, depending on the quality of the iron and the skill of the foundry. The metal would cool and crystallize almost instantly into a silvery white metal almost as hard as diamonds. Chilled iron was for generations the hardest metal commercially available. It could not even be cut with high-speed steel; only the grinding wheel could penetrate its tough hide. Simply stated, the chill is formed by the fast cooling, which occurs so rapidly that the carbon does not have time to separate but remains combined with the iron. This produces a steel-like metal with a very high carbon content of $3\frac{1}{2}$ percent. In the nonchilled portion of the wheel the carbon separates, taking the form of graphite, which results in ordinary soft grey iron.

The origins of the chilled wheel are unknown. The 1888 *Car Builders' Dictionary* said simply, "The name of the inventor is not preserved." Cast-iron wheels were in general use on British mine tramways by 1780, and by 1812 mention is found of "case hardened" cast-iron wheels.[167] Losh and Stephenson obtained a British patent in 1816 that mentions chilled treads. The idea appeared in America a decade later. Richard P. Morgan mentions this style of wheel in his August 29, 1829, patent covering six-wheel cars. During the following year Horatio Allen reported

on chilled wheels in a letter to John B. Jervis.[168] In later years the idea was claimed on behalf of various American inventors. Henry Mooers of Ithaca, New York, maintained that he cast wheels of this type in 1827 for the Ithaca and Owego Railroad.[169] One might doubt his statement, however, because the line did not open until 1834. And of course, although no facts seem to support it, the honor has been ascribed to Ross Winans, the man who invented everything, according to Zerah Colburn.[170]

On July 29, 1834, Phineas Davis received a patent for casting car wheels. Davis hoped to improve the chill and strengthen the wheel by inserting $\frac{1}{2}$-inch-diameter reinforcing rods inside the wheel castings. The rods were located in the tread. After Davis's death in 1835, Ross Winans took up the manufacture of this form of wheel and continued to produce it until at least 1847. An illustrated advertisement appeared in the *American Railroad Journal* during the same year (Figure 7.34)[171]

Jonathan Bonney claimed to have made chilled wheels for the B & O at the Mt. Savage Iron Works in western Maryland in 1829.[172] A few years later he was in the same line of work in Wilmington, Delaware, with a firm that became a leading producer of chilled wheels in this country, Bush and Lobdell. By the mid-1830s many American foundries were producing chilled car wheels. Among them were locomotive builders such as M. W. Baldwin, Thomas Rogers, and Henry R. Dunham. Dunham was not a great success in the manufacture of locomotives, but he did do a lively trade in the car wheel business and held at least one patent.[173]

Car wheel makers were encouraged to expand the production

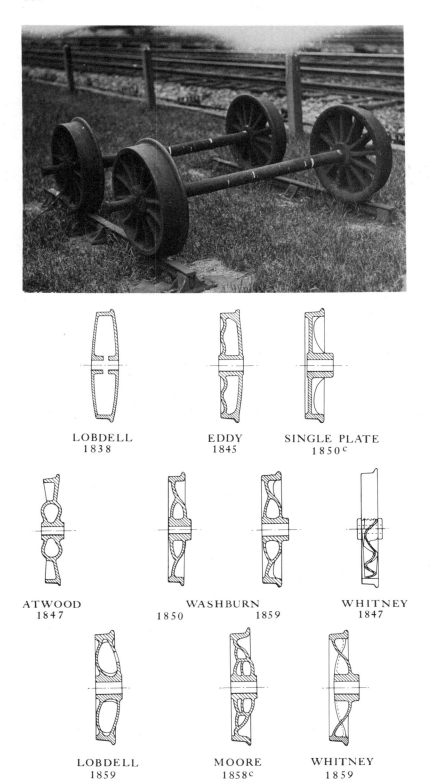

LOBDELL
1838

EDDY
1845

SINGLE PLATE
1850ᶜ

ATWOOD
1847

WASHBURN
1850   1859

WHITNEY
1847

LOBDELL
1859

MOORE
1858ᶜ

WHITNEY
1859

Figure 7.36   A selection of solid-plate cast-iron car wheels. (Traced by John H. White, Jr.)

of the cast-iron pattern by a growing enthusiasm for the product. In its annual report for 1834 the Baltimore and Ohio spoke of the excellent service provided by chilled wheels: They "have stood the test of experience and have come up to our most sanguine expectations."[174] Another indication of their growing popularity was the increasing number of patents: by 1840 a dozen had been granted for cast-iron wheels. They appear to have become the standard form in America by this time.

Spoked wheels were at first favored (Figure 7.35). Some roads, like the Reading, continued using them into the 1850s, but

elsewhere they appear to have been replaced by solid disc wheels during the previous decade.[175] The chill was deficient opposite the ends of the spokes, and these soft spots wore flat, rendering the wheel useless long before the rest of the tread was worn out.[176] It was also found that the rim broke away between the spokes. Moreover, the flat spokes acted as a fan to stir up the dust along the tracks. Because of the uneven contraction of the casting, the hub could not be cast solid; it had to be cast with three or more splits or gaps.[177] Iron bands were shrunk on to bind the hub together, and the gaps were filled with molten zinc. The problem of the broken hub was not restricted to spoked wheels; it was a difficulty experienced with all early forms of cast-iron wheels.

The disc or plate wheel overcame some of the major difficulties of the spoked wheel. It was stronger, and because of the even distribution of material around the rim area, a deeper, more uniform chill was possible. Commercial manufacture of disc wheels began in the late 1830s. Baldwin is said to have produced some as early as 1836, but no patent was taken out; the evidence comes from a patent case some forty years afterwards.[178] On March 17, 1838, Jonathan Bonney, Charles Bush, and George Lobdell patented a double-plate wheel and began production for the trade. The original design was not satisfactory, but it was probably the first widely used plate wheel in the country (Figure 7.36). The design was subsequently refined and remained popular for many years. It continued to be produced late in the nineteenth century but was used only where extra strength was needed, such as for locomotive leading wheels. The original pattern required split hubs because of the uneven contraction between the fast-cooling chilled rim and the slower-cooling hub and plate. Some of this differential was evened out by adding a wave or S curve to the plate. But the Lobdell wheel, as it came to be called, required a large center core, making it expensive to manufacture. Other designs were soon introduced. In 1847 Asa Whitney secured patents on several car wheel patterns after his so-called corrugated design, which carried the wave concept to its ultimate limit. The plate was dished in a series of sharp radius curves, almost like a drapery of cloth, with the idea of relieving the uneven contraction of the castings. In time Whitney was required to simplify his design (Figure 7.36). Other inventors proposed schemes for single-, double-, and even triple-plate wheels that exhibited both graceful and contorted profiles in every imaginable combination of convex and concave curves. By 1860 there were nearly 100 patents covering cast-iron wheel shapes alone.[179]

The single-plate style was simple and cheap to make, because it required no core. But it was not regarded as a first class wheel and seems to have been little used for passenger car service. It was adopted to some extent for freight cars and streetcars before 1880. Double-plate wheels were considered the best form, combining as they did strength and light weight. Triple-plate wheels were a needless complication. Of the various double-plate wheels, none could approach the Nathan Washburn design for popularity (Figures 7.36 and 7.37). Washburn was a practical

*Figure 7.37   Washburn's double-plate car wheel, patented in 1850, was the most popu-
lar cast-iron wheel in America. This scene was photographed in Oakland, California, in
1869. (Society of California Pioneers)*

foundryman whose major business undertakings were in Worcester, Massachusetts. His design, patented on October 8, 1850 (No. 7710), was pleasing in appearance and exceptionally strong. The S-shaped brackets on the rear of the rim strengthened the flange, making it a very serviceable wheel. Wheel makers liked it because of its simple shape and the fact that it was relatively easy to cast. It quickly became the standard style of cast-iron wheel. In 1879 it was declared the strongest and most reliable made; 70 percent of all car wheels in service were said to be of this pattern.[180] Naturally other patentees tried to cash in on Washburn's success. Claiming that he had patented the same wheel in 1848, Anson Atwood wanted a license fee of 25 cents for every Washburn-style wheel made—a royalty that would have amounted to a fortune. But Atwood's bulbous and ill-conceived design bore little resemblance to Washburn's elegant profile (Figure 7.36). Even after the cast-iron wheel disappeared from passenger car service after 1910, it remained the standard form of freight car wheel until 1928, when it was at last displaced by a new design.

Several years before the Washburn wheel was introduced, foundrymen discovered a technique which permitted the casting of solid hubs. The weak and expensive bands and zinc filler

blocks were at last eliminated. Controlled cooling was the solution to the dilemma of uneven contraction. Some foundries simply buried the hot wheels in sand, ashes, or fine charcoal, allowing them to cool slowly.[181] Asa Whitney devised a heated, airtight soaking pit, where the hot castings were kept for three to four days. The temperature was gradually lowered during this period. The process, which did not adversely effect the chill, was described late in 1847.[182] A patent was granted to Whitney in the spring of the following year.

Wheel makers also sought to improve the safety and durability of their product by employing the best metal available. Most early ironmasters had no real idea of the physical characteristics of the metals used; instead they depended on the reputation of the pig iron they bought.[183] Charcoal iron was the only acceptable material for wheels, but again the early wheel men did not really understand why it produced a superior wheel. They selected the right material for the wrong reason, concluding that pig iron free from sulphur and phosphorous made the best wheels, when in fact these very elements were effective in producing a good chill. The undesirable ingredient in a mix was silicon, too much of which had a negative effect on the chill.[184] But even without a scientific understanding of metallurgy, the early-nineteenth-century founders produced good wheels if for no other reason than that excellent grades of cast iron were available to them. Pig iron made by the charcoal process had a tensile strength of 23,000 pounds, compared with ordinary pig iron's tensile strength of 18,000 pounds.[185] Some furnaces produced extra-fine charcoal irons with tensile strengths up to 41,000 pounds.[186] It would be difficult to make a weak casting from such strong iron.

Wheel makers naturally sought the most celebrated brands of iron available and emphasized the trouble they took to secure the very best for their customers. According to Paul R. Hodge, an early locomotive designer, wheel makers of the 1830s and 1840s found that a mix of Baltimore and New Jersey ores gave the best results.[187] Bush and Lobdell of Wilmington, Delaware, imported pig iron from North Carolina.[188] The Ramapo Wheel Company of New York boasted of the quality iron dug from its own mines in Berkshire County, Massachusetts, and Litchfield, Connecticut. By 1880 some makers were mixing a variety of irons for the best alloy. Iron ore from Maine, Michigan, Maryland, and Pennsylvania was stacked in separate piles around the foundry yard.[189] Samples were tested to determine strength and chilling properties. The ironmaster decided on the right balance, taking into account as best he could the adverse effects of the scrap wheels introduced into the furnace. Daily samples were pulled from the furnace and tested to ensure a quality iron. According to the *Railroad Gazette*, much depended on the judgment of the men involved. No textbook formula was enough; it required a knowledgeable foreman whose "care, mature experience and vigilant inspection" could alone produce a first-class product.[190]

Wheel makers were forced to learn more about their art as the price of charcoal iron increased after the Civil War. It was more expensive than coke iron, for even in the early 1870s it was selling for $62 a ton, which was $20 more than coke iron.[191] William G. Hamilton, once superintendent of the Jersey City locomotive and car wheel plant and later employed by the Ramapo Wheel Company, found that poorer grades of cast iron could be greatly improved by mixing in 5 percent of scrap steel.[192] Worn-out steel rail could be purchased cheaply for the purpose. The iron was strengthened by 17 percent. Hamilton patented his "steeled" wheel alloy process in 1868 and 1873. It was widely used, although some makers, such as Whitney, found that the same results could be obtained by adding wrought-iron scrap. Some years after Hamilton's process was announced, other engineers suggested the addition of a trace of manganese.[193] In 1911 Robert C. Totten claimed to have solved the car wheel metal problem by adding chrome and nickel to the alloy.[194] These metals counteracted the bad effects of combined carbon that was present in the chilled portion of the scrap from old car wheels. Totten's wheel was said to be 300 percent stronger than the ordinary cast-iron wheel.

As mentioned earlier, the tread and flange were cast against a smooth iron surface and therefore did not require machining. This portion of the wheel came from the mold like a die casting (Figure 7.38). Unfortunately, a large iron casting stays in a semimolten state for so long that it does not remain absolutely true. Few cast-iron wheels were perfect circles. In 1883 it was reported that car wheels were anywhere from $\frac{1}{16}$ to $\frac{5}{16}$ inch out of round.[195] By 1939, however, techniques and standards had advanced so much that a wheel over $\frac{1}{32}$ inch out of true was not acceptable, and one-third of all wheels were only $\frac{1}{64}$ inch out.[196] Faster pouring (10 to 14 seconds) may have helped increase accuracy.

The safety of cast-iron wheels was another major concern of both the maker and the purchaser. European engineers could not understand the American loyalty to the cast-iron wheel. They found it incredible that this brittle material could provide safe service under hundreds of thousands of cars—cars far heavier than any running outside North America. The Institution of Civil Engineers in England expressed wonder at the extensive use of cast-iron wheels in America. One member stated that they were "but poor expedients" when compared with the wheels used in England.[197]

By the late 1890s cast-iron wheels were outlawed in all European countries except Austria.[198] As already explained, the reason for the success of the cast wheel in America was the availability of high-strength cast irons. And they had a remarkably good safety record in this country: in 1883 there were 5,600,000 cast wheels in service on U.S. lines, and only 28,000 were condemned each year.[199] Many of these were discarded because of cracks and thus cannot be classified as total failures. Wheel failures were controlled in the winter by reducing train speeds. The Boston and Providence said that attempts to maintain normal schedules in cold weather were both reckless and expensive.[200] The line's rolling stock had as many breakdowns in the three winter months as it did in the rest of the year, filling the shops with crippled locomotives and cars. A 25 percent reduction in train speeds proved of some benefit, however.

The introduction of power brakes posed a new threat to wheel safety. The friction of the brake shoes against the tread and flange could raise the temperature of this portion of the wheel to 1350 degrees F.[201] The combination of such high temperatures and rapid cooling by snow and frigid winds crystallized the metal and shelled out the wheel treads.[202] Winter continues to play havoc with car wheels even today because of this problem. It is one reason why some roads have adopted disc brakes.

The most intelligent method of improving wheel safety was to determine the soundness of the wheel before it entered service. Hidden flaws, blowholes, and hairline cracks were difficult to detect visually, but it was possible to uncover a flaw by striking the casting with a hammer. A defective piece would not ring true, and in the case of an extreme fissure it might literally fly to pieces. The sledgehammer was the metal crafter's time-honored testing instrument. At Whitney's plant each wheel was subjected to several heavy blows with a hammer before leaving the shop, and 10 percent were found defective by this simple test.[203] As early as the spring of 1843 drop-hammer tests were conducted in Boston.[204] Several makers submitted their products for examination. A 131-pound hammer with a fall of 10 feet was employed.

American Car and Foundry Co.
St.Louis District,
1918

Figure 7.38   A mold for a cast-iron Washburn wheel. The chilled rim and sand mold for the front tread and flange is suspended by a gimbal on the right side. The bottom or rear mold stands on the floor in front of the men.

Winans's spoked wheel cracked at the rim after one blow, and the entire wheel shattered after eight falls of the hammer. The broken casting revealed an array of blowholes. Several local Boston foundries made only a slightly better showing than their Baltimore rival. Bush and Lobdell, however, put a 496-pound, double-plate wheel on the anvil that survived the hammer admirably. After nine blows the wheel was unaffected; it began to break up only with the tenth.

Late in the nineteenth century, wheel testing became more formal and ingenious. Two wheels from each lot of one hundred were set aside for testing.[205] The sample was expected to withstand fifteen blows from a 140-pound drop hammer. In addition, every wheel in the lot received a heavy blow from a 6-pound hammer. By 1929 a 250-pound ball replaced the old-fashioned hammer.[206] The fall varied from 9 to 13 feet, depending on the size of the wheel being tested. Each piece was expected to withstand a dozen blows.

The second trial was a thermal test to establish the wheel's ability to withstand heating caused by the brake shoes. A ring of molten iron was ladled around the wheel's rim, and the time required for this severe shock to crack the casting was measured. The thermal test was introduced by the Griffin Wheel Company in the early 1890s and was soon adopted by the Master Car Builders Association.[207] At first a wheel that could survive the ordeal for two minutes was considered adequate for normal service, but by the late 1920s the length of the test had been increased to five minutes.[208]

Another check on wheel safety was the record kept on every car wheel in interchange service. The date and the maker or place of manufacture were cast in raised letters on each wheel. If the wheel failed, the nature of the defect was noted and other wheels in the same lot were investigated. In some cases the faulty wheel or its fragments were returned to the foundry for a more careful analysis. The Pennsylvania Railroad began keeping such accounts sometime before 1876.[209] The records specified the part of the cupola charge from which the iron was taken, the depth of the chill, the weight, the size, the wheel's length of service, and its behavior in service. Special entries were made in the event of a failure. The records grew into a prodigious file. In 1883, for example, there were 5.6 million wheels in service, and some twenty-five years later there were 18 million on the books.[210] The clerical time was considered well spent, however. In addition to promoting safety, the records made it possible to enforce the wheel maker's guarantee. For a time the guarantee was on a mileage basis of 40,000 to 60,000 miles, but by the turn of the century the guarantee was limited to a time span of four years.[211]

The yeoman work of wheel inspection was done in the field by a grimy car knocker, who quickly bobbed under the train at every major station stop. His only apparatus was a hand lantern and a long-handled hammer. The uninitiated passenger sat and wondered at the dim light and clanging hammer as the inspector went over every wheel looking for a telltale crack or a muted plate that did not ring like solid metal.

Although the cast-iron wheel was the dominant form of American car wheel during most of the nineteenth century, there was always an undercurrent of interest in tired wheels. The fabricated wheel had several advantages. The entire wheel need not be scrapped just because the tread or flange was worn out; a new tire made it like new. It was salvageable, whereas an old cast-iron wheel could only be broken up for scrap. Supporters' claims that tired wheels were safer and more economical encouraged American railway men to give the tired wheel yet another chance. The engineering trade press was undoubtedly also responsible for periodic revivals of tired wheels. British journals were naturally filled with notes, drawings, and patent descriptions of the latest ideas in this area, and some of the material was reprinted by American magazines. The *Journal of the Franklin Institute*, for example, regularly featured such items even during the 1840s.[212]

Records of actual road tests are also available for tired wheels during the nineteenth century. The Hudson River Railroad made an attempt to use them in 1849.[213] The line was atypical in that it began with an avowed purpose of high-speed passenger service—a necessity if it were to compete with the swift Hudson River steamers. Most American lines of the time were content with operating speeds of 25 mph or less. The new road announced its intention of matching the best English trains with speeds of 50 to 60 mph. In emulation of these trains the line imported a large number of British car wheel sets at the princely sum of $500 each. Even on the relatively straight Hudson River track, however, the flanges were rapidly cut away by the curves. Many years after the road had discarded the British wheels, their remains could be seen at the Thirty-first Street Station in New York City.[214]

Several lines running out of Boston gave the British tired wheels a trial in about 1850.[215] The Boston and Providence devised its own form of cushioned, tired wheel, which had a cast-iron center with 2-inch-thick wooden wedges between it and the wrought-iron tire. It worked well and was put on all the road's passenger cars. In 1857 the B & P's master mechanic, George S. Griggs, patented the same style of wheel for locomotives. In 1853 two lines in New York were also trying wrought-iron tired car wheels.[216] The Utica and Schenectady was disappointed by the experiment, for the wheels cost three times more than the cast variety and their flanges quickly wore thin, causing derailments. Despite the bad experience of the U & S, the Northern Railroad was willing to give the scheme a trial because so many of its cast wheels broke during the winter season.

In England the Mansell wooden-center wheel, introduced in 1848, was receiving great acclaim. The solid wooden center was built up of wedge-shaped pieces of teak, which produced a resilient and quiet-running cushioned wheel. The Camden and Amboy was reported to have used a similar design, with red cedar instead of teak, for some years before (and presumably some years after) 1861.[217] By the late 1860s several American lines were said to be trying Mansell wheels. The Michigan Southern adopted them for the sleeping cars of its Excelsior Line.[218] The Boston and Lowell and the Hudson River railroads reported the experimental use of Mansell's design. Steel tires seemed to solve the problem of rapid flange wear, but the Boston and Lowell abandoned the test because the tires worked loose.[219] Pullman and the Gilbert Car Works also experimented with Mansell wheels during the same period, but no-large scale purchases were made by American railroads despite the exceptional performance of this design in England.

Success came to the tired wheel from an unlikely source and in a most unlikely form. The inventor was a former locomotive engineer, Richard N. Allen (1827–1890), who had drifted from one position and occupation to another.[220] He was persuaded by his brother-in-law to buy into a paper mill at Pittsford, Vermont. The plant produced a common grade of cheap strawboard for which the market was glutted, and Allen soon found himself the sole owner of a profitless business. Rather than close down, he set out to find other uses for strawboard besides its customary role as covers for inexpensive textbooks. Somehow from his railroading experience he thought of adopting the weak paperboard for use in car wheels. How this paradoxical idea came to him is not recorded, but it was greeted with ridicule.[221] A railroad car wheel made of paper? Even after Allen's wheel was in common usage, the term "paper wheel" aroused puzzlement and mirth in most people. Actually the basic idea was nothing more than the substitution of compressed paper for wood at the wheel's center. The paper center was attached to the tire by front and back metal plates securely bolted to the paper disc by twenty-four or more bolts (Figure 7.39). The paper center, even though greatly compressed and nearly as hard as ivory, was spongy enough to cushion the ride and deaden the sound of the wheels grinding over the rails.

Allen began work on his idea in 1869, and after some difficulty he persuaded a local railroad to provide a car for testing.[222] Skeptical of the bizzare experiment, the line was not about to trust a valuable piece of rolling stock to a crank. It provided

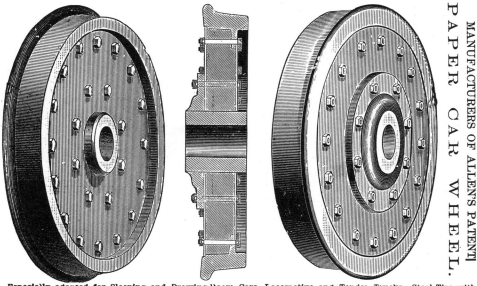

ALLEN PAPER CAR WHEEL COMPANY,
General Offices: 239 Broadway, N. Y.

MANUFACTURERS OF ALLEN'S PATENT PAPER CAR WHEEL.

Especially adapted for Sleeping and Drawing-Room Cars, Locomotive and Tender Trucks. Steel Tire with Annular Web—Strongest, Most Durable, and Most Economical Wheel in use. 74 manufactured in 1877; 60,000 manufactured to date; present facilities, 36,000 per year. Works at Hudson, N. Y., at Pullman, and at Morris, Ill.
W. H. FENNER, JR., President.   J. C. BEACH, Treasurer.   C. H. ANTES, Secretary.

Figure 7.39 The Allen paper wheel, introduced in 1869, had a center core built up from strawboard. Steel or iron plates were bolted on the front and rear. It was widely used in this country between 1880 and 1915.

Allen with an old freight car that carried wood between fueling depots. To everyone's astonishment, the car ran 5,000 miles trouble-free. In the spring of 1870 Pullman purchased a set of Allen wheels for one of his sleeping cars.[223] The wheels performed well, and Pullman bought more and eventually adopted them as a standard. Now that he had the patronage of the rising Pullman empire, Allen was no longer viewed as a foolish tinker. The paper wheel became an everyday fact of luxury railroad travel, making its inventor famous. Allen himself, however, appears to have been squeezed out of the firm by the early 1880s.

The growing interest in tired wheels in this country sprang from several sources. In 1910 George L Fowler, onetime associate editor of the *Railroad Gazette*, recalled that in the early 1870s railroad mechanical men developed "a feeling of insecurity" about the continued use of cast-iron wheels because of the increases in train speed and passenger car weight. These men accepted steel tired wheels from a genuine belief in their greater safety. Others, however, adopted them for more cynical reasons. According to Fowler, some men championed various patented forms of steel tired wheels merely because of their advertising value. This aspect of technical history is one that deserves a study of its own. Pullman himself was perhaps as much interested in the paper wheel's promotional value as he was in its mechanical merits. He knew the value of publicity and pursued it eagerly, particularly in the early years of the sleeping-car business. The paper wheel captured attention, and the early notices of Pullman cars always seemed to mention the wonderful paper wheels which silently and securely transported the car and its occupants across the country.

In one of his master strokes of publicity, Pullman became associated with Frank Leslie's 1877 tour of the United States and may have been one of its backers. The publisher returned the favor with generous coverage of Pullman in *Frank Leslie's Illustrated Newspaper*. One engraving shows Pullman, using his walking stick as an instructor's pointer, explaining the wonders of the paper wheel to an attentive bystander.[224] The accompanying article said:

While our party was viewing the exterior of the vehicle, Mr. George Pullman himself strolled up. Pointing to the wheels, he made the somewhat alarming announcement that they were made of paper! In proportion to its weight, he said, good paper, properly prepared, is one of the strongest substances in the world. It offers equal resistance to fracture in all directions. While the toughest woods are sometimes liable to crack and split under severe trial, and ordinary iron becomes brittle from the constant jarring on the smoothest of steel rails, paper possesses a certain amount of elasticity very desirable in a car wheel. Paper wheels, he said, were subjected to an enormous hydraulic pressure and, when surrounded with a flange of steel, were the most perfect wheels yet invented.

Extravagant endorsements of the Allen wheel were also made by A. B. Pullman, doubtless because of the interest that his brother G. M. Pullman held in the firm.[225] After 1881 Allen's main plant was located on the grounds of Pullman's mammoth Chicago works.

The manufacture of these pasteboard wonders began with circular paper sheets that were glued together with ordinary flour paste.[226] In 1882, 117 sheets were used, but by 1893, 200 were required. The discs were compressed by a 650-ton press for three hours. They were then dried and cured in a warm room for six to eight weeks to ensure the evaporation of all moisture. The seasoned discs were turned in a lathe to size. Bolt holes were drilled, the outer ¼-inch-thick iron plates were put in place, and the steel tire was bolted on. The finished paper core for a 42-inch-diameter wheel weighed 185 pounds. A complete wheel of this size weighed 1,115 pounds.

In its infancy the paper wheel was an insignificant thing, associated almost exclusively with Pullman sleepers. In 1877 Allen produced only 74 wheels.[227] Within a year his company claimed that nearly 1,500 were in service, but this represented only a tiny fraction of passenger car wheels.[228] In 1883 the firm said that

ATWOOD HEMP CAR WHEEL CO.,

MANUFACTURERS OF

ATWOOD PATENT HEMP-PACKED STEEL-TIRED WHEELS FOR CARS AND ENGINES.

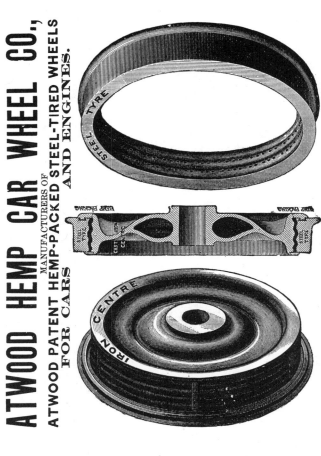

No Bolts.   **Fewest Possible Parts.**   **No Rivets.**

Tires put on cold and held by interlocking of Hemp Packing, which acts as a cushion, cuts off Metallic Connection, and prevents Rigidity

Thoroughly Tested, Safest, Cheapest and Most Noiseless of all Steel-Tired Wheels.

JOS. W. DREXEL, President.   **ROBERT F. SHEPARD,** Sec'y and Treas.

OFFICE: 59 LIBERTY STREET, NEW YORK.

C. W. LEAVITT & CO. SELLING AGENTS, 161 BROADWAY.

PAIGE'S PATENT WROUGHT METAL WHEELS.

Office, 211 Superior St, Cleveland, O.

J. E. FRENCH, President.   **W. S. DODGE,** Secretary and Treasurer.

Adapted for Sleeping and Drawing Room Cars, Locomotive and Tender Trucks. Steel Tires with ½ inch Plates, securely bolted, making it a perfectly Safe, Durable and Noiseless Wheel.

SNOW'S PATENT STEEL-TIRED METAL WHEELS,

WITH INTERCHANGEABLE HUB,

Annular Web to the Tire, Wrought-Iron Plate and Retaining Ring

Manufacture especially for all Classes of Equipment where Safety and Economy are the principal considerations.

—AND—

Durability of Construction, Noiseless in Service

Address: **W. W. SNOW, - - - New York**

RAMAPO, - - - New York.

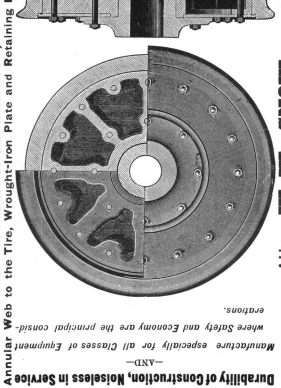

THE ELLIS STEEL-TYRED WHEELS.

MANSELL RETAINING RINGS.

CAST-IRON CENTERS (Double-Plate, Single-Plate or Spoke).

WROUGHT-IRON CENTERS.

W. R. ELLIS, 18 BROADWAY, NEW YORK.

Figure 7.40   A great variety of tired wheels were produced in the United States following the success of the Allen wheel. This collection of advertisements is a sample of the types offered.

30,000 were running on 150 railroads.[229] Since there were roughly 200,000 wheels in passenger car service at this date, Allen's wheels were on less than 15 percent of the fleet. Statistics are not available for the other makes of tired wheels available at the time, but it is clear that cast-iron wheels were no longer the exclusive bearers of the traveling public. During the 1880s more major lines began to adopt paper or some other form of tired wheels. The Santa Fe announced in 1880 that it would equip all its passenger cars and locomotive leading trucks with paper wheels.[230] The Milwaukee Road made a similar commitment in 1882, and the Northern Pacific followed suit the next year.[231] In 1886 Allen's firm announced that 60,000 paper wheels were in service.[232] In seven years the number increased to 115,000.[233]

Allen's success encouraged a band of imitators. Some produced cushioned wheels; others simply offered a demountable steel tired wheel. By the mid-1880s the advertising pages of the Railroad Gazette and other trade journals were crowded with notices by Snow, Pecham, Ellis, Miltimore, Boies, Munton, Allston, Paige, Thurber, and Coopers (Figure 7.40). Some of these were marginal firms that probably produced only a few wheels, but others were sponsored by major railway equipment producers. The Thurber wheel was marketed by the Brooks Locomotive Works, the Boies wheel by the Dickson Manufacturing Company, the Snow wheel by the Ramapo Wheel Company, and the Paige wheel first by the Wason car works and later by an independent firm in Cleveland. Production figures are not available, but it would seen reasonable to assume that even the combined totals would amount to only a fraction of Allen's production. At least one British manufacturer tried to grab a piece of the American market, offering its product through the agency of W. R. Ellis & Company of Boston and New York.[234] It was called the Brunswick Steel Tired Wheel and featured flat metal spokes bent in the shape of a star. The Boston and Albany had been using domestically manufactured tired wheels of this design since 1870 and claimed that they gave excellent service.[235]

A number of Allen competitors tried to capitalize on the idea of the cushioned wheel. Atwood wrapped hemp ropes around a cast-iron center between the tire and rim. A firm in Indianapolis inserted a layer of rubber.[236] The Boies wheel featured flexible center plates. None of these schemes appears to have gained wide acceptance, and in modern times the whole concept of the cushioned wheel has fallen into disrepute.[237] The so-called sandwich wheels, built up from alternate plates of steel and rubber, became fashionable for transit cars in the mid-1930s, but ultimately the industry concluded that they rode no better or more silently than conventional one-piece wheels. They tend to come apart in service and are suspected of creating a "torque effect" that corrugates the rails.

During the era of the tired wheel, its advocates made much of its superior safety, increased mileage, and greater economy, while detractors claimed precisely the opposite. In 1875 the Boston and Albany expected its tires to last for 370,000 miles, or seven times longer than the average cast wheel.[238] The road said that the tired wheel ran for 134,000 miles initially, 186,000 addi-

tional miles after the first turning, and another 50,000 miles after the second turning. Several years later the Grand Trunk said that it had tired wheels running up to 474,000 miles and that it hoped to get 500,000 miles out of them.[239] British lines realized 18 to 20 years' service from a tire.[240] The improved mileage was attributable to the growing popularity of weldless steel tires by the late 1870s.[241] The largest producers were in England and Germany—Allen, for example, bought heavily from Krupp. Such tires were made in the United States as early as 1867 by the Nashua Iron & Steel Company. Some wheel centers wore out one or two tires and were still good for more service.[242]

Critics of tired wheels acknowledged that these mileage figures were impressive. Some were even ready to admit that cast wheels would not last a year in express-train service, even though they were good for seven years under a freight car. However, the longevity of the tired wheels came at a high price. They actually cost more money than they saved. The price of a tired wheel was $80, while a cast wheel sold for $12. One engineer figured that the potential interest from the excess investment on a set of paper wheels would buy a new set of cast wheels every six months.[243] To this must be added the expense of periodic turning and the remounting of new tires. Even the friends of the steel tire agreed that its high first cost was the major drawback. Yet they maintained that the extra cost was more than justified on grounds of safety. They spoke of the tired wheel as "absolutely safe"—an assertion quickly, and rightly, challenged.[244] On November 1, 1876, a paper wheel under the Pullman sleeping car Woodbine broke, derailing the Niagara Express on the North Pennsylvania Railroad.[245] Several passengers were killed. A suit brought against Pullman contended that "paper wheels" were unsuitable for railway service. The suit was disallowed after several years of wrangling, because the plaintiffs could not prove their case. If they had been successful, the paper wheel might have disappeared from the American railroad at an early date. No overall statistics are available, but again isolated reports indicate that tired wheels were not in fact absolutely safe. In 1883 it was reported that three failed in service, although these failures caused no deaths or serious injuries.[246] Ten years later the Wagner sleeping car Alva was thrown into the ditch because of a broken tire.[247] The failure of other wheels used under Wagner cars is documented by a series of accident photographs in the possession of Lee Rogers, a railroad collector of Washington, D.C. One of these views is reproduced in Figure 7.41.

The complaints against tired wheels amounted to more than faultfinding, but the charges, real and imagined, in no way discouraged such enthusiasts as Pullman. Among their other virtues, he claimed, tired wheels produced a substantial reduction in truck repairs because of the cushioning effect of the paper centers.[248] Others clearly shared this faith, for in 1897 all the important manufacturers united as the Steel Tired Wheel Company.[249] The monopoly was composed of nine firms, which included Allen's paper wheel company. The company prospered for a decade and then went into a decline. Steel cars were too great a burden for the fabricated wheels. In the process of brak-

Figure 7.41 *Steel tired wheels had a good safety record, but they could fail, with devastating results. (Lee Rogers Collection)*

ing, the tires became overheated, causing them to expand and loosen.[250] By 1915 they were declared unsafe by a spokesman for the Interstate Commerce Commission.[251] Eight years later the Allen plant at Pullman, Illinois, was abandoned. Production had apparently ceased some time before, and the paper wheel was already spoken of as if it were some almost forgotten antiquity. Apparently no wheels of this type were then on main-line trains, although some were used in and around the Pullman plant as shop trucks. In the mid-1960s paper wheels served the same purpose at the St. Louis Car Company.

The third or modern phase of the railway car wheel began in the first decade of this century, when the rolled- or wrought-steel wheel came into favor. It was a one-piece, single-plate wheel superior to the chilled-iron and fabricated tired style of car wheel. It was stronger and cheaper than the tired pattern, although it could not compete with the cast-iron wheel on the basis of first cost. In 1917 a wrought-steel wheel cost $19, a chilled wheel only $9.[252] It was calculated that $350 million would be needed to reequip all American freight and passenger cars with steel wheels.[253] It was argued that such an extraordinary expenditure was really unnecessary because of the progress being made on the chilled wheel. The wheel makers formed an association in 1908 which opened a laboratory for the sole purpose of improving chilled-wheel design and manufacture. The same group supported a similar program at the University of Illinois.[254] By increasing the wheel's mass by 40 percent, the researchers were able to increase its capacity by 140 percent. The efforts of the Association of Manufacturers of Chilled Car Wheels prolonged the life of the cast-iron wheel for another half century. Its obituary appeared in the February 22, 1901, issue of the *Railroad Gazette*; yet it continued to serve as the standard

freight car wheel for decades. In 1930 it was estimated that 95 percent of all U.S. freight cars, 25 percent of passenger cars, and 66 percent of locomotive tenders were equipped with cast-iron wheels.[255] Faith in the chilled wheel, however, declined permanently following World War II, when larger freight cars and faster schedules proved too much for the old standard. In 1958 the Association of American Railroads Mechanical Division voted to outlaw chilled wheels on new cars.[256] Five years later manufacture of the wheels was abandoned, and in 1968 they were banned from interchange service.

The prolonged acceptance of cast wheels for freight cars was based on what constituted reasonable risk weighed against first cost. They were considered safe enough for this service, and their very low first cost was a definite attraction. But after the coming of the steel car, the chilled wheel was not deemed good enough for passenger service by many engineers. It might carry the weight, but speed was a factor as well, and fast running imposed unsafe stress on the brittle cast-iron disc. The fact that as many as one-quarter of American passenger cars rode on chilled wheels in 1930 (although another source says that only one-fifth were so equipped in 1924) probably means that these cars were either lightweight suburban units or obsolete branch-line cars not in high-speed, main-line service.[257]

Inventors were working on one-piece steel wheels generations before they came into general use. On July 11, 1854, George B. Hartson of New York City received a patent (No. 11,428) for machinery to produce a one-piece rolled wheel. The proposed machinery was not unlike that used in actual production many years later. Another American, Samuel Van Stone, received a patent in 1863 for a different style of machine to roll and forge car wheels. A year earlier Krupp was producing cast-steel car wheels, which were advertised as being available through his

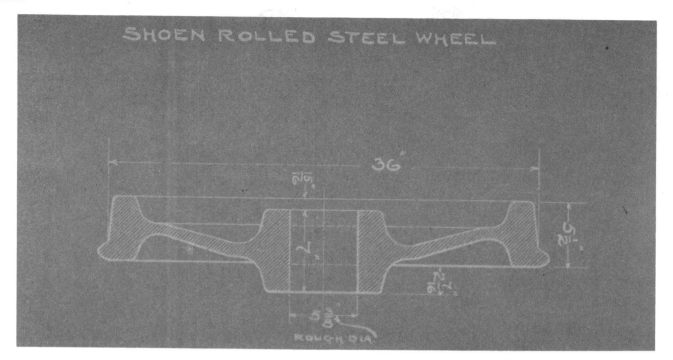

Figure 7.42   Solid rolled-steel wheels were introduced commercially around 1903 and became popular for passenger cars within a few years. Cast-iron wheels continued to be used under freight cars until very recently.

American agent.[258] No reports have been uncovered on the use of these wheels in North America, although by 1881 Krupp claimed to have manufactured 100,000 of them. In 1868 several U.S. lines were reported to be testing cast-steel wheels produced by the British firm of Vickers.[259] It is unknown whether these wheels were literally "cast" or whether they were forged wheels made from cast steel—a term then current to differentiate Bessemer steel from that made in the traditional manner. A year later the Black Diamond Steel Works of Pittsburgh was producing steel wheels for the Pennsylvania on an experimental basis.[260] The first wrought-steel wheel in this country was displayed by a Belgian firm at the U.S. Centennial Exposition of 1876.[261]

A decade later, H. W. Fowler of Chicago patented a process for making a solid, one-piece, rolled-steel wheel.[262] Production began in 1888, and several thousand were turned out for some twenty railroads over the next two years. They were made by heating, rolling, and annealing cast-steel blanks. The blanks, however, proved porous and spongy, resulting in a defective wheel. Most of the wheels were quickly abandoned, but some remained in service for fourteen years and ran as much as 800,000 miles.

There were other experiments during the time of the Fowler wheel, such as a Pittsburgh concern's attempted manufacture of a cast-steel wheel invented by F. W. Webb of England.[263] The steel wheel was cast in a rotating mold, reheated, and rolled. It was said to withstand seventy-three blows during a drop-hammer test, but like Fowler's wheel, it attracted no long-term customers. It would take big money to perfect a practical steel wheel; a rich, patient, and determined entrepreneur was needed who could outlast the period of rejection that all new products must undergo. The man was Charles T. Schoen, a pioneer manufacturer of steel freight cars.

In 1898 Schoen began producing rolled-steel wheels with a process devised by Henrik V. Loss.[264] By 1903 the method had been perfected, and Schoen organized a separate plant for wheel making (Figure 7.42). In 1904 he sold 1,134 of his wheels; in 1905, 22,332; and in 1906, 58,590. Production dropped back to 54,467 in 1907 because of the financial panic, but in 1908 the business was so promising that Carnegie bought Schoen out. The following year the plant produced 269,000 wheels. By this time Midvale, Standard Steel, and the Forged Steel Wheel companies were also in production. In 1912 5 percent of American passenger cars had wrought-steel wheels, a figure that climbed to an estimated 80 percent in twelve years.[265] Today they are still the standard passenger car wheel.

The production of wrought wheels involves the following steps. A thick slab is cut from a 15-inch-diameter ingot. The slab is heated and placed in a 12,000-ton press for rough shaping. It is reheated and undergoes a second forging. The center axle hole is punched out. It is reheated again and placed in a huge rolling machine, where the tread flange and plate are given their final shape. The wheel is put in a press to dish the plate. It is then quenched, reheated, and allowed to cool slowly to relieve internal stress developed by the forging and rolling processes. At last the wheel is machined for final finish.

Beginning in 1930, heat-treated wheels came into favor.[266] The tread and flange were hardened to improve mileage. In this way the body of the wheel could remain as low-carbon steel—a relatively cheap material that was not prone to thermal cracking. Yet the running surfaces were hard and long-wearing. Most roads preferred multiple-wear wheels; even though single-wear wheels were lighter, the multiples were considered more economical over the long run. Machine tolerances were made much more exact (0.003 inches) after about 1940 to ensure a smoother ride and better balance. Ordinary machining in an engine lathe was no longer good enough; wheel sets were now finished by a precision grinding machine.

In 1941 the A.A.R. Mechanical Division began to classify wrought-steel, heat-treated wheels under three broad groups, designated A, B, and C.[267] Class A is the softest wheel of the group, with a relatively low carbon content (up to 0.57 percent).

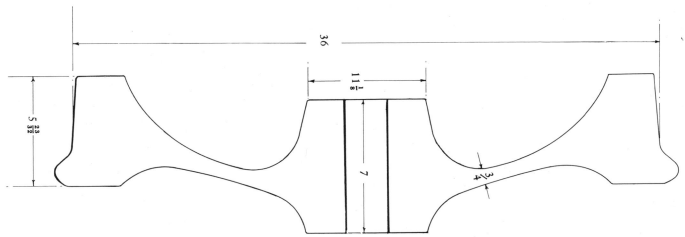

Figure 7.43   Cast-steel wheels are rarely found in passenger service, though they are common on freight cars. This wheel was produced by Griffin around 1970. (Traced by John H. White, Jr.)

These wheels are the fastest-wearing, but they are also the safest, because they are the most resistant to shelling and thermal cracking. They are accordingly favored for service with heavy-tread (brake shoes) braking loads. Class B has a greater carbon content (0.57 to 0.67 percent), wears longer than class A wheels, and is suitable for high-speed service with moderate braking conditions (or disc brakes) and high wheel loads. Class B wheels are preferred for heavyweight cars. Class C has the highest carbon content (0.67 to 0.77 percent) and is the least used because of the tendency to develop thermal cracks, particularly when used with shoe brakes. These are the longest-wearing wheels and could be used successfully with nonthread or disc brakes.

At the time that class A, B, and C wheels were classified, the Mechanical Division attempted to reduce the variety of wrought-steel wheels in production.[268] Prior to 1941 there were about 500 designs in use. This number was cut to 58 in 1942, and by 1967 the number of standard designs was down to 29.

Brakes

In 1882 an English writer observed, "No one can look upon one of our modern locomotives going at the rate of 60 miles an hour without . . . breathing a prayer for the safety of the living freight behind it."[269] Indeed engineers had concentrated on making the train go; swift locomotives and free-rolling cars were the products of their ingenuity. But they had invested less energy and imagination in getting the train to stop. The mechanism for retarding its motion was a miserable collection of light, manually powered rods and levers that was in no way comparable to the elegant machinery of propulsion. Early locomotive superintendents, in fact, wanted nothing to do with brakes. Before about 1875, few American locomotives were equipped with brakes of any kind. The tender had a set of shoes mounted on one truck, but this was mainly for the purpose of holding the engine while the train was at rest in the engine house or at a way station. Brakes, as every good engineman knew, would force the driving boxes and side rod bearings out of line. The reasoning seemed to be that stopping the train was a troublesome job for the car department to handle. In fact, railroad men seemed to feel generally indifferent on the subject. All the major advances in braking technology were made by men outside the field.

The history of railway brakes can be divided into three sections that form a convenient outline of its evolution. These are the hand brake, the continuous brake, and the power brake. Between 1830 and 1875, hand or manually powered brakes were the basic system employed. A lever or hand wheel on the end platforms caused a linkage of chains, rods, and levers to force a block or shoe against the tread of the wheels. The resulting friction retarded the train's motion. The kinetic energy of the train was converted into heat. This technology was directly adopted from highway vehicles. Around 1850 the efficiency of the hand brake was doubled when the brakes on both trucks were interconnected, so that all eight wheels could be braked by turning the wheel at one end of the car.

The second major development, continuous brakes, actually overlaps the eras of both the hand and the power brake. The attempt to devise a continuous brake was an effort to connect the brakes of all cars in the train so that they could be worked in concert. No really satisfactory continuous brake was created, however, until the power brake itself appeared. Various forms of power brakes were introduced at an early date, but none were accepted into everyday practice until about 1870. The schemes offered included steam, electric, vacuum, and compressed-air brakes. In the course of less than a decade the compressed-air system came to prevail, and it remains the standard form to the present day.

When the railroad era opened, brakes were regarded by some managers as an unnecessary appendage. It was claimed that one pioneer line in the South used the following method of pulling up to a station.[270] On approaching, the engineer would signal by raising the safety valves. Slaves then rushed forward to seize hold of the engine and cars and pull back, while the station agent thrust a stick of wood between the spokes of a wheel to complete the stop. The credibility of this story can be questioned, but there is other evidence of the lack of brakes on some infant American railways. In 1840 the Austrian engineer Von Gerstner noted the general absence of brakes on the cars running between Greensville and Roanoke, Virginia, even though the line had some steep grades.[271] Fifteen-car freight trains had brakes on only the first three or four cars. Passenger trains depended entirely on the tender brake. The Boston and Maine followed the British system of using brakeless freight cars with a "brake van" at the end of the train.[272] In the late 1840s the Michigan Central depended on tender brakes, according to the recollections of an

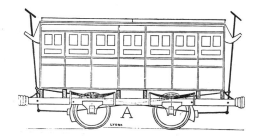

Figure 7.44   A double-acting brake devised by Henry R. Campbell about 1834. Brakemen operated it from roof seats.

old employee.[273] A long lever operated the brakes, and at times it was necessary for the fireman to stand on the lever to exert enough pressure.

Most lines, however, recognized the need for brakes on each car. Illustrations of the earliest American cars uniformly show some form of braking system (for example, Figures 1.2, 1.3, and 1.61). The favored style was the lever brake, copied from road coaches of that day. The B & O, the Camden and Amboy, the New York and Harlem, and other early roads used this form. The brakeman, seated on the roof, exerted his powerful leg muscles to push the long lever. The mechanical advantage was terrific, and for light cars traveling at slow speeds, it was an adequate system. A double-acting arrangement was devised around 1834 by Henry R. Campbell, the engineer for the Philadelphia, Germantown, and Norristown Railway.[274] Cross rods connected both brake shoes on either lever (Figure 7.44). Campbell's plan was copied by the Boston and Providence. But even with a reasonably efficient hand brake, stopping the cars was an uncertain business— as illustrated by an accident that occurred during the opening ceremonies of the New York and Harlem Railroad.[275] The mayor of New York City and other dignitaries were loaded into two horse-drawn cars. To ensure that all went well, the lead car was driven by the railroad's own vice president. A sober and experienced hack driver was engaged to operate the second car. All went well until the vice president made a stop, in part to show the perfection of the car's brake. The hack driver, following closely, pulled back on the reins and called to his team to stop, but in a moment of confusion he forgot the brake. In an ordinary highway coach the horse could draw up and stop the vehicle, but the rail car was heavier and had a nearly frictionless track. It did not stop. The pole of the second car rammed into the rear panel of the lead car with a great crash, and out tumbled the mayor and his friends. No one was hurt, but the railroad had little success afterwards in convincing city officials that street railways were perfectly safe.

The greatest single failing of the manual system was the brakeman himself. He was said to be an indifferent class of worker, reckless, "full of muscle—small brains—inactive" with little or no experience.[276] And where was he when the whistle sounded? Never at his station, but comfortably seated in the cars like a passenger, or loitering around thinking about everything but his duty. In 1853 a writer asked how the railroads could justify hiring such a poor class of men for a job so vital to the public safety. In this "age of lightning and steam," with speed the new god, only the most dependable of men should be manning the brake staff.[277] But the railroads were determined that it should remain a low-paying job. The lines around Boston employed Irishmen exclusively, most likely immigrants fresh from the docks.[278] It was said that native Americans were too enterprising to accept such a poor salary; they aspired to better jobs.

The general ineptness of brakemen was not the sole reason for rough and uncertain stops. Some of the blame is the managers', for they asked too much of the brakeman. He was a janitor as well as a train stopper. If he was frequently away from his sta-

tion, he was often sweeping out the cars, tending to lamps, firing the stove, hooking up the bell cord, calling out the next station, or helping with the luggage. A traveler describes the harried brakeman during a journey of 1865:[279]

Through the aisle scurried the brakeman. Dumping coal into the tall black stove and thumping with the poker—shouting the name of the next station—tearing through the door and slamming it behind him at the whistle's call for "Brakes!", his work was never ended. How interesting it was to watch him set the brakes! His lithe body bending and swaying as he twisted the wheel, one foot pressed against the foot pawl to hold the winding chain—then slinging himself across the gap to the other platform to whirl the wheel there—back again to the first wheel to add a bit more tightening—then like a flash to the other to give equalization.

The passenger train brakeman may have had an impossible job, but it was easy compared with the miserable situation of a freight brakeman. That worker rode atop the cars in all seasons; he had almost no opportunity to get out of the weather. He was expected to jump from one swaying roof to the next, and if it was icy or dark he could easily fall under the wheels or down an embankment. In 1881 it was calculated that at least ten brakemen were killed in this country every day.[280] Many more were maimed. Insurance was not available for such work. And even after the coming of automatic couplers and air brakes, the brakeman's job remained highest on the list of the thirty most dangerous trades, which included the work of electricians, miners, and sea captains.[281]

How well did hand brakes work? Most attempts to assess them quote the biographer of Westinghouse, who contended that the old fashioned "armstrong" method would halt a passenger train traveling at 30 mph in 1,600 feet.[282] Actual tests made during the period, however, indicate that brakemen achieved far shorter stops. An account published in 1855 says that an alert crew stopped a 119-ton train going 30 mph in only 450 feet.[283] A train of equal size going 28 mph was stopped in 376 feet.[284] The popular, 1,600-foot statistic at first seems a gross exaggeration, yet it must be remembered that the smaller figures from the tests are the result of staged stops. The brakemen were at the wheels—tense, alert, and waiting for the whistle to call "Down brakes." During normal operations a brakeman's showing would not have been so good. Suppose that he was not poised over the brake wheel but working inside a car. When the whistle sounded, it might take him 10 to 15 seconds to get out on the platform to man the wheel. At 30 mph the train would already have traveled 325 to 500 feet before the shoes first touched the spinning wheels. That stop would be made in, say, 800 feet. But if the brakeman was half asleep in a seat, or was blocked for a moment by passengers standing in the aisle of the car, it would probably have been 20 or even 30 seconds before he reached his post. Therefore stops of 1,600 feet may not be such an exaggeration.

The number of brakemen per train apparently varied widely, doubtless determined to a large extent by speed and terrain. A report made in 1838 which encompassed most of the major Northeastern lines indicated that it was common to have one

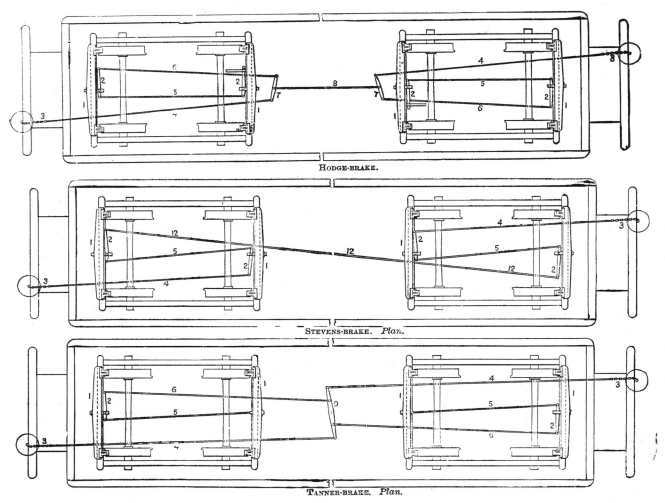

HODGE-BRAKE.

STEVENS-BRAKE. *Plan.*

TANNER-BRAKE. *Plan.*

Figure 7.45   Three popular brake rigs or foundation plans were the Hodge, intro-
duced in 1849, the Stevens, introduced in 1851, and the Tanner (actually designed by
W. I. Nichols), introduced in 1846–1847. (Car Builders' Dictionary, 1879)

brakeman for every two cars.[285] However, a report issued in 1853 concerning New York railroads showed a wider variation.[286] One line had two brakemen on each train, with a conductor who helped when necessary. The number of cars was not indicated. Another line had one man for each car, and another had only one for every four cars. Freight trains were more poorly manned; the Canada Southern employed only two brakemen for each fifty-car train.[287] Of course, the speeds were much less ambitious on freights. The average brakeman could exert 150 pounds' pressure on the brake wheel.[288] The wheel itself multiplied this to 900 pounds, and the levers, depending on their size and arrangement, could compound it to as much as 4,500 pounds. It was this great pressure that pushed the shoes against the wheels.

The effectiveness of hand brakes was materially improved about the middle of the century, when the double-acting foundation plan came into widespread use. It may be difficult to believe, but brakes before then were normally fitted to only one truck. This provided only 50 percent braking ability per car and required the brakeman to run the length of the car to reach the wheel of the next car. The double-acting plan was a relatively simple system of levers and rods that connected the brakes on both trucks to the wheel at either end of the car, permitting all eight (or twelve) wheels to be braked at once. It also allowed the brakeman to easily set the brakes on two cars by stepping between the platforms. The idea was so obvious that it appeared at least as early as the late 1830s, or about the time when eight-

wheel cars were becoming popular.[289] As mentioned previously, the system was already in use on four-wheel cars. Compound or double braking was also an integral part of the earliest continuous braking schemes, such as those tried by Griggs and others. And yet the scheme was resisted on the grounds that such elaborate foundation plans were a needless complication and expense. This argument was reasonable for the first decade or so of American railroading, but by 1850 it was simply an excuse to justify a false economy. Double brakes had become a necessity because of the increase in train weights and speeds.

While many mechanics had considered the problem, the first to devise a long-term solution was Willard I. Nichols, car superintendent on the Hartford and New Haven Railroad. Nichols, who thought of the double brake more as a laborsaving than a lifesaving device, perfected his foundation plan in 1846 and 1847.[290] He died not long afterwards without patenting the invention. Nichols's brake, which became known in later years as the Tanner brake, was widely used (Figure 7.45).

Meanwhile other inventors were active in the field. Charles B. Turner of Buffalo obtained a patent in November 1848 for a bumper brake that included connecting brakes for both trucks. The bumper feature was a failure, but an associate of Turner's, Henry L. Tanner, saw the potential profit in the scheme for double-acting brakes.[291] Tanner, who described himself as a "general trader," was not a mechanic or a railroad man but a shrewd speculator who knew a good thing when he saw it. He modified Turner's patent and began selling rights to a double-

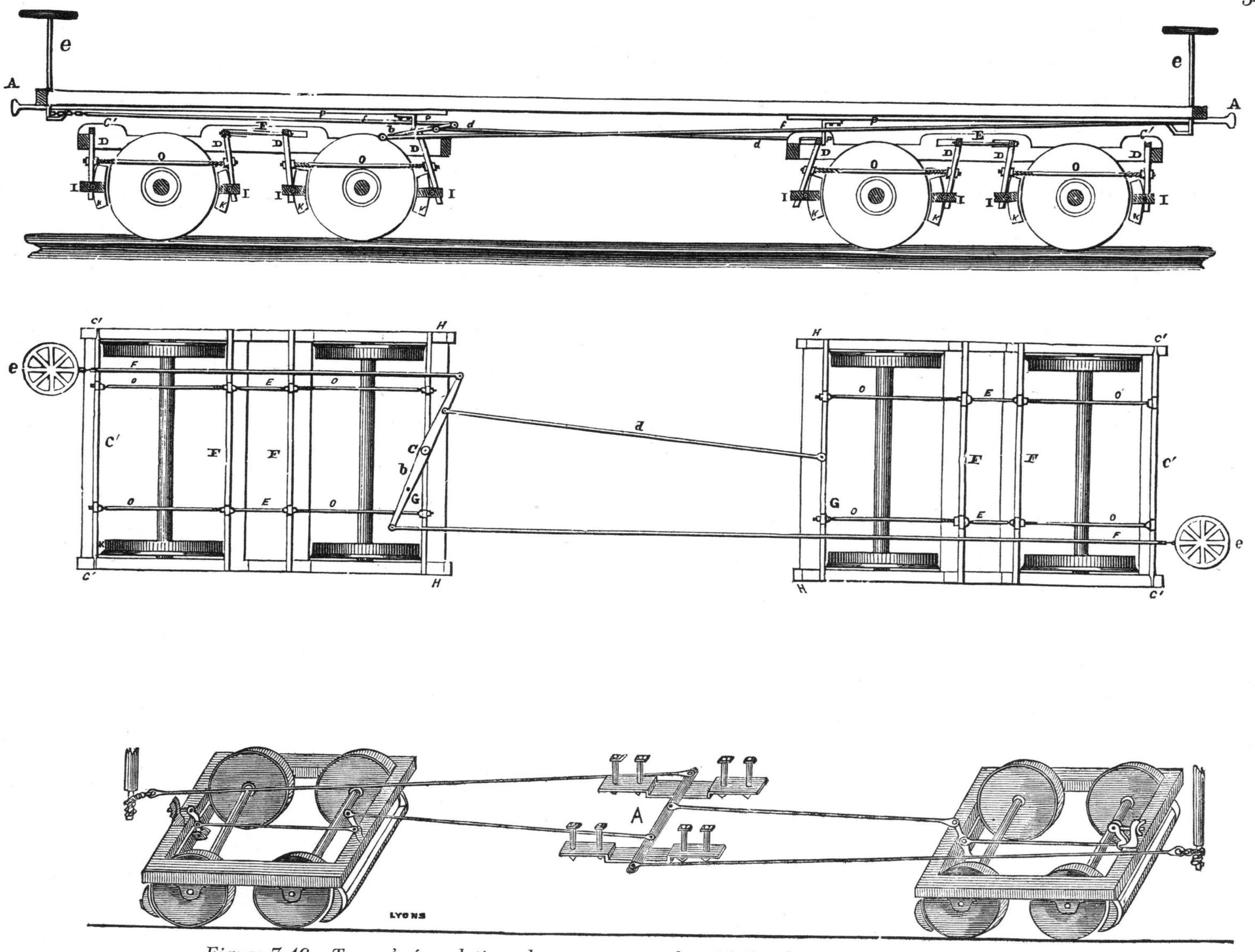

Figure 7.46 *Turner's foundation plan was patented in 1848. The two upper drawings show the original design; the lower perspective sketch indicates Tanner's modifications in 1849–1850.*

acting brake totally unlike the device described in the patent (Figure 7.46). During his travels in the Midwest he saw the Nichols brake and realized that it was a foundation plan far superior to the reformed Tanner brake. Because it was common property, Tanner decided to take Nichols's plan for his own and reissue the Turner patent to cover the idea. By then both inventors were conveniently dead. For unexplained reasons (perhaps due to resistance on the part of the Patent Office), Tanner changed his tactics on reaching Washington. Instead of revising the Turner specifications, he bought rights to a brake patent that had been pending since 1847. Lafayette F. Thompson and Asahel G. Batchelder, both of Massachusetts, had applied for a patent on a bumper brake described as a wretched contrivance that refused to work. Apparently its novelty was also in question, because the government would not issue a patent. (Several buffer or bumper brake patents were already in existence.) Thompson and Batchelder, one a mechanic, the other a carriage maker, were probably open to offers after five years of waiting. Tanner canceled their drawings, specifications, and model and substituted new ones in Thompson and Batchelder's names to cover the Nichols designs. The Patent Office cooperated nicely, granting Tanner a patent in the name of Thompson and Batchelder on July 6, 1852. Tanner now set out to sell licenses for the use

of his brake, his customary fee being $5 per mile of line and a free pass. Where he met resistance, he entered suit, usually seeking a small claim. A victory against the Erie and the Hudson River railroads in 1853 helped to establish the validity of his claim. It was the same technique that Winans had used so effectively.

In 1855 Tanner sold his rights to Thomas Sayles of Lansingburg, New York. By this time double-acting brakes were common on passenger cars, though it would be some years before they were installed on much freight equipment. Tanner acted as Sayles's agent, and his vigorous pursuit of license fees soon made him anathema to the industry. Roads that did not pay were taken to court. In 1907 Angus Sinclair mentioned that Tanner's name was of "litigious memory," but Tanner and Sayles were interested in extortion, not in being kindly remembered. During the first term of their patents they collected over $53,000 in fees. During the period of the first renewal (1866–1874) they took in another $80,000. At least, these are the available figures; the exact total, which was thought to be far greater, was never revealed. Even after the patents expired and a second renewal was denied, Tanner pressed old claims amounting to 90 million dollars. In 1877 he was involved in some two hundred suits. In 1881 his partner Sayles died. Tanner's attorneys were suing the

Chicago and North Western for 15 million dollars, but by this time the case was going poorly and he settled for 500 dollars. Finally in 1882 the U.S. Supreme Court put an end to the swindle, ruling that no equity could be claimed after a patent had expired.

Tanner was not the lone purveyor of double-acting brake foundation plans. He had competition from two other major patentees, Nehemiah Hodge of North Adams, Massachusetts, and Francis A. Stevens of Chicago. Hodge's gear was patented in 1849 and became the most popular single form of double-acting brake in the United States. By 1874 it was said to have been used by 75 percent of the members of the Master Car Builders Association. On first inspection the rig looks overly complicated, with its six levers and seven rods, but it combined well with the air brake and had good compensating characteristics which equalized brake shoe pressure between both trucks (Figure 7.45). Stevens devised a much simpler form of rigging which was patented in 1851.[292] It employed no floating or center levers but depended only on brake beam mounted levers. It was popular in the West and, like the Hodge brake, remained in favor until modern times.

The braking of all wheels on each car was accomplished by the double-acting brake. This great advance was apparently in wide use by 1855. But railroad men envisioned an even better form of brake that would not depend on the uncoordinated efforts of several brakemen but would control every brake on the train through a single mechanism. Such a brake would be operated by the engineer. It would respond more certainly and quickly in the event of danger. It would produce a more even stop through relatively light pressure on all wheels rather than hard braking of a few. This would cause less wear and tear on the machinery and largely avoid the problem of flattened wheels. Inventors' first ideas on how to power such a brake centered on capturing the momentum of the train itself. In theory it was an appealing plan, because it conserved energy and required no auxiliary power source, such as steam or compressed air. The idea fascinated railroad mechanics throughout the nineteenth century, and countless plans were tested. But like so many alluring concepts, the notion of using the train's own power was to remain an unfulfilled dream.

The momentum brake went under several names: buffer, bumper, and compression brake. Most depended on the motion of the draft gear. Every railway man had observed the gathering in of the slack when the engineer shut off steam preparatory to a stop. Harnessing this enormous energy through a contrivance of levers and rods leading to the brake gear seemed beautifully simple. Stephenson was probably the first engineer to try to create a buffer brake. His plan, tested on the Liverpool and Manchester Railway in 1832, was a failure. Three years later, John K. Smith of Port Clinton, Pennsylvania, patented a buffer brake that was tested on the Little Schuylkill Railroad during the same year.[293] Reportedly it worked well, but no other roads are known to have used it.

In 1839 a novel form of momentum brake appeared on the Boston and Providence Railroad.[294] It remained in service there for at least a decade and was tried on several neighboring lines. Part of its longevity must be credited to the fact that it was the invention of the presiding master mechanic, George S. Griggs. Under the watchful eye of its creator, almost any mechanism can be nursed along indefinitely. Griggs did not employ the motion of the buffers; rather, he utilized the motion of the train to drive a windlass which wound up a rope or chain. A control rope ran the length of the train. A tug on the rope moved a lever on each car that engaged a clutch on a windlass, or drum, which was belted to one of the wheel axles (Figure 7.47). Thus all brakes on the train could be set at one time. Griggs's brake was said to be six times as powerful as the ordinary hand brake. During one test a sixteen-car freight train going downgrade at 15 mph was stopped in its own length, yet only one car on the train was fitted with Griggs's brake.[295] The cost of the apparatus was estimated at $25 per car. This momentum brake was one of the few that enjoyed early and prolonged success, but it was hardly widespread.

In the 1850s many inventors offered plans for buffer brakes. Turner's patent, as well as the patent of Thompson and Batchelder, have already been mentioned. In 1852 a Philadelphian, T. G. McLaughlin, convinced the Camden and Amboy to try his new self-acting buffer brake.[296] Seven years later another advocate of compression gear, W. R. Jackson, made tests on the B & O.[297] These and similar designs all failed; the universal defect was inability to control the brake's action. When the train speed slackened for any reason, the brakes came on whether they were wanted or not. This made running slowly a tricky affair, and switching a near impossibility. It was necessary to disconnect the brake to back up. In addition, most brakes took hold unevenly. The slack running in at the head of the train would jam on the front brakes hard, while the rear cars were rolling free. Joseph Olmsted, a Chicago inventor, attempted to master the control problem by introducing an electromagnetic clutch with a basic mechanism much like Griggs's old plan. Olmsted's momentum brake was tried here and in England in 1869 and 1873, and then like so many others, it disappeared. After this date there appears to have been no more interest in momentum brakes for passenger service. Power brakes had already proved themselves, but some optimistic mechanics seemed to believe that a workable buffer brake was possible for freight cars. In the 1880s when the industry was debating over the desirability and the possible form of power brakes on freights, the ancient compression scheme was revived as an economic alternative to the costly Westinghouse apparatus. These schemes raised many false hopes and resulted in as many patents, but the inherent defects of the compression brake were not overcome by the new generation of inventors.

At the outset of the new wave of compression brake fiascos, simpler forms of continous brakes were being employed. Some of the coal roads in eastern Pennsylvania were having success with manually powered continuous brakes that involved little more than tying together the brake levers on a series of small four-

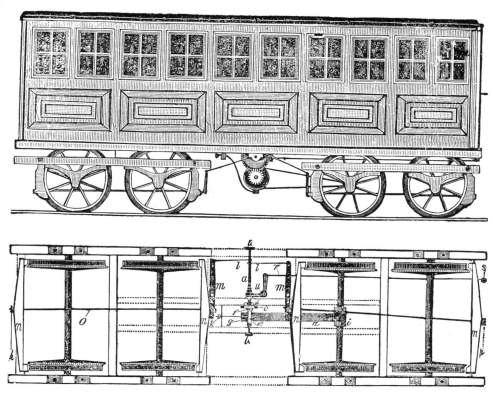

Figure 7.47 Griggs's momentum brake of 1839 was the first continuous, mechanically powered brake to be used in everyday service. In the lower figure, C is the gear drive, d the clutch, e the gear-drive frame, f the pinon gear, g the pulley, and h the drive belt. (American Railroad Journal, *September 15, 1841*)

wheel cars with a long wooden rod. The train of cars was under the control of a single brakeman.[298] The gravity road at Mauch Chunk was running fourteen car trains in this fashion in 1838. A more sophisticated scheme had been worked out for passenger cars by the B & O three years earlier. A tall brake staff at one end of the car was actuated by an overhead rope that ran forward to the engine. When the rope was pulled, the lever was jerked off a perch, allowing a counterweighted lever to fall and apply the brakes on all eight wheels. The device is pictured in the German drawing of the B & O Washington branch car reproduced in Figure 1.70. The plan was credited to the brother of the B & O's first president, Evan Thomas. A history of the company printed in 1853 claims that a patent was issued, but no record of it can be found.[299]

These primitive continuous brakes must have suggested the possibility of even better train-stopping systems. Many mechanics believed that a chain connecting all the car brakes together would be the solution. English mechanics seemed particularly attracted to the chain brake idea and began trying to perfect it in the 1840s.[300] They were moderately successful in adapting the chain brake to everyday operations, and a few lines, such as the London and North Western, had some in service as late as 1892. The earliest such brake recorded in an American patent was invented by Lucius Stebbins of Hartford and patented in April 1848. The chain was drawn taut by a friction drum powered by the locomotive driving wheels. No record exists on the adoption of Stebbins's design, but an identical system, patented in 1855 by William Loughridge (d. 1890) of Weverton, Maryland, was in rather wide use for about fifteen years (Figure 7.48). Loughridge's apparatus was in effect a form of momentum brake, because the power drum was driven by the motion of the

train. A pulley fastened to a countershaft was pressed against one of the locomotive's rear driving wheels. The inventor claimed that a series of movable pulleys overcame problems of correct chain tension and brake shoe pressure. He called his system the graduating car brake. Within two years it was in use on the Cincinnati, Hamilton, and Dayton; the Ohio and Mississippi; and the Mad River and Lake Erie.[301] In 1859 the Pennsylvania adopted the Loughridge brake—such a triumph that the inventor reproduced the letter of acceptance in an advertisement.[302] A few years later the railroad claimed that Loughridge's lifesaving device made the P R R the safest route possible, and that after three years of use and the transport of over 3 million passengers, not a single fatality had occurred.[303]

But as critics of the chain brake pointed out, while the inventor claimed that brake shoe pressure was uniform on all cars and that no wheels could slide, in actual service the wheels locked on the front car and there was almost no pressure whatever after the third car.[304] Loughridge's pulleys and movable bearings could not compensate for the varying length of the train, a measurement which fluctuated considerably because of the huge amount of slack allowed by link and pin couplers. The friction wheel was cut away by the flange and developed flat spots, making its operation unreliable.[305] Stops tended to be rough. Also, in extreme winter weather the chain would freeze around the clusters of pulleys and become inoperable. Even Loughridge must have recognized the defects of the chain brake, for in the 1860s he began to experiment with steam and air. Commenting on the problems of the chain brake, a British author said that after so many efforts to perfect the system, "it would have become, ere now the universal brake of the world" if that had been possible.[306] In spite of its deficiencies, the Pennsylvania did not

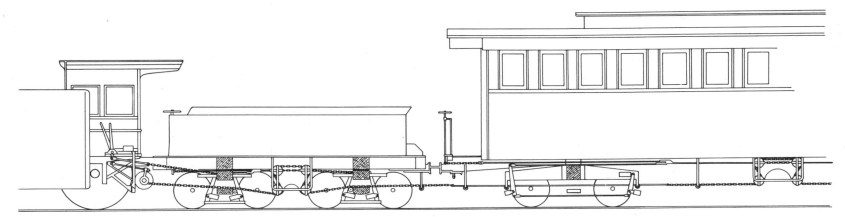

Figure 7.48 *William Loughridge's chain brake was patented in 1855. A drum, powered by the rotation of the rear driving wheel, wound up the chain to apply the brakes. (Smithsonian Neg. X2401)*

abandon the chain brake until late in 1870. For nearly fifteen years the constant jangling of heavy chains supplied a noisy counterpoint to the grinding and clicking of the wheels on many of the road's better trains.[307]

Most continuous braking systems were designed to function for both service and emergency stops. Thus most were intended to replace the brakeman, or as one Victorian writer observed, to make him "a victim of evolution." But William G. Creamer of New York City was less ambitious; he introduced a spring-powered brake meant only for emergency stops. The brakeman was to make regular or service stops as before, but in a crisis an auxiliary system would be available that could instantly apply all the brakes for a fast halt. Creamer selected a large spring to power his apparatus (Figure 7.49). The spring and its case were geared to the brake wheel shaft. A rope ran the length of the train to the engine. A pull of the rope lifted the catches on the spring cases, thus releasing the springs and setting the brakes. The engineer, conductor, or any trainman (passengers were advised not to touch the apparatus) could set the brakes anywhere on the train by pulling the control rope. Once the rope had been pulled and the Creamer brake had stopped the train, it was necessary to rewind the springs. The brake was not a laborsaving device, but it did solve one of the gravest problems not met by the manually powered brake—the occasional need for an emergency stop.

Creamer patented his device in December 1853; he was issued a second patent in November 1856. In 1855 he organized the U.S. Railroad Car Brake Company to exploit the scheme.[308] An enthusiastic first customer was the Hudson River Railroad. The spring brake added a measure of much needed security for the line's rapid express trains. During one test a train moving at 24 mph was stopped in 277 feet.[309] Creamer's brake was adopted by other lines such as the New York Central, the Lake Shore, and the Central Pacific. Contemporary photographs dating from the late 1860s clearly show the apparatus on the C P's passenger equipment. Creamer brakes were installed on the coaches and Silver Palace cars. Even the Central Pacific's emigrant sleepers were so equipped, as illustrated in Figure 6.27. Precisely how widely used Creamer's brake was cannot be determined, but as early as 1861, 1,000 were reported in operation.[310] In the following year the inventor attempted to sell the spring brake to British railways.[311] He persisted for at least five years, but apparently only the South Eastern Railway purchased it.[312]

With the coming of power brakes, Creamer's brake was obso-

lete. Heavier cars pushed train weights beyond the capacity of spring power. When the Creamer brake was introduced, trains were light enough that the springs worked as well as or better than the hand system, but by 1870 this brake was doing well to stop a train in 1,000 feet.[313] Even so, some lines still believed that the Creamer brake was a desirable extra precaution in the days before the automatic air brake. The Lake Shore, for example, advertised in its 1873 calendar that it was using the Creamer brake. Creamer himself believed in the future of the spring brake, for as late as 1870 he applied for a patent extension. Patent officials denied the request, stating that the estimated $80,000 profit realized by the inventor was adequate compensation.[314] Some railroad officials were already disenchanted with the Creamer brake as a safety device. Brakemen often failed to rewind the drums after the brake had been used, thus rendering it useless for the next emergency.[315] Ironically, what finally discredited Creamer's invention was a terrible wreck—exactly the sort of disaster that it was intended to prevent. A train on the Hudson River line plowed into a derailed freight train, killing twenty-one persons. It was declared a "historic failure," and much criticism was directed at both Creamer and the railroad for depending on so absurd a system as a collection of clock springs to stop an express train.[316]

In the last major epoch of the railway brake, mechanical power replaced human effort. The introduction of the power brake was a direct outgrowth of increasing speed and traffic density. With the rise in traffic came the additional revenue to pay for more elaborate safety apparatus. Many systems seemed suitable to drive the power brake: steam, compressed air, vacuum, electricity, and hydrostatic pressure were all suggested and tried. All were intended to replace the brawny but brainless brakeman with a more powerful and dependable means for retarding travel. All, except for certain forms of the steam apparatus, were meant to be continuous brakes.

Naturally steam was the first power source tried. When any engineering problem was being considered in the nineteenth century, steam was always offered as a sure solution. If it could drive the wheels of commerce, why could it not stop them as well? A good supply of steam was available from the locomotive; it need only be harnessed to a cylinder and the brake shoe. In 1833 Robert Stephenson patented a steam brake for locomotives. It was an intelligently arranged device, but it was a locomotive and

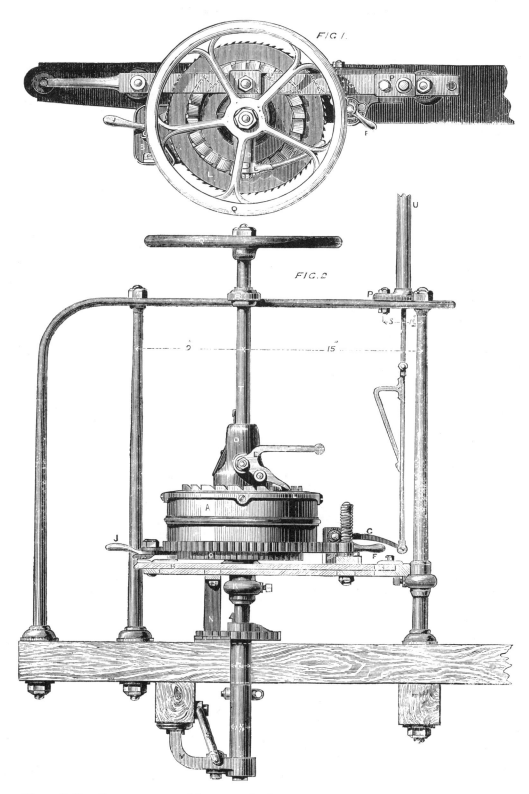

Figure 7.49 Creamer patented his spring brake in 1853. It was an emergency brake that could be activated by the engineer. (Engineering, July 12, 1867)

not a train brake.[317] Fifteen years later steam brakes were tried on the Boston and Providence, but a boiler explosion gave them a bad name, even though they were in no way connected with the accident. In 1855 Henry Miller of New York City offered a steam brake with cylinders under each car and a rubber hose between the platforms to draw steam from the locomotive.[318] His device was tested on the Michigan Central and the New Haven. It stopped a 104-ton train running at 30 mph in 700 feet. Other inventors active in this area included Wendell Wright,

who patented a brake in January 1855 that could work on steam or air. The following year Thomas W. Smith, a partner in the Alexandria, Virginia, firm of Smith and Perkins, introduced a steam brake that was given flattering testimonials by several local and Northeastern lines.[319] Among the lively succession of inventions at this time, apparently none could overcome the defects of steam for train brakes. High-pressure steam lines between and under the cars were dangerous. Low-pressure steam did not have enough power and would condense into water that could freeze inside the line and block passage of the vapor (as it also did in heating lines, mechanics later discovered). Steam brakes on the locomotive were simple, efficient, and widely used around 1880 to 1900, but how to convey steam to the train of cars was a problem never to be solved, at least so far as braking systems were concerned.

Water was another familiar power source in nineteenth-century America. It propelled grist mills, elevators, cotton presses, and washing machines. As early as 1859 Lewis Kirk, onetime master mechanic for the Reading, suggested a hydrostatic railway brake. His patent, granted in May of that year, specified a water cylinder under each car. Pressure was maintained by a buffer-actuated pump, giving each car a self-contained system. In 1874 two Philadelphians, W. H. Henderson and Thomas McBride, independently devised hydrostatic systems, both of which were tested on the Philadelphia and West Chester, a suburban line with many stops.[320] Henderson used a large tank under the engine cab floor with a head of steam to maintain the pressure. He thus claimed to have avoided the expensive and troublesome steam pump necessary for other systems, such as those of his competitors, McBride and Westinghouse. Hoses and cylinders much like the equipment for the air brake carried the water back under the cars. An obvious objection to the water system was that the braking fluid would freeze in cold weather, but the inventors suggested a glycerine mixture. The water brake made little progress, however, and was never used in regular service so far as is known. The French or Le Chatelier water brake, which was adopted fairly widely in the American West late in the century, was not a water brake in the true sense. It was a locomotive compression brake that overcame the usual problems of braking the train by reversing the locomotive and allowing the cylinders to work as an air pump. Le Chatelier closed off the exhaust passages so that cinders were not drawn into the cylinders. He also introduced a spray of water to lubricate the cylinders. However, his was strictly a locomotive and not a car or train brake.

The air brake was to sweep away all rival braking systems, though it did not seem like such a sure thing during the first years of the Westinghouse brake. At the time a competing form of atmospheric brake was being vigorously promoted, with considerable success. It was efficient, reliable, and cheaper than the compressed air brake. It required no pump and could maintain full power even with frequent stops, an advantage not offered by the air brake. It provided a quick release and a fast application. This was the vacuum brake, a product that initially showed great promise but was widely purchased only in Great Britain, where it remains a national standard.

Simplicity was the great virtue of the vacuum brake. A vacuum was created in the train line(a system of pipes and hose running the length of the train) by a steam ejector on the locomotive. The pressure of the atmosphere then pushed a piston or caused a diaphragm to collapse, depending on the design employed. The motion was transmitted through a system of rods and levers to the brake shoes. The apparatus required no compressor or stor-

age tanks. The negative pressures involved in its operation made seals a relatively small matter.

Vacuum brakes also had some serious shortcomings. Before 1883 there was no American vacuum system with an automatic safety feature, such as the air brake had incorporated for a decade. If a hose broke or an ejector failed, the system was useless. The problem was solved in time, but meanwhile Westinghouse enjoyed an enormous advantage during a critical period in the history of power brakes. Once a railroad had adopted the air system, it was not likely to scrap the equipment just because an equally efficient vacuum system was made available. Another and perhaps even more important defect was the vacuum brake's lack of power. The ejector could create an effective pressure of 8½ to 11½ psi.[321] This figure probably reflects the vacuum's maximum efficiency at sea level. It would fall off with each foot of elevation, reducing its effectiveness on mountainous routes. The air brake, with pressures originally of 70 and later of 110 psi, was a much stronger brake, and power obviously became more vital as train speeds and weights increased. For short or light trains, the vacuum brake was very satisfactory, and it consequently flourished briefly during the first years of the power brake in America.

Vacuum brakes were patented as early as 1844 in England. Sixteen years later Nehemiah Hodge, already mentioned in connection with foundation gears, obtained a U.S. patent for a vacuum brake. Hodge created a vacuum with a steam ejector on the locomotive. Accordion-like rubber bags mounted under the cars would pull together or collapse, creating the motion to apply the brakes. It is not known if Hodge's brake was actually used. It was a precedent, and Westinghouse thought enough of it to purchase the patent and have it reissued in 1879.

The vacuum brake idea remained dominant until the success of the air brake prompted a general upswing of interest in the subject of power brakes. John Y. Smith, a partner in the locomotive manufacturing firm of Smith and Porter, lived in Pittsburgh, the city where Westinghouse introduced and perfected the modern air brake (Figure 7.50). Smith, an Englishman, had been active in American railway mechanical affairs since the time of the Civil War.[322] He was familiar with the requirements of the industry and must have recognized the potential market for power brakes that Westinghouse's system had only begun to satisfy. Smith selected the vacuum system and began public tests in 1871.[323] It is impossible to say with certainty whether or not he was aware of Hodge's 1860 patent, but his apparatus copied that scheme part for part. Beginning in July 1872, he obtained several patents which provided details and improvements, but the basic scheme remained Hodge's.

Smith did well in selling his apparatus. By the spring of 1873 it was in use on the Boston and Lowell, the New Haven, and the Central Railroad of New Jersey.[324] Within a year several more lines had found it satisfactory. A test on the Central of New Jersey showed that it was equal to the air brake: a twelve-car train with a 35-ton engine going 35 mph was halted in 555 feet.[325] The *National Car Builder* recommended the Smith

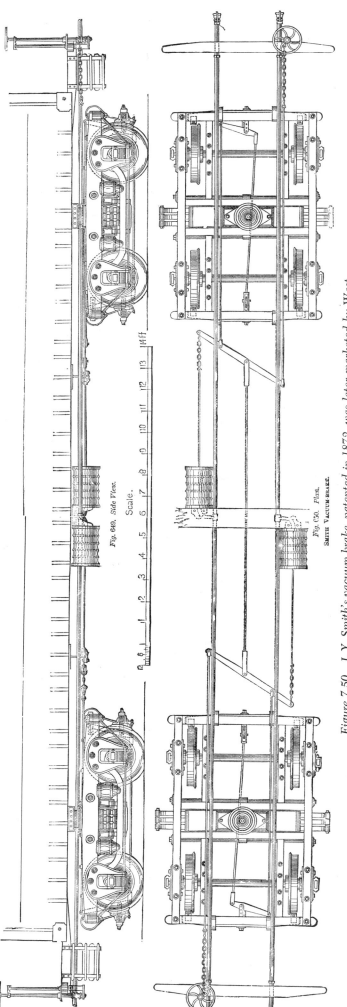

Fig. 649. Side View.

Scale.

Fig. 650. Plan.

SMITH VACUUM BRAKE.

Figure 7.50   J. Y. Smith's vacuum brake, patented in 1872, was later marketed by Westinghouse. (Car Builders' Dictionary, 1879)

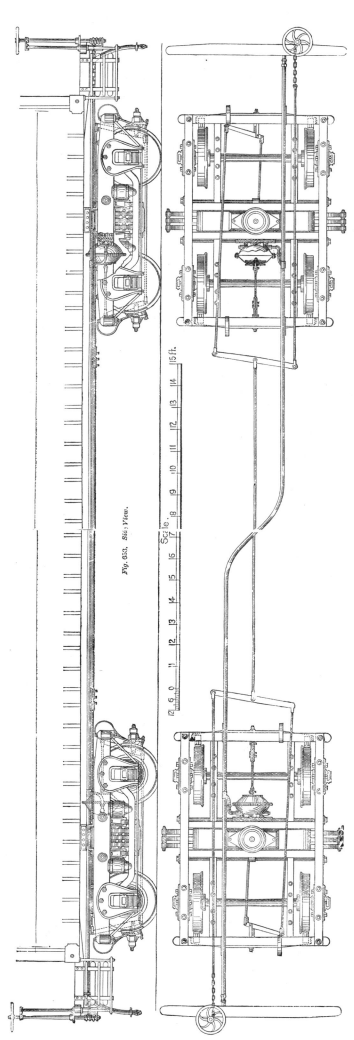

Fig. 653. Side View.

Scale

Figure 7.51   The Eames vacuum brake, introduced in 1874, had a moderate success in North America. (Car Builders' Dictionary, 1879)

apparatus, saying that while it was not as powerful as the Westinghouse brake, it was adequate for the requirements of most U.S. lines.[326] Its first cost was lower, and it did not require a "mint of money and a portable machine shop to keep it in good order." This statement was an exaggeration, but it contained enough truth to prompt Westinghouse to buy rights to Smith's patents, apparently in 1875 or 1876.[327] Westinghouse produced vacuum brakes on Smith's design for several years under his own name.

Smith continued to market his apparatus in England, with enormous success. In 1876 he was said to be taking over the British market, where his system was more widely used than the Westinghouse brake.[328] A few years later nearly 1,900 British cars were equipped with the Smith brake.[329] However, the fortunes of the Smith brake began to decline almost as rapidly as they had risen. A modern British railway historian has declared it "a pernicious thing" and has blamed Smith's apparatus for a series of accidents, culminating in the 1889 Armagh disaster on the Great Northern of Ireland in which eighty people died.[330]

The Smith brake was on the market for only a year before another major vacuum brake was offered by Frederick W. Eames of Watertown, New York. Eames received his first patent (No. 153,814) on August 4, 1874. His brake incorporated no remarkable features, although he did use a cup-shaped brake cylinder (Figure 7.51). The large cast-iron cup was about 30 inches in diameter. A rubber diaphragm was bolted to the periphery of the cup and collapsed into it when the vacuum was created. Like Smith and Hodge, Eames produced the vacuum with an ejector.

Eames took over an old water-powered stone mill building on Beebee's Island in the Black River at Watertown and began to manufacture his apparatus around 1876.[331] He had already run tests on the nearby Rome, Watertown, and Ogdensburg Railroad. Within three years, twenty-nine U.S. railroads were using Eames's brake. By 1881 sixty-seven different railroads in nine countries had adopted it.[332] The brake was particularly well received in South America; in Brazil, for example, nine out of eleven lines installed it.[333] But the true showcase of the Eames brake was the New York Elevated railways. Some 200 locomotives and 500 cars hurried up and down Manhattan Island; 3,500 trains made 70,000 stops daily, demonstrating the efficiency of the Eames brake before an enormous audience.[334] Here the vacuum brake made a good showing, because the trains were extremely light and the stops frequent. It had plenty of power to stop the diminutive locomotives and cars, and unlike the air brake, the vacuum system had no difficulty in recovering from a series of stops.

Eames knew that British railway operations were ideally suited to the vacuum brake. Many manufacturers, including his American rival, J. Y. Smith, were already demonstrating their wares in England. Seeking a way to dramatize his product abroad, Eames found a flashy express locomotive built in 1880 by the Baldwin Locomotive Works. The high-speed, single-axle engine had been repossessed by its maker and was awaiting a buyer.[335] Eames had the engine refurbished and made ready for road trials. (He

named it in honor of his father, Lovett Eames, a fact that has caused some confusion as to the inventor of this form of vacuum brake.) After a test on the New Haven, the demonstrator was shipped to England. In 1881 it appeared on several British lines and at the Exhibition of Life Saving Appliances. British railroads were under pressure from the Board of Trade, a powerful government agency, to adopt some form of continuous brake.[336] Legislation seemed imminent that would require such action, and Eames hoped to capture at least a portion of the windfall orders for power brake gear. But his expectations were not realized, for only two lines, the Lancashire and Yorkshire and the Rhymney railways, seem to have chosen his gear.[337] The English were perfecting their own vacuum brakes and did not need outside help. After all, they had gifted mechanics such as John Aspinall, who had perfected an automatic vacuum brake in 1879, years before any such device appeared in America.[338]

Eames returned home to develop an automatic vacuum brake. In order to be competitive with the Westinghouse and Aspinall systems, it must incorporate the fail-safe principle; that is, if the train parted or the ejector failed, the brakes must come on automatically. His first effort, a complicated two-pipe system devised in 1881 or 1882, was not dependable.[339] Early in 1883 he tested an improved duplex-automatic brake on the New Haven and acquired new capital for its manufacture.[340] Meanwhile a strike occurred at the plant and Eames, attempting to enter the works, was shot and killed by a striker on April 20, 1883.[341]

His successors in the company realized that the vacuum brake was not likely to survive in America. The export trade was doing well, but domestic sales fell off as more lines adopted the Westinghouse system. As late as 1886 the Eames company claimed that 200 railroads were using its equipment and reported one new customer, the Indianapolis, Decatur, and Springfield Railroad.[342] Two years later, however, the *Car Builders' Dictionary* stated that vacuum brakes were used only to a very limited extent in this country. The Smith brake was nearly extinct, being found only on the Long Island, and Eames's last stronghold was the New York Elevated.[343]

In 1885 the Eames company's new chief mechanical engineer, A. P. Massey (1842–1898), devised an air brake.[344] Westinghouse naturally sued, but the Watertown firm continued to push into the air brake field, encouraged by several railroads who wanted to block Westinghouse's growing monopoly. In 1890 the Eames company was reorganized as the New York Air Brake Company, and since the earliest Westinghouse patents had now expired, the new firm was free to copy their designs.[345] As air brake technology advanced, and as the need for complete interchange of cars became acute, Westinghouse and New York Air Brake were again in the courts. The litigation was at last resolved in 1912, when Westinghouse and its competitor agreed on a cross-licensing accord which covered all air brake patents.[346] Under this arrangement New York Air Brake's business expanded to a yearly capacity of 50,000 brake sets. Among its customers were such major roads as the New York Central, the Seaboard Air Line, and the Union Pacific railroads. Of the total air brake market, New York Air Brake held—and probably still holds—one-quarter; the great bulk of orders go to Westinghouse.[347]

The air brake is an important bench mark in the history of railroad technology. It came into being simultaneously with the opening of the transcontinental railroad, the introduction of the dining car, and Janney's first coupler patent—events which marked the opening of a new railroad era. There is a tendency to ascribe the invention of the pneumatic brake to George Westinghouse, Jr., alone. He was unquestionably an outstanding engineer and businessman; it may not be an exaggeration to call him a genius. Nevertheless, some admirers have overstated his contributions. In the dedication of *Development of the Locomotive Engine* (1907), Angus Sinclair eulogized Westinghouse as "The preserver of the Travelling Public whose great invention of the Automatic Air Brake has preserved more human lives than any military general or tyrant ever succeeded in destroying." Westinghouse was not a latter-day Prometheus who seized a lifesaving miracle from the heavens to save mankind. The modern railway brake was the work of many men. It evolved through countless successes and failures. Westinghouse benefited directly and indirectly from the experiments of his predecessors, although he claimed to have been unaware of their efforts and to have repeated past mistakes before arriving at the belief that compressed air was the answer.

The air brake was not Westinghouse's idea, nor was it a novelty when he began to work on his system in 1868. Numerous British and American inventors had previously patented and tested such apparatus. In fact, compressed air as a power source was almost as old as the locomotive, for compressed-air engines and vehicles had been proposed as early as 1800.[348] An Englishman, D. Crawford, patented an air brake in 1845; a fellow countryman, S. C. Lister, did likewise in 1848. Twelve years later John McInnes actually tested an air brake on a British line.[349] Meanwhile U.S. mechanics and inventors were trying to develop the same idea. In 1852, for example, Cutler and Rapp of Buffalo proposed a compressed-air brake that included all the basic features of the first Westinghouse system except for the steam pump on the locomotive.[350] Their pump was driven by the motion of the train. Several other designs were patented before Westinghouse even began to study the problem.

His attraction to the subject of railway brakes after he had been in a train wreck, his chance reading of a magazine article about digging railway tunnels with compressed-air machinery—these stories about Westinghouse have been often told. The important facts about Westinghouse are his background and his relationship with railway officials of the day. Westinghouse was barely twenty when he began to think about railway brakes, but he already had a good knowledge of practical mechanics. His father was a manufacturer of farm machinery, and George had worked in the family shop since he was a boy. A serious and ambitious youth, he took to the road in his late teens to promote a car replacer he had patented. During his travels he became acquainted with several Eastern railway officials. They liked and trusted the sober young man from Schenectady, and when he ap-

proached them later with an idea for a pneumatic brake, they did not dismiss him as foolish or impertinent. Two local railway superintendents, W. W. Card of the Pan Handle and Robert Pitcairn of the Pennsylvania, figured large in the early success of the Westinghouse brake. Their help was not forgotten; both became officials in the Westinghouse Air Brake Company.[351]

After he had considered chain and steam brakes, the idea for a pneumatic brake matured in Westinghouse's mind by July 1868, when he applied for a patent.[352] A reciprocating steam pump on the locomotive supplied compressed air to a tank located under the cab floor. When a valve was opened in the cab, air was sent back to the cars through a pipe, with flexible hoses between the platforms. The pipe directed the air into cylinders attached to the brake rigging. Although the scheme was hardly original, no one had managed to translate it into a dependable, working system. Westinghouse found backers in Pittsburgh, Robert Pitcairn among them. A Worthington feed water pump was rebuilt into an air compressor, and the other necessary parts were fabricated in a local machine shop. W. W. Card gave Westinghouse permission to fit his apparatus to locomotive Number 23, with four coaches, and make a road trial on a local train, the Steubenville Accommodation.[353] The test run began in September 1868, and the train had barely cleared the Pittsburgh terminal when the brake was thrown into emergency.[354] A team and wagon straddled the track; they were spared only because of the fast stop possible with the air brake.

Although the brake worked, it needed refinements before it was ready for the market. The steam pump valve gear was a troublesome detail that required reworking. While Westinghouse struggled to perfect the mechanism and refine the design for commercial manufacture, his patent (No. 88,929) was issued on April 13, 1869. More tests and demonstrations were needed to introduce the brake to the industry. Just a year after the first trial, the brake was shown to members of the Master Mechanics Association. A special train was run from Pittsburgh to Altoona, where the efficiency of the air brake could be exhibited at its best over the mountainous divisions of the Pennsylvania Railroad. In November of the same year, a ten-car train demonstrated the merits of the air brake to the directors of the railroad. It was then sent to Chicago, where its performance inspired enthusiasm. During one trial the train, moving at 30 mph, stopped in 19 seconds, or 380 feet.[355]

Orders began to flow into Pittsburgh from the West, New England, and elsewhere. In September 1870 the Westinghouse brake was declared beyond the experimental stage and was regarded "as a decisive solution" to the problem of brakes on passenger trains.[356] Eight railroads were now using the apparatus on 355 cars. Of these, 200 were on the Pennsylvania. In less than two years 85 railroads and 4,000 cars were equipped with the Westinghouse brake.[357] By 1874 the number of cars so fitted had nearly doubled. At the time of the Centennial, nearly three-quarters of the passenger equipment in this country had air brakes.[358] So rapid a conversion is almost without precedent in an industry famous for its wait-and-see attitude.

Several factors explain Westinghouse's success. These were the industry's need to improve the safety and speed of railroad service and each road's need to outperform its competitors. The need for a dependable power brake was obvious. Traffic density and train speeds had advanced markedly since the Civil War. The public was calling for greater safety measures. In a letter written in 1875 to John Garrett, president of the B & O, Mrs. Ella M. Williams said that she and some friends were returning to Baltimore from the West, and that before reaching Chicago their train stopped short of what might have been a major collision, thanks to the wonderful new brake invented by "Mr. Westinghouse."[359] Her party had planned to take the B & O for the final portion of their journey, but on learning that the Westinghouse apparatus was not employed by that road, they selected an alternate route. Mrs. Williams was from Baltimore and wanted to support her home-town railroad, but considerations of personal safety ruled out such loyalty. She asked Garrett, "Oh, why don't you make your cars safe too!" The message was clear: to keep passengers it was necessary to adopt the latest safety apparatus. When competing lines advertised such safety measures as Miller platforms, Baker heaters, and air brakes, there was nothing to do but follow their lead. Some political pressure was also brought to bear. In 1873 the Michigan legislature made air brakes compulsory on all passenger trains.[360] Other state governments showed similar inclinations, although state laws appear to have had much less influence on the railroads' decision to patronize Westinghouse than the pressure of public opinion itself. The passenger car fleet was rapidly and voluntarily converted to air brakes. It was the far larger and more costly outfitting of freight cars that required Federal legislation.

Service was another major reason for Westinghouse's success. He did not merely devise a workable mechanism; he instituted a systematic process of manufacture, quality control, and service in the field.[361] Repairs were simplified because parts were precisely made. Westinghouse was a meticulous businessman whose careful handling of orders and accounts was a relief to railroad managers weary of dealing with brilliant but disorganized inventors. Westinghouse was a marketer as well as an inventor. He made sure that dependable field representatives were available to answer technical questions, advise on repairs, and see that parts were available. He was careful not to sell apparatus that had not been absolutely proved. Devices that worked well on the test rack might fall to pieces under the rigors of everyday service or in the hands of unskilled workers. He ran continuous experiments to test the durability of his apparatus even after its efficiency seemed irrefutable. The triple valve, for example, was subjected to 460,000 applications before Westinghouse was satisfied that it was ready for production.[362]

Even during the early years of his success, Westinghouse knew that the original straight air system was defective. Conceptually it was pedestrian, a system that almost any mechanic might devise. In practice, however, it suffered from several serious problems. Because air had to be pumped from the locomotive the full length of the train, it took too long to reach the rear cars. So

much time elapsed that the front brakes were pulling hard before the rear ones began to take hold. This was an inherent problem of the air brake. The slow response of the rear brakes was due to the slow rate of travel of compressed air, which will move at 1,100 feet per second in the open but only 900 feet per second inside a pipe. There was no real way to resolve the difficulty, but the situation could be improved by tinkering with the mechanism, or by increasing the pipe size to speed the air flow. On short trains, the slow response was not critical, and because the air brake was used almost exclusively on passenger trains for its first fifteen years, Westinghouse could well have ignored the problem. But he was not satisfied; he wanted the braking to be uniform. This meant regulating the air volume and pressure at both ends of the train, something that could not be achieved with the old straight air system.

There was another and more pressing reason to revise the air brake: it was not really all that safe. If a hose burst, if the train broke in two, if the main air reservoir burst, or if the pump failed, the train would be without brakes. Westinghouse wanted a "fail-safe" system, so that if any of these events occurred, the brakes could still be applied. Moreover, in an ideal system they should come on automatically. At this point Westinghouse broke away from conventional thinking and began to evolve a truly epochal invention: the automatic air brake.

There is possibly even a third reason that Westinghouse felt a new design was necessary. The straight air system was hardly patentable; competitors could easily point to prior inventions. A growing business such as Westinghouse's would attract rivals, and he needed a specialized design to protect his valuable monopoly. Late in 1871 Westinghouse filed for a patent to cover his first work on the automatic system.[363] It involved an independent air apparatus for the cars, with an auxiliary tank and control valve under each car. If the train broke in two, the brakes would automatically set for an emergency stop. The cars might even have an axle-driven pump. The engineer would have no control over this secondary system, and the normal service stops would be made, as before, with the straight air system. Two patents were issued to Westinghouse in 1872. But he was not satisfied with the emergency system; he wanted the safety feature to work with the normal braking operation. And then in 1872–1873 came his inspiration for an air-actuated brake. In operation it was exactly opposite to the old system. The brakes were activated by discharging rather than charging the train line. A brain was needed: a pressure-sensitive valve that could admit air from the train line into the reserve tank, readmit air from the same tank to the brake cylinder, and exhaust the air to release the brake. Each car must have such a valve as well as its own air tank (Figure 7.52). This lively, sophisticated valve was called the triple valve because of its three basic duties. It was a remarkable chunk of cast iron no larger than a grapefruit, and so amazing to mechanics of its day that some of them half expected it to speak.

Thus the triple valve made two contributions. First, it achieved more uniform braking, because full pressure was available in

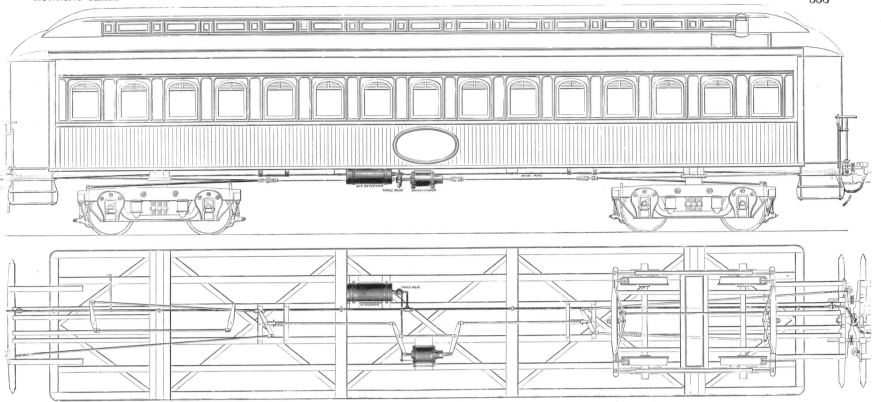

Figure 7.52   The Westinghouse brake, put on the market in 1868–1869, is shown here as it was improved in 1873, with a reserve tank under each car and the triple valve which automatically applied the brakes in case of an air failure. (Railroad Gazette, April 14, 1876)

each car's reserve tanks. Second, it provided a fail-safe operation, because any major reduction in the pressure of the train line (as would happen if it broke in two) caused the triple valve to pop open to full emergency. Normal service stops were made by small, controlled reductions in the train line pressure: the triple valve was designed to respond in $2\frac{1}{2}$ ratio. That is, if the engineer exhausted 10 pounds from the train line, the triple valves would feed 25 pounds of air into the brake cylinders. Before about 1900, the standard line pressure was 70 psi.

The first triple-valve patent was issued to Westinghouse in February 1873, and an improved version was patented on May 13, 1873, (No. 138,827). All subsequent air brake developments were an outgrowth of the 1873 automatic system. It was, of course, refined over the years to improve its transmission speed and power. The triple valve itself disappeared, to be replaced by a more complex and sensitive control valve. But even the most recent form of air brake follows the basic 1873 plan.

Acceptance of the automatic system, despite its olympian superiority over the old form, was not swift. The Pennsylvania managed to convert all its passenger cars by 1879, but the New York Central did not adopt the air brake in any form until 1880.[364] It was the last major railroad to do so and was criticized for this tardiness. There is a frequently told anecdote about the old Commodore scoffing at the idea of stopping a train with a whiff of air. Westinghouse specifically denied the tale in his 1910 recollections, but the Central's slow acceptance does lend plausibility to the story. At the national level, less than half of all air brake equipment was of the automatic design by April 1880.[365] At that time 1,770 locomotives and 5,218 cars were equipped with the new or automatic system, while 2,677 locomotives and 10,244 cars had the old straight air equipment. In 1881, 12,270 passenger cars had automatic brakes, while 11,389 still had the

old system—a great increase in use of the new brake in only one year.[366] But progress was not fast enough. The dangers of the old system were tragically demonstrated by three wrecks on the B & O in 1887. The first and most serious occurred at Republic, Ohio, in January, when an express running at 65 mph crashed head-on with a freight, killing thirteen people. The straight air brake was said to be partially responsible.[367] A few months later, part of a passenger train broke away because of a defective coupler and rolled downgrade on the most mountainous portion of the line, near Terra Alta, West Virginia. Again the straight air brake was blamed. In November an express trying to make up time ran out of control and derailed in Washington, D.C., an accident also attributed to the old-fashioned brake.[368] (It is possible that the B & O was still using the Loughridge air brake, which did not incorporate the patented automatic features of Westinghouse.)

If certain lines were negligent about buying the latest apparatus available, at least they were using some form of power brake. By the early 1880s it appears to have been universal on American passenger cars. In 1882 Westinghouse claimed that 36,218 cars of all types had his equipment.[369] During the next year he claimed 55,000 cars on a worldwide basis, but said that the majority were running on domestic lines.[370] To accommodate this booming trade, he purchased a cotton mill in Allegheny City (north Pittsburgh) for conversion into a new air brake plant.[371] Two years later the facility had to be enlarged, and by the late 1880s the Allegheny works was too small to handle the increasing orders for freight brakes. Westinghouse decided to build an industrial complex in Turtle Creek Valley, east of Pittsburgh. He had ridden countless miles through this valley while testing the triple valve on a local train called the Walls Accommodation. He located his signal company, the Union Switch and

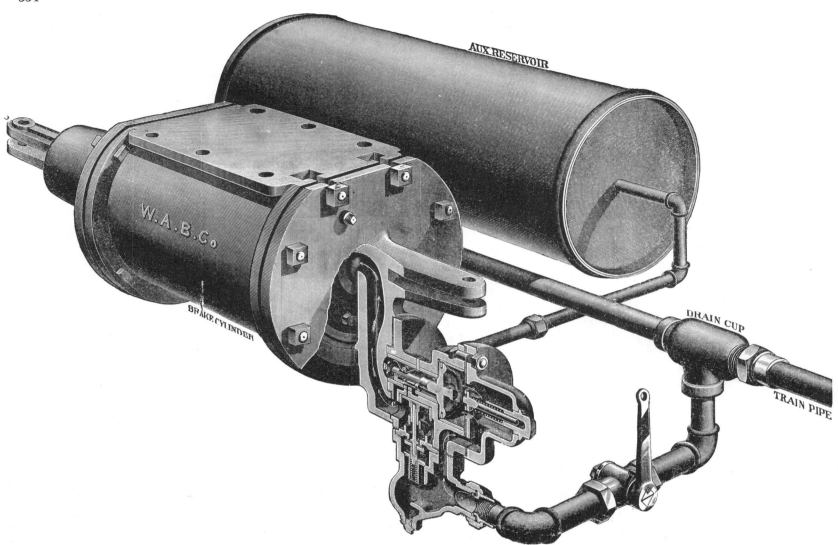

Figure 7.53   The Quick Action brake was introduced in 1887–1888 to speed the braking on long freight trains. It was also used for passenger equipment.

Signal, in the valley town of Swissvale and erected a new air brake plant in Wilmerding. Money was no problem, for the firm paid a 400 percent dividend in 1886.[372] The new plant opened in 1890 and has remained the home of the business, which currently calls itself WABCO.

Westinghouse had a Pullman-like compulsion to dominate the field; he sought to buy out competitors and develop foreign sales. Many of his rivals were self-eliminating but others could only be overcome by absorption. His purchase of the Smith vacuum brake patents was the first acquisition. In 1898 he bought the Boyden Air Brake Company of Baltimore, and not long afterward the American Brake Company of St. Louis.[373] Only the New York Air Brake Company managed to withstand his patent suits and stock offers. In 1872 Westinghouse began to push for overseas sales, organizing an export branch and establishing a shop in England.[374] Seven years later he opened a Paris branch, and in subsequent years formed companies in Germany and Russia.

Meanwhile Westinghouse was also working on the air brake. The railroads at last seemed ready to make the tremendous capital outlay necessary to apply power brakes to the freight car fleet. This would involve not just thousands of units, but more than a million cars. The Master Car Builders scheduled a test near Burlington, Iowa, in July 1886. Of the brakes tested, only the West-

inghouse air brake and the Eames vacuum brake proved successful. But they were hardly perfect. The air brake's old problem of delayed action proved a near-disaster on the fifty-car test train. It took 22 seconds for the rear brake to take hold in an emergency stop, and the rear of the train piled up against the lead cars with a thunderous crash.[375] Westinghouse was aware of the need for quicker action in uniformly reducing the air throughout the system. The solution was found in venting the air at each car for emergency stops. Westinghouse developed a new triple valve that achieved serial venting: rather than bleeding the air off at one point, the apparatus now discharged it at fifty points. During the second Burlington tests of 1887, the action of the brake was improved from 22 seconds to only 6 seconds. Westinghouse was determined to do even better. He made other changes in the quick-action triple valve and provided a larger train line—$\frac{1}{4}$ inch greater in diameter than the old standard 1-inch pipe. These improvements helped speed the action on a fifty-car train to a miraculous $2\frac{1}{2}$ seconds. The story of the quick-action brake is normally told exclusively in relation to freight cars, but it was also the successor to the original 1873 automatic brake for passenger cars (Figure 7.53). The development of this design was the last creative association that George Westinghouse had with air brake technology. Hereafter his associates, such as W. V. Turner, took charge of design improvements.

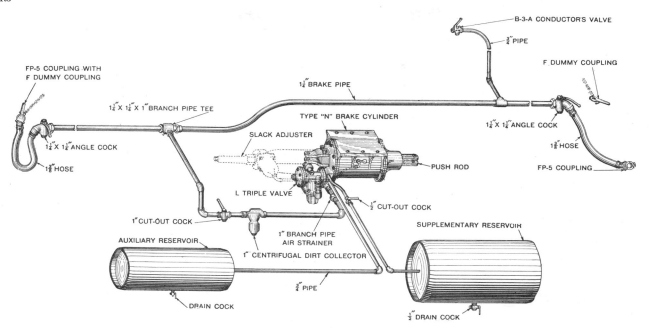

Figure 7.54   The LN brake, which appeared in 1908, was one of a series that improved the power and responsiveness of the air brake. (Car Builders' Dictionary, 1919)

The industry's languid application of the quick-action brake to the freight car fleet is well documented in the railroad trade press and subsequent writings on the 1893 safety appliance act. But the applications to passenger cars are less well recorded. Some indication of the pace of conversion may be gained from the example of the Pennsylvania Railroad, then one of the most progressive in the country. Late in 1893 it was reported to have nearly completed the conversion from automatic to quick-action equipment.[376] If it took the most liberally disposed roads five years to make the change, how long must it have taken the most conservative?

The quick-action apparatus had no sooner become a standard than the industry discovered that a more powerful system was needed for high-speed passenger trains. In the early 1890s some very fast trains were introduced, such as the Empire State Express. But the speed of these trains was no advantage unless they could be safely controlled. In 1894 Westinghouse introduced the high-speed brake, which increased train line pressure from 70 to 110 psi.[377] A reducing valve was attached to the brake cylinders. In an emergency stop it permitted very high-pressure air to work the brakes; as the train slowed, it gradually cut back on the pressure so as not to lock the wheels. This device was based on a discovery made in 1878 by George Westinghouse and Douglas Galton during a series of railway brake tests conducted in England. They found that the friction between the brake shoe and wheels is considerably *less* when the wheels are revolving rapidly than when they are revolving slowly.[378] Hence initially, a very high brake shoe pressure is important to stop the train in the shortest possible distance. However, the pressure must be carefully and automatically controlled to avoid rough stops, flattened wheels, and skids. Sliding wheels adversely affect the braking process. The new high-speed brake functioned well and registered stops 450 feet shorter than the quick-action brake.[379]

From this time forward, passenger and freight car brake apparatus went separate ways. The K (1905) and later the AB (1933) brakes became the standards for freight service. During the same period passenger cars went through a half-dozen major changes in brake styles, each representing a refinement upon a refinement. All aimed for greater power and smoother stops.

Moreover, all these systems had to be designed to work together, and while they might be quite different in details, all had to be capable of functioning interchangeably, considering the mix of cars possible. That is, an aging baggage car at the head of the train might have a PM style of brake, while the club car would have the latest HSC gear.

In 1902 an improved version of the high-speed brake was placed on the market.[380] It featured a new style of triple valve and air cylinder; otherwise it was much the same as the 1894 design. In 1908 the LN brake was introduced, again with a new triple valve and cylinder (Figures 7.54 and 7.55).[381] It featured a quick recharge that permitted successive brake applications with no material loss in the braking force. A second air tank aided in this operation.

With the coming of the steel car, trains suddenly outgrew their brakes, for train weight tripled. In 1890 a 280-ton train going 60 mph could be stopped in 1,000 feet.[382] In 1914 a 920-ton train traveling at the same speed took 1,760 feet to stop with the 1890 style of brake. Westinghouse was called upon for a quick solution. Increasing the line pressure much above 110 psi was not considered practical, nor was an increase in cylinder size, which was already 18 inches in diameter.[383] Multiplying the leverage of the foundation gear beyond the 9 to 1 ratio resulted in dragging brake shoes, slow releases, and added maintenance. Clasp brakes were of some benefit but did not really solve the problem. The best interim solution was to mount two brake cylinders per car. Thus in 1910 Westinghouse introduced the PC brake, which could stop a train of steel cars traveling at 60 mph in 1,100 feet (Figure 7.56). This was an improvement of almost 40 percent over the older forms of high-speed equipment.

In 1914 the PC brake was replaced by a more powerful successor, the UC brake, which incorporated three air tanks and a control valve so radically different that it was called a universal rather than a triple valve (Figure 7.57). This brake could stop a heavyweight train traveling 60 mph in 860 feet.[384] The UC was a longtime standard for passenger equipment in this country.[385]

Just before the opening of the lightweight era, a new electrically controlled brake, the HSC, was placed on the market. Electric pneumatic brakes had appeared at the Burlington tests in

*Figure 7.55   A better understanding of the brake rig and apparatus can be gained from
this rare 1913 underbody photograph. LN apparatus is shown.*

1887–1888. For a time Westinghouse had reluctantly believed that electric control was a necessity, but he did everything in his power to avoid the introduction of a secondary system that might fail and jeopardize the basic operation. His company successfully avoided electricity for many years, but at last it became apparent that the relatively slow-acting, purely pneumatic system was not good enough. To guarantee safety, however, the HSC brake was not dependent on the electrical control. Without electricity its action slowed down, but it could still function pneumatically if

the circuit failed.[386] The HSC was first used in 1932. Its superior power and extra-smooth stopping characteristics made it a standard for the lightweight streamline trains then coming into service. Electric control improved stopping capability by 40 percent.[387]

An even better form of electropneumatic brake, the D-22, was introduced in 1935. It broke away from traditional running-gear design by mounting the brake cylinders on the trucks rather than under the body of the car (Figure 7.58). This eliminated the

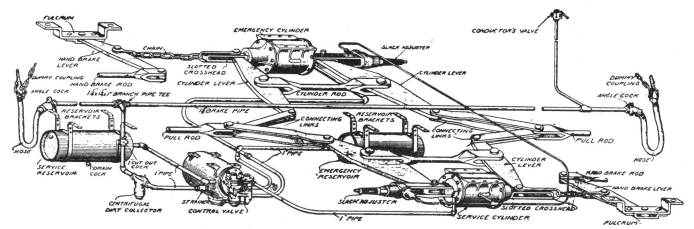

Figure 7.56 *In response to the need for more powerful brakes on steel cars, the twin-cylinder P.C. brake was developed in 1910.* (L. K. Sillcox, Mastering Momentum, *New York, 1941*)

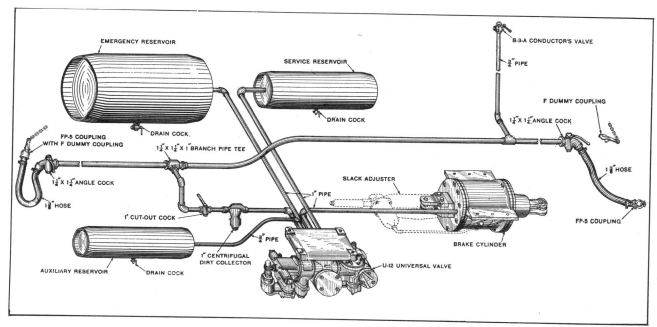

Figure 7.57 *The UC, placed on the market in 1914, became an industry standard.* (Car Builders' Dictionary, 1919)

foundation gear. Two or four small brake cylinders attached to each truck's side frame were connected directly to minimal brake rigging of the truck. This saved weight and the lost motion of the long levers and rods typical of the old system. The smaller cylinders also used less air.[388]

Because of the decline in railway passenger travel, recent developments have been sparse in this area of railroad technology. One of the latest improvements was a pneumatic-hydraulic brake introduced by the New York Air Brake Company in 1969.[389] It features a high-pressure hydraulic system (800 psi) controlled by, and interchangeable with, conventional air brakes. The apparatus has been applied to over 120 electric suburban cars built for the Illinois Central Railroad in 1971. It is possible that this brake or some equally radical departure will at last displace the present standard D-22.

Brake Shoes

Train stopping involves some basic principles of physics. Because energy cannot be destroyed, the kinetic energy or motion of the train must be converted into some other form of energy. When we think about energy conversion, we tend to imagine the transformation of heat into energy. In braking, however, just the opposite occurs. The heat generated by the brake shoes bearing against the wheels dissipates the train's forward motion. The brake shoes and, to a lesser degree, the wheels are ground away in the process. The shoes, being more expendable, are made of a comparatively soft material such as cast iron. The ideal brake shoe is also an efficient radiator that can dissipate large amounts of heat without rapid self-destruction. Again, cast iron proved well suited for this purpose. Practical as it was, however, cast iron did not produce the perfect brake shoe, and car builders tried a variety of other materials.

Cast iron appears to have been used only rarely during the first two generations of American railroading. Wood was the principal brake shoe material then. A half-dozen pioneer mechanics, including Wilson Eddy, Reuben Wells, and John Kirby, were interviewed on the subject in 1891.[390] None could recall the use of cast-iron shoes much before 1855, though after that time they became an industry-wide standard. In the early period white oak was favored, but beech and hickory were also used. Some lines

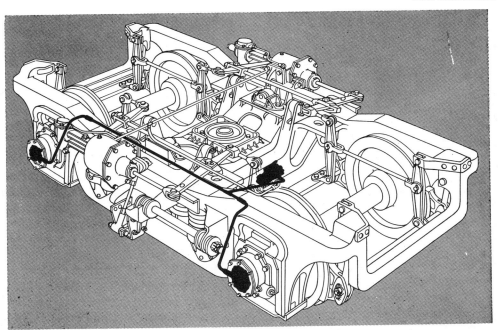

Figure 7.58   Lightweight equipment and faster trains called for more powerful pneumatic brakes, such as the model D22 shown here. The black silhouette is a control system called a decelostat that prevented wheel skids by automatically adjusting the air pressure. (Car Builders' Cyclopedia, 1953)

nailed leather to the brake heads or blocks, as shoes were often called at this early date. The great defect of the wooden shoe was its tendency to wear or char. As train speeds and weights increased, wooden shoes would simply burn off. The smoke was frightening to passengers, and the flaming shoes endangered the entire carriage and its occupants during the wooden age. The burnt-off shoes posed an even greater threat, because they left the train without brakes.

To overcome this problem the Boston and Providence tried stone shoes in 1853, but they proved unsatisfactory, presumably because of excessive tread wear. The Madison and Indianapolis, which had a surplus of strap rail after rebuilding its tracks with T iron, began shoeing its wooden brake heads with the short pieces of the discarded rails. The straps became so hot, however, that the blocks charred, the shoes loosened in time, and the experiment was abandoned. The road then adopted cast-iron shoes. And what happened in Indiana happened elsewhere, with very few exceptions. The B & O installed wrought-iron shoes and continued to use them on passenger cars until 1882, when the line at last made a tardy conversion to cast iron. The Hudson River Railroad alone remained loyal to wooden shoes into the 1870s.[391] As late as 1894, efforts were made to revive the wooden shoe.[392] The Wharton Switch Company placed a cast-iron shoe on the market that had end grain oak inserts. The Wharton shoe was reported to be widely used on street railways, where it accumulated mileages up to 6,000 miles, but it never seems to have been accepted by any steam lines.

Cast shoes were a national standard before 1870. They were a simple article that any foundry could produce, but it was not so easy to make a really good shoe. Indifferent makers turned out shoes with hard chilled surfaces or poorly mixed irons that produced hard spots and cut grooves in the wheel tread.[393] George M. Sargent, a major shoe manufacturer, admitted that after fourteen years in this specialty he was unable to determine the optimum mixture for the perfect brake shoe. He knew that uniformity of the metal was paramount, and that it called for a degree of quality control not found in any but the best-managed

foundries. So long as purchasing agents sought the cheapest supplier, he said, the railroads would be plagued with poor shoes.

The rapid wearing away of the shoes required frequent and troublesome renewals. The "car knocker" had no more miserable job than to crawl under the trucks and fit on a new set of shoes. In the wintertime especially, he could think of the most imaginative excuses for putting it off. Some roads were inclined to stay with wrought-iron shoes because they lasted five or six times longer than cast iron.[394] But wrought iron tended to warp and curve when heated. The metal rolled and bunched up at the ends, and the surface became hardened and slippery. Wrought-iron shoes might outlast cast iron, but they were less effective in stopping the train.

It occurred to Isaac H. Congdon (1833–1899), master mechanic for the Union Pacific, to combine wrought and cast iron in a brake shoe. The body was made of cast iron, with wrought-iron segments cast in the front wearing surface. These composite shoes were said to outwear the common pattern by three to four times.[395] They cost 4 cents a pound, or only 1 cent more than conventional shoes. Congdon patented the idea in March 1876. Manufacture began the same year in Chicago by George M. Sargent, who developed Congdon's invention into a widely used variety of brake shoe.[396] Congdon shoes were still being produced in 1909.[397]

A variation on the Congdon shoe was developed by James Meehan, master mechanic on the Cincinnati, New Orleans, and Texas Pacific. Meehan embedded pieces of hardened tool steel in the face of the shoe to cut away the surplus metal rolled over the wheel tread's outer edge. Such shoes helped to maintain the correct tread profile on locomotive and steel tired passenger car wheels. They partially eliminated the costly procedure of the wheel turning lathe. It was still necessary to turn the wheel sets after extended mileage, but the Meehan shoes were effective in maintaining the profile for long periods of time. Meehan patented the idea in 1884. In later years it was known as the Ross-Meehan shoe.

More important than these improvements was the introduction

Figure 7.59 *The Christy brake shoe, patented in 1865, permitted quick replacement by removal of the key that held the shoe to the brake head.* (*William Voss,* Railway Car Construction, *New York, 1892*)

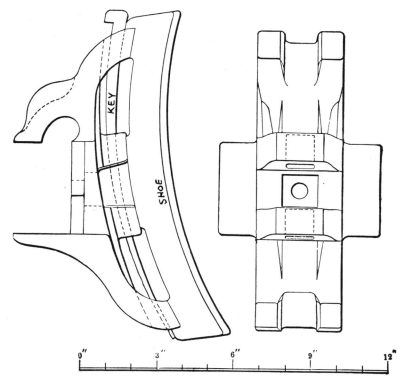

of demountable brake shoes by James Christy of Philadelphia in 1865. Christy offered a quick-change shoe held to a bracket or head on the brake beam by a long, tapered spike-like key (Figure 7.59). It greatly simplified the changing of shoes—the job could be done in one minute. By 1885 fifty to sixty railroads were using the Christy shoe. A year afterwards it was adopted by the Master Car Builders Association as a standard. The Christy patent involved another very desirable feature: the back of the shoe and the head were separated so that air could circulate and cool the shoe.

Ordinarily only one brake shoe was used per wheel. This remains true for freight cars, but during the first decade of the twentieth century the need for greater braking power prompted experiments with double or clasp brakes. A shoe was placed on the opposite sides of the wheel. Their vise-like grip not only doubled the braking surface but also helped to hold the axles and boxes in alignment. The idea was suggested at an early date, but no serious efforts to perfect it were made until 1887, when the Pennsylvania tried it on two cars.[398] The test failed because of a poorly designed brake gear. The clasp brake made little progress until 1909, when Thomas L. Burton of the Reading began to devise a better foundation plan.[399] A car fitted with the double brake stopped in two-thirds of the distance that a conventional car required. The moderated pressure resulted in less shoe wear and added to wheel life by lowering tread pressure. Shoes lasted for 25,000 miles in some cases. Shoes were so cool after emergency stops at 80 miles per hour that they could be touched with the bare hand. By August 1912 clasp brakes were standard on the Reading and the Central of New Jersey. As the heavyweight cars grew in number, so did the acceptance of the clasp brake.

A more radical departure was offered as railroads moved into the lightweight era. The time-honored shoe bearing against the running edge of the wheels was discarded. In its place was an axle-mounted drum brake not unlike that used for automobiles. It was felt that the wheels were doing enough to guide and support high-speed trains; they should not be required to stop the train as well. High-speed braking was destructive to car wheels, because the resulting high tread temperatures resulted in thermal cracks and shelling. These bad effects could be entirely avoided by installing a separate "wheel" or drum for braking. As described in Chapter 1, drum brakes had been tried in the 1830s on the Mohawk and Hudson Railroad (Figure 1.61). It was a scheme forgotten for nearly a century, until in 1936 Edward G. Budd asked one of his engineers, C. L. Eksergian, to develop a modern car brake.[400] Eksergian's first idea was to adapt the automotive drum brake, but the shoes, being inside of the drum, could not dissipate the heat quickly enough. The design was not practical for so large a vehicle as a railroad passenger car. An open disc with shoes pressing on both sides offered the cooling characteristics needed. The disc was actually a thick cast-iron drum with ribs or blades at its center (Figure 7.60). It was a combination drum and blower, whose 7,000 square inches proved effective as a heat exchanger. The drum was bolted to the wheel hub. Small air cylinders, one for each drum, operated tong levers

fitted with asbestos composition shoes. After a series of test-stand experiments, the disc brake was tried on a Burlington coach early in 1938. During the next year the lightweight motor train General Pershing, also on the Burlington, was fitted with the new form of brake. The results were so encouraging that the Santa Fe and the Milwaukee railroads ordered several sets. Budd claimed that the disc's cooling action was so effective that no checking of the drum or dusting of the brake linings could be detected.[401] Stops were made quietly—the usual screeching was eliminated. This was of great benefit to the passengers, for tests showed that they often complained about a train's poor ride when their discomfort was actually caused by noisy brakes.[402] Budd gave the following figures on the brake's efficiency in train stopping:

60 mph	Train stopped in less than 1,000 feet.
80 mph	Train stopped in less than 1,600 feet.
100 mph	Train stopped in less than 2,500 feet.

It was claimed that the shoes lasted 100,000 miles, while ordinary metal shoes wore out in 6,000. Wheel life was increased by 70 percent. And the dead weight of the conventional brake rigging was eliminated.

Remarkable as these advantages appear, there was no rush to adopt the disc brake. In 1946, nearly eight years after its commercial introduction, only 200 cars were equipped with it.[403] More interest was evident in the postwar years. By 1949, 500 cars carried the Budd brake, and by 1960 the number climbed to 2,200, a sizable portion of the passenger car fleet.[404]

The resistance to the disc brake can be explained in part by the railroads' normal rejection of new ideas, but it was also due to improvements being made in the conventional brake shoe. In the mid-1950s Westinghouse introduced a high-friction composition shoe composed of rubber, asbestos, and powdered iron. It offered faster stops and more mileage than cast-iron shoes. Several years later the Griffin Wheel Company marketed a similar shoe that outlasted iron shoes by four to five times and required only one-third to one-half the braking force for a stop.[405]

*Figure 7.60   Shoe or tread brakes were standard on American lines until 1938, when
disc brakes were introduced by the Budd Company. (Pullman Neg. 64008)*

Couplers

If railroad cars were permanently fastened together, the simplest form of attachment would be possible. But cars are constantly being coupled and uncoupled in the normal course of railway operations. Trains are made up at the terminals before departure. Cars are then picked up and set out during the run, and the train itself is generally dismembered upon arrival at the last station. Despite the obvious need for an efficient and convenient coupling device, American railways were long content with the most primitive methods for connecting cars. As late as 1870 the link and pin coupler remained common on passenger cars. And this primeval device predominated on freight cars until almost the end of the nineteenth century.

Not much thought was given to the very first car couplers; a few links of chain were considered good enough. Figure 1.2 in Chapter 1 shows such rudimentary couplings. The Mohawk and Hudson cars of 1831 had chain links that were dropped over pins fastened to the car's frame. A hasp-like iron strap was hinged (or possibly sprung) over the pin to keep the chain from losing its purchase. The plan was simple, but the links provided too much slack, resulting in whiplash starts and thunderous stops. The shortcomings of the chain coupler were recorded by a passenger

who described the inaugural trip of the Mohawk and Hudson:[406]

When he [the conductor, Mr. Clark] finished his tour, he mounted upon the tender attached to the engine, and, sitting upon the little buggy-seat, as represented in our sketch, he gave the signal with a tin horn, and the train started on its way. But how shall we describe that start, my readers? It was not that quiet, imperceptible motion which characterizes the first impulsive movements of the passenger engines of the present day. Not so. There came a sudden jerk, that bounded the sitters from their places, to the great detriment of their high-top fashionable beavers, from the close proximity to the roofs of the cars. This first jerk being over, the engine proceeded on its route with considerable velocity for those times, when compared with stagecoaches, until it arrived at a water-station, when it suddenly brought up with jerk No. 2, to the further amusement of some of the excursionists. Mr. Clark retained his elevated seat, thanking his stars for its close proximity to the tall smokepipe of the machine, in allowing the smoke and sparks to pass over his head. At the water-station a short stop was made, and a successful experiment tried, to remedy the unpleasant jerks. A plan was soon hit upon and put into execution. The three links in the couplings of the cars were stretched to their utmost tension, a rail, from a fence in the neighborhood, was placed between each pair of cars and made fast by means of the packing-yarn for the cylinders, a bountiful supply being on hand. . . . This arrangement improved the

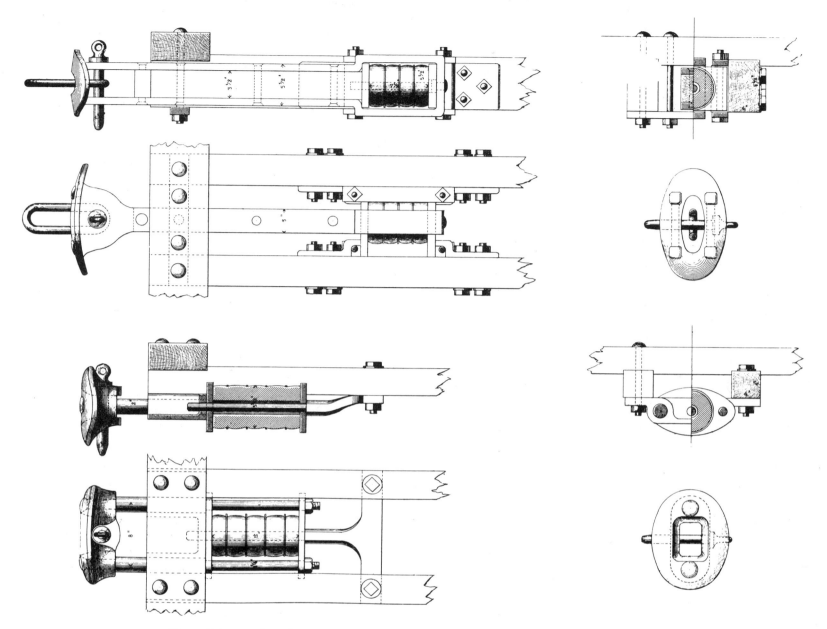

Figure 7.61   Link and pin couplers were the basic passenger car union until the 1870s, and they prevailed in freight service into the 1890s. The top four drawings are of a typical coupler; the bottom four show the New Jersey Railroad and Transportation Company's standard. The drawings date from 1859. (A. Bendel, Aufsätze Eisenbahnwesen in Nord-Amerika, Berlin, 1862, plate 11)

order of things, and it was found to answer the purpose, when the signal was again given, and the engine started.

A better scheme was soon evolved that remained a standard on American railways for many generations. A hollow iron tube about 5 inches square was fastened to the underside of the end platform. The end that projected out a few inches from the platform had a face plate. A hole was bored through the top and bottom of the drawhead to receive a pin. The pin held a link inserted through the face plate. It was a loose, rudimentary sort of fastening that a country blacksmith might have devised (Figure 7.61). Pictorial evidence indicates that this form of coupling, commonly called the link and pin, was in use by the mid-1830s. In the drawing of the B & O's Washington branch car of 1835, the coupler is placed well back under the platform, the pin being dropped through the floor (Figure 1.70). A very long link would be required. It should also be noted that the drawheads were

connected to an iron drawbar running the full length of the car. This device protected the body and frame from the draft strains. A more conventional link and pin coupler is shown in the Boston and Providence eight-wheel car drawing of 1838 (Figure 1.75). It was the type commonly used until the coming of the automatic coupler.

An ordinary link was made of 1½-inch-diameter bar stock, slightly flattened. It measured about 13 inches long by 5 inches wide. Links weighed 6 to 8 pounds each. The pin was made of the same-sized wrought-iron bar and also generally measured just over 1 foot in length. These rugged chunks of iron were intended to withstand the heavy draft strains as the train pulled and surged along the tracks. Some mechanics thought that it served its office too well under emergency conditions, and that it might be better if the link were designed to break under an unusually heavy strain such as a derailment. If the link snapped at the juncture of the derailed car, the wreck would be con-

Figure 7.62   An ironbound wooden coupler link used on the Phila-delphia, Germantown, and Norristown. Links of this type were intro-duced in the 1830s. (Smithsonian Neg. 16204)

tained. The rest of the train could remain on the track instead of being pulled off to the uncertain end that awaited the errant car or locomotive. Sometime before 1840 the B & O began using coupling links made from wood, of all improbable materials.[407] They functioned admirably by snapping off at the critical moment, leaving the nonderailed cars standing unharmed on the tracks. One report stated that this safety coupling saved a train of cars when a cow derailed a locomotive near Cumberland. It was regrettable, the *American Railroad Journal* said, that so few lines were using the superior wooden fastening. The Reading and its subsidiary, the Philadelphia, Germantown, and Norristown, also adopted the wooden link.[408] A link from the P G & N is in the Smithsonian Institution's collections (Figure 7.62). It is made from an oak plank measuring 15½ inches long by 4 inches wide by 2 inches thick. Two ¼-inch iron bands are riveted around its edge, but note that they do not meet in the center. The link is weakest at the point of separation. The pinholes are lined with small iron bands to increase their service life. Such links were in use until the early 1870s, when the air brake's slow releases put such a strain on the wooden couplings that they would break when the engine started up.

The link and pin coupler remained popular because of its low first cost and simplicity. But was it actually economical? Certainly not when its record of human misery is considered. In maiming and killing it was without equal; crushed fingers, hands, and ribs were its specialty. Trainmen could recognize one of their fraternity anywhere by his missing finger or mutilated hand. And the link and pin was also uneconomic from an operational point of view. Even without the accident claims, railroad managers began to realize that the coupler was expensive because of the prodigious loss of links and pins. In 1875 an official of the New York Central said, "Millions [of pins] are made every year but no one knows what becomes of them."[409] A relatively small Indiana trunk line found that it lost over 43,000 links and pins annually.[410] The Lake Shore calculated the cost of such losses at over $52,000 a year, or $5.20 a car.[411] This material did not just innocently disappear; it was stolen. The Atlantic and Great Western found that couplers were being sold to local junkmen.[412] In one yard investigators discovered twelve barrels of links and pins that had been labeled as old iron. In a nearby lumbering camp they saw a huge logging chain, worth $1,500, made from coupling links.

Railroads sought ways to overcome this financial drain, but there seemed no efficient way to police the thieves. Cars were exposed in yards and on sidings, and much of the thievery was probably carried out by railroad employees. The Pennsylvania, which retained the old-style coupler for passenger cars long after most roads had switched to Miller couplers, solved the problem by securing the link to the drawhead with a secondary rivet (Figure 7.63).[413] Other roads searched for ways to lower the cost of manufacturing links and pins. The Central Pacific devised a machine that turned out 1,000 links a day, whereas only 200 could be made in a day by the old hand method.[414] The process was further automated by the nation's leading maker of such

materials, the Pittsburgh Forge and Iron Works. In 1880 it had a machine that would cut, bend, and scarf (bevel the ends of) 6 to 7 links a minute.[415] The weld was formed under a steam hammer in ¼ minute. Another forging machine pounded out a matching pin every 1¼ minutes. Six years later a new pin-making machine was producing 35 pins a minute, and an improved link maker, about ready for use, was scheduled to complete 25 links a minute.[416]

A better way of reducing the financial drain was to eliminate the link and pin coupler itself. An automatic coupler would have no loose pieces to be pilfered, and the savings thus realized would ultimately pay for the new device. Furthermore, there was a positive passion among inventors during much of the nineteenth century to create an automatic railroad coupler. In 1897 1 percent of all U.S. patents involved car couplers.[417]   Of 28 patents related to railroad devices issued during the week of September 5, 1882, 11 were for car couplers. The total number climbed year by year. In 1873 there were 800 car coupler patents; by 1875, 900; by 1884, 2,300; by 1887, over 4,000; by 1897, 6,500; and by 1930, long years after the Janney coupler's adoption, 11,813.[418] Coupler patentees became a nuisance that railroad men avoided at all cost. Mechanical officers were exhausted by their entreaties, models, visits, and prospectuses. The enthusiastic inventors were said to represent "baffled hopes and misdirected labor" more fully than any other single segment of the population. The scheme of one rustic mechanic was likely to be mistaken for a threshing machine component. Another resembled a torpedo, and the car coupler of Syrus Washington Wright of Eden, Alabama, was construed as an unduly complicated steam engine valve gear. Unfortunately, sensible ideas were often buried in the overwhelming debris of foolish plans. It seems fair to say that the very intensity of interest in the subject delayed rather than facilitated the introduction of automatic couplers.

Suggestions for a self-acting coupler appeared as early as the late 1840s. In 1847 A. G. Heckrotte of Washington, D.C., began to promote a device that a Baltimore paper felt would "no doubt supersede the old fashioned bolts and chains."[419] The drawbar was a box, open at one end, with a spring trundler much like that of a gun lock, and a roller inside. A side lever released the trundler. A solid iron bar in place of the usual link made the connection. It had notches at both ends which snapped in between the jaw formed by the trundle and roller (Figure 7.64). The device was actually tested on the B & O but was not adopted.

David A. Hopkins (1824c.–1889) of Elmira, New York, patented a semiautomatic coupler on August 1, 1854 (No. 11428), That drew some favorable attention (Figure 7.64). It involved a modified link and pin drawhead with a block inside that prevented the pin from falling in place until the block was pushed back by a link from the opposing coupler. The pin was then free to drop. The scheme was simple, and because it closely resembled the standard form of coupler instead of some outlandish contraption, it was rather well received. The Hopkins coupler was used on the Erie for nearly five years.[420] The inventor claimed that it was not more widely adopted because of defec-

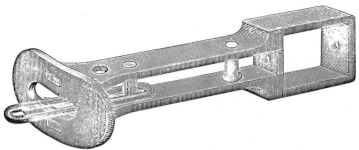

Entire length, 18½ inches.   Iron, 4 by 1 inch.   No. 1 Face Plate.

Figure 7.63 The link in the coupler here was permanently attached to prevent its removal or loss. The engraving dates from about 1875. (DeGolyer Foundation Library)

tive manufacture rather than poor design.[421] It is also probable that the extra cost discouraged sales. As one railway manager said, "Your device would be more expensive than Paddies are."[422]

Just a year after Hopkins's coupler appeared, J. T. England, a minor official of the B & O, unveiled a semiautomatic coupler that had some limited use. England followed the same general idea of the Hopkins coupler, but instead of a spring-actuated block he employed an iron ball. The ball blocked the pin, holding it up. When the link entered the drawhead, the ball was pushed back into a cavity out of the way and the pin was free to fall in place (Figure 7.64). England received a patent (No. 13869) on December 4, 1855. Endorsements were offered by several railway mechanical men, but the device seems to have been an ingenious curiosity rather than a standard product.[423]

Literally thousands of other schemes were patented; the few that have been described represent some of the better-received early automatic coupler designs. Their lack of success can be ascribed as much to an indifferent market as to their mechanical defects. But now that the railroads were becoming the dominant public carrier, railroad operations and safety records were ever more important and ever more a matter of public concern and criticism. By 1860 railroads were no longer an infant industry but a major force in American transportation.

At this juncture an ebullient figure, Colonel Ezra Miller, came forward with a solution to one of the railroad's worst safety problems. In the mid-1860s Miller devised a combination trussed end platform, draft gear, and coupler that diminished the likelihood of telescopes and eliminated the link and pin coupler. Within a decade it was widely used on American passenger cars.

Miller had already had careers in politics, the New York State Militia, and civil engineering. He was described by one admiring biographer as "a cultivated and experienced" engineer notable for his purity of character and animated conversation.[424] Miller had just the right combination of talents to create and sell a new form of car coupler. While traveling in the Midwest to survey land, he was impressed not so much by the inferiority of the car couplers on the trains he rode as he was by the weakness of their end platforms. Miller rightly felt that car builders had made a major error by dropping the end platforms below the floor level. The main frame and the platform should be in line and strongly fastened together to withstand end buffering. Miller also decided that the ends of the car should be drawn tightly together to diminish the likelihood of one car mounting the end of the car in front. With the loose, slack link and pin arrangement, cars bounced so freely that they could jump up onto an adjacent platform—a sure invitation to the much dreaded telescoping accident (Figure 7.65). The old system of loose links also permitted the cars to sway so much that it added substantially to the general misery of travel, except for passengers who happened to be sailors.

Miller first began to think about the problem during his travels in 1853. The standard biographical accounts of his life say that the idea took nearly a decade to mature. It is recorded that he re-

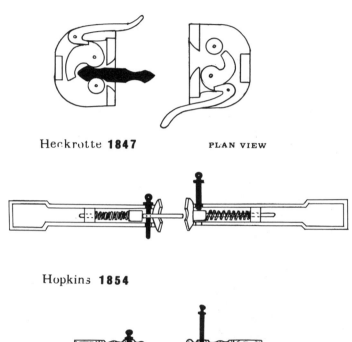

Heckrotte **1847**   PLAN VIEW

Hopkins **1854**

England **1855**

Figure 7.64 Some early efforts toward developing an automatic coupler. All were tested, and the England device was put into rather wide use on the B & O. (Traced by John H. White, Jr.)

ceived his first patent on March 31, 1863 (No. 38057), and two others in January 1865 (No. 46126) and July 1866 (No. 56594).[425] However, in 1887 it was revealed that he may have been inspired by the work of Joseph Miller, believed to be a cousin, who patented a hook style of coupler in 1854.[426] Four other hook couplers were on file at the Patent Office before Miller received his first patent. Their principle was simple: two drawbars with spear-shaped ends were sprung at their ends so that when they passed by each other, they would snap together and hold fast (Figure 7.66). A lever on the end platform was used to pull them apart for uncoupling.

Miller wisely included a trussed platform and a central buffer in his general design, thus giving the railroad a neat package that solved several problems at once. The scheme was so attractive that it met with almost immediate success. The Erie adopted it in 1866, the year that Miller's third patent was granted.[427] The Burlington was using it on many trains by the following year.[428] In 1869 the Philadelphia, Wilmington, and Baltimore was re-

Patented March 31, 1863, January 31, 1865, and July 24, 1866.

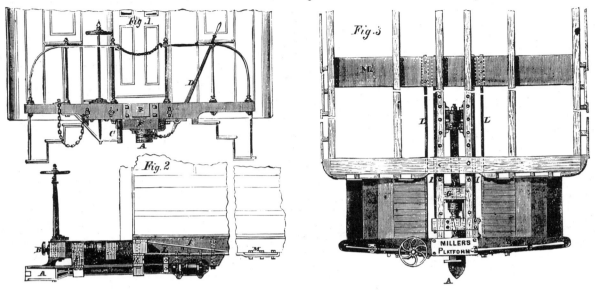

Figure 7.66   Miller's hook-style coupler was widely adopted in this country for passenger cars between 1875 and 1900. For most of that period it was a national standard.

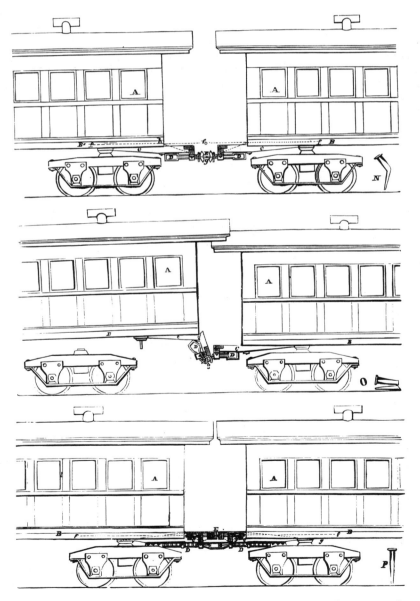

Figure 7.65   Miller combined a strong antitelescoping platform with his passenger car coupler. The weakness of the conventional drop platform is illustrated in the upper two details of the drawing.

ported to be putting it on all new cars.[429] By 1870 it was extensively used in the Middle West and in the Western states, although the more conservative New England lines showed little interest. But railroads in all regions were soon persuaded to install Miller's platform by the reports of its miraculous performance. In May 1870 a day express on the Allegheny Valley Railroad rammed into a slide at 40 mph, but the sturdy Miller platforms prevented telescoping.[430] A more spectacular accident on the Boston and Providence two years later again proved the effectiveness of Miller's lifesaving invention.[431] The train struck a wagon and team, the remains of which fell under the cars and sheared off the trucks. The wheelless cars slid along the track, then went off the right of way and down a small embankment into a field. But the Miller hooks remained fast, holding the train together until it came safely to a stop amidst the weeds. One nervous couple on their way to the 1876 Centennial Exposition were reassured by a fellow traveler:[432]

"Now, madam," continued our interlocutor, "there is no possible danger of the accident you fear, the ordinary oscillation, by which a car is swung from the rail, being very nearly quite removed; and, my friend," addressing me, "there is equally small danger, in the middle cars, of 'telescoping;' the train is made up of cars provided with the trussed platforms of Colonel Miller, which absolutely prevent both the kinds of casualty you have feared."

What possible excuse could be offered for not adopting the Miller platform and coupler? So many lines began using it that by December 1874 Miller could claim that 85 percent of all U.S. and Canadian railroads were carrying his apparatus.[433] Of the 110 railroads that had not yet converted, only the Pennsylvania was a major line; the others were small roads with little passenger traffic. The passenger car fleet had abandoned the link and pin coupler in something like eight years. It was a swift and deliberate transition, free of the painful delays and deliberations that attended the conversion of the freight car fleet. Of course, the freight changeover involved more than a million cars and a vast expenditure compared with the relatively small passenger car fleet. There was also more public pressure to make passenger trains safe. Fatalities and injuries among trainmen were viewed as a secondary evil; the statistics were even separately reported by the I.C.C.

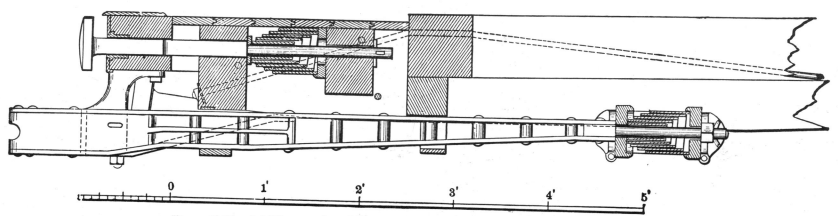

Figure 7.67 *A Miller coupler of about 1885. The single buffer, immediately above the hook, maintained tension between the cars so that the couplers would not separate. (Voss, Railway Car Construction, 1892)*

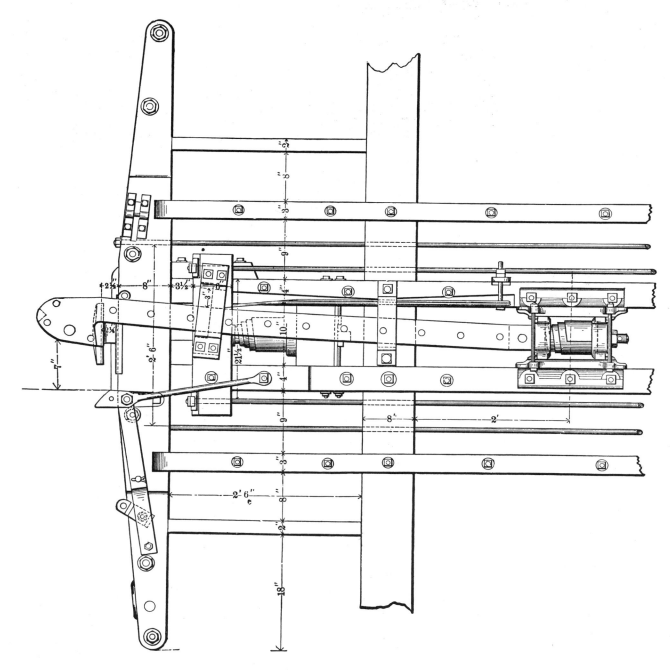

Figure 7.68 *This drawing represents an underside view of a Miller coupler and platform of about 1885. (Voss, Railway Car Construction, 1892)*

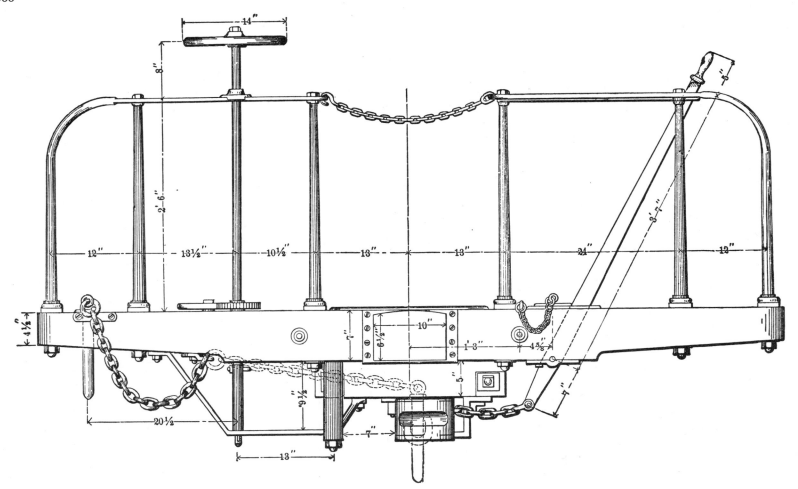

Figure 7.69   An end elevation view of a Miller coupler and platform. The large lever on the right is used to pull the drawhead to one side for uncoupling. Note that the end of the drawhead is slotted to receive a conventional link and pin. (Voss, Railway Car Construction, 1892)

But in fact, the often criticized pace of installing automatic couplers on freight cars was not all that slow, for it was accomplished in little more than a decade. Tourists in Russia are told with some pride that the Soviet Union completed its conversion around 1957. And in 1971 the large-scale application of automatic couplers was just getting underway in Western Europe, according to some firsthand reports.

The success of the coupler made Miller a rich man. He was said to be one of the few inventors who personally realized a fortune from his patents.[434] In one of his earliest catalogs he discussed the question of his growing net worth.[435] "Is my charge too high? I am told that the improvements will earn for me a fortune! I often hear this remark. Granted. Should I not have it? If I earn *hundreds* of fortunes for the railroads of America, should I not have *one* for myself and family?" He claimed that his basic license fee was $100 per car, but in practice it proved far less. As it worked out, the actual fee fell between $25 and $53.[436] Some roads got by more cheaply by paying a flat amount. Among the Burlington papers is an agreement with Miller dated June 11, 1879, that grants use of the patents until they expire for a mere $800.[437] Miller also profited from the sale of his gear, though he does not seem to have been directly engaged in its manufacture. In a letter of September 14, 1877, he quotes the following prices: coupling hooks, $24.50; buffers, $11; uncoupling lever, $2.[438]

Miller's design was praised by William Voss, an expert in railroad car construction, who said around 1890 that the coupler was little changed from the original pattern introduced nearly twenty-five years before.[439] The parts were admirably proportioned in both size and shape. Stronger materials had been substituted in some instances, however. The original hooks had been fabricated from $5/8$-inch flat bars, the skeleton filling pieces being of cast iron (Figures 7.67 to 7.69). But now hooks were made from solid wrought-iron forgings or steel castings. Voss went on to say that Miller's apparatus was used on the majority of American passenger cars, but that it faced an uncertain future because of the recent (1888) adoption of the Janney coupler for freight cars. Voss admitted that the Miller hook had some notable failings. It was difficult to uncouple, particularly on curves.[440] The single central buffer was ineffectual in stabilizing the car's rock and roll, because the single faces were located at the axis of the motion. The hooks tended to uncouple unexpectedly, because they were only 6 inches deep, and variations in the level of cars, height of the draw bars, or condition of the track surface could cause them to slip apart. Such occurrences were more common than the public knew. The deeper jaws of the Janney coupler solved this problem. The introduction of the vestibule created another difficulty for the Miller coupler: there simply was no space for the long throw of the uncoupling lever. It was also necessary to slam the cars together in order to couple the hooks.

These failings were one reason for the Miller's gradual decline, but the main cause was the adoption of the Janney coupler for freight cars. The operation of mixed trains was common on branches, but even on the main line it was occasionally necessary

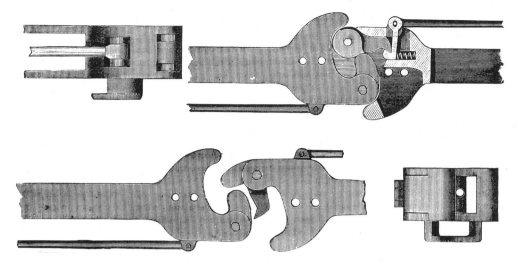

Figure 7.70   *In 1873 Janney patented what has become the basic coupler of today. This drawing, published in 1874, shows the device in its first form.* (National Car Builder, November 1874)

to connect all varieties of rolling stock. And at many terminals, switching engines handled both freight and passenger cars. A common standard was desirable. Since the Pennsylvania had never used the Miller Hook, its subsidiaries and connecting lines made the conversion to the Janney at an early date. In the spring of 1890 the Burlington changed from Miller to Janney.[441] Ironically, in the following year the I.C.C. noted that the Miller was at its peak: of some 28,000 passenger cars, 21,350 had Miller hooks.[442] Within two years, however, the number was down to 17,830; and by 1899, the last year for which statistics were compiled, it had dropped to 11,600.[443] Even this figure represents a sizable portion of the American passenger car fleet. Some lines obviously were not bothered by the interchange of passenger cars and saw no need for a hasty conversion. It is also true that a combined Janney-Miller coupler which had been developed sometime in the 1880s extended the useful career of the obsolete hook coupler. Just when the last Miller hooks disappeared from regular service does not seem to have been recorded, but it is safe to assume that some were in operation into the early twentieth century.

The device that replaced Miller's coupler was invented by Eli H. Janney of Alexandria, Virginia. By all odds Janney should have remained one of the thousands of anonymous amateur inventors whose couplers rested untried in the files of the Patent Office. He had no mechanical or railroad background. He came from a farming family, served in the Confederate Army, and worked as a dry goods clerk during the hard times following the Civil War. Superficially he was the very stereotype of the "coupler crank." But Janney somehow stumbled upon the plan for a rotating jaw coupler that in time became the American standard. He tinkered and whittled his way toward a major advance in railroad technology. Just why or how he was so motivated is not really known; presumably he was drawn to the coupler problem by the distressing stories of injuries to trainmen in the nearby yards. In April 1868 he received a patent for a drop-hook coupler. It was a poorly conceived plan that justifiably received little attention, although some popular writers have mistakenly assumed that it was the basis of the modern coupler. Five years later, on April 29, 1873, Patent No. 138,405 was issued to Janney for an open-jaw or knuckle coupler that incorporated the elementary design of the present-day style. A clenched fist may have

suggested the idea: when closed, it forms a hook; when opened, its grasp is broken. The coupler had a bifurcated head, representing the hand, and a hinged or revolving jaw, representing the fingers. The jaw was locked closed by a pin to effect the coupling. When the pin was released, the jaw could swing open to release the coupling (Figure 7.70). Because the connection (or jaw) was on edge, it was called the "vertical plane coupler," a phrase contrived by the M.C.B. to characterize the general style of coupler devised by Janney and his imitators.

In 1874 Janney formed a partnership with two local backers to promote his 1873 patent. Late that year five railroads were already testing the device.[444] Of these lines the Pittsburgh, Fort Wayne, and Chicago, a subsidiary of the Pennsylvania, was the most significant for Janney's future. From the initial performance of the coupler on eight coaches in 1874, the road was so impressed that within four years it equipped all 152 passenger cars with Janney's device.[445] The parent company watched with interest. Its allegiance to the archaic link and pin was an embarrassment, but for some reason the Pennsylvania continued to reject the Miller hook. Janney's device was at that time "in the inchoate model stage," but the Pittsburgh, Fort Wayne, and Chicago's good experience convinced the Pennsylvania that the design could be perfected.[446] In 1876 the road applied Janney's coupler to 55 cars and showed an inclination to install it more widely. With such an endorsement, Janney lost his amateur status.

In 1877 Janney broke off with his original partners and made an agreement with William McConway and John J. Torley of Pittsburgh to manufacture his coupler.[447] These men were experienced mechanics who specialized in malleable iron castings. Janney now had the expert help he needed to manufacture and market a practical coupler. The Pennsylvania made good its promise and applied the Janney coupler to its passenger cars sometime before 1883.[448] Progress was slower elsewhere, however, because of the prevailing loyalty to Miller.

Meanwhile Janney worked to improve his product (Figures 7.71 and 7.72). In a third patent of 1879 he altered the shape or contour of the knuckle in order to smooth its engagement and disengagement. Three years later Janney developed a coupler specifically for freight cars. The spring catch that was suitable for passenger cars was considered too weak for freight trains, so he

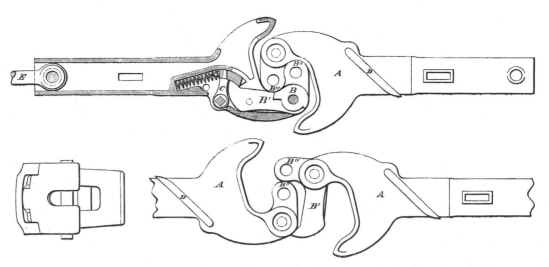

Figure 7.71   Janney improved his coupler with the help of practical advice from McConway and Torley, malleable-iron makers of Pittsburgh. (Railroad Gazette, January 10, 1879)

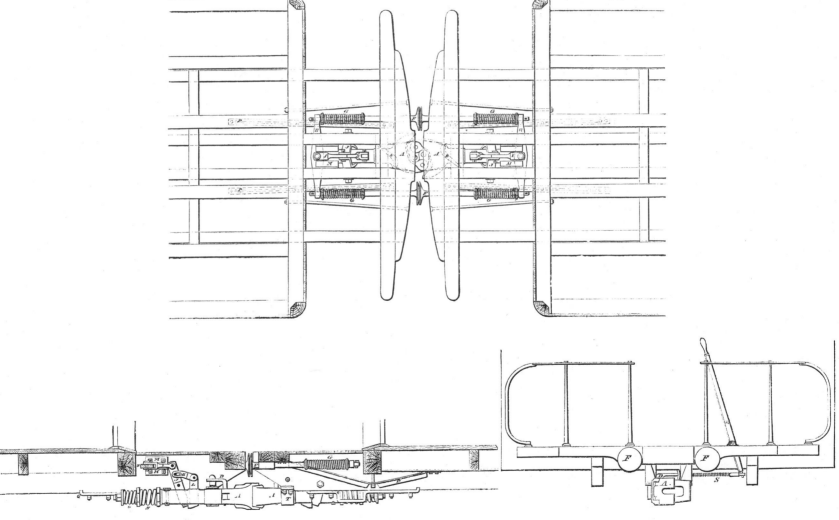

Figure 7.72   Janney combined his coupler with a safety platform and buffer arrangement much like that offered by Miller. (Railroad Gazette, January 10, 1879)

used a drop pin to hold the knuckle closed. This arrangement was soon applied to both freight and passenger car couplers, although many spring catches remained in service as late as 1905. By the end of the 1880s the Janney coupler had evolved into a rugged, serviceable piece of apparatus.[449] Less the draft gear, it weighed 210 pounds. It measured 33¼ inches long. The shank or

barrel was a hollow casting 5 inches in diameter and ⅞ inch thick. The head and barrel were a malleable iron casting, and the knuckle was forged iron or cast steel. The drop pin was malleable iron, while the knuckle pin was crucible steel 1⅜ inches in diameter. A pair of passenger car couplers sold for $190.

The merits of Janney's design slowly gained recognition during

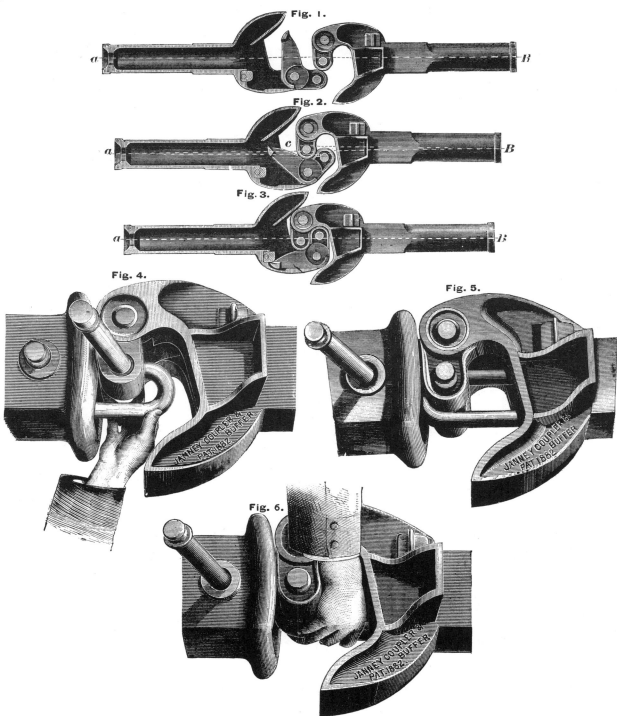

Figure 7.73   *By 1882 Janney was offering couplers with pin catches in place of the original spring latch. The three lower engravings show how the slotted jaw could accommodate the old-fashioned link and pin couplers.*

the eighties. By 1888 it was acknowledged the best possible form of freight car coupler: in that year the Master Car Builders adopted it as their standard after years of debate and testing.[450] Janney won the M.C.B. adoption by a large majority, but he was forced to waive his rights to the eighth and ninth claim of his patent 212,703 (Feb. 5, 1873), which covered the knuckle profile. This permitted other manufacturers to produce a vertical plane coupler that would couple with those manufactured by McConway and Torley. The M.C.B. did not ask Janney to abandon patent claims on other details of the coupler, such as the drop pin lock. But none of Janney's patent covered the general plan of the open-jaw coupler, and other makers were thus free to produce devices of a very similar pattern. During the year of the M.C.B.

ratification, 150 American railroads were already using the Janney coupler.[451]

The early years of the coupler were not without difficulty, for the knuckles broke too easily. The Boston and Albany claimed that 296 fractured in just two months, and the line's master car builder, the quarrelsome Fitch Adams, said that this dismal record vindicated his objections to the Janney coupler.[452] A broader survey showed that Adams's contention was incorrect, because the annual national breakage rate for Janney knuckles was only 3 percent, and overall Janney coupler failures ran no more than 1 percent.[453] Much of the problem was attributable to the necessity for cutting a large slot in its face and drilling a hole through the knuckle so that the link and pin connections

Figure 7.74   The adoption of the Janney-style coupler in 1888 did not usher in an age of standardization. Each manufacturer had an individual plan for the jaw and latch pin assembly. While all could couple together, parts were not interchangeable. (Railroad Car Journal, July 1899)

could be made (Figure 7.73). By 1899 the transition period was nearly over, most cars had automatic couplers, and stronger, solid knuckles would soon be universal.[454]

Maintenance and road repairs were another grave problem during the early years of the M.C.B. coupler. In its concern to avoid a monopoly, the association had created an even more undesirable situation by encouraging a proliferation of coupler makers, each with his own pattern. All were of the same general dimensions and all would work together, but each make was unique insofar as individual parts were concerned. No two made the same locking pin or knuckle. In 1891 there were nineteen major coupler makers producing Janney-type couplers.[455] Every division repair shop was obliged to carry parts for Smillie, Gould, Hinson, Trojan, and the other brands. By 1896 there were twenty-

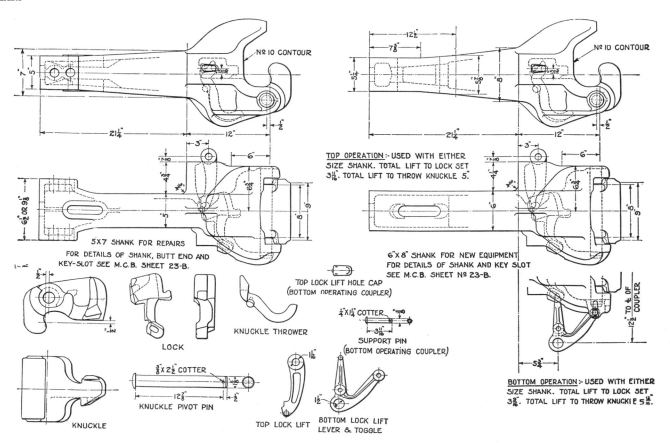

Figure 7.75   The type D Janney coupler was introduced in 1916. Basically intended for freight service, it remained the industry standard until 1931. (Car Builders' Cyclopedia, 1928)

four makers, and by 1899 nearly eighty styles of knuckles were in interchange service.[456] A chart of this bewildering variety of parts is shown in Figure 7.74. It has been claimed that the problem was solved in 1916 when the type D coupler was adopted by the Master Car Builders (Figure 7.75).[457] This coupler was designed essentially for freight cars and in time did bring a measure of standardization to that portion of the car fleet. However, the same was not true for passenger cars. The *Car Builders' Cyclopedias* showed a variety of coupler styles made by Symington, Gould, Sharon, and other makers (Figure 7.76).

If complete uniformity in design was beyond the reach of the car masters, they did succeed in strengthening the Janney coupler. Cast steel replaced those parts formerly made of malleable iron. All dimensions were enlarged except for the contour of the knuckle opening. In 1904 the Master Car Builders voted to increase the thickness of the knuckle's hub by 1 inch.[458] Because so much metal was added at this critical point, it was necessary to change the standard profile that had been in effect since 1888. In 1911 the Car Builders decided to restudy coupler design and develop something beyond the piecemeal reforms made during the past twenty years. The heavy steel cars that were then coming into favor required a basic overall improvement in couplers. Finally in July 1916 the M.C.B. voted to adopt the new pattern of coupler, which was designated as the type D.[459] It weighed 400 pounds, or one-third more than its predecessor. Its strength was improved by 100 percent, and its designers claimed that it would extend service life by 300 percent.

The type D incorporated an improvement devised by Eli Janney the year before his death. It was the knuckle pin protector, which strengthened the knuckle, relieved the knuckle pin from side strain, and reduced leverage on the knuckle by moving the fulcrum point forward. A patent was issued posthumously to Janney in 1912.

The type D was perfect for freight cars, but it was adopted for passenger cars as well.[460] Yet the type D was hardly accepted before planning was initiated for a successor. The new coupler, type E, was a modification of type D. Again the individual parts were slightly enlarged to increase overall strength.[461] Knuckle depth, for example, was increased from 9 to 11 inches. The type E was adopted as the new standard in 1931.

Meanwhile efforts were underway to perfect a distinctive coupler design especially suited for passenger cars. The conventional Janney style of coupler effected a relatively loose connection between the cars. This was satisfactory for freight equipment, but the resulting end play adversely affected the ride. A tighter, more secure coupler was required for passenger cars. This need led to the introduction of the tightlock coupler (Figures 7.77 and 7.78). It had a massive head with a side-wing pocket. A mating, wedge-shaped lug on the opposing coupler fitted into the pocket and held the coupler heads rigid. The actual connection was made, as before, by a Janney-type open-jaw coupler. Charles H. Tomlinson (1869–1938) pioneered this style of coupler.[462] He had some experience in the field through his association with the Ohio Brass Company of Mansfield, Ohio, a producer of electric railway apparatus. Tomlinson designed some of the company's most successful traction car couplers. In 1928 he interested the B & O in testing a tightlock coupler that he hoped to perfect for main-line railroad service.[463] Tomlinson envisioned a fully automatic device that did not simply couple the cars but also connected the steam heating, air brakes, air signal, and electric lines. The industry was not ready for a fully automatic coupler, but it did like the tightlock feature, and over the next few years this

Figure 7.76 · A passenger car coupler of 1910. Cast steel had replaced malleable iron by this period.

design was revised for commercial use. Its extra strength and its antitelescoping and antijackknifing characteristics were regarded as adding significantly to the safe operation of fast, lightweight cars. By 1937 the tightlock coupler was an alternate standard.[464] Its popularity grew in the prewar period, and in 1954 it was made an industry-wide standard for passenger cars.

In recent years interest in a fully automatic coupler has revived. Manual handling of the brake hoses, signal hoses, and steam connections is labor-intensive. It is also dangerous. The coupler itself is often a problem when it is not positioned properly, for it tempts foolhardy trainmen to step in toward an approaching car and kick one coupler. Occasionally the pin, when lifted, will not cause the knuckle to swing open fully. Again, a trainman will sometimes step inside the rails and pull it open by hand. These mechanical failings, together with the carelessness of operatives, have led to the death and injury of many men. As recently as 1965, 7 trainmen were killed and 944 injured in coupler-related accidents.[465]

Nearly a quarter century before Tomlinson's efforts to introduce a fully automatic coupler in 1928, Westinghouse attempted to market a combination car, steam, and air coupler.[466] The device was not widely adopted, but this did not discourage others from trying to interest the railroads in similar contrivances. The *Car Builders' Cyclopedias* issued in the 1920s contain illustrations of many such designs. In 1969 the Midland-Ross Corporation reintroduced the scheme and ran road tests on the Union Pacific.[467] More recently the Association of American Railroads began a research effort toward this end in response to a strong expression of interest on the part of railroad managers.[468] However, the prospects for fully automatic couplers have received a setback because of the failure of the control line on the Metroliner cars to stay in contact. Trains occasionally broke in

two while running. The problem was solved by installing conventional jumper lines between the cars. But even if a practical design were perfected, it is not likely that the industry would attempt a rapid conversion. The cost would be prohibitive, at least as far as freight cars are concerned, although the conversion of the now relatively modest passenger car fleet would be more likely.

Draft Gears

Couplers are not rigidly fastened to the cars; the mounting includes a spring or friction mechanism to serve as a shock absorber. This apparatus, attached to the rear of the coupler's shank, is called the draft gear. Its main function is to protect the car structure from coupling impacts. It is expected to absorb, or at least cushion, the forces between colliding cars and to equalize the speed of moving cars when coupled together as a train. Draft gears perform well in impact speeds up to about 5 mph. Conventional gears can handle blows up to 300,000 pounds, but beyond this point (2¾ inches of travel), the springs are totally compressed and the car frame must absorb the collision. If the force is great enough, of course, the frame will give way.[469] The compression of the draft gear is called buff.

Draft gears serve a second major function: the stretching, or slack action, considered necessary for train operations. This was regarded as essential to the starting of heavy trains because the engine was not required to pull all the cars at once. When stopping, the slack would bunch in, so that when the locomotive started up, it was pulling against one car at a time until the slack pulled out. By the time the draft gear on the last cars was elongated, the forward part of the train was rolling at a relatively

Figure 7.77 *The tightlock coupler, first tried in 1928, eventually became standard for passenger cars because of its greater strength and antitelescoping characteristics.* (New York Central)

fast gait. If too fast a start was made, the couplers on the rear cars were endangered. Inexpert handling resulted in many draft gear failures. It was the major car repair expense; estimates of repair costs arising from draft gear and coupler failures ranged from 38 to 70 percent of total car repairs.[470] The great majority of these problems, however, were freight-car-oriented. Passenger cars were relatively free from the failures because trains were short, and they received more careful handling. Also, they were not run through hump yards, where much of the buff damage took place. One car designer compared passenger equipment to a pampered house pet and freight cars to a stray dog.[471] The analogy is a good one, for passenger cars rarely left the home road, while freight cars wandered about the country on foreign roads, enduring a rough-and-tumble life in interchange service. So long as a freight car could roll, no one seemed to care whether the draft gear functioned or not.

While passenger car draft gears did not undergo the abuse that freight gears received, they were an important adjunct to passenger train operations because they could dramatically affect the ride. A defective draft gear could cause a surging, or rapid forward jerking, of the car that a nonexpert might well blame on the trucks.[472] In such a case the springs were not necessarily broken; they might simply have grown soft with age, so that the changeover between the full compression of the springs and the friction gear was sharp rather than smooth. As the train's speed fluctuated, which was inevitable because of grade and power variants, the cars developed an annoying jerking motion. The draft gear had to be well maintained to perform optimally.

The desirability of draft gears was recognized during the first decade of American railroading. Charles Davenport, a pioneer car builder of Boston, patented a device on September 9, 1835, to ease shocks and cushion the motion of the train at the drawbars. A full leaf spring was attached to the end of the coupler shank. This style of draft gear is illustrated by the English drawing of an American passenger car of 1845 which is reproduced in Figure 1.78. In 1838 P. Alverson, car builder for the Hartford and New Haven, began testing a spiral spring draft gear.[473] The arrangement was shown in a contemporary engraving, which also included the wooden style of coupling link described in the previous section (Figure 7.79). The inventor said that he pre-

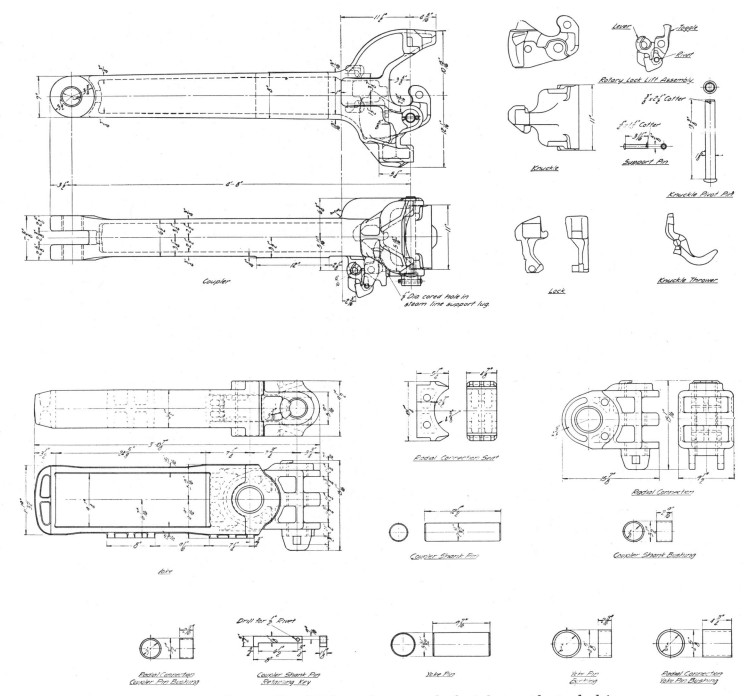

Figure 7.78   The type H tightlock coupler was made the industry-wide standard for passenger cars in 1947. (Car Builders' Cyclopedia, 1953)

ferred this form of coupling link and suggested that it be used in combination with the spring draft gear. It is not directly stated that he was the originator of the wooden link, but he may have been. Alverson's gear was said to have performed well after two months' service, and the *American Railroad Journal* expressed the hope that the elastic connection would entirely replace the old fashioned stiff ones. The editorial went on to say: "We hope that after this, we shall not hear of those unpleasant jerks—one of the greatest sources of nuisance in Railroad travelling—and the more reprehensible, because they can so easily be remedied."[474]

Both Davenport and Alverson offered the industry a simple solution to the draft-cushion problem that was largely followed in succeeding generations of railroad cars. Steel springs were about as direct a means of dealing with the matter as could be

found, but other inventors envisioned more complicated methods. As early as 1835 air cylinder draft gears were tried on the Boston and Worcester.[475] The scheme was not successful, however, for one of the air springs failed on the Whitmore car, leaving the remainder of the train behind. A few other roads, notably the B & O, insisted that the solid, inelastic draft gears were best after all. Those of this persuasion championed the continuous drawbar. The scheme was introduced on the B & O in 1835 by Jacob Rupp.[476] A heavy rod or flat bar ran the full length of the car between the couplers, thus entirely relieving the frame of draft and buff. The New Jersey Railroad and Transportation Company tried the scheme around 1840.[477] It put a spring in the center to cushion the draft, but found the scheme heavy and unnecessary. The Lehigh Valley tried a massive continuous drawbar measuring 3 by 7½ inches, but even this huge plate

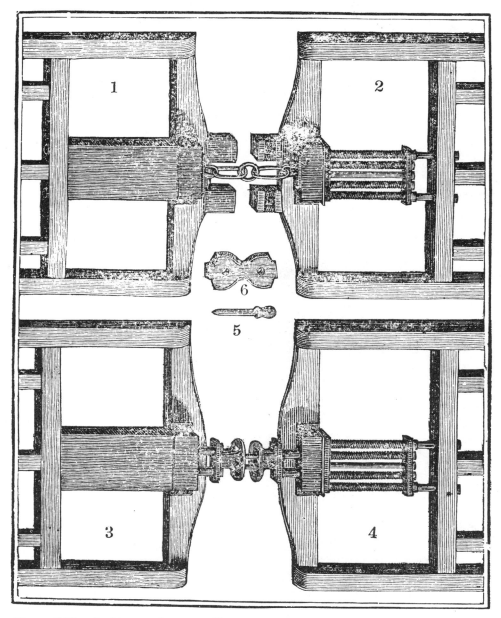

Figure 7.79 Cushioning devices, called draft gears, fasten couplers to the car frames. The 1838 scheme shown here used coil springs. Note also the wooden link in detail 6. (American Railroad Journal, January 1, 1839)

buckled in service, throwing the couplers so far out of line that they could not be joined. The dead weight was another disadvantage. At last even the B & O conceded that the scheme was not such a good idea, and by 1875 the road was using continuous drawbars only on freight cars.[478]

As already mentioned, rubber springs were introduced for truck suspension during the mid-1840s. They were also applied to draft gears in place of conventional steel springs (Figure 7.80). They undoubtedly exhibited the same advantages and deficiencies that they showed in truck service. As Figure 7.61 suggests, they appear to have remained popular for draft gears until the 1860s. In 1929 the idea was revived by the Waugh Equipment Company for passenger cars and became so popular that after 1940, rubber springs were applied to freight cars as well.[479]

During the last quarter of the nineteenth century, coil steel springs were the basic form of draft gears. Often a double spring was used. Some were made with a smaller spring inside the main one, but in other instances the springs were placed in tandem on a common shaft.[480] Miller used volute springs on the buffer and coupler hook, while Janney used large coil springs in his

more elaborate buffing system (Figures 7.67 and 7.71). The matter of drawsprings was so important that the M.C.B. established a standard capacity for them as early as 1879. In that year a drawspring had to have a capacity of 13,000 pounds. As car weights increased, the limit was increased: to 18,000 pounds in 1884 and to 22,000 pounds in 1893. Such increases did much to save cars from destructive buffing shocks, but as springs grew in strength, so did the problem of recoil. The kickbacks produced jolts almost as great as the shocks that produced the kickbacks. How to dampen the recoil was a major problem to be overcome.

Westinghouse was attracted to the problem during the course of the Burlington, Iowa, brake trials.[481] He worked out a plan with a triple wedge that compressed an interlocking set of friction plates (Figure 7.81). The action of the wedge and plates dissipated the spring's recoil. The friction draft gear also increased the overall capacity of the device, well beyond what the ordinary coil spring gears could offer. A patent (No. 391,997) was issued on October 30, 1888, just about the time that the device was placed on the market. The first models were rated at 60,000 pounds, although some individual units were said to ab-

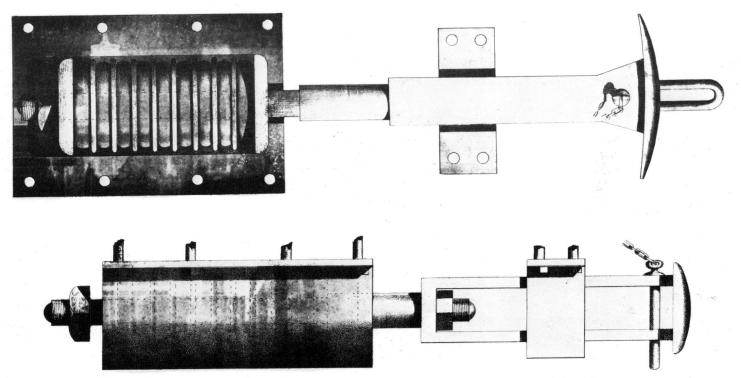

Figure 7.80   Rubber springs were used for draft gear cushions during the 1840s. This engraving is from an advertisement appearing in the American Railroad Journal, *September 16, 1848.*

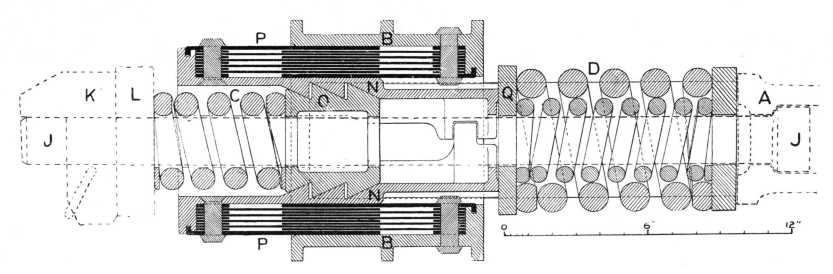

Figure 7.81   A Westinghouse friction draft gear patented in 1888 offered an alternative to older designs which depended entirely on springs. O-N is the wedge and P-B the friction plates. (Engineering News, *March 1, 1890)*

sorb as much as 100,000 pounds.[482] The Westinghouse apparatus made a good showing during demonstration tests. Fourteen couplings were made with two cars fitted with friction gear at Altoona early in 1889.[483] The speed was gradually increased after each coupling. By the twelfth coupling, with the speed at 28 mph, some of the blocking gave way behind one of the buffers, and one truck broke loose and stripped away a body bolster. When the speed went up to 30 mph, even heavier damage was sustained by the cars, but the Westinghouse gear was retrieved from the wreck undamaged. Despite such good results, the industry was slow to adopt friction gears. By 1899 Westinghouse had boosted the capacity of his apparatus to 170,000 pounds, but only 3,000 freight cars and few locomotive tenders were using them.[484] The extra cost was one reason for the rail-

roads' resistance. There were also complaints about the gear's stiffness; under ordinary draft and buff movements, it tended to act more like a solid block of iron than a spring.[485] The introduction of steel cars made friction gears a necessity, and despite their failings they became the basic forms of draft, at least for freight cars, around 1920. By 1943 85 percent of the U.S. freight car fleet had friction gears.[486] They were extensively used for passenger cars as well.

In 1927 a radically new form of draft gear was introduced by Otho Duryea.[487] Conventional draft gears, he observed, moved a scant 2 to 3 inches and then went solid. Duryea envisioned not only a longer travel of 7 inches but also an independent draft gear that would, in effect, float within the car's center sill.[488] A large bank of coil springs absorbed the impacts. Duryea's cushion underframe ran the full length of the car inside the center sill and had couplers attached at both ends. It was something of a super, continuous drawbar, with a capacity of 64,000 pounds. Duryea intended the device primarily for freight cars, but it was used for passenger cars as well. Apparently the first passenger use was on the B & O's lightweight trains of 1935, the Abraham Lincoln and the Royal Blue.[489]

As already discussed, draft gears posed a relatively minor problem for passenger cars. In the area of freight cars, however, they amount to an obsession among car designers. The problem has been partially solved in recent years by various forms of hydraulic cushioned gears. The first appeared commercially in 1955, and it is assumed by many that they are a modern improvement. But hydraulic draft gears were used in 1893 on some passenger cars built for the New York Central Railroad.[490] Perhaps if railroad managers were better students of their own history, the long-standing problem of damage claims might have been resolved generations ago.

CHAPTER EIGHT

Self-Propelled Cars and Motor Trains

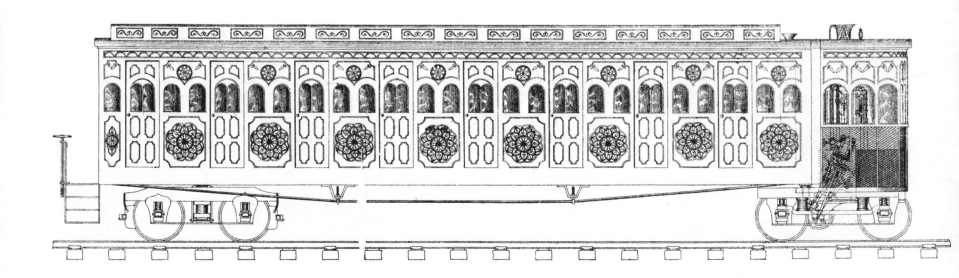

THE HISTORY OF THE MOTOR CAR revolves about the basic question of cheap transportation. It originated in the search for small, self-propelled units to service marginal runs that did not require a full passenger train. Traffic on many branch and suburban runs would barely fill a single coach, yet it was necessary to operate a coach, a baggage car, and a locomotive. Even such a minimal train required a full crew, and costs easily outran revenues. Conversely, on heavily traveled suburban lines conventional steam trains could not efficiently move the traffic. Because of the poor acceleration inherent in the steam locomotive, the frequent stops slowed the flow of commuters to and from the great cities. Multiple-unit, self-propelled electric cars were better adapted to handle high-volume traffic economically. On both the lightly traveled branch line and the crowded commuter run, the story of the rail car is dominated by economics rather than mechanics. Its major theme is the hard facts of operating costs.

Self-propelled, or motor, cars are at both ends of the passenger traffic spectrum. One tends to think of them only in terms of the branch-line shuttle, but the electric multiple-unit cars of the New Haven and Illinois Central are also self-propelled cars. And they are engaged in moving vast crowds along the heavily patronized commuter routes leading out of New York and Chicago.

The great majority of self-propelled cars are electrics, although there have been some quixotic experiments with steam and internal combustion over the years. In general the motive power methods that have been tried for self-propelled cars form a rough chronology, with steam first, then gasoline or gas-electric, straight electric, diesel and diesel electric, and lately, gas turbine. The chapter will follow this general outline, keeping the focus on standard, main-line developments. Street railways, rapid-transit cars, inspection vehicles, and experimental or home-made branch-line cars will receive only brief mention.

Self-propelled cars were tested on American railroads before the Civil War in an effort to reduce operating costs for lightly patronized or marginal trains. The attempt to introduce such equipment was persistent but largely unsuccessful during the nineteenth century. The railroads faced no serious competition in this period, and the abundance of obsolete or surplus equipment, which had little book value, was suitable for secondary runs. The expenditure of large sums for expensive motor cars did not seem necessary. After the new century opened, however, the situation began to change rapidly. Street railways and interurbans had already cut into suburban and branch-line traffic. The automobile promised to draw away even larger numbers of short-haul passengers.

Railroad men began to worry. They became more receptive to radical cost-cutting methods. If the compact internal combustion engine was so efficient for the highway, why not use it for light rail vehicles as well? Several hundred gasoline rail cars were purchased by American railroads between 1905 and 1915. The cars worked fairly well, but they were sensitive and proved cantankerous unless given special treatment. The mechanism was alien to the average railroad shopman. The operating crews distrusted the motor car's springy, uncertain, "I am about to go off the track" ride. The raucous, evil-smelling engine added little to the charm of the gasoline car. The prejudice against such vehicles was evident in the scornful names given them by railroaders and the traveling public: "doodle bug," "galloping goose," and "wooden shoe" do not describe conveyances of much style or comfort. There was a general retreat from the rail car; after a brave beginning it was left to wither and die.

However, the economic crisis following the end of World War I persuaded railroaders to reexamine the question. The war had demonstrated the reliability of the internal combustion motor. Spark plugs and carburetors had become everyday words and devices. The transport of the future, the airplane and automobile, was gasoline-powered. If the railroads were to hold any portion of the short-haul market, they must do so with economical, small-capacity units. This ruled out conventional steam trains with their large operating crews. One- or two-man motor cars were seen as the solution. Between 1921 and 1930 the railroads purchased approximately 800 such units, but the Depression and World War II curtailed new orders. During the 1950s the diesel rail car became popular, although it was cut short of its full potential by the general decline in rail travel. After 1960 nearly all motor cars were in suburban commuter service.

The general trend of main-line electric cars is somewhat different. Mechanically, they were a direct outgrowth of the electric street railway. The rolling stock was far heavier—on a scale with existing main-line trains. Main-line electric, self-propelled cars were also normally outgrowths of terminal electrifications, such as the New York Central's Grand Central project. A spin-off of this scheme was the installation of multiple-unit cars in 1906 on the Harlem division to service heavy commuter runs in and out of Manhattan. Unlike the free-ranging motor car, the electrics were tied to short lengths of suburban trackage and rarely ran more than 50 miles from the home terminal. The self-propelled electric car reached its apogee in 1969 with inauguration of the high-speed Metroliners between New York and Washington. The multiple-unit vehicle had become part of first-class, long-distance train travel.

Steam Cars

Beginning around 1850, there was an intermittent effort spanning eighty years to introduce steam-powered self-propelled passenger cars on American railroads. It seemed to run in ten-year cycles, with occasional short periods of intense activity. After a series of failures, the idea would fade away. But after a short hiatus it would be revived, presented as a fresh solution to an old problem, and rejustified by claims for lower fuel and labor costs. The memory of past failures was forgotten or ignored; the steam car taught no lesson except the moral that it is unwise to disregard the lessons of history. Over a hundred steam cars were built for service in this country, and of the total only a few could be classified as more than a marginal success.

Steam cars were first tried in England on the Eastern Counties

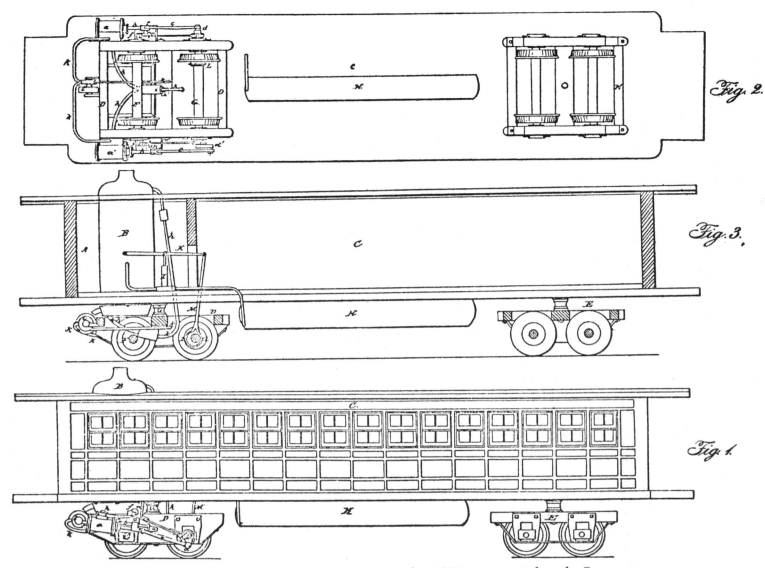

Figure 8.1   Moore and Parrott's steam car, patented in 1851, was operated on the Boston and Worcester. Similar cars were later used on the Pittsburgh, Fort Wayne, and Chicago Railroad. (U.S. Patent Office)

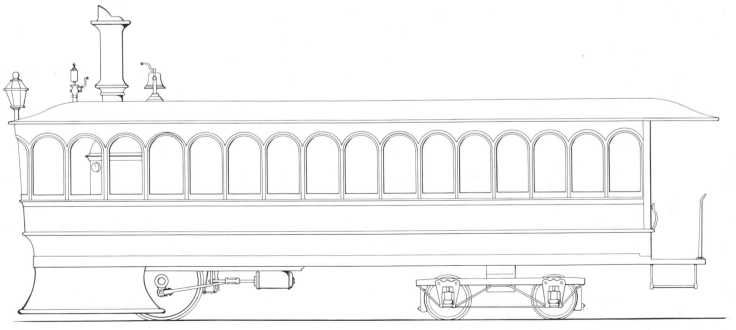

Figure 8.2   The Baldwin Locomotive Works built the steam car Novelty in 1860 for the Pittsburgh, Fort Wayne, and Chicago. The driving wheel was a solid, plate-style wheel. (Traced by J. H. White, Jr.)

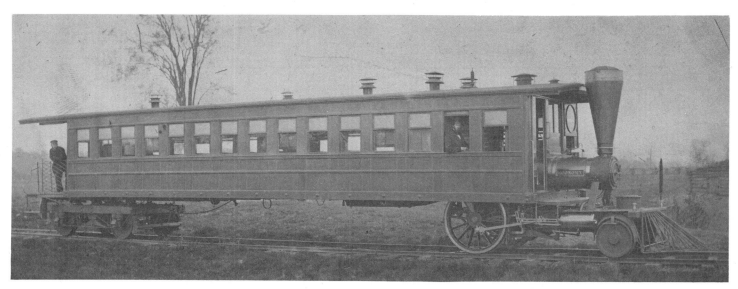

Figure 8.3   *William Romans constructed the* Economy *in the Columbus, Piqua, and Indianapolis repair shops in 1861.* (*Railway and Locomotive Historical Society*)

Railway. Late in 1848 W. Bridges Adams built the *Fairfield*, a six-wheel, 31-foot-long combined locomotive and passenger car, at the suggestion of James Samuel, an official on the same line.[1] A year earlier the two men had produced a very small, open steam car, but apparently it was an inspection vehicle not intended for revenue service. In 1849 Adams fabricated an eight-wheel car, the *Enfield*. These units performed acceptably but failed to spark much enthusiasm elsewhere in England, for it was almost a quarter century before another steam car was produced there. The pioneering experiments of Adams and Samuel, however, may have encouraged M. W. Baldwin to build similar machines in the United States. Positive information is not available on the point, but in 1846 and 1847 Baldwin did produce some extremely small power units for the Vermont Central and the Georgia railroads.[2] They were so small, even for this early time, that they could not properly be called locomotives. They were about the right size to power a self-propelled car, although no further inference can be drawn.

The first American steam car for which there is positive information was produced in 1850 or 1851 by Joseph H. Moore and William P. Parrott of Boston. On December 2, 1851, the partners received a patent (No. 8561) in which they stated that they had already built a 12-ton test car (Figure 8.1). The eight-wheel model had a vertical boiler, with cylinders mounted on the outside frame of the front truck. A flexible steam pipe connected the boiler and engine. Its weight was less than one-half that of an equivalent locomotive, tender, and coach. A light locomotive and tender were priced at $6,000 and a coach brought $2,000, while the steam car cost only $4,000 and was equal to the full-sized train. The inventors reported speeds of 40 mph with the test car but admitted that there were some mechanical problems. These difficulties were overcome, for a later report states that Moore and Parrott cars operated successfully on several branches of the Boston and Worcester.[3] They were economical and well suited to the light traffic patterns of the tributary lines.

A few years after the Boston cars appeared, J. H. Moore moved west to become superintendent of the Pittsburgh, Fort Wayne, and Chicago. His interest in steam cars remained strong, and in 1860 he took over a car plant in Massillon, Ohio, to produce an improved model of his 1851 patented design.[4] The new, all-iron car was under construction in June 1860. It would seat 94 and carry 3 tons of baggage in its long, 77-foot body.[5] A

separate apartment for mail was also provided. The cylinders measured 8 by 14 inches, the wheels 42 inches. Its weight was light: computed at 16 tons, with fuel and water adding another 3 tons.[6] It cost $6,000.[7] The designer claimed that a cord of wood would propel the vehicle 125 miles, and that the total operating cost would amount to one-third that of a regular train. In the fall of 1860 the car carried the president of the railroad from Pittsburgh through to Chicago, demonstrating its ability to make long runs when necessary. Two sister cars were produced to assist in servicing the New Brighton accommodation run. Moore's success was short-lived, however, for in 1862 the U.S. War Department requisitioned the cars for use in the South. Photographs exist showing them in service in the Atlanta and Nashville area during the war years.[8]

At the same time that Moore was building his huge iron locomotive-car, M. W. Baldwin was preparing a more modest steamer of his own design (Figure 8.2). The six-wheel wooden car measured just under 32 feet in length and weighed only 7½ tons.[9] The front compartment, housing the vertical boiler, occupied only 30 inches. The 5- by 12-inch cylinders propelled a solid-plate, 42-inch driver. Longitudinal seats accommodated 38 passengers, and long tanks under the seats stored 100 gallons of water. Exhaust steam heated the passenger compartment. In June 1860 the car, named the *Novelty*, made a test run over the Norristown line at 30 mph. A month later it was in service on the Pennsylvania out of Pittsburgh, exhibiting good fuel economy. A bushel of coal was sufficient fuel for 50 miles. The car was delivered to its consignee, the Pittsburgh, Fort Wayne, and Chicago, which intended it for suburban operation.[10] However, no subsequent accounts of its service life can be found. It must have been a failure, since Baldwin built no more steam cars for another fifteen years.

Late in 1861 the Erie's master car builder, C. A. Smith, produced a 60-passenger steam car of his own design.[11] It had a wooden frame, iron gas pipe posts, and a sheet-metal body. The composite body was powered by two 8- by 12-inch cylinders and an upright boiler. The firebox exhausted out of twin smokestacks. It could pull one car with ease, but with two, it had some trouble starting. Although it ran for a time between Paterson and Jersey City, it was not found suitable for suburban service and was sent west to work between Binghamton and Hornellsville, New York. But its jerky, uneasy ride annoyed passengers, so that the ma-

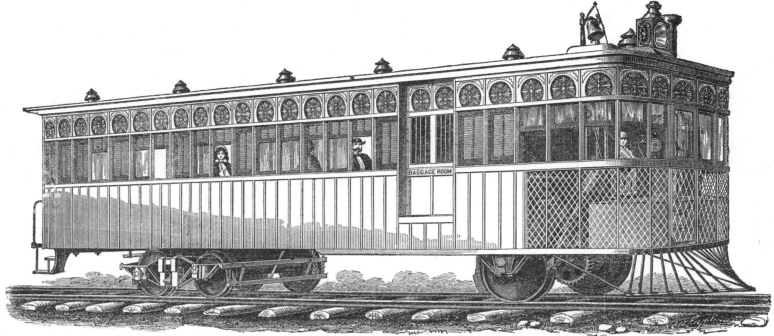

GRICE & LONG,
114 Walnut Street, Philadelphia, and 94 Wall Street, New-York,

Call the attention of all interested in cheap Railway transportation to their STEAM PASSENGER CAR. A speed of 25 miles per hour, and a traction equivalent to a train of 80 passengers, is attained.   The consumption of fuel and running expenses are very light; carries fuel and water for 25 miles, and weighs less than half of the Locomotive.   For City Railways the heaviest grades may be overcome, and, by the use of their patent VIBRATING TRUCK, the shortest curves can be turned.

GRICE & LONG'S STEAM PASSENGER CAR.

Reference is made to Messrs. J. W. Van Rensselaer, Sup't West Jersey R. R. Co., W. H. Gatzmer, Camden and Amboy R. R. Co., James Mornell, Philadelphia and Trenton R. R. Co.

Figure 8.4   Grice and Long of Trenton, New Jersey, built a number of steam cars for both street and steam railways in the 1860s. Larger units, such as the one shown here, were intended for branch and main-line service. (Scientific American, October 26, 1861)

Figure 8.5   The Vulcan Iron Works of San Francisco completed the Napa *in 1865 for the Napa Valley Railroad. It is shown as it was relettered in later years for service on the San Joaquin Valley Coal Mining Company. (Locomotive Engineering, October 1891)*

chinery was eventually removed and the car was retained as a smoker. Years later it was dropped beside the tracks near Attica, New York, to serve as a section-gang shelter. It was reported at this site in 1898.

William Romans was another railway mechanic smitten by steam car fever in the 1860s. At the time he was master mechanic for the Columbus, Piqua, and Indianapolis Railroad, with headquarters in Columbus, Ohio. The road was obliged to operate a lightly patronized train to Piqua, 73 miles from Columbus. Romans saw the steam car as the best way to cut costs, and he designed and built one in the railroad's own shop during the

summer of 1861.[12] Appropriately named *Economy*, the 16-ton, 70-foot-long car seated 48 passengers (Figure 8.3). A 6-foot-deep rear platform provided extra baggage space. The rear truck was made so that it could be pulled forward to reduce the overall wheelbase for short turntables. Romans built a miniature self-contained 2-2-0 locomotive for the power unit. It had 7- by 18-inch cylinders, 54-inch driving wheels, and 28-inch leading wheels. The *Economy* was designed to operate at 25 mph, the normal running speed of the time.

Sometime after the *Economy* entered service, Romans produced a second car, the *Express*. Anticipating that others would

Figure 8.6 *Joseph P. Woodbury's steam truck of about 1865 was attached to the car body by a ball-bearing ring plate. The smokestack is not in place in this construction photograph. (Railway and Locomotive Historical Society)*

want duplicates of the Columbus-built cars, he patented his design on January 28, 1862 (No. 34,270). Romans must have been disappointed by the reaction of other railroads, for it was not until 1865 that he received an order. The Minnesota Valley Railroad commissioned him to build a car for a 10-mile suburban run between St. Paul and Shakopee.[13] The *Shakopee* made four trips a day and became a great favorite with engine crews because of its warm, roomy cab; it was an ideal working station during the bitter Minnesota winters. When the railroad's western terminal moved to a more important junction, however, the steam car could not handle the traffic, even with a trailer car. The *Shakopee* was stored for a time and used only for an occasional inspection or paymaster's run. Then it was leased to the Northern Pacific for similar duties. Late in the 1870s it was dismantled, and the body was salvaged for a peanut stand.

Other hopeful inventors tried to promote the steam car. Prominent among these were Robert H. Long and Joseph Grice, car builders of Trenton, New Jersey.[14] Long obtained a patent in January 1860 for a six-wheel steam car that featured an offset boiler and gear drive. The design was intended for street railway service; however, difficulties in selling the idea to the transit industry led Grice and Long to turn to main-line steam roads for a market. Approximately ten railroads bought one or more cars. One of the first was the Camden and Amboy. The *Scientific American* of October 26, 1861, described the car as 37 feet, 6 inches long, with seating for 36 passengers (Figure 8.4). A small baggage room separated the engine and passenger compartments. The 11-ton steamer consumed 7 pounds of coal per mile. Similar cars were sold to the West Jersey, the Jacksonville and Lake City, the Huntington and Broadtop, and other lines, including one in Mexico. Around 1863 Long and Grice introduced a larger eight-wheel car. The Portland and Kennebec Railroad bought one of these 12½-tonners for $8,500. That price may well suggest why steam cars were not more widely used.

The far West also contracted the steam car fever of the 1860s. A number were produced for the street railways of San Francisco, but only two seem to have been built for a steam line. In November 1865 the Vulcan Iron Works of San Francisco completed the 21-ton car *Napa* for the Napa Valley Railroad.[15] It was powered by a hefty 2-2-0 locomotive unit with 9- by 18-inch cylinders and 48-inch drivers (Figure 8.5). The locomotive car was capable of pulling as many as eighteen freight cars, and it could move along smartly with eight.[16] The passenger compartment seated 26. The water tank held 750 gallons, the coal bunker 800 pounds. In later years the *Napa* had several owners. It finally came into the possession of the Southern Pacific, which stored it outside the Sacramento repair shops. Around 1925 the old veteran was junked. The *Napa's* sister car, the *Calistoga*, had a much shorter career. It was completed in 1867 by Vulcan but in 1875 was rebuilt into a conventional locomotive by the Napa Valley Railroad.

Although the excitement created by the steam car during the 1860s led to no widespread adoption, this failure was certainly not due to lack of enthusiasm by the steam car's promoters. This

group was filled with zealots, of whom the most active was Joseph P. Woodbury of Boston. His belief in the steam car was absolute; he felt that it was essential to the success of both street and steam railroads. He contended that his cars could do the same work as conventional equipment for half the cost, and at a saving of one-third to one-fifth in dead weight.[17] Woodbury's devotion to the perfection of a superior self-propelled rail car is shown in the five patents that he took out between January 1865 and February 1871. Several ideas described in these papers were gear and direct-drive power units, center door entrances, and roller-bearing mounting of the engine-boiler to the car body. The latter plan was perhaps Woodbury's most original contribution. It permitted the driving machinery and a vertical boiler to be fastened to a four-wheel truck and yet stay free to swivel beneath the car's floor (Figure 8.6). A double iron ring turned easily on iron ball bearings.

In the fall of 1865 Woodbury exhibited a demonstration model built to his designs by the Salem Car Company.[18] The 30-foot car, which could seat 40 passengers, was tested on the Eastern Railroad and other New England lines. The steam, or at least a portion of it, was exhausted into the water tanks both to muffle the car's running noise and to preheat the feedwater.[19] Next Woodbury produced a large, 75-passenger car, its machinery presumably made by the Manchester Locomotive Works. In 1866 Woodbury contracted with the New England Steam Car Company to build cars of his patented design. It cannot be determined which, if any, railroads bought them. But occasional reports, one appearing as late as 1876, indicate that Woodbury was still actively pursuing the idea.[20] A photograph taken around 1890, now in the private collection of Thomas Norrell of Silver Spring, Maryland, shows an aged Woodbury car in switching service at the Bellows Falls Brewing Company, Walpole, New Hampshire.

If the 1860s can be described as the most active period of the steam car in America, the next twenty years were the most inactive. The 1880s in particular were a decade of almost complete stagnation. In the 1870s, however, there were a few feeble attempts to push steam cars. The most interesting was conducted by William Baxter of Ilion, New York. Baxter was active in many areas of mechanical engineering and was associated with the

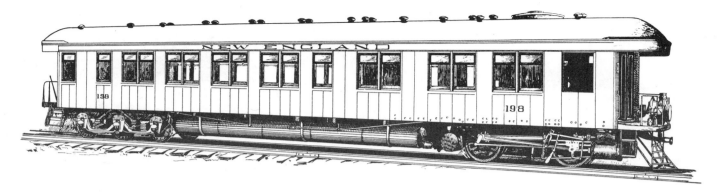

Figure 8.7 *The New England Railroad's steam car of 1897 was powered by a steam truck fabricated by the Schenectady Locomotive Works. Note the large water tanks under the car body.* (Railway Age, *October 15, 1897*)

Figure 8.8 *The Erie produced a steam car in 1897 by combining a Schenectady steam unit with an old emigrant car body.* (Alco Historic Photo, *National Railway Historical Society*)

Remington Arms Company, which produced his compact, portable steam engine. He introduced a new concept in steam car practice, for it used a compound engine—one of the first such applications in American railway service. Remington built the machinery for the test car, which was ready by the fall of 1872.[21] The equipment was mounted on a 30-passenger horsecar with a vertical boiler on the front platform and the engine directly connected to the wheels without benefit of gearing. The 5- and 8- by 12-inch cylinders were arranged so that the engine could operate as either a simple or a compound unit. When extra power was required, a transfer valve was opened to direct high-pressure steam directly into both the high- and low-pressure cylinder. The car eventually entered service on the Worcester and Shrewsbury Railroad, a short suburban line that connected Worcester, Massachusetts, with a lake resort.[22] It remained on the line until almost 1890. But for all the merits of the car, Baxter appears to have had no success in selling his design to the railroad industry.

In the far West, the Central Pacific's Sacramento shops turned out a new steam car for service on the subsidiary Northern Railway between Woodland and Williams, California.[23] The car did not include Baxter's compound-engine idea or any other notable mechanical features, but it was described as a very neat and

compact 30-passenger car. It had cylinders that were 8 by 14 inches and 42-inch driving wheels, and it measured 61 feet in length. This car, like so many of its predecessors, appears to have been an experiment that did not spawn any offspring.

By the late 1890s the past failures of the self-propelled cars had been forgotten, and the idea was basically so appealing that it revived among a new generation of railway managers. The high cost of branch-line train operations and the savings possible with self-contained cars again seemed irresistibly attractive. It was a reenactment of the 1860s—a flurry of activity that in the end produced few lasting results. This late-Victorian revival can be credited to C. Peter Clark, general manager of the New England Railroad and, not so incidentally, the son of the road's president.[24] Clark was disturbed about the diversion of traffic by a new trolley line that paralleled one of his branch lines. In a single year 350,000 passengers abandoned the New England for the trolley.[25] Hardly enough patrons remained to justify continuance of service. Clark made a general study of the problem and found, to his consternation, that local trains running in the Boston area averaged only 47 passengers per trip, hardly enough to justify a locomotive, two cars, and a full crew. A combined coach and locomotive could do the same job for a fraction of the cost.

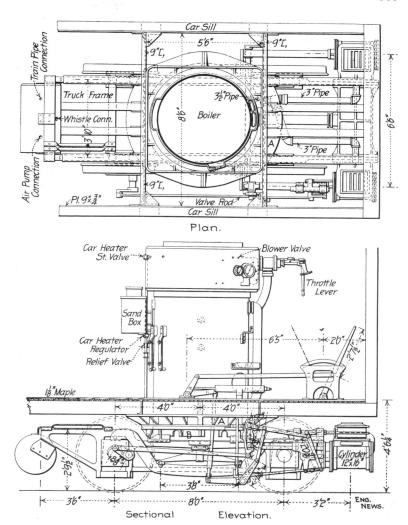

Figure 8.9 *The steam power truck furnished by the Schenectady Locomotive Works for the cars shown in Figures 8.7 and 8.8* (Engineering News, *February 6, 1902*)

Clark, who appears to have considered the idea original, named his new concept in transport the Composite car. In the spring of 1897 he sought the advice of several locomotive builders, among whom A. J. Pitkin, general manager of the Schenectady Locomotive Works, was the most responsive. Pitkin took a personal interest in the novel project and devised and patented a ball-bearing, circular, iron-frame boiler mounting, apparently not realizing that the same scheme had been invented thirty years before by Woodbury.

In October 1897 the Composite car was ready for service. To hold down costs and speed delivery, the machinery was installed in an old dining car. The boiler and engine truck were placed in the kitchen end of the car; the original six-wheel truck was removed (Figure 8.7). The compartment was insulated and fireproofed with asbestos board and galvanized sheet iron. In the passenger compartment the tables were replaced with side seats, boosting the capacity to 60 passengers. The boiler, cylinders, and driving wheels were built as a unit. The power truck was attached to the car by a circular iron frame which could turn on a nest of one hundred and twenty-five 1½-inch steel balls.[26] The 52-inch-diameter boiler contained three hundred and eighteen 1¼-inch-diameter fire tubes and was built for a working pressure of 200 psi, which was rather high for the time. The cylinders measured 12 by 16 inches; the drivers were 42 inches in diameter. The 57-ton Composite was a full sized, powerful unit that could be operated by a two-man crew. Oil fuel had been considered, but in the end the firebox was designed for coke or hard coal.

The Composite began service on a 20-mile branch, but the grade proved too much for its limited boiler.[27] It was next tried on Cape Cod, whose dead-level profile did not tax the Composite's steaming capability. Here it pulled six freight cars without difficulty. It often hauled a coach, and on one occasion moved eleven freight cars. It registered speeds up to 60 mph, but this was recognized as too fast for its small drivers, and it was limited to 45 in regular service. The fire could go for 12 miles without attention, and the 1,500-gallon water tanks were good for 60 miles of running. Yet for all these glowing reports, the Composite was not duplicated for the line. The negative decision may have been made by the New Haven, which took over the New England Railroad during this period. The car was not the success that Clark envisioned, but it did remain in service longer than the typical main-line steamer. It worked on the Dedham branch of the New Haven until October 1904.[28] After two years in storage, it was sold to a small line in northern New York.

The Erie also faced a decline in traffic on certain branches because of the encroachment of electric street railways, and like the New England Railroad, it sought ways of cutting costs and increasing the frequency of service. In the spring of 1897 it tried a small steam storage car produced by the Kinetic Power Company.[29] The car seated only 20 passengers and was undoubtedly too small for the job. And then news came of Clark's full-sized steamer. The Erie quickly asked the Schenectady Locomotive Works to build a duplicate power truck (Figure 8.8).[30] It was delivered in October 1897, just weeks after the New England

unit was ready. An old emigrant coach provided the body (Figure 8.9). The assembly cost the Erie $10,000. It was considered a bargain, however, because operating costs were pared from 27 cents a mile for a locomotive, combine car, and crew, to 16 cents a mile for the Composite and its crew. The road calculated that the steamer would pay for itself in just over three years. The Erie talked of buying another five or six, but in the end it acquired only one more.[31] Purchased in 1898, the car carried the road number 680 and was lettered for the Erie subsidiary, the New Jersey and New York. The anthracite coal burner was fitted with a power unit of the same size and design that Schenectady had produced for the New England Composite and for the Erie 1897 emigrant conversion unit. But the Erie's enthusiasm was short-lived, for even though the car ran fourteen trips a day over the 4-mile branch, breakdowns were frequent and the ride was rough, more like a locomotive's than a coach's.[32] In addition, the boiler was too small to produce the necessary quantity of steam for the schedule. The car was retired and dismantled sometime before 1902.

During the same period the Baldwin Locomotive Works decided to enter the steam car competition. Its main rival was Schenectady, and it could not allow its opponent to enter a new field without challenge. Also, Baldwin's chief, Samuel Vauclain, was confident that he could build a superior steam car. At the time, he was pushing his own compound locomotive patent, and a successful steam car might be a novel way to advertise the system. Compounding was then regarded as a cure-all for the steam locomotive's poor fuel economy. It might prove a double benefit for steam cars, since a smaller boiler would be possible because the steam was used twice. Vauclain also felt that a con-

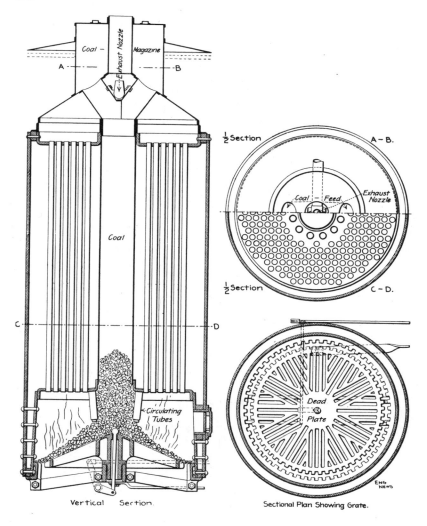

Figure 8.10   Baldwin's self-stoking vertical boiler was designed in 1897 to eliminate the need for a fireman on steam cars of its manufacture. (Engineering News, *February 6, 1902*)

Before the *Woodsdale*'s disgrace, Baldwin managed to sign on two more Midwestern customers for larger versions of Vauclain compound steam cars. The Detroit and Lima Northern's Number 100 was more on the scale of a full-sized coach, measuring 58 feet in length.[36] Mechanically it was similar to the *Woodsdale*, except for the lack of a condenser (Figure 8.11). Another variation was the enclosure of the valve gear to protect it from dust. But apparently the Number 100 never even entered regular service on the D & L N; it was sold to the Fairfield Traction Company of Lima, Ohio, which subsequently remodeled it into a combine car.[37] The steam unit was used as a small switcher.

The third Vauclain car was produced for the Pittsburgh, Cincinnati, Chicago, and St. Louis Railway in 1899.[38] It incorporated the compound-engine, condenser, and magazine-boiler features in a 58-ton car. The 30-foot passenger compartment seated 40. The frame was a composite of steel channels and timber. The magazine held 2,500 pounds of coal. According to the Baldwin order books, the car was intended for service on the Cincinnati, Lebanon, and Northern branch, but its only record of service was on the Xenia to Springfield, Ohio, branch. There it ran for a month, only to be withdrawn and sold to the Clearfield Coal Company of Pennsylvania. Poor steaming and problems with the coal magazine were responsible for its retirement.[39]

Thus Baldwin and Schenectady both failed in their effort to introduce a successful steam car. Neither was ready to admit defeat, and both would try again within a few years, but for the time being the self-propelled car was again discredited. Many reasons were given for its failure. The vertical boiler was a major problem because of its limited steaming capacity, at least in sizes practical to enclose within the confines of a car body. Also, deposits settled on the lower tube sheet of such boilers, causing them to overheat and thus loosening the tubes.[40] To maintain steam, small boilers were worked very hard, resulting in rapid deterioration. Too much time was required to water, coal, and fire up a cold boiler. And there was the inherent defect of the small steam plant, where efficiency seems to be directly related to size. The most productive steam units are ocean vessels and power plants, which do not have width, height, or weight limitations. Steam cars are relatively small units that proved high in maintenance expense and low in thermal efficiency.

The huge water tanks under the car were a weight problem in themselves, but the heavier floor frames needed to support them added substantially to the car's dead weight. To propel this extra tonnage required a larger engine and boiler—again more unwanted weight. Condensers partially solved the problem, but in themselves they were heavy enough to make the cars sway because of the rooftop mounting. Axle loadings were heavy, requiring outsize bearings and extra care in maintenance to avoid hotboxes. The cars rode poorly, being part locomotive and part car. It was in general an unhappy combination, with all the poorest qualities of both components. The "Little Wonders," "Economies," and "Lilliputians" failed to outperform conventional steam trains even in operating costs, for when depreciation and maintenance were figured in, their record was unimpressive. It

---

denser would materially reduce the size of the water tanks, thus cutting dead weight. It also offered the side benefit of muffling the exhaust. Of course, these auxiliaries amounted to something of a trade-off because of the extra weight, cost, and complexity of the subsidiary systems. A peculiar self-firing, vertical boiler was another part of the Baldwin plan (Figure 8.10). Coal was housed in a central magazine, which was loaded from a rooftop hopper. The fuel fed itself into the firebox by gravity and by the car's motion, thus freeing the engineer to concentrate on his operating duties. At least, that was Baldwin's intention.

In June 1898 the first Baldwin car went into service on the Cincinnati, Hamilton, and Dayton's Middletown, Ohio, branch.[33] Named the *Woodsdale*, the 24-ton steamer looked like an oversized streetcar (Figure 8.11). It seated 28 passengers and had a 6-foot-long baggage room. The 48-inch-diameter boiler supplied steam at 180 psi to cylinders measuring 5½ and 9 by 12 inches.[34] During tests, the car ran for 38 miles with no attention to the fire and at speeds of 42.5 mph. The C H & D paid $6,000 for the *Woodsdale* and put the car in service between Hamilton and Middletown, Ohio, on a schedule of six trips a day, with stops spaced out at 2 miles. After a few months, however, the *Woodsdale* was withdrawn from service and placed in the storage line outside of the shops. The road said that many problems had forced this early retirement.[35] The boiler was too small. The water capacity was inadequate. The coal magazine was difficult to load. The fire needed frequent clearing. The water capacity of 300 gallons was insufficient, even with the condenser. In addition to these difficulties, the car rode poorly.

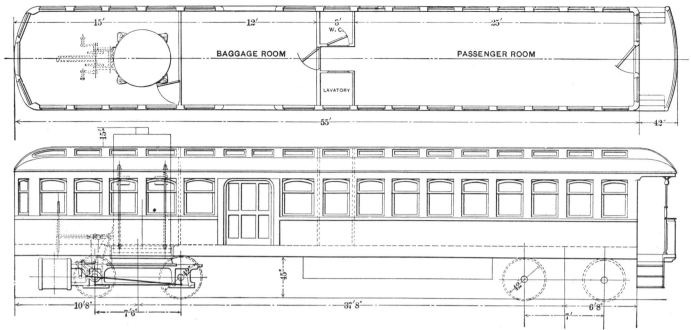

Plan and Elevation of Car for the Detroit & Lima Northern.

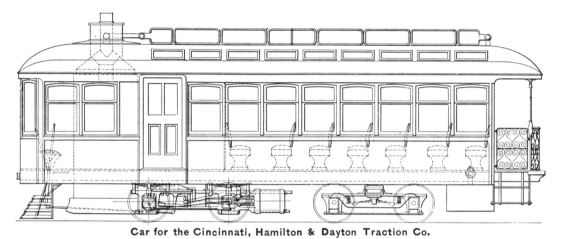

Car for the Cincinnati, Hamilton & Dayton Traction Co.

Figure 8.11   Two Baldwin steam cars of 1897 and 1898. Both cars were retired early. (American Railroad Journal, *April 1898*)

was also contended that when a conventional locomotive fails, all is not lost: a replacement engine can get the train underway again. With a steam car, however, the entire train is out of commission when the motive power fails. These criticisms were applicable to steam cars in all periods and were so basic that no one succeeded in solving them. Even so, it would be nearly forty years before the last steam car appeared in America.

During the early years of the twentieth century, the success of the electric interurban and the automobile renewed interest in self-propelled rail cars. Most efforts were reasonably directed toward perfecting a gasoline-powered vehicle. But despite its record of repeated failures, a few railway engineers insisted that the locomotive-car had the best potential. This belief was nurtured by reports from Europe that a superior steam car was being manufactured in sizable numbers by the Ganz Company of Budapest. The past failures in America were blamed on a want of inventiveness rather than any inherent defect in the steam car itself.

In 1904 the Rock Island began a study of train operating costs.[41] It found that the daily operating cost for a branch-line

train amounted to $13.75, while that of a self-propelled car would equal $9.25. A 65-ton steam car cost an estimated $12,000, while a coach and a baggage car ($5,000 ea.) and a small locomotive ($7,000) represented a total investment of $17,000. Any reasonable manager would opt for the steam car. But there was a serious flaw in the logic of the study that apparently went unrecognized by the Rock Island's managers. This was that the existing rolling stock available for branch-line operations represented only a very small percentage of the investment necessary for new equipment. In addition, the report underestimated the actual first cost of a new steam car by nearly 50 percent.[42] But the optimistic conclusions were accepted, and the Rock Island placed orders for two cars. One was a coal burner; the other was oil-fired.

The coal burner was ready for service first. It was an enlarged, Americanized version of a Ganz car built by the Baldwin Locomotive Works to the designs of the Railway Auto Car Company of New York, Ganz's American agent. The Rock Island's Number 2551 was delivered in the spring of 1907.[43] Ganz's largest European model was an 80-horsepower rail bus, but the Rock Island unit was a 250-horsepower, steel-bodied, 53-ton unit (Figure 8.12). The 55-foot-long body was divided into engine, baggage,

Figure 8.12   A Baldwin-built steam car incorporating the European Ganz designs as enlarged and modified for North American service. The Number 2551, completed in 1907, was a coal burner. (H. L. Broadbelt)

smoker, and passenger compartments. Mechanically it followed the plans of its Hungarian designers. A fully enclosed compound-engine, valve gear, and gear box were attached to the front truck wheels.[44] A compact boiler produced superheated steam at 270 psi. The 2551 could pull a trailer at 35 mph with a fuel consumption of 16½ pounds per mile.

The second car, Number 2552, was an oil burner produced by the American Locomotive Company in 1908.[45] It looked like a modern version of Romans's locomotive car (Figures 8.13 and 8.14). The husky power truck was a grotesque 2-2-0 locomotive with a horizontal, return-flue boiler. Mellin's cross-compound system was used to propel the single pair of 38-inch driving wheels. The 2552 was nearly identical in size, floor plan, and weight to the Baldwin car, despite its very different mechanical arrangement. Truss rods were added to stiffen the frame, to reduce vibration, and to help support the massive 1,000-gallon water tanks and 100-gallon fuel oil tanks.

The 2552 was placed in service on the Salina to Herrington, Kansas, branch, where it had an operating cost of 15.1 cents a mile—some 5.9 cents above the original estimates.[46] The Baldwin car, Number 2551, was a far greater disappointment; it was returned to Philadelphia for a new engine, boiler, and superheater.[47] Although neither car fulfilled the Rock Island's hopes, neither could be classified as a failure. Both operated for a number of years as steam cars. In 1917 both power plants were removed, and the cars were converted into combine trailers for commuter or branch-line service.[48] The Baldwin car was retired in 1943, and the Alco unit followed five years later.

Other roads were also studying costs and testing steam equipment. Several exhibited a strong interest in the performance of the Ganz cars. The Budapest firm had built its first car in the late 1890s, and during the next decade it completed 200 cars for lines in Hungary, Austria, Germany, and Russia.[49] By the time the automotive age began, the Ganz products could hardly be called experimental. Ganz felt that it had overcome the basic defect of

the steam car: its boiler. Boiler failures and maintenance were the chief cause of downtime. Ganz offered a quick-change boiler that could be pulled out and replaced by a standby unit in less than one hour. It seemed logical that this proven stock design could be successfully imported for use in North America. The Florida East Coast and the Intercolonial (Canadian) railroads bought units in 1906 and 1907.[50] Other lines felt that the design was not suited for operations here and proceeded to enlarge and revise the Ganz car. The Erie acquired a large wooden Ganz car, the Number 3000, in 1907.[51] Ganz produced a somewhat larger car in 1911 for the Santa Fe.[52] It had an oil-fired Jacobs-Shupert boiler with feedwater and superheaters. The power truck was overburdened, causing the journals to overheat. Additional leading and trailing wheels did not correct the problem, and the car was retired after only three months' service. The body was remodeled as a conventional car. Why the Ganz car should have been so successful in Europe and so unsuccessful here is a question that no one seems able to answer.

During the same period a few railroads tried to produce their own steam cars. Around 1903 Theodore H. Curtis, newly appointed superintendent of motive power for the Louisville and Nashville, began to build experimental units.[53] Little is known about these, but a description of a later Curtis car is available. In 1909 an eight-wheel, steel-body car of his design was running on the Hocking Valley Railroad. It was a large 76-foot-long, 700-horsepower unit designed to pull one or more trailers. One of Curtis's associates blamed the demise of the project on the railroad's overly conservative president, who not only condemned the idea of self-propelled cars but wanted nothing to do with electric car lighting.

The Great Northern tried a car designed by its former locomotive superintendent, Maximilian E. R. Toltz.[54] The oil-burner steamer was built at the Burlington's Aurora shops. The Great Northern was unimpressed with the Toltz car, however, and it was returned to Aurora for storage.

Figure 8.13 The American Locomotive Company completed this oil-burning steam car in 1908. (Alco Historic Photo, National Railway Historical Society)

Figure 8.14 The grotesque power unit of the Rock Island's steam car Number 2552 was hidden inside the front compartment.

Figure 8.15   *This formidable steam car was produced by the St. Louis Car Company in 1906 to the designs of W. G. Wagenhals.* (Scientific American Supplement, *December 22, 1906*)

Surely the most spectacular steam passenger car of this period was the Wagenhals-Kobusch car built by the St. Louis Car Company in 1906 for the Missouri Pacific.[55] It was a Leviathan among steam cars, measuring over 82 feet long and weighing 89 tons. Its designer, W. G. Wagenhals, apparently believed that an increase in size would solve the steam car's problems. Wagenhals had been general manager of a street railway in Cincinnati before taking the same job with the Ohio River and Columbus Railroad. A man who liked mechanics, he had built a gasoline automobile in 1891 and a small inspection car for the O R & C. With this background he recognized the potential of the self-propelled rail car. In 1905 he ordered a steam car from the St. Louis Car Company. That firm's president, George J. Kobusch, was impressed with the idea and encouraged Wagenhals to develop the ultimate steam car.[56] The result of this partnership was the mighty Number 1000 (Figure 8.15). The wooden body was supported by a steel frame with truss rods for extra stiffness. The forward portion of the body was metal-sheathed to help fireproof the engine room. An oil-fired, fast-steaming, torpedo-boat style of boiler furnished steam at 250 psi. The cylinders were mounted inside the wheels on the front truck, with a gear drive to one pair of drivers. The second set of wheels was driven by connecting rods. The 1000 was scheduled for a cross-country tour before delivery to the Missouri Pacific. The demonstrator won only two orders for the St. Louis Car Company: two smaller cars, presumably duplicates, were produced in 1908 for the St. Joseph and Grand Island and the Crystal City and Uvalde (Texas) railroads.[57] The 40-footers seated 41 and weighed 33 tons. The machinery was similar to the 1000's, except that the side rod drive was eliminated. The service history of these cars has not been uncovered. They could not have been notably successful, because no more were produced.

After 1910 the gasoline engine was the unrivaled motive power unit for small vehicles. By this time it had been developed into a compact, dependable power source, as demonstrated by the growing numbers of automobiles and trucks in everyday service. The lesson was obvious, and the railroad industry began to accept gasoline rail cars. Yet an undercurrent of resistance was evident among railroad men. They knew and trusted steam and saw little to admire in "doodle bug" gasoline cars. They still hoped that the past failures of the steam car might be reversed by a more refined and scientific design. Much had been accomplished in steam automotive design; perhaps these techniques could be applied to the self-powered rail car. The Stanley brothers of Newton, Massachusetts, were leaders in the steam auto field. They were the only domestic manufacturers to mass-produce steamers over an extended period. They had a genius for light steam power design, and they devised reliable, practical vehicles. The Stanleys were confident that they could redeem the failure-ridden steam car. Late in his life, Francis E. Stanley took personal charge of the project.[58] He was about to retire from the automotive business, and he apparently viewed the rail car as a challenging retirement project.

Francis Stanley took established designs of the Stanley Company and adapted them directly to the rail car. They included a high-pressure (up to 1,000 psi) fire-tube boiler wound with piano wire for extra strength and light weight. The engine, a faithful copy of the standard auto design, was a simple noncompound, two-cylinder unit. Kerosene was used to fire the boiler. A test car, the Number 100, was produced around 1915 for service

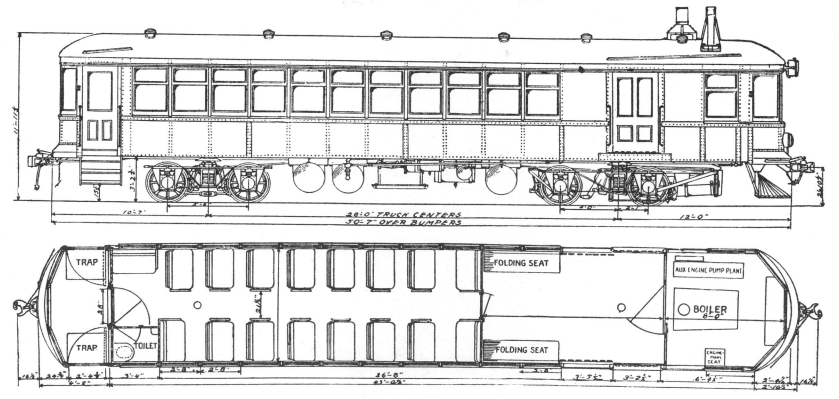

Figure 8.16 *Several Unit steam cars were built between 1915 and 1922 after designs prepared by the famous steam automobile manufacturers, the Stanley Brothers. The car here was made for the Canadian National in 1922. (Railway Review, September 30, 1922)*

on the Boston and Maine. In 1916 a second car, the Number 101, was produced for the White River Railroad of Vermont.[59] The 50-foot, 44-passenger car body was built by the Laconia Car Company. The Stanley Company provided the machinery. The squat vertical boiler contained 600 tubes ¾-inch in diameter. The fire was automatically regulated by the rise and fall in boiler pressure. The 15-ton car could operate at speeds of 45 to 60 mph—a rate more than adequate for any branch line. A roof condenser was installed to avoid the need for large water tanks or frequent water stops. The manufacturers claimed that 1 gallon of fuel (kerosene or heating oil) would propel the car 5 miles, but in actual service fuel economy was only about one-quarter of this figure.[60] In 1922 the car was reported to be very satisfactory and dependable. It often pulled a trailer, and on occasion it was harnessed to a freight car. Since the 101 could develop up to 1,000 horsepower, at least for short intervals, it was capable of maintaining a satisfactory schedule on hilly lines.

The project was just getting established when Francis Stanley was killed in an auto accident on his way to a meeting involving the rail car. The Rushmore Company of Boston then acquired control of the business.[61] The new owners adopted the trade name Unit Car. They built cars for the Boston and Maine, the Canadian National, and the Uruguayan government in South America. These cars resembled the 100 and the 101, except that they were heavier. The 1922 Canadian National car, for example, weighed nearly 30 tons.[62] The piano-wire boiler was replaced by a larger watertube unit (Figure 8.16). The engine design remained unchanged. However, the Unit Car failed to revolutionize the industry and seemed to have little effect on the growing popularity of its gasoline-powered competitors. The reasons for its demise are unknown, but no more cars were produced after the early 1920s. The original demonstrator was not scrapped until 1941 or 1942, after being stored for several years.[63]

At the time that the Unit Car was introduced, the International Harvester Company began to study the possibilities of a steam car.[64] The scheme matured slowly, for it was not until 1927 that the first pilot model entered service. The Locomotor, as it was called, was a massive 73-foot combine car seating 63 passengers. It had a forced-circulation watertube boiler furnishing steam at 600 psi. Two 8-cylinder uniflow engines, mounted under the floor, powered the trucks through automotive-type drive shafts. The roof condenser's work was accelerated by power-driven fans visible from the car's end. The Locomotor moved 1½ miles per gallon of fuel oil—a respectable figure considering its size and weight. In 1929 the Milwaukee Road purchased a combined baggage–Railway Post Office Locomotor.[65] The 58-ton unit had no passenger compartment; trailers were used for this purpose. The B & O and the New York Central were reported ready to purchase similar units. However, only the Milwaukee ordered a second car. It was produced in 1931 by the Ryan Car Company of Chicago, which had meanwhile taken over the Locomotor's assets.[66] This car weighed 63 tons and had two 225-horsepower, eight-cylinder uniflow engines (Figure 8.17). The two Locomotors covered Milwaukee runs, mainly in Iowa, before being retired in 1938 and 1939.[67]

The Locomotors were working out their last years when George D. Besler, son of a former Central Railroad of New Jersey president and owner of the Davenport Locomotive Works, began to develop the ultimate steam railway car. Around 1934 Besler was experimenting with an advanced, high-pressure, steam power package. He gained considerable notoriety by producing several vehicles, including an airplane, powered with his engine. The New Haven decided that the young engineer had created more than a popular science novelty and agreed to cooperate in the fabrication of a two-car train (Figure 8.18). Two semilightweight commuter cars were selected for the conversion,

Figure 8.17   The Locomotor Company delivered this powerful eight-cylinder steam car to the Milwaukee Road in 1931. Except for the roof-mounted condenser, it looked like a conventional gas-electric car of the period. (William Baldridge, Cedar Rapids, Iowa)

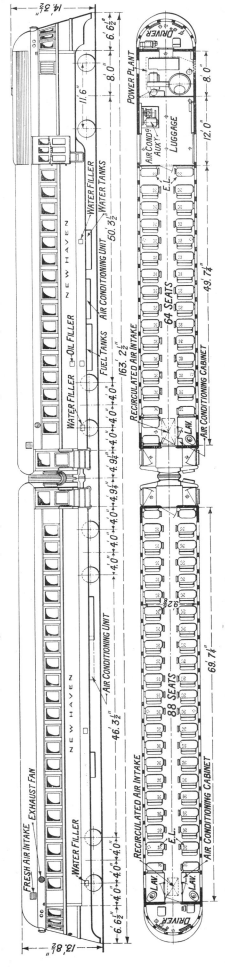

Figure 8.18   Bester's high-pressure steam train was placed in service on the New Haven in 1936. It represented the last effort to produce an American steam rail car. (Railway Gazette, March 12, 1937)

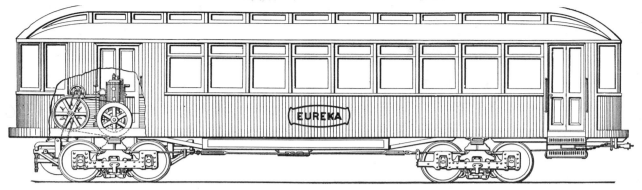

Figure 8.19   A three-cylinder gasoline engine powered the belt-driven Eureka. *It was tested on the New York Central and Pennsylvania railroads in 1899.* (Railroad Gazette, May 5, 1899)

and the work was carried out by the Budd Company.[68] Streamlined front and rear cabs were installed. Body skirts and a blue paint scheme helped create a more modern appearance. The interior was refurnished as well, and new seating was provided for 152 passengers. The power plant featured a forced-circulation boiler with a continuous tube tapering from ¾ inch to 2¼ inches in diameter.[69] Set for 1,500 psi, it was unquestionably the highest pressure ever used for a steam car. The long-wheelbase power truck had a cast-steel frame with two pairs of compound cylinders cast integrally. They measured 6½ and 11 by 9 inches. The drive was directly connected to cranks on the axle ends; there were no gears or transmission. A roof condenser was used. The 1,000-horsepower Besler train cruised comfortably at 70 miles per hour. It could be operated from the rear cab by pneumatic controls.

Besler's car entered service late in 1936, running six trips daily between Bridgeport and Waterbury, Connecticut.[70] The train remained in service for about eight or nine years. It was reasonably satisfactory, but the fact that it was never duplicated indicates that it was no competition for diesel rail cars of the time. It was heavy, weighing nearly 152 tons. Although much of the excess weight can be blamed on the use of old cars and the added burden of air conditioning, the power machinery alone weighed 45,200 pounds—far more than a comparable internal combustion plant. The 14-foot, 6-inch wheelbase lead truck must have caused heavy flange wear. And because everything about the power plant was exotic and custom-made, maintenance was itself a custom process. Perhaps the Besler car was too sophisticated to succeed, or perhaps it was not given a fair trial. Whatever the reasons for its failure, it was the dramatic final chapter in the history of the American steam car.

Internal Combustion Motor Cars

Long before the final steam car entered revenue service, railroad mechanics had begun to explore other power sources that might prove suitable for rail car service. Actually it was the street railways, rather than the railroad industry itself, that showed the most willingness to test exotic types of propulsion units. Perhaps their need was greater, since horsecars were slow and inefficient. The search for a compact mechanical drive led to experiments with steam, soda, compressed air, electric batteries, and gasoline cars. Steam failed miserably, even though it was familiar and highly successful in other applications. Its deficiencies caused mechanics to try out power units then bordering on the visionary, such as the internal combustion engine. Gas and gasoline engines

were being installed on horsecars during the middle and late eighties. In Europe such engines were being successfully applied to road carriages, but America was slow to accept the automobile. The street railway was the gasoline engine's first proving ground on this side of the Atlantic. Early in 1886 the Connelly Gas Motor Company began work on its first car, which was tested in Brooklyn in January of the following year.[71] At least two more Connelly cars were produced and tested during the next two years.

Even more important experiments were started in 1888 by William H. Patton in Pueblo, Colorado.[72] The first car had a mechanical transmission that proved to be its weak point. Patton recognized this defect, and while seeking to overcome it, apparently became the first person to discover the happy marriage of the internal combustion engine and electric transmission. He embellished the idea by interposing batteries between the engine-generator and the traction motors, because he believed that a "flywheel" effect was necessary. This part of the scheme proved extraneous, but the basic concept later came into widescale use with the advent of the diesel-electric locomotive. Between 1890 and 1898 Patton built several gas-electric streetcars and small locomotives. The apparatus was impressive enough to gain the backing of George M. Pullman. By the time the first car began sputtering around the yard of the Pullman plant, however, the street railway industry was firmly committed to electric traction. One branch line in Maine purchased a Patton car, but the inventor was unable to interest the larger steam railroads in trying one.

Other mechanics also attempted to introduce gasoline rail cars during this period, but only one is known to have built a unit that a steam railroad considered worthy of testing.[73] In 1899 the Vimotum Hydrocarbon Car Company of Chicago sent an eight-wheel passenger car produced by the Jewett Car Company to Indianapolis for tests on the New York Central and the Pennsylvania railroads. The 41-foot-long unit was more on the order of an interurban car than a full-sized steam road coach. A three-cylinder Wolverine engine generated 45 horsepower to propel the 21-ton car at speeds up to 41 mph. The transmission included a variable-speed pulley with a huge flat belt carrying power to the wheels on one truck. The car does not appear to have measured up to its name, *Eureka*, for the Vimotum Company retreated into an early and probably well-earned obscurity (Figure 8.19).

The twentieth century was destined to be the internal combustion age. Many progressive engineers and businessmen understood why the compact, oil-burning engine would prevail, yet few seemed ready to try it on a rail vehicle. One was Edward H. Harriman, a financier and railroad executive of remarkable en-

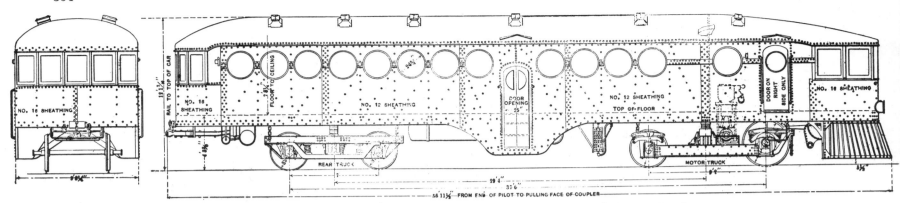

Figure 8.20   *William R. McKeen, superintendent of motive power for the Union Pacific, began producing gasoline-powered rail cars in 1905. This, his seventh car, was completed in 1906. (American Railroad Journal, May 1906)*

Figure 8.21   *McKeen's six-cylinder engine powered only the rear wheel of the front truck through a chain drive.*

ergy and vision. His forte was corporate and financial planning at the highest level, but he also seemed to find time for mundane management problems, such as how to cut costs on marginal passenger operations. During the first years of the new century, Harriman was impressed by the performance of high-speed torpedo boats being tested by the Navy.[74] It occurred to him that the light, yet powerful engines used in such vessels could be placed in a passenger coach. Or better yet, a railroad could build a special car on the order of the sleek, knife-edge torpedo boats. He turned the idea over to William R. McKeen, Jr., superintendent of motive power for the Harriman-controlled Union Pacific Railroad. McKeen was not an old-guard master mechanic but one of the new, college-trained breed of engineers, still in his

early thirties. This in itself did not guarantee an openness to new ideas, but McKeen also happened to be an independent and free-thinking man who reveled in the unorthodox. Although he had no experience in the field of gasoline engines, he was intrigued by the novelty of the assignment and took to it with enthusiasm.[75]

The resulting McKeen car represented the real beginning of the main-line gasoline rail car. It was the first to be built in any significant number, and it marked the origin of a serious interest in such rolling stock by an American railroad. It was also one of the most distinctive and imaginative rail vehicles ever produced. McKeen literally followed Harriman's idea for placing a torpedo boat on the rails (Figure 8.20). The nautical look of the slender, prow-shaped body was later reinforced by the use of porthole windows. The steel body was novel if not exactly a pioneer. The wind-splitting, aerodynamic design was futuristic, albeit scientifically incorrect. Later experiments showed that the wedge-shaped prow should trail the car, not lead it.

McKeen's power train was as novel as the body plan (Figure 8.21). The engine (200 to 300 horsepower) was carried directly on the front truck. A 5-inch-wide Morse silent chain powered the forward or large-diameter wheel set; only one set of wheels was driven. The six-cylinder engine was separated at the middle into two three-cylinder units. The chain drive sprocket was mounted on the crankshaft at this central point, so that the chain could run to a corresponding sprocket on the forward axle. A friction clutch could disengage the chain and the low gear. The gear was used to start the car, but at a speed of 10 mph the chain was shifted to direct drive.[76] The direction of travel was changed by reversing the engine. The engine was started by compressed air stored in tanks, which were replenished by an air compressor driven by the main engine.

Designs for the first car were ready early in 1904. It was the only single-truck, wooden-bodied car built by McKeen. The Number 1 was produced at the Union Pacific's Omaha shops, as were all the other McKeen cars, totaling approximately 150. The 31-foot experimental model seated 25 and was powered by a 100-horsepower engine.[77] It made a trip from Omaha to Portland, Oregon (about 2,000 miles), and returned without difficulty, or so read the official news releases. Years later an old U P mechanic recalled that the long journey involved many problems, and that Motor Car Number 1 was pushed into more than one terminal.[78] After its return the Number 1 was assigned to a branch line, where it pulled a mail car. It eventually caught fire (thus demonstrating one of the hazards of gasoline vehicles) but was rebuilt with a box cab.

A second, larger car was completed in September 1905.[79] It was mounted on eight wheels, measured 55 feet in length, seated 57, and weighed 28 tons. The steel body was scientifically patterned after a bridge truss, and the diagonal members of the frame necessitated small windows—one drawback to the plan. Since light weight was a paramount consideration, the trucks and wheels were designed to eliminate as much excess metal as possible. The interior was beautifully finished, with antique mahogany paneling and a cream-white ceiling embellished with gold and sepia designs. The seats were covered in leather, and the car was brilliantly lit with acetylene lamps. The exterior was painted a rich maroon with gold stripes and lettering. The trucks and undercarriage were olive green.

McKeen was now ready to tool up for production, and four cars similar to the Number 2 were quickly rolled out of the shop. In 1906 came the Number 7—the first McKeen car to display the distinctive porthole windows and drop-center door. With the next car, McKeen began to manufacture his own gas engines. In 1908 a separate corporation was set up for the production and sale of gasoline motor cars. All work continued to be performed within the Union Pacific shop complex. Orders were solicited and

received from other railroads. The curious torpedo cars began to appear on the Erie, the Norfolk Southern, the Alton, and numerous Western lines (Figures 8.22 and 8.23). By November 1909 fifty-six McKeens were reported in service on U.S. railroads (mostly on the Union Pacific), with one in Mexico.[80] Some had 70-foot bodies, and nearly all were powered by 200-horsepower engines.

McKeen presented papers and furnished data to railway clubs in an effort to promote his car.[81] His material was, of course, laudatory and intended to allay fears that such units would always be experimental. He portrayed them as reliable, economical, and safe, and pointed out that they were smoke-free, easy on the track, and instantly ready for service. McKeen documented his claims with specific examples. The cars on the Union Pacific ran mainly on branches in Nebraska; some traveled 390 miles a day, and many hauled trailers. During its first year's service, car Number 19 had had only two engine failures. The cars were as reliable as a steam locomotive, and were available twenty-three days out of the month. One unit on an isolated branch with no repair facilities ran for six months without a single failure. It performed splendidly during a blizzard that froze the steam locomotives at the same terminal. And for economy the cars had an outstanding record, according to McKeen. In regular running they averaged 3 miles to the gallon. Total operating costs, including normal running repairs, were figured at 12 cents a mile. And the two-man crew saved $195 a month in wages.

McKeen also boasted about the wedge-shaped front end which, he explained, had been suggested by earlier experiments in Germany and by a demonstration electric car shown at the St. Louis World's Fair in 1904.[82] He claimed that a car with a flat front was subjected to wind resistance equal to 87.5 horsepower, while resistance to the wedge front amounted to only 21.9 horsepower. Just how McKeen arrived at these figures is not known, but in light of later streamlining experiments, their validity is questionable. In addition, the normal top speed of the cars was 50 miles an hour, and even at this velocity, it is unlikely that the prow-shaped nose would offer any material saving of fuel. The wedge front did have at least one practical, if unintended, advantage: it served as a collapsible buffer in accidents. In one case, two McKeen cars hit head-on with a combined impact speed of 75 to 85 mph.[83] Although both front ends were demolished, the engines were undamaged and no one was seriously injured or killed. The cars were not even derailed. In another incident, a McKeen car hit a steam locomotive at 30 mph with no damage beyond a dented front end.

The reports issued by McKeen were not entirely objective. His critics admitted that his design was original and distinctive, but not everyone found the cars' appearance pleasing or their performance satisfactory. A contemporary observer objected to their "warlike-looking submarine design." Ralph Budd, onetime president of the Burlington and no enemy of motor cars, recalled that the arched-roof McKeens were called potato bugs.[84] Aesthetic objections aside, the design was also attacked for mechanical shortcomings. The engine was balky and could be operated

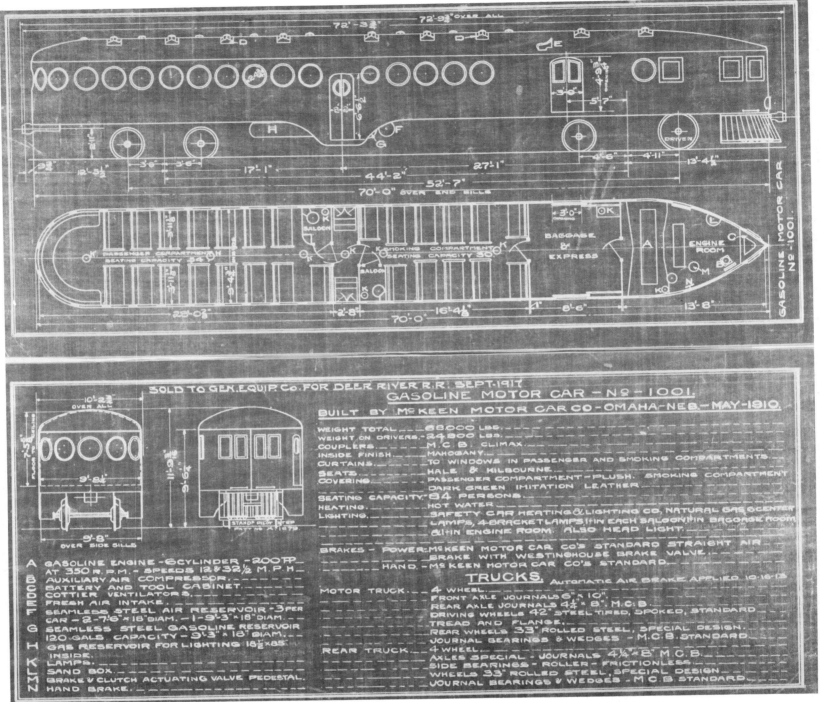

Figure 8.22   McKeen built approximately 150 cars, the majority of which operated on the Union Pacific and its related roads. The example here was produced for the Baltimore and Ohio.

only by expert workmen. The camshaft and timing parts were prone to rapid wear.[85] The compressed-air starting system was troublesome, and the storage tanks were inadequate. There was no reverse gear. To back up, it was necessary to stop the engine, reverse it, and start again—a cumbersome process. The engine, even when functioning well, was noisy, particularly in climbing a grade. Mounting the engine directly on the front truck was considered a poor idea, because it was subject to every shock and jar that the roadbed had to offer.

These defects did not encourage sales. Except for the closely associated Union Pacific, few repeat orders were placed. Most lines bought one or two cars as an experiment, and after 1917 very few new McKeens were built.[86] Three years later, the com-

pany was liquidated. But McKeen's demise may have been occasioned by more than engineering defects, for it is possible that the market was simply saturated by the 250 cars produced by McKeen and General Electric. The industry may have felt that only so many runs were suitable for motor cars and that no more were required. This speculation is supported by the fact that both manufacturers suspended production simultaneously.

A few McKeen cars remained in service thirty years after the U P's Omaha shops suspended production. Some were rebuilt with squared-off front ends, new front trucks, and gas-electric drives. The Chicago Great Western rebuilt three into a deluxe motor train, the Blue Bird.[87] The lead unit was given a new 300-horsepower engine, and trailing cars were remodeled with buffet,

Figure 8.23   Because of its unorthodox appearance and troublesome power plant, most McKeen customers bought only one or two experimental cars. (Railway and Locomotive Historical Society)

sleeping, and lounge sections. McKeen cars remained most conspicuous on the Union Pacific, which had thirty-two in service in 1929.[88] The majority had run more than a million miles by then. In later years some were rebuilt as gas-electrics, and two were in use as late as 1947. Possibly the last McKeen to operate was the unit belonging to that museum of American railroading, the Virginia and Truckee. The car was in service as late as 1946 and looked very much as it did in 1910, when it was sold to the Nevada short line for $24,000.[89]

General Electric decided to enter the self-propelled rail car market just months after McKeen began design studies for his first car. The project was officially launched late in 1904, but G.E. officials had been considering it for several years.[90] They were motivated by two considerations: the success of gasoline cars in Europe, and the prospect of a new outlet for electrical apparatus. They envisioned a self-energized trolley car that could operate on ordinary steam railroads without the colossal capital cost associated with conventional electrification. Patton had explored an idea that G.E. engineers now hoped to perfect. They felt that a major drawback to the larger gasoline rail cars was the mechanical transmission. The gearshift and clutch might be suitable for light highway vehicles, but it could not withstand the strains of a large gas engine and a heavy rail vehicle. An electric transmission could handle a prime mover of any horsepower. Moreover, it would relieve the engine of torsional strains and shocks caused by crossings, switches, curves or rough track that were ordinarily transmitted by a direct mechanical connection between the engine and the wheels. Because there was no fixed relation between the engine and the car's speed, the engine could be turning over at its optimum revolutions per minute. Hence the engine's maximum power could be utilized when the car was starting up or when it was climbing a grade. Conversely, the engine-generator

set could be loafing while the car sped over a level portion of the line. The electric transmission offered the possibility of an infinite number of speed changes. It was easily reversed. Double end controls were readily installed. If desired, every wheel could be powered by a separate motor. In short, G.E. felt that it had a salable engineering concept. It also produced a stock line of generators, motors, controllers, and their associated electric street railway apparatus which could be adapted for gasoline rail cars.

The first G.E. car was something of a makeshift (Figure 8.24). A 65-foot-long wooden body with arched windows and a steel underframe was furnished by Barney and Smith. A six-cylinder, 140-horsepower engine was delivered from England in mid-1905. Assembly proceeded slowly, for the car was not ready for testing until January 1906. Subsequent trials on the Delaware and Hudson were disappointing. The 68-ton gas-electric was heavy, underpowered, and undependable. It was returned to its maker, where the engineers began on it afresh. Two years would pass before a second G.E. car was ready for a trial.

Meanwhile McKeen was already in production. His lightweight steel cars proved a model for his competitor. General Electric turned to the Wason Car Company for assistance. The Massachusetts car builder came forward with a body design that in effect translated McKeen's thinking into a more conventional form. The sharp prow gave way to a rounded snout, and the porthole windows were replaced by conventional sash. However, the general concept of a light, arched-roof steel body was retained, as was the center door entrance and the rounded rear end. G.E.'s own design team took over the planning of the new engine. The result was a compact V8 weighing less than one-third of the imported English engine. The combination of Wason body and new engine resulted in a lithe, 50-foot motor car that weighed only 31 tons, or less than half as much as the first test car.

Figure 8.24   *In 1905–1906 General Electric built its first experimental gasoline-electric car, which was fitted with an English engine.* (Jim Shaughnessy Collection)

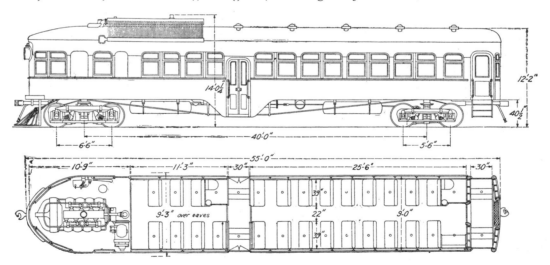

Figure 8.25   *After constructing several experimental units, G.E. introduced a production-model gas-electric car in 1909. The two cars for the Southern shown here were completed during the following year.* (American Railroad Journal, *February 1910*)

Although the second G.E. car was vastly superior to the original wooden demonstrator, it did not fully meet the expectations of the designers, for a third demonstrator was soon underway. It was not ready until early in 1909. General Electric had clearly lost the race with McKeen, who had already produced thirty-five cars and had another twenty underway. But G.E. doggedly pressed on. The Number 3, somewhat larger than the Number 2, began a series of road tests that covered 50,000 miles. The 35-ton demonstrator toured every railroad that would allow it on the property and made an appearance at the Master Car Builders Convention in Atlantic City.[91] Its 125-horsepower plant delivered about 2 miles to the gallon and could propel the 52-passenger car at a steady 60 mph. Its good performance finally proved that General Electric was ready to produce self-propelled cars on a commercial basis. In October 1909 the Southern ordered two 55-foot-long cars that were near-duplicates of the Number 3 (Figure 8.25). Three other railroads placed orders as well, and deliveries began the following year.

Figure 8.26   Interior of a G.E. gas-electric of about 1910. Most of the bodies were built by the Wason Car Company (General Electric Company)

The initial performance reports were encouraging. Exploiting the testimonials of its customers, G.E. reported a dazzling series of successes in a presentation before the New York Railroad Club.[92] The 107-mile-long Minneapolis-based Dan Patch Line was serviced exclusively by gasoline motor cars. The cars ran the full length of the road in only three hours and twenty-five minutes, including stops. They used the cheapest fuel (naphtha at 6 cents a gallon), even on days when the temperature was 16 degrees below zero. They ran 3,900 miles a month at an average cost of 18 cents a mile. The equivalent cost for a steam train was figured at 33 cents a mile. The gross yearly saving came to $6,970, or roughly 25 percent of the car's original cost. Repair figures were equally impressive. The gas-electric's repairs cost 4¼ cents per mile, a locomotive's 8½ cents. The cars were rugged and could be pushed well beyond their rated capacities. They were used to switch at terminals and could pull as many as twenty freight cars. The Frisco, another large-user of G.E. equipment, pulled two trailers and 415 passengers with one of its gas-electrics.

The industry reacted enthusiastically, during the year of these sensational reports, G.E. delivered 31 cars—the greatest number produced in any twelve-month period (Figures 8.26 to 8.28). Several models were offered, the largest being a 71-footer weighing 49 tons and seating 91 passengers. But product variety and customers' praises were not enough to sustain purchases. Orders fell off after 1912. Deliveries slowed to sixteen in 1913, and then to only two by 1916. No definite explanation can be found for the abrupt decline that both G.E. and McKeen experienced after

1915. The possibility that the market was saturated has already been suggested. In addition, there was prejudice against gasoline motor cars in railroad circles. Although such attitudes are difficult to document, R. S. Zeitler does mention the "disparaging view of the internal combustion engine" held by railroaders at the conclusion of his paper on the self-contained railway motor car appearing in the *Cyclopedia of Engineering*.[93] Under pressure, many old-line master mechanics could be persuaded to buy a motor car for testing. But even if management was ready to give it a fair trial, a hostile operating crew could sabotage the interloper. Only a few embarrassing road failures were needed before its enemies proclaimed that the damfool thing was a failure, just as they had predicted. There was no question that motor cars could succeed only under favorable conditions. Repair and operating crews had to be trained in their care and handling. Shops, tools, and parts must be available. A gas-electric car could not be maintained properly by a blacksmith or a boilermaker.

There were other more direct reasons for the gas-electric's initial failure. G.E. expected that the advantages of electric transmission would sell the motor car, but the system was not without faults. It was heavy and expensive: electric transmission cost $10,000 to $12,000 more than a mechanical drive and weighed 10 to 20 percent more.[94] The first cost of the cars, ranging from $20,000 to $30,000, discouraged larger sales. The same money would buy a locomotive, which represented a profit-making tool, while the gas car was viewed as a marginal cost-cutter. In addition, G.E. management shifted policy. Rather than promoting gas-electric cars, it decided to aim for a larger market by en-

Figure 8.27   Engine-room compartment of a G.E. gas-electric car of about 1910. The V-8 engine developed 125 horsepower. (General Electric Company)

couraging main-line electrification. This would involve the sale of generating, substation, and overhead-wire hardware in addition to motor and control equipment. In 1917 the decline in orders and the war emergencies provided good reasons to close out the gas car division.

While G.E. and McKeen dominated the first generation of motor cars, a number of smaller firms also attempted to introduce them. For example, the Kuhlman Car Company of Cleveland, Ohio, built an eight-wheel gasoline motor car for the New York Central Railroad in 1906.[95] The 43-foot, 9-inch car was powered by a 220-horsepower Chase engine and registered speeds up to 65 mph. Another streetcar builder, Brill, constructed several heavy gas-electric-battery cars between 1906 and 1909 for the Strang Gas Electric Car Company of New York City.[96] They were tested in the Midwest but apparently did not seem worth serious study to any major railroad. Fairbanks, Morse, & Company, a major producer of handcars and other railway supplies, began to make four-wheel streetcar-like gasoline cars around 1906. Of the limited number built, most went to short lines, but one was purchased in 1909 by the Pennsylvania for a branch in Delaware.[97] In the Far West the Hall-Scott Company of Berke-

ley, California, began rail car production in 1909.[98] The firm was already established in the gasoline engine business. Steel bodies, 150-horsepower, six-cylinder engines, and automotive-style drive shafts to the rear truck were typical of the Hall-Scott design (Figure 8.29). Most of the cars were built for Western customers, but at least one unit was sent as far east as Indiana and a few were turned out for Canadian lines.[99] Rail car production apparently continued until 1926, when the firm became associated with the American Car and Foundry Company. After that time it produced engines for Pullman and Brill rail motor cars.[100]

Besides the cars built by small commercial suppliers, all kinds of home-made gasoline rail cars were fabricated by American short lines. Documentation is scarce, but enough pictures have been printed in the railroad hobby publications to indicate that the practice was fairly widespread. Although the variety and crudeness of these contraptions defy any general analysis, all were born from a "root hog or die" necessity. The Rio Grande Southern, a poverty-stricken Colorado narrow-gauge line, constructed a distinctive group of rail busses between 1931 and 1937 that were locally known as the Galloping Geese.[101] Awkward in

Figure 8.28   G.E. produced one of its larger cars in 1914 for the Illinois Central. The 71-foot-long car seated 86 passengers and included a separate compartment for Negroes. (DeGolyer Foundation Library)

appearance and bumpy in ride, these homely vehicles helped keep the line in business during the worst years of the Depression (Figure 8.30). Elsewhere, local mechanics combed the junkyards in search of just the right piece to complete a rail car. The parts available—a truck chassis, a tractor engine, or perhaps a discarded streetcar body—dictated what form the car would take. Most had simple mechanical drives, but at least one rather elaborate gas-electric was assembled in 1919 by a Kentucky short line, the Flemingsburg and Northern. A 90-horsepower gas engine powered a 55-kilowatt generator.[102] The body appears to be made from two four-wheel streetcars.

Technically the improvised short-line motor cars were of no importance. In the general field of transportation they were also fairly inconsequential. They are not represented in I.C.C. statistics, which recorded only railway motor cars that were operated by class-one lines. Yet it is worth noting the existence of this motley fleet of home-made cars, because at one time hundreds of them picked their way over the shaky, weed-grown rail lines of rural America (Figure 8.31).

The motor car was an idea that had been introduced ahead of its time. By 1915 it was considered a failure; almost no new units were being ordered, and most railroads seemed only too pleased to return to traditional methods of operation. Yet just five years

later motor cars were again warmly received. Much had changed during this short period. World War I had profoundly affected national attitudes as well as the national economy. The shift to an automotive society was accelerating, with a corresponding erosion of short-haul rail passenger traffic. At the same time, inflation was boosting wages and the price of materials. Operating costs soared. Some roads figured that branch-line steam train costs per mile were as high as $1.50, while revenues amounted to no more than $1 and sometimes as little as 20 cents per mile.[103] Formerly, branch-line trains could be expected to make a small profit or at least break even. A small loss could be absorbed or charged off to good will, but in the postwar period a 50-mile branch could lose $15,000 a year per train.[104] One Western line calculated its annual losses on short-haul passengers at 7 million dollars. In 1923 the B & O reported that 30 percent of its passenger trains (all were short runs) failed to cover expenses and that only the main-line expresses showed a profit. The railroads were caught in the double squeeze of declining revenues and rising costs. Every cost-cutting method was being investigated. Now even the despised motor car began to look rather inviting.

At first the industry tended to overreact by adopting extremely small units. These cars were actually city busses outfitted with flanged wheels. They were a sensible investment for the most lightly traveled branch lines, but they did not have the capacity

Figure 8.29   The Hall-Scott motor company of Berkeley, California, built gasoline rail cars for the Nevada Copper Belt Railroad in 1910. The 36-ton car seated 32 passengers.

Figure 8.30   The Rio Grande Southern fabricated a gasoline rail car powered by using a 1936 Ford V-8 truck engine. This car and its sisters were locally known as the Galloping Geese.

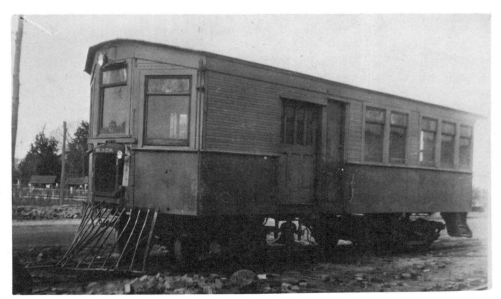

Figure 8.31   Some poor short lines produced ungainly home-made gasoline rail cars. This one was photographed on the Collins and Glennville Railroad in 1926.

Figure 8.32   *J. G. Brill, a major streetcar and motor bus builder, entered the rail car field in 1921. Its first model was an adaptation of a standard 30-passenger bus. (George Krambles Collection)*

or stamina for main-line service. There was no question, however, that the rail bus could make a better showing than the steam train. In 1921 Brill began to produce diminutive, 6-ton rail busses that ran 14 miles per gallon and cost no more than 30 cents a mile to operate.[105] The 29-foot-long vehicle seated 31 passengers and cruised at speeds up to 30 mph (Figure 8.32). It was the ultimate compact, which could provide rail transportation at the minimum cost. Comfort and safety were also minimal, however. The featherweight bus bounced and rolled and offered its human cargo little protection in a smashup. Yet setting a production-line highway bus on flanged wheels was so cheap that several firms were soon engaged in their manufacture. Mack, White, and F.W.D. (Four Wheel Drive), already established as truck and bus makers, entered the rail bus market in the early 1920s. Brill, the nation's leading producer of streetcars, became so enthusiastic about rail cars that it organized an automotive car division in 1923 to specialize in their production.[106] These firms were challenged by such unknowns as Edwards, Russell, and Sykes, all of whom perished after building a few units except for the Edwards firm of Sanford, North Carolina. Edwards produced nearly 100 cars, some for such major roads as the Burlington, before it too expired in the late 1940s.[107]

Most rail busses were produced for short lines, but a few big roads found employment for them. The New Haven purchased one of the first Macks to come off the assembly line in 1921. The 32-footer seated 35, and it remained in service until 1939.[108] Edwards's first production car, built in 1922, was sold to the B & O for use on a branch in West Virginia.[109] It was a 34-passenger, 60-horsepower car that pulled a tiny, 9,350-pound, eight-wheel trailer car (Figure 8.33). According to the *Scientific American* of June 1922, the Edwards car was trimming costs beyond any reasonable expectation. The steam train servicing the Green Springs branch cost 43 cents a mile, the rail car

only 8. This included wages, fuel, maintenance, and depreciation.

Highway-type rail cars continued to be supplied to main-line roads in later years. In 1933 the Cotton Belt purchased a 10-ton Astro-Daimler rail bus.[110] Called *Eagle of the Rails*, the 38-foot-long car seated 40 and was said to give a smooth, silent ride at speeds up to 80 mph. At the same time the Cotton Belt converted a highway bus for rail service in its own shops. Fifteen years later, six more rail busses appeared in Texas on the Houston North Shore branch of the Missouri Pacific.[111] These Twin Coach units remained in service until about 1960. A more modest example of a late-model rail bus was the 28-passenger unit built by the Superior Coach Company of Lima, Ohio, in 1948 for the Mississippi and Skuna Valley Railroad (Figure 8.34). The stock school-bus body was fitted with a small Railway Post Office compartment.

As already noted, the rail bus never really flourished on the main line, but some units of this type were supplied to the New Haven in the 1950s. In 1951 Mack furnished a 21-ton double-ender mounted on P.C.C. (President's Conference Committee) streetcar-style trucks.[112] The body resembled a standard city bus of the time. The 41-foot, turquoise and cream bus did well in its tests, and another nine units were ordered in 1953 for service on the New Haven's Cape Cod lines. Deliveries were made during the following year, but the railroad management had changed and the new administration put the busses in storage. Few ever turned a wheel in revenue service. Two were sold to the Sperry Company in 1958 for conversion to rail inspection vehicles. A few years later others were sold to Spain, but they were lost at sea when their cargo ship sank during the crossing.[113] The ill-starred Mack rail busses were never really able to demonstrate their worth.

Under certain conditions, the ideal bus would be a combination highway-railway vehicle. It would be able to use the

Figure 8.33   *The Edwards chain-drive rail bus of 1922 was powerful enough to haul a trailer. The motor train was retired in 1936.*

Figure 8.34   *Twenty-eight passengers, as well as mail and express packages, were jammed into the 27-foot, 8-inch length of this rail bus completed in 1948 by the Superior Coach Company of Lima, Ohio. (Superior Coach Company)*

Figure 8.35 *The Evans Auto-railer began operation in the northern Virginia suburbs of Washington, D.C., in 1936. The bus could run on railroad tracks and on the highway with equal ease.* (Horace Thorne)

public streets for access to the central city and yet thread its way through the suburbs on a private right-of-way. The scheme was tried as early as 1932 by the Twin Coach Company.[114] Two years later the Public Service Company of New Jersey outfitted three city busses with retractable flange wheels for use between New Brunswick and Trenton. Meanwhile the Evans Products Company of Detroit, a major railroad supply firm, produced a vehicle that it named the Auto-Railer. Its chief selling points were its flexible operations and its smooth, resilient ride, a result of the fact that the pneumatic tires actually carried the bus on the rails. In 1936 sixteen passenger coach Auto-Railers were sold to the Arlington and Fairfax Railway for operations in the Washington, D.C., area (Figure 8.35). The stubby cars, resembling light delivery trucks, were unable to save the Arlington line, nor did they convince the industry that combination rail and road vehicles offer any special advantages. The pneumatic-tire-on-steel-rail scheme did have a wider appeal, and a few such units were produced during the early 1930s by Fairbanks, Morse in cooperation with the Chrysler Corporation and the Budd Company.

Rail busses had only a limited market on the main line because they were simply out of scale for the job. Larger cars with generous seating and baggage space were required. And they had to be substantial enough to withstand the buffing and occasional collisions that were inevitable on a heavily traveled class-one road. Seating for 40 to 50 passengers, together with a 15-foot express baggage space, called for a body somewhat smaller than a full-sized coach but at least equal to a commuter car. An engine of 100 to 200 horsepower was about the minimum requirement if the car was to keep up with the prevailing schedules. The undersized, pigmy-powered rail bus could not meet these demands, and the supply industry soon developed an enlarged style of gasoline car especially suited to the needs of the main line.

In 1923 and 1924 Brill introduced several standard models for branch- and main-line service. The model 55, the smallest unit offered, was a 43-foot, 38-seat car powered by a four-cylinder, 68-horsepower engine (Figure 8.36).[115] The 14½-ton model's normal running speed was 38 mph. The model 65 was essentially the same car as the 55, except that it had a larger six-cylinder engine for extra power (Figure 8.37). The model 75 featured a 55-foot-long body, seating for 59 passengers, a 190-horsepower engine, and a total weight of 25 tons (Figure 8.38).[116] All the cars used a mechanical transmission with six speeds forward and three in reverse. An auxiliary transmission was incorporated in the cast-steel bolster of the forward truck. Only the forward truck was powered. The engine was mounted to the car frame, with a short drive shaft connecting the main transmission to the secondary gear box on the truck. These cars were not made from highway busses; the design was specially developed for railway service. The trucks had inside bearings, a feature uncommon in American rolling stock. The wheels were held on by large nuts so that they could be removed to allow access to the auxiliary gear box for servicing.

At the same time that the gas mechanicals appeared, Brill

offered a 250-horsepower, 60-foot, 37-ton gas-electric car (Figure 8.39). Mechanical transmission was not considered adequate to power plants over 200 horsepower. Some model 250s had elongated 73-foot bodies with seating for 84 passengers.[117] Between 1921 and 1937 Brill built approximately 250 motor cars in all styles, from miniature rail busses to heavy diesel-electrics for domestic railroads.[118] Trailers and foreign orders swelled production by another 70 units. In general the Brill cars performed well, but they rode poorly and tended to "nose," or sway noticeably back and forth, even on a straight track.[119] This swaying was characteristic of most rail cars, because the engine was placed at the extreme front of the body, adding as much as 10 tons to the forward end.

The Edwards Railway Motor Car Company, already mentioned in connection with light rail busses, began to offer a larger car better suited to heavy branch- or main-line operations. In 1922, just as Brill was tooling up for large-scale gas car production, Edwards built the Burlington a 9½-ton car that was 32 feet, 7 inches long.[120] The 39-seater was propelled by a 60-horse-power Kelly-Springfield engine with a normal top speed of 45 mph. During the next four years the Burlington purchased another eight cars from Edwards. All were larger than the first unit, and most conformed to the maker's popular model 20. This design measured 43 feet long, seated 39, and weighed just under 21 tons (Figures 8.40 and 8.41). A 100-horsepower engine and four-speed mechanical transmission made up the power plant. One of the best and most original features of the cars was the truck mounting of the engine. This placed the engine under the floor, freeing the forward compartment for more baggage space. In 1925 Edwards produced an extra-large car, Number 552, for the Burlington. It measured 65 feet, 7 inches in length and had a 100-horsepower Buda engine on each truck.[121] The Colorado and Southern, a subsidiary of the Burlington, considered buying a model 20 for $16,500, but decided instead to purchase a second-hand unit from the C B & Q.[122] The car performed splendidly at first, paying for itself during a single year's operation. But it was a mechanical disappointment, with a particularly troublesome transmission. The C & S also found the car too small and too light for the run from Fort Collins to Greeley, Colorado. It was sold in 1929. Presumably the parent corporation also found the Edwards cars unsatisfactory, for all were sold or retired between 1931 and 1939. These units were better suited to short-line operations, where they gave good service. One of them is still running on the California Western Railroad.

The most astute effort to reintroduce the motor rail car was made by Harold L. Hamilton (1890–1969).[123] His experience as a railway mechanic and later in the engineering and sales offices

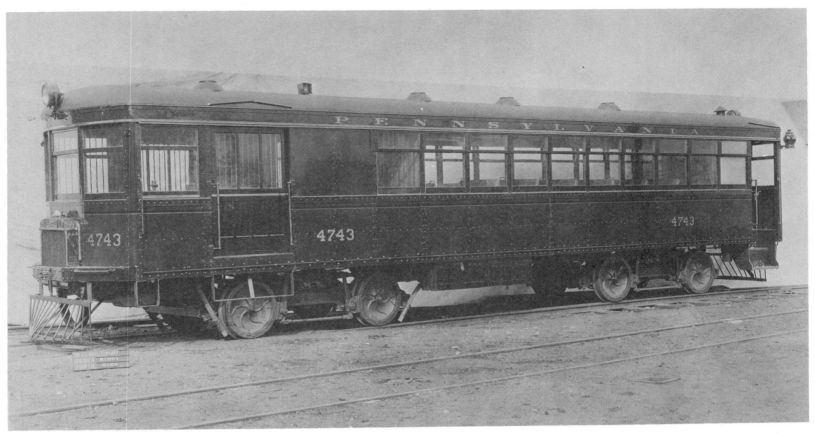

Figure 8.36 Brill built this gasoline mechanical-drive Model 55 around 1924. The 15-ton car seated 38. (Historical Society of Pennsylvania)

of the White Automobile Company suggested to him the potential of the railway motor car. Hamilton admired the G.E. gas-electric car and believed that the giant firm had foolishly abandoned a splendid engineering concept. He decided to conduct a survey to discover exactly what features were most wanted by railroad managers.[124] Around 1921 he began gathering data, which revealed a clear pattern of railroad requirements that was eventually translated by Hamilton's small staff into a practical rail car. His chief engineer, Richard Dilworth, was a former G.E. gas-electric mechanic, who would later play a major role in the perfection of the diesel-electric locomotive.

During the course of the survey and design work, Hamilton organized the Electro-Motive Corporation with offices in Cleveland. Again Hamilton moved carefully. Rather than rushing to acquire a plant, machinery, and staff, he avoided large capital outlays and the obligations of an expanded payroll by jobbing the manufacture of the cars out to subcontractors. G.E. supplied the electrical gear and the Winton Engine Company the prime mover. The St. Louis Car Company fabricated the body and trucks and performed the final assembly. Electro-Motive's small sales and engineering staff entered the uncertain rail car market with little more than a post office box and a roll of blue prints. The first car was ready for testing by the firm's first customer, the Chicago Great Western, in the summer of 1924.[125] The 57-foot, 38-ton unit was powered by a six-cylinder, 175-horsepower engine specially designed for railway service by Winton (Figures 8.42 and 8.43). The passenger compartment seated 44; folding seats in the baggage room could accommodate 8 more. The engine room was very compact, measuring only 8 feet long. Traction motors powered only the front-truck wheel sets. The car was no beauty, but its square, rivet-encrusted steel body had a rugged, utilitarian look that was somehow rather appealing. It

performed as efficiently as its no-nonsense appearance promised, and without the noise and vibration characteristic of most rail cars. At 4 miles to the gallon, its fuel consumption was considered satisfactory.

The Northern Pacific also took delivery of an E.M.C. car in 1924. Orders began to flow in, stimulated by reports of good service. The Colorado and Southern, for example, saved $14,000 per year by replacing a steam train between Cheyenne and Wendover, Wyoming, with an economical E.M.C.[126] The car itself was not cheap—$50,022.32, or as much as a moderate-sized steam locomotive—but any piece of capital machinery that can pay for itself in less than four years is a bargain. During its second production year, Electro-Motive sold 36 cars. The business continued to grow during the prosperity of the twenties (105 units were sold in 1928), and by the end of the decade E.M.C., which continued to depend on outside contractors, absolutely dominated the rail car market (Figure 8.44).[127] The Depression ended Electro-Motive's boom, but before rail car production was suspended in 1932, 500 cars had been completed. The firm's potential was so promising that General Motors purchased it and its chief supplier, the Winton Engine Company, in 1930.[128] G.M. channeled its new subsidiary away from rail cars and into diesel locomotives. Electro-Motive's golden touch continued, and the firm came to control the domestic motive power market—an exceedingly rich prize during the years when American railroads converted from steam to diesel on a wholesale basis. From E.M.C.'s humble origins in the gas-electric rail car field, both an industrial giant and a revolution in railroad operations had evolved. If only in this transitional role, the motor car achieved real significance.

The many marginal producers of rail motor cars even included the Pullman Company. As the nation's largest producer of pas-

MODEL 65 GASOLINE CAR

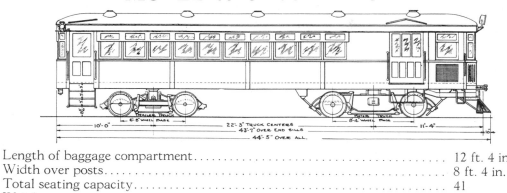

Length of baggage compartment.................................... 12 ft. 4 in.
Width over posts... 8 ft. 4 in.
Total seating capacity... 41
Weight.. 31,000 lb.

MODEL 75 GASOLINE CAR

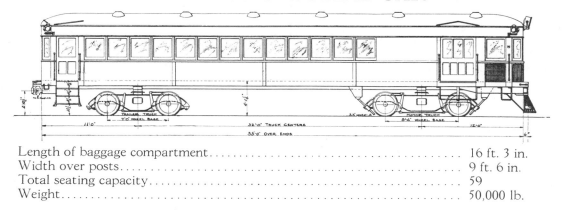

Length of baggage compartment.................................... 16 ft. 3 in.
Width over posts... 9 ft. 6 in.
Total seating capacity... 59
Weight.. 50,000 lb.

GAS-ELECTRIC CAR

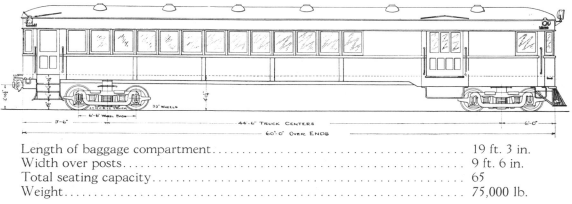

Length of baggage compartment.................................... 19 ft. 3 in.
Width over posts... 9 ft. 6 in.
Total seating capacity... 65
Weight.. 75,000 lb.

Figure 8.37   Three Brill gasoline rail cars from an advertisement of 1925.

senger cars, Pullman had dabbled in motor car construction since the time of the Patton gas-electrics in the late 1880s. During the twenties the firm built gas-electrics in limited numbers for the Santa Fe, the Burlington, and other lines. Many of these were produced for Electro-Motive. Two Pullman rail cars were notable for their size and mechanical arrangement. They were built in 1926 for the Detroit, Toledo, and Ironton.[129] The 72-foot, 6-inch-long units weighed a moderate 66 tons because of the liberal use of aluminum in the body structure. The engine-generator mounting was below the floor. Each car carried two 150-horsepower Hall-Scott engines, one of which turned clockwise and the other counterclockwise in an effort to dampen vibrations. The big cars rolled along at speeds up to 60 mph. Pullman's share of the market was relatively small, but it soon acquired two subsidiaries, Standard Steel and Osgood Bradley, that were more active in this field.

All gas-electrics tended to look alike: in the tradition of the heavyweight era in steel car architecture, they were boxy and plain. This superficial standardization was further reinforced by Electro-Motive's dominance. The surface uniformity was undermined, however, by a trend to increased horsepower and to the introduction of extra-large units and diesel power. By the late 1920s horsepower had advanced to 300 or more, and electric transmission had become common. Rail cars were designed to pull trailers for peak loads, but the original units of 175 horsepower were generally not powerful enough for such assignments. In addition, more than one trailer might be needed on certain runs. The Lehigh Valley, for example, wanted a motor car capable of towing three trailers, and late in 1926 Brill produced four 500-horsepower motor cars to meet this requirement.[130] The 70-foot, 65-ton cars had two engines apiece. Although two engines were common for the larger gas-electrics, a few triple-engine cars

Figure 8.38 Brill's model 75, delivered in 1925, was a gasoline mechanical car. The wooden trailer was an ex-Staten Island Railroad commuter car. (Baltimore and Ohio Railroad)

Figure 8.39 Brill delivered this gas-electric rail car in 1926. (Westinghouse Electric Corporation)

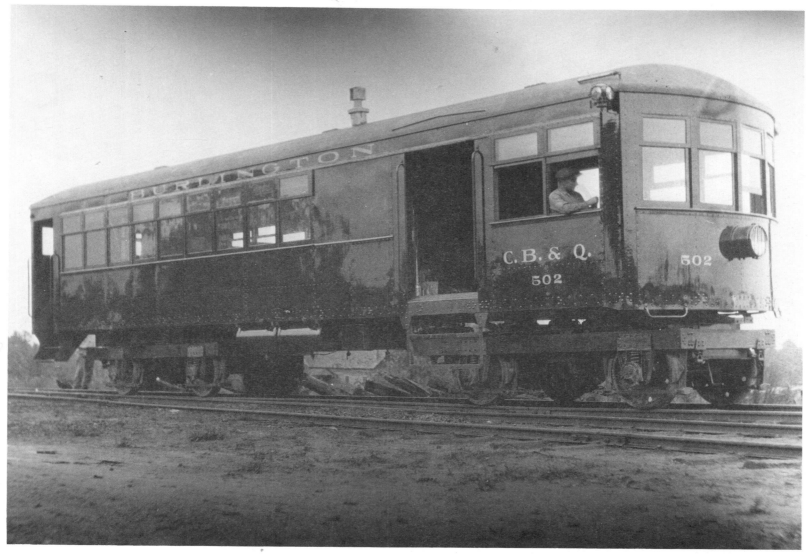

Figure 8.40   Edwards delivered this Model 20 gasoline mechanical car in 1924. It measured 43 feet in length and weighed 17 tons. (Horace Thorne)

were built by Mack. In 1928 Mack sold the Reading Company a 76-foot, 7-inch triple motor car that had three 120-horsepower motor generator sets. Two years later the B & O purchased a 600-horsepower car from Standard Steel (Figure 8.45). It was really more locomotive than car and represents, as do the other high-horsepower rail cars, a move to replace steam trains with motor trains. In 1930 Brill built a 550-horsepower gas-electric car and three cars (one of them an observation coach) for the Gulf, Mobile, and Northern. The Santa Fe, the New York Central, and other roads also acquired powerful rail cars. The trend to greater power was evident in cars produced during 1931: of thirty units, all gas-electrics, that were delivered for domestic service, twenty-three had 400 horsepower, one had 900, and the remaining six had less than 400.[131] Rail cars were being used for faster and more heavily traveled runs, but internal combustion power was coming to the railroads surreptitiously. Rather than being accepted directly in a locomotive, it was introduced for hauling trains in the disguise of the rail car.

Nearly all the rail cars constructed during the twenties were powered by conventional gasoline engines, but builders were also looking for engines capable of utilizing cheaper fuels. It was possible to burn a low-grade petroleum by-product, called distillate, in a gasoline engine with at least moderate success. By the late twenties gasoline was up to 15 cents a gallon, while distillate cost only 4 cents.[132] Taxes and the growing demand created by highway vehicles were responsible for the escalating gasoline prices. But distillate could not be efficiently burned in a low-compression gasoline engine. Combustion was incomplete, smoke abundant, and power low. The railroads were urged to try diesel engines, which were well suited to the consumption of cheap fuel. Diesel engine design had not been perfected for vehicle use, however. The engine remained a heavy, bulky mechanism suited to stationary power plants and little else.

Even so, some attempts were made to harness the diesel to rail cars. The Canadian National was probably the first North American line to try the diesel-powered rail car. It acquired the first unit in 1925 and during succeeding years purchased more of the cars, which were fitted with English engines, while the vast majority of rail lines remained loyal to the conventional gas-electric.[133] In the United States the Rock Island appears to have been the leader in testing diesel-electric power plants, for in 1925 and 1926 the road repowered four McKeen cars in this fashion.[134] In 1927 the Milwaukee Road installed a Foos diesel in an aging General Electric rail car.[135] During the following year four diesel-electrics were ordered by the Pennsylvania and the Reading railroads.[136] The 300-horsepower engines were built by

Figure 8.41   Edwards' best design feature was in mounting the engine and transmission on the power truck. (Horace Thorne)

Westinghouse. The car bodies were produced by Pullman for the Pennsylvania and by Bethlehem Steel for the Reading (Figure 8.46). Ingersoll-Rand also tried to sell the railroads on diesel-electric motor cars by producing a 600-horsepower demonstrator in cooperation with the St. Louis Car Company.[137] The car could pull four trailers and still maintain a schedule, and it achieved the mileage of a 400-horsepower gas-electric with a cheaper fuel. But the railroads were still skeptical, and the gas-electric remained the clear favorite through the early 1930s.[138]

Gasoline motor cars were not confined to one region of the country. They sputtered and smoked through the pines of New England, the wheat fields of the Great Plains, and the mountains and deserts of the Southwest. The New Haven was one of the first major lines to seriously consider equipment of this type. In 1921 the road began to purchase light rail busses, but it soon realized that larger motor cars were required for most trains.[139] By 1931 the New Haven had thirty-six motor cars in operation.[140] By 1927 the neighboring Boston and Maine was covering 130 schedules with motor cars on 60 percent of its system.[141] Twenty-four cars ran a total of 3,000 miles daily, with such success that the line hoped to replace more steam trains with gas-electrics. They were popular with the public, whose only reported complaint concerned the smell of the exhaust. The Penn-

Figure 8.42   Electro-Motive did more than any other firm to revive the gas-electric rail car. Its first rail car, produced in 1924, is shown here. (Electro-Motive Division, General Motors Corporation)

sylvania also found the motor car a necessity and acquired over fifty such units by 1930. The Burlington had the largest fleet (fifty-seven cars), which covered 27 percent of its passenger mileage.[142] The Burlington's motor cars rolled up 3.3 million miles annually. The Southern Pacific carried its Texas-Louisiana passengers in gas-electrics that paid for themselves in twenty-two months.[143]

Some smaller lines abandoned all steam passenger operations in favor of motor cars. The Gulf, Texas, and Western netted $22,555 from gross passenger and mail revenues of $35,558.[144] Its Brill rail car, which cost about $17,000, paid for itself in less than a year. In 1925 the Cincinnati Northern, a New York Central subsidiary, replaced all its steam trains with motor cars and thereby saved 29.3 cents per mile.[145] The year before, the Gulf, Mobile, and Northern had begun a similar program on a far grander scale. The 800-mile railroad was a major trunk line that operated the full range of passenger service, from plush limiteds to seedy locals. At first, motor cars were assigned to the lesser trains, but in 1930 the St. Tammany Limited was reequipped as a four-car motor train. The new train was hardly super-deluxe, but it had nice furnishings and was used for through service. In 1935 the last steam passenger operations on the Gulf, Mobile, and Northern were abandoned in favor of motor trains.[146]

During a single decade the motor car had grown from an unloved infant into an admired adult. Since the early 1920s it had proved a reliable transportation tool, able to run for millions of passenger miles at remarkably low cost. Much of its success is attributable to the perfection of the internal combustion engine. Such power plants could now operate for 6,000 hours, or 150,000 miles, between major overhauls.[147] The fussy, undependable

motors that had so frustrated the efforts of McKeen and the other motor car pioneers were now an antique memory. The growth of the motor car fleet is outlined in Table 8.1.

Table 8.1   Self-Propelled Rail Cars on Class-One Railroads

Year	Electric	Gasoline	Gas-Elect.	Diesel	Not specified	Trailers & misc.	Total
1915	—	—	—	—	—	—	1,647
1925	1,462	←	(112)*	→	12	293	1,879
1930	2,541	227	581	6	6	2,233	5,594
1935	2,451	277	611	19	9	1,178	4,545
1940	2,290	192	550	28	6	1,041	4,107
1950	2,178	134	252	137	0	944	3,645
1960	—	—	—	—	—	—	2,975
1970	2,444	←	(174)*	→	15	—	1,879

* Breakdown by gasoline, gas-electric, and diesel motive power is not given. The figure covers all three categories.
Source: I.C.C. statistical annual reports for the years noted, except for the 1915 figures, which are from *Railway Mechanical Engineer,* June 1915.

The motor car's triumph was cut short in 1930 by several factors. The Great Depression following the 1929 panic dried up passenger traffic. In 1920 each American traveled an average of 440 miles a year by rail; in 1934 that figure was down to 146 miles.[148] The most noticeable decline was on short-haul trains, where traffic decreased by 84 percent. Thus the motor car's chief market had nearly vanished, since the people who continued to travel were going by highway. Automobile registration continued

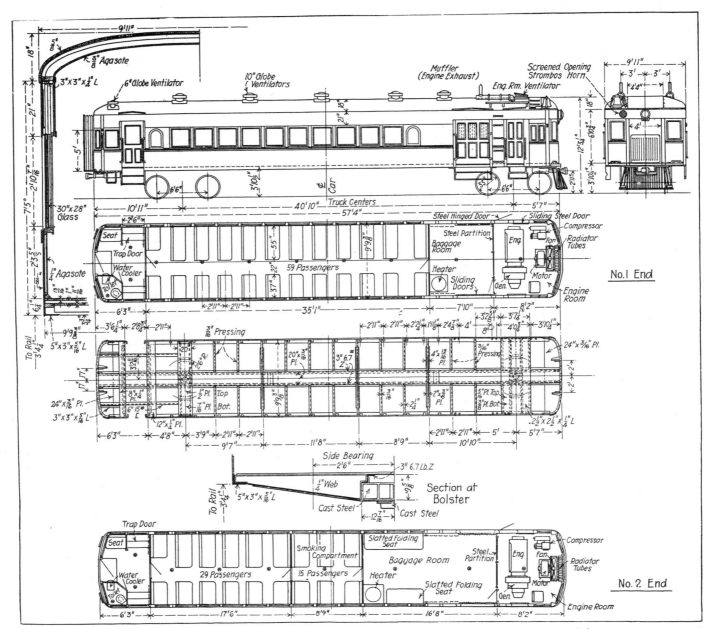

Figure 8.43 Electro-Motive's early rail cars weighed 38 tons and were powered by 175-horsepower Winton gasoline engines. (Railway Review, September 6, 1924)

to climb; even during the worst years of the Depression, 1 to 3 million new cars crowded onto the highways annually. In addition, intercity buses expanded their operations. By 1930 the Greyhound Corporation operated over 5,000 miles of routes and contemplated forming a national network.[149] Within three years that goal was more than realized, for Greyhound had expanded its route mileage to 40,000. Other independent bus operators were expanding just as aggressively—all to the detriment of the railroads. Some roads decided to join the competition rather than fight it; the Pennsylvania, for example, bought a large block of Greyhound stock, while the Burlington formed its own motor bus division.

The railroads were forced to admit that although rail cars could reduce operating expenses, they could not reverse America's trend toward the highway. Motor rail cars proved to be only an interim solution. As branch-line and secondary trackage was abandoned (over 15,000 miles disappeared during the 1930s), more and more surplus motor cars appeared on the storage tracks.[150] The orders for new cars dried up: in 1929, 132 were ordered; in 1933, 11.[151] Yet just when the motor car seemed

destined for extinction, an unexpected revival took place in the form of the glamorous motor trains for the long-distance traveler. A series of sleek, radical, lightweight self-propelled cars opened a new era in American railroading and made what was once a dowdy necessity into a star. The coarse utility of the traditional gas-electric gave way to the racy, glossy look of the streamliner. In the words of Ogden Nash: "The Iron Horse has run its course and we ride a chromium weasel."

The streamline motor car era began with an experimental unit designed and built by W. B. Stout (1880–1956), an aeronautical engineer. Stout's turtle-like *Rail Plane* was an ultra-lightweight vehicle which incorporated airplane materials and construction techniques. Its welded tubular frame and aluminum skin helped hold the weight of this 60-foot vehicle to only 12½ tons, or 500 pounds per passenger. A conventional steel coach weighed nearly 5,000 pounds per passenger. The *Rail Plane* was ready for tests by September 1932.[152] It registered speeds of 90 miles an hour and ran briefly on the Gulf, Mobile, and Northern Railroad.[153] Pullman had sponsored the project in the hope that the radical design might lead to a commercially viable product. Indirectly

Figure 8.44   This Electro-Motive car of 1926 developed 220 horsepower. The body was produced by the St. Louis Car Company. (Railway and Locomotive Historical Society)

this goal was realized, for Stout's experiment attracted the attention of W. Averell Harriman, the board chairman of the Union Pacific Railroad. Twenty-five years earlier his grandfather had been inspired by torpedo boat experiments to order the studies that led to the McKeen cars; now the younger Harriman encouraged the development of Stout's lightweight design for everyday service. The project was headed by a member of the Union Pacific's engineering staff, Everett E. Adams. Pullman's own designers and W. B. Stout were also involved.

Preliminary studies quickly elminated the single-unit self-propelled car and concentrated the design effort on a self-propelled car with two trailers permanently attached. The Union Pacific wanted a multiunit, articulated motor train for high-speed main-line service. An order was placed with Pullman in May 1933 for a three-car train, referred to as the M10,000.[154] The first, or power, car contained the engineer's cab, a 600-horse-power distillate engine-generator, and baggage and mail compartments. The trailing cars offered coach and buffet seating for 116. The 200-foot-long train was built largely of aluminum alloys and weighed 85 tons (Figure 8.50). It was 3 feet lower than conventional equipment, and its sheathed underside cleared the

rails by just 9½ inches.[155] The framing, riveted together from plate, channels, and square tubes, was reminiscent of the antique schemes proposed by LaMothe and the other pioneers associated with tubular car construction. But Stout's goal of lightweight construction was partially frustrated by the safety requirements. Heavier framing members were needed to match the strength of conventional equipment, because the smaller cross section of the motor trains resulted in less stiffness.[156] In addition, overall weight could not be reduced much more than 25 percent, even with high-tensile-strength metals. Because thin sheets tend to buckle or deflect, actual strength is below the theoretical figures. For these reasons the remarkably low weight per passenger achieved in the *Rail Plane* could not be duplicated in the M10,000.

When delivered in February 1934, the M10,000, or *City of Salina*, was a sensational brown and yellow caterpillar on rails. It looked like a monstrous airplane fuselage without wings, tail, or wheels. It was heralded as "Tomorrow's Train Today" during a 13,000-mile, twenty-two-state tour. President Roosevelt personally inspected the train during its Washington stop, and it was said that the Federal government shut down so that the humblest

Figure 8.45   Standard Steel Car Company constructed this powerful gasoline rail car in 1928 for the Baltimore and Ohio. It was converted to diesel power in 1935 and was retired in 1950. (Baltimore and Ohio Railroad)

civil servant might retrace F.D.R.'s steps. Finally in January 1935 it entered regular service.[157] It continued to operate until late 1941 when, after nearly 900,000 miles, it was sold for scrap. Its limited capacity had always been a handicap, and now that the engine was worn out, installing a new diesel was not considered economic. In addition, the demand for scrap aluminum was great because of the war emergency. For all these reasons the M10,000 was condemned to early retirement.

Even before the first motor train was completed, Harriman began to order more aluminum streamliners from Pullman. Each one grew in consist and in overall physical size until 1936, when the last of the motor trains, the City of Denver, was delivered (Figures 8.51 and 8.52). The equipment had swollen almost to the full configuration of conventional rolling stock. All were named for principal municipalities on the Union Pacific system: the City of Portland, the City of Los Angeles, the City of San Francisco, and the City of Denver. The City of Denver had two locomotives totaling 2,400 horsepower, ten cars with seating for 182, and a combined weight of 666 tons. This train, which was

equal to full-sized equipment, ran an impressive total of 5.6 million miles before being retired in 1953.[158]

Pullman produced only one more motor train on the general design of the M10,000. The Green Diamond, a five-car articulated unit, was delivered to the Illinois Central in the spring of 1936.[159] The 230-ton train was propelled by a 1,200-horsepower diesel-electric plant. Unlike the Union Pacific automotive trains, the Green Diamond was made from Cor-Ten steel instead of aluminum, which accounted for its greater weight. The bodies were riveted rather than welded fabrications. The first car was devoted entirely to the machinery of locomotion. The second was a combined Railway Post Office–baggage car. The third and fourth cars were coaches, while the trailing observation had a kitchen as well as dining and lounge compartments. The 328-foot, 6-inch-long train seated 146, including the occupants of the diner-observation car. The Green Diamond represented the Illinois Central's effort to capture a larger share of the Chicago–St. Louis traffic, for which it competed with the Alton, the Chicago and Eastern Illinois, and the Wabash. Mileage and speed on all

Figure 8.46   The Reading Company's Number 74 was among the first diesel-electric rail cars in the United States when it entered service in 1928. The 73-foot car weighed nearly 69 tons. (Westinghouse Electric Corporation)

Figure 8.47   Plain interiors were characteristic of gas-electric cars, because they were normally assigned to short secondary runs.

Figure 8.48   The engineer's cab of gas-electrics was compact and noisy. The engine is visible to the left.

Figure 8.49 *The B & O modernized a twenty year-old Brill gas-electric in 1946 with a fancy paint job and a 250-horsepower diesel engine. The Number 6040 was retired in 1953.*

Figure 8.50 *The Union Pacific's* City of Salina *was the first streamlined motor train to enter service in the United States. Pullman completed it in 1934.* (Pullman Neg. 37922)

*Figure 8.51   Because the City trains were articulated, car bodies spanned a common pair
of trucks at each end. (Pullman Neg. 38747)*

roads was about equal, but the Illinois Central hoped to lure
patrons away from its rivals by capitalizing on the public's in-
fatuation with streamliners. The ploy seems to have worked very
well, for the train ran full, and a second section was often needed
to accommodate the disappointed passengers who could not
crowd aboard the Green Diamond. Reservations had to be made
a week in advance, and business was so good that the I C re-
gretted not having bought a ten-car motor train. In the first six
months of 1938, the train was netting $110,000, or nearly 40
percent of gross ticket receipts.[160] This was an excellent return,
considering its first cost of roughly $425,000. The line's experi-
ment with motor trains was amply repaid before the Green
Diamond was retired in 1949 and scrapped one year later.

Pullman's rivals were eager to gain a share of the streamline
motor train market. Budd, in fact, had entered the car-building
business by this means even before the City of Salina's delivery.
But an even more daring attempt to introduce a new dimension
in railroad passenger equipment was made by the Goodyear-
Zeppelin Corporation of Akron, Ohio. The firm was looking for
work, because the Depression and the disaster to the airship
Macon had left its plant idle.[161] The company decided that it

had the technical experience and the staff to design and build a
radical motor train by applying existing aircraft materials and
techniques. The bankrupt New Haven showed an interest in the
idea, and with a $250,000 loan from the Public Works Adminis-
tration, the road signed a contract early in 1934 with the airship
manufacturer. The general plan followed that of Pullman's
M10,000 for a three-car articulated aluminum train (Figures 8.53
and 8.54). It was about equal in size to the M10,000, for the
Comet, as the New Haven named it, seated 160 and had a power
plant at either end of the train, which contributed appreciably to
the total weight of 126 tons.[162] The train was equipped with 400-
horsepower, six-cylinder Westinghouse diesels that were used in
connection with the usual electric transmission. The body was
nearly square in cross section, and the underbelly was fully covered for increased strength and better aerodynamics. The lower
exterior body panels were machined in a decorative whorl or fish-
scale pattern. Light- and dark-blue bands and a gray roof added
to the exterior finish.

The *Comet* was ready for testing in late April 1935. Its high-
speed run of 110 mph between New Haven and Boston beat the
fastest regularly scheduled train by fifty-five minutes. But the

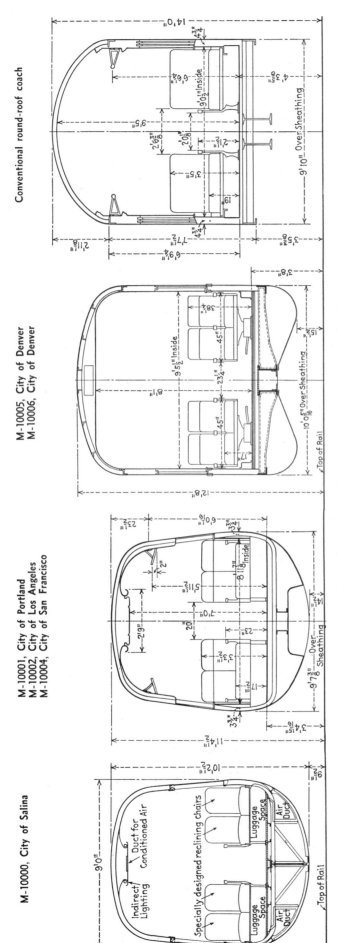

Figure 8.52 The undersized Union Pacific motor trains were compared with a conventional coach in this 1936 drawing. (Railway Mechanical Engineering, September 1936)

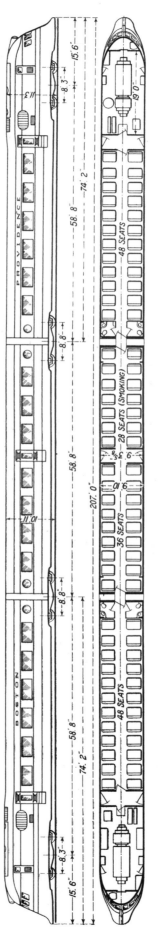

Figure 8.53 The Goodyear-Zeppelin Corporation completed the motor train Comet for the New Haven in 1935. (Railway Gazette, February 1937)

Figure 8.54   The aluminum, diesel-powered Comet *of 1935 was a mechanical success,
but it was too small for main-line needs. (Association of American Railroads Neg. 659)*

Comet had been purchased for the more prosaic operation between Boston and Providence, which did not include such a
record-breaking pace; the average speed was 60 mph. Patronage
was not as good as expected, and additional local stops were
added that further slowed the schedule. In all, the New Haven
seems to have purchased an elegant racehorse for draft-horse
service. Even so the road had no complaints, for the Comet was
available 89 percent of the time and produced a respectable net
revenue. After three years of service and 405,000 miles, a general
overhaul showed that it was in excellent condition.[163] There was
some engine wear and the train had to have new wheels; otherwise few repairs were required. The silvery, lizard-shaped rail
car rolled along satisfactorily until its retirement in 1951; yet it
remained the sole rail car produced by Goodyear-Zeppelin.

If Goodyear-Zeppelin offered the most exotic design, American
Car and Foundry produced the most conservative. In shape,
general arrangement, styling, and materials A.C.F.'s first motor
train, the Rebel, was the opposite of the trains designed by Pullman, Goodyear-Zeppelin, and Budd. A.C.F. was the last to join

in the streamline craze and may have concluded that the time for
extravagant designs was past and that more utilitarian plans
were needed to save railroad passenger traffic. Furthermore, the
purchaser, the Gulf, Mobile, and Northern, had no illusions
about extraordinary service or high-speed operations. Aerodynamic design was hardly needed for a train that would stop
every 10 miles and would average only 35 mph. The Rebel trains,
as they were called, were intended to replace standard steam
trains on existing schedules. It was hoped that they would attract
new passengers because of their fresh appearance and air conditioning. The road also expected, in view of the good record of
gas-electric cars, that the motor trains would reduce operating
expenses. The Rebels might be called the sensible man's streamliners—solid, practical, devoid of foolish experiments.

The Gulf, Mobile, and Northern ordered two complete four-
car Rebel trains for delivery in 1935 (Figures 8.55 and 8.56).
A fifteen-year loan from the Public Works Administration made
the purchase possible. Construction was undertaken at A.C.F.'s
Berwick, Pennsylvania, plant, the former Jackson and Woodin

Figure 8.55 The Gulf, Mobile, and Northern's Rebel trains were built by A.C.F. in 1935. Unlike most of the other early streamlined motor trains, they were not intended for high-speed service.

Car Company. The 73-foot motor car of the Rebels had a baggage and mail compartment.[164] Two coaches measured 75 feet long; each seated 62 and 71 passengers respectively, and each had separate sections and toilets for white and colored passengers. A small buffet in the first coach offered light food service, but no separate seating for diners. The 77-foot-long observation car had six open-section sleeping berths, a single stateroom, and a lounge seating 18. The cars were not only slightly shorter than conventional cars, but also somewhat lower and narrower. The eight-wheel cars weighed about 62 tons each—an unimpressive achievement, considering that most full-sized lightweight cars weighed about 50 tons. The blunt-nosed Rebels had less flare and style than other streamliners, but at least they had smooth sides. The exterior Cor-Ten steel sheets were welded to framing members. The paint scheme was also a plain, straightforward silver, with two broad red and gunmetal stripes.

Actually, the Gulf, Mobile, and Northern had every reason to be happy with the Rebels, for they were well suited to the job at hand. Traffic grew while expenses declined. The double-truck cars proved much more practical than the articulated trains. An extra coach operated between Jackson, Mississippi, and New Orleans, which would have been impossible with an articulated train. The equipment worked so well that the road bought a third train in 1938. The Rebels continued to run into New Orleans until 1954.[165] Then, after the trains had stood for a while in storage, the Alco engines were removed from the power cars and sold to a shipping company, and in 1962 the cars were scrapped.

Even before the time of the Rebel, A.C.F. had been considering the manufacture of a more advanced style of motor rail car. The general planning called for a single-unit, air-conditioned streamlined car with a hydraulic transmission and power plant under the floor. Quantity production was also contemplated, but the number of orders received never permitted it. Late in 1934 two cars were delivered to the Norfolk Southern by A.C.F.'s Brill

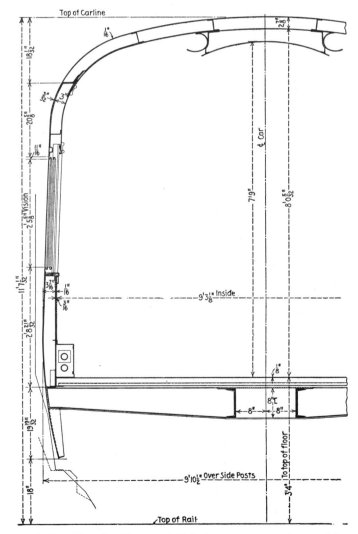

Figure 8.56 The Cor-Ten steel Rebel cars were undersized but not remarkably lightweight. (Railway Mechanical Engineer, July 1935)

Figure 8.57   A.C.F.'s motor rail car Number 2025 seated 57 and weighed 26 tons. It was delivered in 1935. (Seaboard Airline Railroad)

Figure 8.58   A.C.F. hoped that its streamlined Motorailers would develop quantity sales, but relatively few were sold. The Miss Lou was delivered in 1940. (Alan R. Lind Collection)

plant.[166] The low, well-shaped bodies were not unlike the streamlined electric suburban cars produced by the same plant for the Norristown suburban line. Two more were acquired by the tidewater line soon afterwards. The good service rendered by these units prompted the Seaboard Airline to purchase three A.C.F. rail cars.[167] These were slightly larger than the Norfolk Southern units: the body was stretched out to 64 feet, the total weight registered 26 tons, and seating was provided for 57 (Figure 8.57). Because the cars were not intended to pull trailers, no center sill was required. Couplers were provided, however, for emergency towing in the event of a breakdown. The bolsters were a box-girder weldment. The use of high-tensile steel alloys and aluminum, together with the light frame, helped trim unwanted pounds from these units. A.C.F. supplied similar cars to the Chicago and Eastern Illinois and the Missouri and Arkansas.

In 1940 A.C.F. introduced a larger, more powerful streamlined diesel rail car: the Motorailer. Again quantity production was envisioned, but again only a limited number were made—fewer than a dozen for domestic use. During the first year of production, two units were delivered to the New York, Susquehanna, and Western, a short line operating through several New Jersey bedroom communities in the environs of New York City.[168] The cars were 76 feet long, weighed 38 tons, and seated 80. They featured controls at either end of the car (eliminating the need for a turnaround at the end of the run) and center door entrances. Later in 1940 two similar Motorailers named Miss Lou and Illini were placed on the Illinois Central (Figure 8.58). Like the Susquehanna cars, they were built of Cor-Ten steel. Because they were intended for intercity rather than commuter service, baggage and buffet facilities were included. The Illinois Central also acquired a two-car Motorailer, the Land O'Corn, for service between Chicago and Waterloo. Seating for 109 passengers was supplemented by 16 places at the lunch counter situated in the trailer car. None of the trio proved entirely successful, and in 1943 and 1947 they were sold to the Susquehanna. They ran on that line until the 1950s, when they were replaced by Budd cars. Not long after the Land O'Corn entered service, another Motorailer was produced for the Missouri Pacific. Except for a large order of 22 cars sold to the South American republic of Colombia in 1941, A.C.F.'s Motorailer was indifferently received. Production was, of course, cut short by World War II, but the company's failure to reintroduce the design afterwards indicates that the product was not in demand. Moreover, A.C.F.'s management had become enchanted with another fleeting jack-o-lantern, the Talgo train, and possibly saw no need to refine the Motorailer for the postwar market.

While Pullman and A.C.F. were producing their first lightweight motor trains, a Philadelphia auto parts manufacturer decided to enter the rail car field. The Budd Company had been associated with the French Michelin firm for many years, and the success of its overseas partner in introducing rubber-tired rail cars encouraged Budd to try a similar course. The Depression had dried up normal business, and any new product line was worth investigating. Michelin produced its first rail car in 1929

and had nine in service within three years.[169] The use of pneumatic tires distinguished these vehicles from all previous rail cars. The ultimate in cushioned wheels was said not only to give a superlative ride but to increase traction by 35 percent. An internal aluminum safety ring allowed the wheel to fall only $\frac{5}{8}$ inch in the event of a puncture. A flange was also applied to each wheel to guide the vehicle in case of a failure, but under normal conditions the flange did not touch the railhead, because the soft rubber tire hugged the rail. Budd hoped to make the rubberized rail car even more appealing by offering a strong, light, rustproof body made of stainless steel. The advantages of this material and of the flawless method of fabrication, Shotwelding, were described in Chapter 2.

The first car, ready in early 1932, was a boxy twelve-wheeler seating 40 passengers. It was powered by an 85-horsepower Junker diesel engine. A smaller car, La Fayette, was next built for testing in France.[170] In November of the same year a 50-foot, 12-ton car was delivered to the Reading Company for use on its New Hope branch.[171] The 47-seater could rarely be coaxed to speeds beyond 45 mph. It was prone to derail, a failing attributed to its bouncy ride. These deficiencies, coupled with its price of $75,000, discouraged any widespread interest in the Budd-Michelin rail car. Only two more were built for use in this country, although Michelin produced thirty rubber-tired cars in 1937 for the Eastern Railway of France. Such vehicles have also been successfully used for subway service in Europe, Canada, and Mexico in recent years.[172] In 1933 the Pennsylvania acquired a two-car gas-electric, which generally followed the Reading car in overall size and design. But the pneumatic wheels worked no better than they had on previous Budd cars, and they were replaced with conventional steel wheel sets. The Number 4688 and its trailer were a balky and never-to-be-repeated experiment on the Pennsylvania, but they did remain in service longer than any of the Budd-Michelin units. In 1943 the Pennsylvania leased them to a Virginia short line, the Washington and Old Dominion Railroad. The second owner came to regret the deal, however. The units were retired after a few months, and finally were scrapped in 1948.

Late in 1933 the final Budd-Michelin car was delivered to the Texas and Pacific.[173] The two-car, air-conditioned train measured 140 feet in length and weighed 52 tons. The power car had a knife-edge leading prow, two 240-horsepower gasoline engines, and conventional steel wheel trucks. The extremely light 12-ton trailer seated 76 passengers and was carried by sixteen rubber tires. It was Budd's largest rail vehicle to date, and would prove the biggest failure. Before any revenue passengers were carried on its projected Fort Worth–Texarkana route, the Silver Slipper derailed on a test run.[174] The trailer was fitted with new trucks, but the power car proved unreliable, and after only a year the Silver Slipper was scrapped. The costly failure was extremely embarrassing to Budd and forced the firm to abandon its touted rubber-tired rail vehicles.

Yet the failure of the Michelin scheme did not persuade Budd to quit the rail car field, because the firm rightly felt that its

Figure 8.59   The Pioneer Zephyr under construction at Budd's plant in 1934. The stainless-steel motor train remained in service until 1960. (Budd Company)

stainless-steel car bodies were superior to anything available in the field. Its next train, the Zephyr, confirmed Budd's faith in his product. It was such an outstanding success that the previous disappointments were quickly forgotten and Budd became a leader in the passenger car field. The Zephyr was based on an optimistic belief that rail passenger service could be saved by a fresh approach. The name itself means the west wind of springtime promise. The Zephyr was a three-car motor train ordered in 1933 by the Burlington as a countermeasure to the Union Pacific's growing fleet of streamliners (Figures 8.59 and 8.60). The entire train was only 196 feet long and seated 72 passengers.[175] Food service was limited to a buffet. Yet because of its small size a 600-horsepower engine could propel it, and do so at a fraction of a conventional train's operating cost. Like the other pioneer streamliners, the Zephyr was an effective traffic generator. From the beginning of regular service late in 1934, the seats were well filled. Business was so good that there was a waiting list. In three years the Zephyr ran thousands of miles with very few failures.[176] The train paid for itself in thirty months, and misgivings about its safety and durability were silenced by its performance. The Zephyr struck several animals and was involved in four grade-crossing accidents (in two of

these the vehicles were trucks) without any major damage.[177] It was predicted that the train would derail easily, but the accidents proved this fear unfounded. However, a head-on collision with a steam locomotive in October 1939 did cause heavy damage to the power car.[178] It was collapsed for 28 feet, and the engine was driven back into the mail room, resulting in two deaths and thirty-eight injuries. But the train was rebuilt and ran another 3.2 million miles before its retirement in 1960.

The Zephyr's success encouraged the purchase of more Budd motor trains. In 1935 the Boston and Maine bought the Flying Yankee, a slightly enlarged articulated train. Like the Zephyr, it was an extremely durable first-generation streamliner; it ran 2,735,600 miles before retirement in 1957.[179] To cover more intensively patronized routes, the Burlington decided to duplicate the Zephyr on a larger scale, with heavier, bigger, and wider cars, and more cars per train. The Mark Twain, the Twin Cities, and the Denver Zephyrs followed in quick succession. Although these motor trains reflected great credit on their maker, Budd's prewar motor rail car manufacture was to end in failure. Actually, Budd had dropped motor car production to concentrate on fabricating conventional cars, but in 1941 it received an order from the Denver and Rio Grande Western to produce a pair of

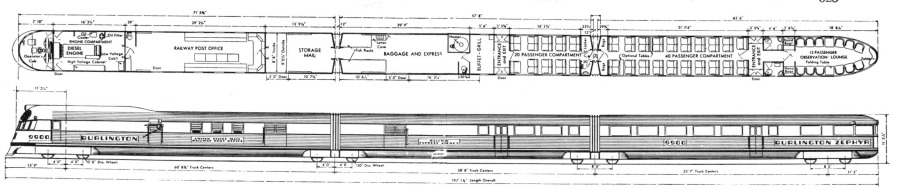

Figure 8.60   The Pioneer Zephyr began service on the Chicago, Burlington, and Quincy Railroad in 1934. (Railway Mechanical Engineer, May 1934)

two-car self-propelled trains.[180] It built the Brigham Young and the Heber C. Kimball, which weighed just under 132 tons each and measured 153 feet in length. Two 192-horsepower diesels powered each train. But Budd had grievously miscalculated, because the cars were underpowered for the stiff grades between Denver and Salt Lake City. The engines were so overworked that breakdowns were frequent and a steam locomotive was on standby service to tow the trains in. After only seven months, the Budd twin units were retired.[181]

Perhaps the humiliating failure of these prewar trains eventually prompted Budd's engineering staff to prove that they could design a dependable rail car. Whatever the reason, the company decided to produce one for the postwar market. The project seemed ill-advised, for no such cars had been built for years, and the industry's rejection of the Motorailer appeared to demonstrate that even a modern rail car was not wanted. Streamlined motor trains proved a short-lived phenomenon as well. American railroads chose instead to buy full-sized lightweight locomotives and cars. Rail cars and motor trains were a vanishing breed. But Budd felt that it could succeed where others had failed, because a dependable lightweight rail car could provide a cost-cutting tool that would attract the railroads. A high-horsepower unit built with economical off-the-shelf components would be more than a streamlined gas-electric. In fact, electric transmission would be replaced by a hydraulic transmission to save weight and first cost. Two 275-horsepower G.M. diesel engines, designed for rugged duty in army tanks, supplied the snap and ginger unknown to the old gas-electrics. Horsepower was figured at 9 per ton, or about double that of the older generation of rail cars.[182] The engines were placed below the floor; a jack shaft transmitted power to one pair of wheels on each truck. The engines were mounted in a cradle that could be quickly rolled out from under the car for inspection or servicing. If one should need major repairs, a skilled crew could replace it in just over an hour with a standby engine.

The 59-ton cars were designed for multiple-unit operation, so that two or more might be coupled together and operated by a single driver. Disc brakes made stops possible in 2,300 feet at 85 mph.[183] Generous 90-passenger seating made the cars ideal for commuter and accommodation trains. And they were attractive, with shiny stainless-steel bodies and bright modern interiors. The entire package was thoughtfully conceived and well executed. The original selling price of $128,750 was not much greater than the charge for a lightweight coach.[184] Operating costs were equally attractive: in 1950 they were calculated at 56 cents per mile. Diesel fuel was cheap at 8 cents a gallon, and the thrifty Budd car realized 2 to 3 miles per gallon. In later years the selling price crept up to $160,000 and operating costs rose to "less

than 72 cents per mile," but all other costs had gone up as well, and the Budd rail car remained a bargain.[185] Maintenance cost 26 cents a mile, or about one-quarter that of an equivalent steam train.

Budd named its new product the RDC for Rail-Diesel-Car. A demonstrator was ready in July 1949, and the first production car went to the Boston and Albany in May of the following year.[186] Within three years 116 were in service on railroads in all regions of the country, as well as in Australia, Cuba, and Saudi Arabia. When the last RDC left the Red Lion plant in 1962, a total of 398 had been produced. Of the four production models offered, the RDC-1, a simple coach, was the most popular (Figure 8.61). It accounted for two-thirds of the cars produced for domestic service. The RDC-2 and RDC-3 had baggage and mail space, making them suitable for long-distance, main-line, or branch-line service where head-end business was a factor. The RDC-4, the least popular model, had no passenger space, for it was devoted entirely to baggage and mail (Figure 8.62). It was 11 feet shorter than the 85-foot standard length of the other models. In addition to these four, an odd class, the RDC-9, was produced for the Boston and Maine in 1956.

The RDC's performance on the Western Pacific was an early and much publicized success.[187] The railroad operated a steam train, mainly for the accommodation of its employees, between Oakland, California, and Salt Lake City. The route ran 924 miles through sparsely settled country that was a traffic manager's nightmare. The train cost as much as $310,000 a quarter to operate, and because most of the passengers were deadheads, the best the line could hope to do was cut its losses. Two RDC-2s purchased in 1950 achieved that goal by trimming operating costs to a manageable $76,900 a year. Thus a net loss of $60,000 became a profit of $207,000. The cars proved economical and dependable, performing without failure through subzero weather and driving snowstorms so fierce that the glow of the headlight was obscured.

Eastern lines made good use of the RDCs as well. With their help the B & O cut operating costs on local trains between Baltimore and Washington by 51 percent. Later the road operated RDCs in over-the-line service. The cars were rarities in that they offered dining facilities which went beyond minimal buffets. Six tables seating 24 passengers were an unexpected luxury for patrons of the normally bare-bones rail motor car.[188] The B & O also demonstrated the Budd car's capacity for speed by staging a fast run between Washington and Chicago that took twelve hours and twenty-nine minutes. This was not typical, however; most RDCs operated on leisurely schedules.

The two major roads serving Boston were the largest operators of RDCs. In 1953 the New Haven had forty of them running 1.25

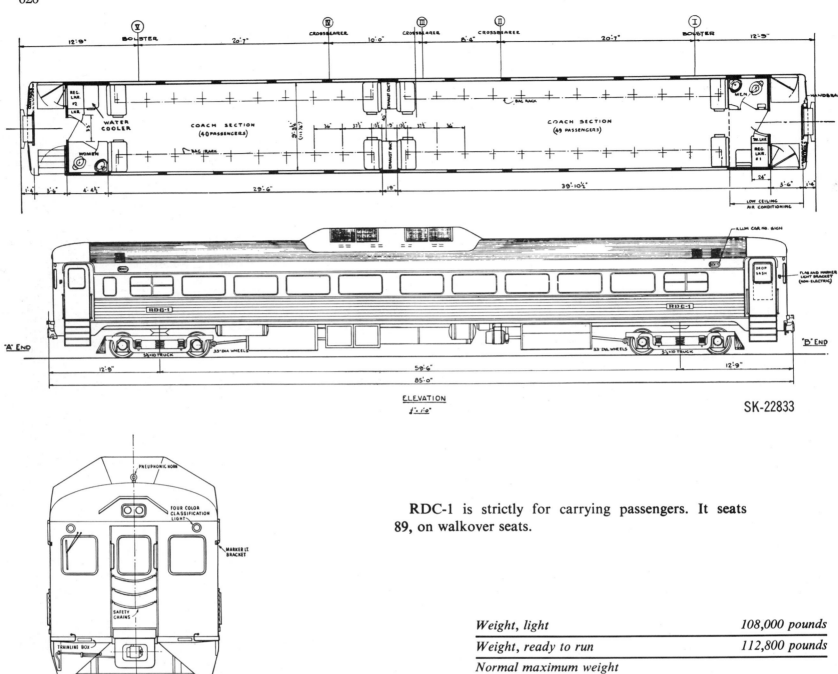

RDC-1 is strictly for carrying passengers. It seats 89, on walkover seats.

Weight, light	108,000 pounds
Weight, ready to run	112,800 pounds
Normal maximum weight (including 89 passengers)	126,600 pounds

Figure 8.61 Budd's RDC, a diesel mechanical-drive unit, proved one of the most popular and successful rail cars produced in the United States. Nearly 400 were turned out between 1949 and 1962. (Budd Company)

million train miles a year. They replaced sixteen locomotives, sixty-four cars, and four old-model rail cars. But the Boston and Maine had an even larger fleet—the largest in the country. How and why the road came to acquire 109 Budd cars between 1952 and 1958 was explained by a former member of its mechanical department, Ellis Walker.[189] He said that the state public utilities commission had pressured the line for years to retire its aging fleet of wooden commuter cars. The railroad pleaded poverty but finally agreed that if certain trains, all notorious money losers, could be abandoned, the old cars would be taken off. The deal was in the public interest, and after some negotiation apparently an unofficial understanding was reached. The B & M bought secondhand steel cars as well as new Budd rail cars, both of

which finally replaced the last wooden cars serving the American traveler.

The last RDC was produced in 1962 for the Reading Company, yet hundreds of them continue in daily service. Today most are assigned to commuter runs. Many have been scrapped, others have been sold to Canadian roads, and in the case of the Rock Island, several have had their engines removed and been converted into ordinary locomotive-hauled coaches. But even as they drop away one by one, the RDCs are remembered as one of the most successful styles of self-propelled rail car ever constructed.

The other rail car designs of the postwar era present a marked contract to the RDC. Fanciful concepts were promoted in place

Figure 8.62   This RDC-3 seated 48 and carried 5,000 pounds of mail and 8,250 pounds of baggage. It was delivered in 1950. (Lawrence L. Williams)

of sound engineering. The low-center-of-gravity panacea became fashionable despite an embarrassing series of failures. Recent history was ignored as this superficially appealing scheme was resold to another generation of railroad presidents, whose ignorance of the technical problems led them to waste millions of dollars. The Talgo, Xplorer, Train X, and their various cousins, sisters, and aunts were costly failures that briefly captured public attention because of their radical appearance and their promoters' hard-sell tactics. But they failed to deliver safety, comfort, or reliability. Their modest operating savings were more than offset by their high first costs, large repair bills, and operating difficulties.

The leader of the low-center-of-gravity revival was the American Car and Foundry Company. Once a major builder of passenger equipment, the firm had lost its number-two position in the industry to Budd even before Pearl Harbor. A.C.F. apparently hoped to regain its position by marketing a revolutionary concept in passenger train design that had been advanced by a Spanish engineer. It was a low-slung, ultra-lightweight design that A.C.F. hoped would complete the transition to the automotive train. Some staff members may even have viewed the success of the lightweight car as only an intermediate phase in railroad development. Perhaps they hoped to lead a large-scale conversion to automotive trains before their rivals could perfect their own designs. Whatever the real motives of A.C.F., the firm spent ten frustrating years trying to sell the Talgo concept in America.

As early as 1837 W. B. Adams, an English mechanic, had proposed low-slung, articulated, two-wheel railway carriages. The idea was reintroduced more than a century later by A. A. Goicoechea, a Spanish engineer who named his design Talgo as an acronym for train, articulated, light, Goicoechea, and Oriel—the last being the name of the family who helped finance the project. The inventor proposed a two-wheel triangular frame or running gear, coupled together so as to lead the trailing units into curves. A prototype was built in Spain for test purposes, but it was realized that a serviceable train must be produced by an established car builder. The job was given to A.C.F., and so began America's misadventure with the Talgo.[190] By April 1949 the first of three duplicate trains was ready (Figure 8.63). Two were sent to Spain, and the third was retained by A.C.F. for testing and display.[191] The 167-foot-long train was made up of an 810-horsepower diesel electric locomotive, a baggage car, and four coaches. Except for their width, the squat coaches were equal in weight and size to a nineteenth-century horsecar. The basic construction was aluminum, which helped to hold the 20-foot-long cars to a weight of 3½ tons. The wheels were of the sandwich-core type made of rubber and steel, much like those used on P.C.C. streetcars. The air-powered hydraulic brakes had automotive-type brake drums. The two trains that were shipped to Spain entered regular service in July 1950 after prolonged testing. They cut 3½ hours from a 396-mile run that traversed mountainous territory between Madrid and the French border. They were found to be somewhat underpowered and difficult to operate in reverse except at extremely low speeds.[192] An American writer reported that the ride was poor, with a pronounced fore-and-aft motion, but that the trains went around curves smoothly and with little roll. The Spanish were pleased with the Talgo and eventually purchased more, though not from A.C.F.

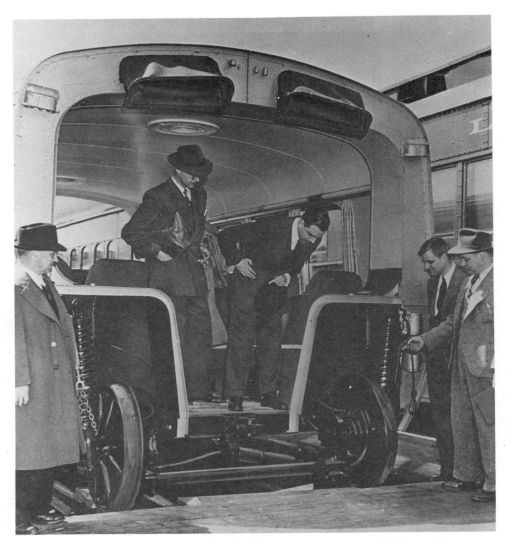

*Figure 8.63   The low-slung, articulated Talgo train was a Spanish design first manu-
factured in this country by A.C.F. in 1949. (National Archives)*

Meanwhile A.C.F. was unable to create much enthusiasm among American customers. In 1953 the company built a second, improved demonstrator in which the individual coach units were lengthened from 20 to 34 feet.[193] Three units were coupled together to form a six-wheel car that could be cut in or out of the train, thus answering the usual objection to articulated trains by American operating men. The running gear was redesigned to facilitate operating in reverse, and the framing was strengthened to meet the 400-ton buff load requirements. Yet the cars were still low-slung and undersized. The new Talgo was indifferently received at first. But a shake-up in top railroad management brought men into power who were attracted to innovative schemes such as the Talgo. Robert R. Young had gained control of the New York Central, and one of his admirers, Patrick B. McGinnis, was running the New Haven. Both faced critical fiscal problems because of passenger train losses. In former years, Young had expressed faith in exotic equipment as a solution to the passenger train dilemma. He envisioned a vast new fleet of cars for the Chesapeake and Ohio and dreamed of taking over the Pullman Company. None of these plans were realized, but he did finance experiments with a Talgo type of design in 1947.[194] The C & O was a coal road and passenger traffic was a minor consideration, but the Central was a major passenger carrier, and its heavy losses made desperate measures appropriate. Young, who was a showman, hoped to reverse the irreversible by introducing a futuristic train.

It seemed that the Talgo would shed its Cinderella role, for orders began to appear in 1955. The New Haven purchased two, the John Quincy Adams and Dan'l Webster. The Adams was delivered by A.C.F. in 1956.[195] It had five three-unit cars seating 480 passengers. The Webster was produced by Pullman and seated only 392. The Boston and Maine ordered a train similar to the Adams but did not name it. The Rock Island purchased another A.C.F. Talgo, also in 1956, and called it the Jet Rocket. The New York Central showed good faith by purchasing a Pullman-built Talgo named the Xplorer. These trains cost anywhere from $788,000 to $1,200,000 each. They were scheduled for main-line, long-distance travel, and according to their publicity, would usher in a new era of railroad travel.[196]

But once the press parties and champagne excursions were over, the super-lightweight bubble burst. The Talgos really offered nothing new. A low center of gravity provided no measurable advantage in the ride or in passenger comfort. Just the reverse—passengers complained about the poor ride and high noise level.[197] The low ceilings, small seats, tiny toilets, and box-lunch dining service were decidedly substandard, and the public was not impressed. Both the Jet Rocket and the B & M Talgo were quickly withdrawn from through-train service and downgraded to commuter runs. The New Haven trains were disaster-prone. The Dan'l Webster caught fire on an inaugural press run, and it and its companion cars regularly broke off their third-rail shoes when entering the Grand Central electrified trackage.[198]

Figure 8.64   *The United Aircraft Turbo Train of 1967 was a space-age fizzle. All U.S. units were retired in 1976. (Amtrak)*

They were retired after several years, and the Xplorer was laid off after two years.

The Talgos were another costly failure for the railroads, but other industrial giants also took losses on the extra-lightweights. General Motors thought it could build a cheap lightweight by adapting stock components for the purpose. EMD (the Electro Motive Division) built two Aerotrains in 1956. The 1,200-horse-power locomotives were duplicates of the unit fabricated for the Jet Rocket. The cars were standard 40-foot G.M. bus bodies widened by 18 inches.[199] They were carried on four wheels and suspended on motor-bus air springs. The 16-ton cars seated 40 passengers. The trains had ten cars in their consist. One train was sent East for a year's trial on the Pennsylvania. The second train operated on several Western lines. The Union Pacific leased it for a time, but rapidly became disenchanted with its poor performance and high repair costs.[200] After a brief Mexican tour, the Aerotrain was sold in 1958 to the Rock Island for commuter service. At low speeds the train worked fairly well, but passengers complained about the annoying sideways motion that sometimes became violent.[201] The Aerotrain was retired in 1965 and was subsequently given to the National Museum of Transport in St. Louis. There it now stands, a forlorn monument to a well-intended but ill-conceived effort to revive passenger train travel in America. The Talgos and their second cousin, the Aerotrain, were clearly experiments introduced after their time rather than before it. In the words of David P. Morgan, editor of *Trains*

magazine, they "proved spectacularly unsuited" to the needs of the railroad industry.

The most recent development in rail motor cars is the gas turbine car. Actually its roots are in the Train X, Talgo era, though it has been represented as something very new. The publicity once again claimed that the car had achieved a marriage of aircraft and railroad technology. Supporters spoke of lightweight aluminum construction, low centers of gravity, wind-tunnel tests, and other supposedly revolutionary ideas that had in fact been knocking around for more than forty years. The designer of the Turbo Train was Alan R. Cripe, an engineer who had been involved in the planning of Train X and its successors. He held patents on many components used on these trains and was a firm believer in guided axles and controlled banking systems.[202] In 1963 the prospects for more experiments in this area seemed remote, but Cripe had faith in the concept and delivered a paper on it that was heard by an official of United Aircraft. The U.A. executive was impressed and convinced others in his company to support Cripe's research. The decision proved timely, for the high-speed ground transportation act was passed in 1965 and Federal funding for experiments of this nature was now available. A contract was signed the following year, and a test train was rushed through production and completed in May 1967. The double-ended, welded-aluminum train, reminiscent of the motor train built nearly thirty-five years earlier, measured 203 feet long and weighed 105 tons (Figure 8.64). It seated 144 passengers and stood 35 inches lower than a regular train. Four 400-horse-power turbine engines propelled the train; a fifth turbine ran the generator for heat and lighting. The train was complex, sophisticated, and admittedly experimental, and understandably it proved troublesome. It was not ready for regular service until almost two years after delivery, and even then mechanical problems kept it out of service 54 percent of the time. This was a dismal record compared with an ordinary diesel passenger locomotive, which registered a downtime of 22 percent.[203] The transmission, lubrication pumps, air suspension, and axle guidance system were particularly temperamental.[204] A ride on the train in 1970 was memorable for the tangle of loose wires, gauges, and electrical test gear in the passageways of the power units, the entire maze being monitored by a worried young man who was presumably a United Aircraft engineer. The train rode poorly at slow speeds but smoothed out remarkably as the pace quickened. The seats were full, and according to the crew, the train was very popular with the public.

By the end of 1970 the Turbo train had logged 500,000 miles. Better insulation had reduced the noise level, and an additional car was being added to boost seating.[205] It was kept running, though not without difficulty, and the pitiful breakdown record of the five turbos that had been built for Canada did little to restore faith in United's ability to produce an economical or dependable rail car. When it came time to order more turbine rail cars in 1972, Amtrak selected a French contractor. These units were straightforward double-truck rail cars, unencumbered by exotic suspensions, that had operated successfully in France since 1967. When a second order was given to the French in 1974, United Aircraft decided to quit the rail car field. It seems regrettable that no American manufacturer developed a practical turbo rail car to replace the RDC. Still, it is more than likely that the Talgo–Train X scheme will snare yet another sponsor, for like the monorail, it is an alluring engineering concept whose charms never seem to fade.

Electric Self-Propelled Cars

The growth and concentration of suburban travel to and from major American cities led to the adoption of electric self-propelled cars on several North American railroads. (It will not be possible to discuss all these installations here, nor is it necessary to do so, since they have been well covered in a volume by William D. Middleton.[206] Also, this section will not treat interurban and rapid transit lines even though some, such as the Chicago, South Shore, and South Bend, operated equipment of main-line proportions.) Beginning as early as 1835, Boston became a center for commuter traffic.[207] Within three years four railroads were hauling commuters into the hub; some even offered season tickets. By 1860, 143 daily trains were serving the outlying towns. Twenty years later, 403 trains were necessary as Boston jobholders sought housing in Newton, Lynn, and other nearby villages.[208] The outskirts of New York, Philadelphia, and

Chicago were growing in the same way. In time this traffic swelled to overwhelm the big-city terminals: the locals were crowding out the main-line trains. In the early years of the twentieth century suburban traffic was growing at 20 to 25 percent a year, and while four or five tracks could handle the through trains even in major cities, another twenty-five tracks were needed to accommodate the suburbans.[209] By 1925, 364,000 commuters—more than the entire population of Arizona—moved in and out of New York every day.

Some lines built auxiliary downtown stations to handle the traffic; others continued to expand their main terminal. After a time, however, no more property was available, or else it was not available at a price any railroad could afford. Managements made various efforts to improve the efficiency of suburban trains. The Boston and Albany and the Illinois Central purchased special double-ended locomotives with high tractive power to hustle the trains in and out of the station more quickly. Because they could run equally well in forward or reverse, no time was lost in turnarounds. But such measures were not enough in the more extreme cases of traffic density, and some railroads decided that electrification was ultimately cheaper than endlessly expanding terminal trackage. What the electrics offered was rapid acceleration—three times that of a steam locomotive. The effect on terminal capacity is obvious. But electrification involved more than cars and trolley wire: expensive substations and the mechanics of power distribution added materially to costs, which even in the 1920s could reach nearly a million dollars per mile. Sometimes the cost of electric suburban operations was justified on a road because of its traffic volume, yet its management could not or would not raise the capital to do the job. The lines running out of Boston were cases in point. Elsewhere, electrification was adopted only after municipalities passed smoke abatement laws, for clean air was another major advantage of electric suburban service.

Electric railroad technology was worked out in all details by the street railway industry almost a decade before any main-line electric operations took place. Traction motors, gearing, controllers, and power generation and distribution were refined and proved for dependable everyday use when the Baltimore and Ohio opened its Baltimore tunnel line in 1895. Although some additional problems were involved, main-line electric railroading was essentially trolley engineering on a larger scale. Sensibly, the steam railroads borrowed as many ideas as possible from the city systems. They took the 600-volt DC plan and its appurtenances, and they even employed engineers, such as Frank J. Sprague or George Gibbs, who had made their reputations in the transit field. Sprague offered what was possibly the single most important contribution to electric railway engineering: the multiple-unit system, which enabled one man to operate two or more motor cars. The M.U. system saved more than labor alone. Because the motors were under each car, power was spread throughout the entire length of the train. And because the ratio of power to weight was generally higher than that of a locomotive, acceleration was greater. This was a prime requirement on

Figure 8.65   The cluttered underside of a standard-era, multiple-unit electric rail car is shown here. Most of these pipes are electric conduits. (Pullman Neg. 29720)

lines where traffic was congested; the faster that trains could get in and out, the greater the capacity of the railroad. Sprague's system was first used on the Chicago elevated in 1897. It quickly established itself as a basic transit tool and is used today by every electric railway that runs multiple-unit equipment.

Electric cars have other common characteristics. They are heavier, more expensive, and more complicated than an ordinary coach. Their seating capacity is greater, their luxuries fewer than their locomotive-hauled counterparts. Toilets are small or nonexistent. Interior decor is spartan and the seating elementary. They are arranged to carry the maximum loads with the minimum comfort. Because the trips are short, the fares substandard, and the prospects of a profit ludicrous, most patrons bear the miseries of commuting with resignation. Mechanically the cars are a forest of wiring and underbody gear (Figure 8.65). Where to hang all the controls, breakers, resistors, transformers, air compressors, and air brake gear are problems that baffle successive generations of engineers. To keep the body space free for the maximum number of passengers, everything possible is placed under the floor. The operator's cab is a tiny closet crowded into one side of the vestibule. Electrical gear easily adds 10 tons to the weight of the car, which makes it necessary to use heavier axles to carry the load and withstand the torque of the traction motors. Because of all the extra mechanical-electrical apparatus, M.U. cars cost roughly 25 percent more than regular coaches.

Multiple-unit cars were commonly coupled together semipermanently as married pairs—sometimes as motor-trailer combinations, but also as twin motor cars. It was considered prudent to have controls in the trailer car to facilitate switching operations. Large entrances and sliding doors were also common. The newer cars were patterned ever more on subway vehicles and often had center entrances and no vestibules. Toilets were eliminated wherever local health regulations did not interfere, and separate toilets for the sexes were rare. Good ventilation was a necessity because of the heavy loads carried at rush hours. Air conditioning came last to the suburban commuter, and it arrived in a big

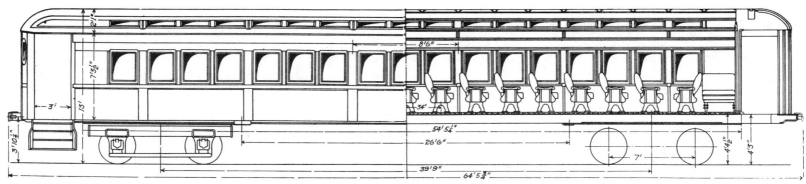

Figure 8.66   The Pennsylvania Railroad designed its class P-54 suburban coach in 1907
for conversion to electric service. Figure 8.72 shows a later example. (Railway Review,
June 8, 1907)

way only after government subsidies were offered for new equipment.

The first large-scale use of M.U. electric rail cars in America began in 1905 on the Long Island Railroad.[210] Traffic was booming and plans were underway for a direct connection into the Pennsylvania's Manhattan station, which was then under construction. The program was started none too soon, for between that date and 1919, passenger volume climbed by 353 percent. Eventually about 40 percent of the line was electrified, and these nearby sections carried over 70 percent of the traffic.[211] The first cars were designed by Gibbs and duplicated those planned for the New York subway. In 1905–1906, 134 were built. They measured only 51 feet long and were made decidedly undersized in all dimensions in order to fit the subway loading gauge. Sometime afterwards it was decided to build a separate full-sized tunnel, and the next cars were modeled after the parent road's standard steel suburban coach, the class P-54. Because the electric cars were self-propelled rather than locomotive-hauled, they were classed as MP-54 (Figure 8.66). In 1909 and 1910 A.C.F. turned out 222 MP-54s for the Long Island.[212] They measured 64 feet, 6 inches long, seated 72 passengers, and weighed 54 tons. Two 225-horsepower motors drove each car. Around 1915 the Long Island adopted a Gothic arched roof in place of the squat clerestory. The design was not particularly pleasing and gave the cars a more subway-like appearance (Figure 8.67). Hundreds more were built on this plan; by 1930 the road had 1,000 MP-54s in service.[213] This was the largest M.U. fleet in North America and represented a triumph in standardization.

In 1932 the Long Island acquired an experimental double-deck car, the Number 200, produced in the parent road's Altoona shops. The MP-54s were dependable workhorses, but in view of their limited seating and heavy weight, it seemed as if a better design must be possible. The prototype had a split-level, or seat-over-seat, arrangement; it was not a double-decker in the sense of the modern gallery car. The Number 200 was a trailer measuring 72 feet long and weighing 36 tons.[214] Two more were constructed in 1937, but further production was delayed until 1947 because of the war. Sixty were built in Altoona in 1947–1948 at a cost of $143,000 each.[215] They were 80-foot-long motor cars equipped with air conditioning (Figures 8.68 and 8.69). They seated 132, or nearly double the number of the old MP-54s. But they were not popular with the public; the steps to the upper seats were a hazard, the many nooks and corners made them difficult to clean, and women disliked to occupy the upper seats for fear of encouraging peeping toms.[216] The railroad decided that the cars were too expensive and that single-level cars with

two-and-three seating were a more economical alternative. No more bilevels were ordered, and by 1971 a number had been retired. During the period 1953–1955, 160 new M.U.s were purchased that featured two-and-three seating. These larger-capacity cars permitted the M.U. fleet to be reduced to 718.[217] At the same time, hundreds of old cars were given a thorough mechanical refurbishing, modern interiors, and new seating. Air conditioning was considered too expensive, but powerful fan-ventilating systems were installed.

In 1906 as part of its Grand Central improvement program, the New York Central began the second main-line electrification to make large-scale use of M.U. cars. The third rail was eventually extended 34 miles to Croton and 24 miles to White Plains.[218] Electric locomotives hauled through trains in and out of Grand Central, but the suburban service was to be maintained by 125 motor cars furnished by A.C.F. These short, arched-window cars had a distinctly interurban look (Figure 8.70). They measured 62 feet long over the couplers, seated 64, and weighed 53 tons. The trucks were fabricated by the American Locomotive Company. The side frames were forged from square bars much like a miniature steam locomotive. There were 600-volt DC motors that produced 200 horsepower per car. The cars had both electric and gas lighting. Steam and electric heating were provided as well. The dual systems were necessary in the event that the cars operated outside of the electric zone. Between 1910 and 1929 the Central acquired over 200 additional motor and trailer cars. These units were about 3 feet longer than the original cars and looked more like conventional coaches. Express trains consisted entirely of motor cars, ranging from two to ten per train. Locals ran with a mix of two motor cars to each trailer, with three to twelve cars per train.

In 1926–1927 the motor cars were cut in half for the addition of an 8-foot section, which increased the weight by 15 tons and the seating by 12.[219] The art-glass arched windows in the older cars were plated over. Many of the trailers were motorized, and the electrical systems of all units were reworked so that trains could be operated in multiples of fifteen cars. Except for ten new cars delivered by Standard Steel in 1929, the New York Central M.U. car fleet remained static until 1950, when St. Louis began deliveries of 100 new motor cars.[220] The 85-foot-long, smooth-sided, welded steel cars weighed 75 tons and carried 130 passengers in two-and-three seating (Figure 8.71). These arched-roof cars followed lightweight car design. They were a welcome supplement to the Central's aging M.U. fleet, although technically they were unremarkable. The old-fashioned 600-volt DC system was dependable but obsolete. In 1962 fifty-three new motor cars

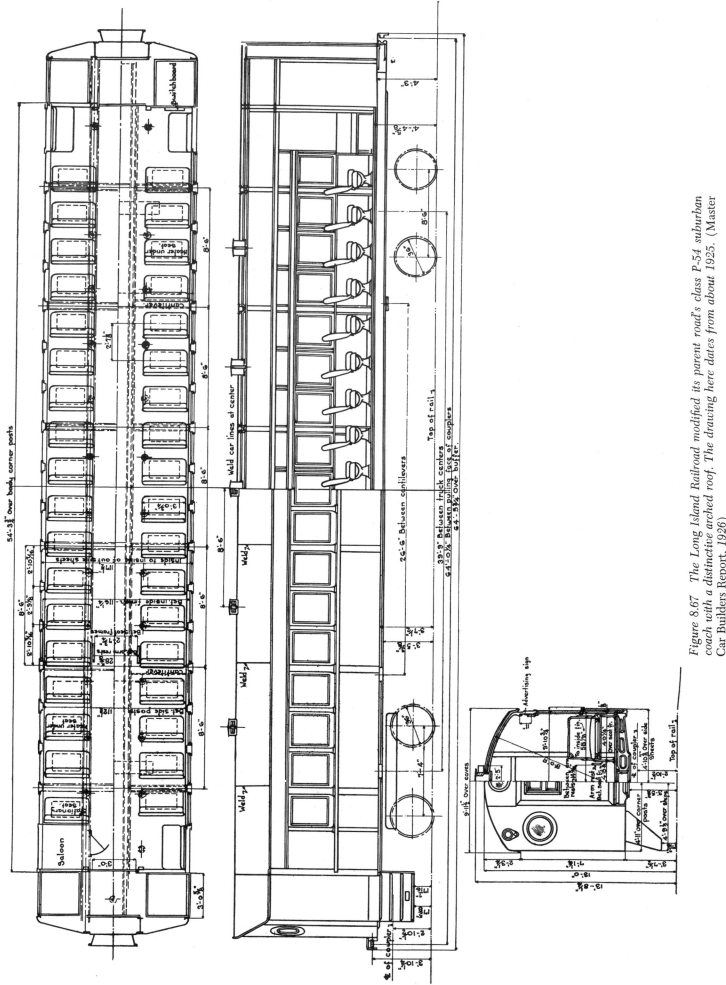

Figure 8.67 The Long Island Railroad modified its parent road's class P-54 suburban coach with a distinctive arched roof. The drawing here dates from about 1925. (Master Car Builders Report, 1926)

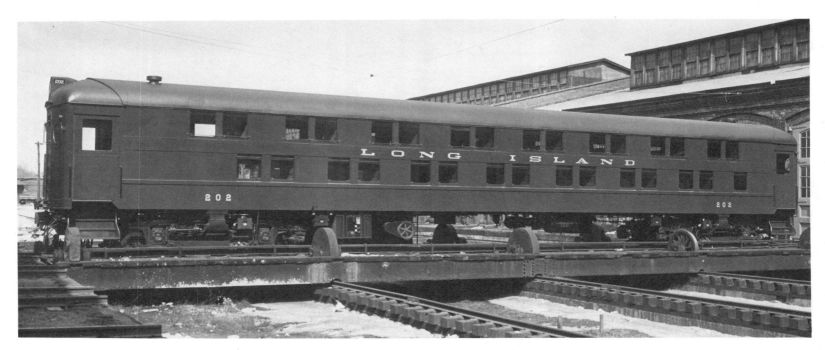

Figure 8.68   The Altoona shops produced this electric double-decker in 1947. The 80-foot car seated 132 passengers. (Pennsylvania Railroad)

Figure 8.69   The interior of a Long Island Railroad bilevel commuter car. (Pennsylvania Railroad)

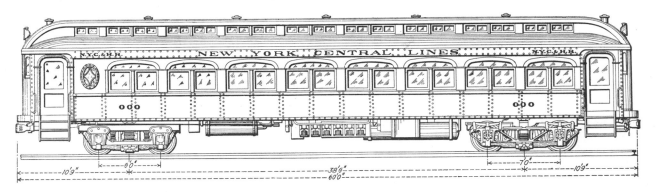

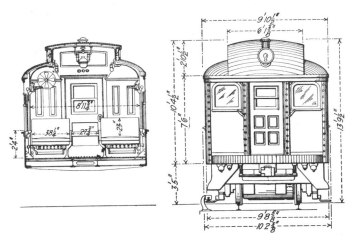

Figure 8.70   The New York Central's short electric cars of 1906 were among the first steel cars used in main-line service in this country. (Car Builders' Dictionary, 1906)

Figure 8.71   The St. Louis Car Company produced this electric M.U. car for the New York City suburban service out of Grand Central Station in 1950. (St. Louis Car Company)

Figure 8.72   A late-model class MP-54 electric M.U. car built at the Altoona shops.
(Altoona Public Library)

made possible the retirement of a large portion of the older cars.[221] Only the Bronx locals were serviced by the prewar heavyweights after this time.

Some historians have contended that the New Haven pioneered electric self-propelled rail cars in the United States. It is true that the road began to operate electric cars on their Nantasket Beach line in 1895.[222] Several other New Haven branches in Connecticut were electrified in the period 1896–1898. The experience gained from these operations was valuable, but it did not lead to any immediate main-line electrification, nor was it technically more advanced than anything that was being done on hundreds of suburban and interurban electric lines elsewhere in the country. Actually, the New Haven had no main-line M.U. operations until 1910, some five years after such service began on the Long Island and the New York Central. Again the scene of operations was the New York City area. The road's four-track main line from Manhattan to Stamford, Connecticut, was the most heavily traveled in the nation. Electric locomotive-hauled trains began running in 1907, but commuter traffic was so dense that the advantages of M.U. trains were obvious. In February 1910 motor cars began operations between Grand Central Station and Port Chester.[223] The first four motor cars were delivered by the Standard Steel Car Company. Their open platforms and clerestory roofs gave them an antique appearance, but electrically they were the most advanced electric cars in America because of their 11,000-volt AC electric system. The circuitry was further complicated by secondary wiring for 600 volts DC, so that the cars could operate on the Grand Central's trackage.

With their four 150-horsepower motors, they were undoubtedly also the most powerful electric cars in the country and could pull two trailers. They measured 70 feet long, seated 76, and weighed just under 87 tons.

More cars were acquired as traffic doubled during the next decade. By 1922 the New Haven was transporting 16.3 million passengers a year over this short section of railroad.[274] Nine years later 129 motor cars and 92 trailers shuttled to and from Grand Central. The New Haven, like most other steam roads with electrified sections, acquired little new rolling stock that might be described as distinguished until after the end of World War II. In 1954, for example, they received a number of lightweight electrics from Pullman's Worcester, Massachusetts plant.

The Pennsylvania Railroad cautiously studied the M.U. systems of the other Eastern lines. Its Penn Station electrification, opened in 1910, remained a locomotive-powered tunnel operation, but during the next year the road began to consider its chaotic Philadelphia suburban service.[225] Local traffic had overwhelmed the Broad Street Station. When it was opened in 1881, the eight-track terminal comfortably handled 160 trains a day. But the mounting number of commuters required two enlargements in 1890 and 1894 that resulted in a sixteen-track station. By 1910, 574 trains a day created a traffic snarl beyond its capacity. Delays were frequent during bad weather, yet further expansion of the facility was considered impossible because of the downtown setting. It was decided to electrify the two most heavily traveled commuter routes, Paoli and Chestnut Hill, and thus speed up train movements. Faster in and out speeds, and

Figure 8.73   The Illinois Central began large-scale electric suburban operations out of Chicago in 1926. The cars were constructed by Pullman. (Illinois Central Photograph, Trains magazine collection)

the elimination of the turnaround switching movements that were necessary with steam locomotives, improved the situation immeasurably. With no additional tracks, the terminal was thought to have an actual capacity of 1,000 trains a day after electrification was completed.[226]

In 1915 the first M.U.s began to run out of Broad Street. Ninety-three class P-54 steel suburban cars were converted, and indeed had been designed for conversion to M.U. service. The low overall height permitted the mounting of pantographs. Two 225-horsepower motors were fitted on the front truck. The 64-foot-long cars seated 72 passengers. As the electric operation expanded from Philadelphia to Norristown, to Wilmington (and later to Washington), and to Harrisburg, hundreds more MP-54s joined the roster (Figure 8.72). Eventually more than 500 were

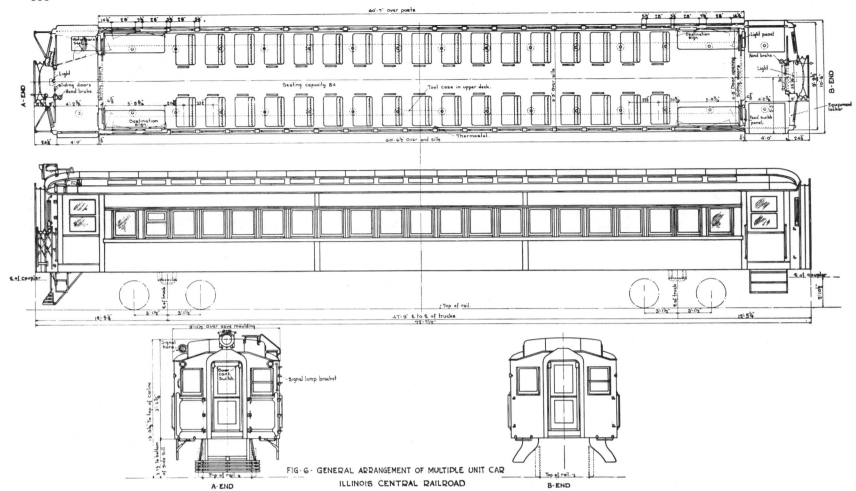

Figure 8.74   *In all, 280 steel motor and trailer cars were built for the Illinois Central's electric service. The design followed conventional steam railroad practice.* (Master Car Builders Report, 1926)

placed in service.[227] Standardization was carefully maintained, but some improvements, such as roller bearings, were applied to the newer units. In 1950 a major program of remodeling was started at the Wilmington shops, which installed new trucks, windows, frames, and interiors. Eight years later the Pennsylvania belatedly began to replace the MP-54s. (The new cars will be discussed later in this chapter.)

Outside the Northeast, the only main-line M.U. operation of any size was the Illinois Central's Chicago suburban line. Historically it was an early and increasingly important route, serving the residential-industrial communities to the south of Chicago. It began in 1856 with four daily trains between Chicago and Hyde Park.[228] By 1880 traffic had grown so much that two separate tracks were built and reserved exclusively for suburban trains. Double-ended locomotives and light cars, the latter modeled after those used by the New York Elevated, were purchased. Sixteen trains a day were run, but the *Railroad Gazette* predicted that "there is plenty of room for growth and the growth is sure."[229] Thirteen years later 62 locomotives and 380 cars carried an astonishing 9 million passengers to the Columbian Exposition grounds in the space of five months. Yet the traffic continued to grow. In 1925, 24 million passengers were carried; between 1900 and 1929 the traffic increased by 350 percent. One-third of the total Chicago suburban service was funneled into the Illinois Central, and the lakefront sky was blackened by the daily passage of more than four hundred trains.[230] The city had forty railroads with 1,700 steam locomotives working within its border,

but the Illinois Central's suburban operation was the most exposed and hence received the most criticism.

In 1891 talks began on the advantages of electric operations. Definite planning was initiated in 1911, but the project languished until the city passed an act in 1919 requiring electric operations by February 1927.[231] The conversion involved a major reconstruction of the entire 30-mile line (actually 250 miles of track) which included grade separations and new stations. The total cost of the project came to 24 million dollars.[232] Electric operations began ahead of schedule: the first M.U.s were running in August 1926. The electrics cut as much as 30 percent from the old schedules and were credited with an acceleration rate six times that of the steam trains. In addition to clean and rapid service, the electrics proved themselves a prudent investment, for the steam trains were losing $337,000 annually while the M.U.s made a profit of $400,000.[233]

In 1921 when the electrification program was getting underway, the Illinois Central purchased its first steel cars.[234] They were intended as prototypes for the electrics and were used later as trailers. A worldwide study was made of suburban rolling stock. This information was used in designing cars that set the basic pattern for 280 motor and trailer units built between 1921 and 1929 (Figures 8.73 to 8.75). They were conventional-appearing, heavyweight steel cars, arranged in married pairs of a motor and trailer car. They measured 72 feet in length and seated 84. The first group of trailers weighed 94,800 pounds, but in the second order a total of 7,600 pounds was saved by substi-

Figure 8.75   The clean interior of an Illinois Central electric suburban car just before it entered service in 1929. Rattan seats were durable but not very popular with passengers. (Pullman Neg. 33068)

tuting aluminum for the roof sheets, doors, and interior sheathing. An additional saving of 6,100 pounds was made in subsequent orders by using aluminum for the conduits and junction boxes. The motor cars weighed 141,200 pounds. The bodies and frames were made from copper-bearing steel. The interiors were considered bright and spacious because of the large windows, high-gloss white ceilings, wide aisles (31 inches), and the absence of luggage racks. Wide, 48-inch doors and electric "door engines" speeded loading and unloading, as did the platform-level stations. In rush hours the rated seating of 84 was augmented by an additional 109 standees. The motorman was protected by safety glass in the end windows. The exteriors were painted dark olive green, with black roofs and gold lettering. Four 250-horsepower, 1500-volt DC motors per motor car pushed the trains along at 60 mph. They accelerated at 1.5 mph per second. The traction type of Tomlinson coupler was used. For forty-five years these Pullman heavyweights unfailingly carried millions of Chicago commuters. In 1971 new lightweight cars began to replace the Pullmans, but a number of the old units remain in service today, a testament to their durability and solid construction.

The first generation of American M.U. cars actually lasted fifty years. The emphasis was on compatibility of electrical equipment, which made good economic sense but retarded technical advancement in the area. This philosophy was engendered by the long life characteristic of traction equipment and by the economics of rail commuter operations. Poor earnings do not encourage substantial new investment. Body styling and interior appointments changed slightly with each new order of cars, but the basic electrical systems remained largely unaltered. The new cars purchased in 1949 by the Reading for its Philadelphia suburban lines were equipped to duplicate the original cars delivered in 1931.[235] The Long Island, an even more extreme example, was receiving cars as late as 1963 with performance levels equal to those of its 1908 equipment.[236]

The advantages of the AC-DC electrical sytem had been considered for many years, and when it was finally adopted after 1950 it represented a major technical advance in this conservative field. Alternating current had excellent characteristics for long-distance power transmission, while direct-current motors were superior to AC motors. Because 600-volt DC was an early standard on street railways, this type of traction motor became highly refined and mass-produced. It was adopted for diesel-electric locomotives as well, which established its position even more securely. In addition to being cheaper than AC motors because of quantity manufacture, it was more powerful at low speeds—by as much as 47 percent, one estimate claimed.[237] What was needed was a compact, dependable unit aboard each car to convert AC to DC. Motor generators were far too bulky. In 1913 Westinghouse tested a mercury-vapor tube rectifier on a M.U. car.[238] However, the rectifier interfered with telephone service, and the tube itself had to be replaced every three months. In 1949 a more durable ignitron mercury-arc rectifier was tested on a Pennsylvania M.U. car. This trial was so successful that the New

Haven specified ignitron converters on 100 new electric cars built by Pullman's Worcester plant.[239] Deliveries of the stainless-steel-sheathed cars began in February 1954 (Figures 8.76 to 8.79). The 86-foot-long coaches seated 120 and were powered by four 100-horsepower motors. Because of this larger seating capacity, 150 old M.U.s were retired. Four years later the Pennsylvania purchased six Pioneer III-style coaches with ignitron rectifiers.[240] These were the lightest (45 tons) and fastest (90 mph) multiple-unit cars produced up to that time. The trucks were of an advanced inside-bearing design. The small, 30-inch-diameter wheels, the air springs, and the gear box motor drive were all new features (Figure 8.79).

The electrified Pioneer IIIs were such a success that Budd modified the design slightly for the first of the Silver Liners that have since become the standard electric car in the Philadelphia suburban lines of both the Pennsylvania and the Reading railroads. Rather than carrying mercury-arc rectifiers, the new cars were equipped with rectifiers of silicon diode, which were more compact and did not require firing circuits or cooling systems. The first 55 Silver Liners were delivered in 1963 at a cost of 14 million dollars.[241] They weighed 50¾ tons and seated 124 passengers in cars with toilets and 127 in those without. No interior paint was used; sheet and molded plastic shapes covered the entire surface. By 1974 over 200 Silver Liners were in service and another 144 were under contract.[242] The last group was produced by General Electric's Erie, Pennsylvania, plant. Each car had 8,600 parts and 3½ miles of wiring.

The success of the Silver Liners encouraged the development of a more modern style of car for 600-volt DC commuter lines. In 1966 Budd began designs for Long Island's new Metropolitan class. These 85-foot-long cars had side doors and no vestibules. Their curved-side bodies helped create the image of a giant subway vehicle. They weighed 46 tons and had molded plastic interiors.[243] The exteriors were smoothed (rather than fluted)

Figure 8.76   A major advance in electric car technology was the rectifier, or AC-DC
equipment. The Number 4674 and its sisters were the first production models to be so
equipped when delivered by Pullman in 1954. (E. B. Luce Corporation Photograph)

Figure 8.77   A New Haven electric rectifier car interior of 1954. (E. B. Luce Corporation Photograph)

stainless sheets accented by blue stripes. It was one of the largest
orders on record: 620 cars, at a cost of 130 million dollars. Despite the size of the order, it was not a happy one for the supplier
or the purchaser. The cars did not perform well at first, and this
proved an embarrassing and costly problem for the Budd Company. In fact, the firm resolved to quit the car-building field,
although new orders which were subsequently received caused
management to reverse its decision.

The latest innovation in electric self-propelled cars appeared
on the Illinois Central in the spring of 1971.[244] Planning that
was begun in 1967 produced designs for double-deck or gallery
cars similar in their general arrangement to those running on
other nonelectrified Chicago suburban lines. The presence of
overhead wires inhibited the use of high cars. However, the Illinois Central justified its choice of galleries, rather than single-
level cars with two-and-three seating, on the grounds of passenger comfort. Two-and-two seating was preferable and allowed space for wider aisles. An order for 130 cars was given to
the St. Louis Car Company. The cars cost $307,000 each and, like
most recent commuter equipment purchases, a large share of the

Figure 8.78 Most M.U. cars were outfitted as coaches, but a few, such as this 1954 New Haven club car, had more luxurious furnishings. (E. B. Luce Corporation Photograph)

funding was public. In this case two-thirds of the money was Federal.

The 85-foot-long units stood 15 feet, 10 inches high, weighed 67 tons, and seated 156 passengers (Figure 8.80). The front portion of the roof was depressed 22 inches to accommodate the pantograph. Center and end entrances were provided. The Cor-Ten steel bodies were painted silver gray with orange and black highlights. Vinyl-covered, foam-cushion seats replaced the old-fashioned rattan of the heavyweight cars. Tinted safety glass eliminated the need for shades. Air conditioning and electric heat ensured passenger comfort, while four 160-horsepower motors guaranteed a rapid acceleration rate of 1.9 mph per second. New York Air Brake furnished the latest electropneumatic-hydraulic brake equipment. The old Tomlinson style of coupling was retained, however. The trucks were similar to those used on the Metroliners.

The new cars, named Highliners, were on all counts a splendid improvement over the old equipment, but not enough were acquired to fully service the line. During peak rush periods it was

necessary to operate some heavyweight trains, and the dangers of this practice were soon tragically demonstrated. On October 30. 1972, one of the new gallery trains overshot the 27th Street Station.[245] The crew then backed into the station—a flagrant violation of the Standard Code. Meanwhile, a heavyweight train following behind had already passed the signal guarding the station and was thus unaware that another train was in the same block. The old train tore through its modern replacement, telescoping the rear car and causing 45 deaths and over 300 injuries. The new cars met the 400-ton buff standard, but the end framing was not sufficient to withstand the impact of a 400-ton, seven-car train at 30 to 40 mph. The difference in end profiles also facilitated telescoping, and once through the end stanchions, the light sheet-metal body of the Highliner offered little resistance to the terrible progress of the heavyweight train.

The Metroliner, America's most spectacular electric rail car, is essentially a multiple-unit car despite its sophisticated engineering, its claims to space-age technology, and its interiors that are surely the ultimate in plastic aircraft decor. It is simply the fast-

Figure 8.79   Budd's Pioneer III–style truck was adopted for electric M.U. cars in 1958. Note the disc brakes and massive cup-shaped mounts for air springs. (Lawrence S. Williams)

est, most powerful, and best-furnished electric self-propelled car yet constructed. The Metroliners were introduced when the North American passenger train seemed to have no future. They did much to restore faith in the railroad's capacity to move passengers efficiently—at least over moderate distances—but technically they proved troublesome and may be the last of their breed.

The Pennsylvania's New York–Washington line was one of the most heavily patronized in the world. Even after the World War II peak, it carried 130,000 passengers daily. But after 1946 business began to fall away as travelers returned to the highway. New highway networks stimulated the decline. And in 1961 Eastern Airlines began the low-cost air shuttle between New York and Washington. Frequent schedules and a flying time of under one hour drew thousands of commercial travelers away from the Pennsylvania. The railroad trimmed its schedule by half an hour and installed some rebuilt cars, but it was a losing battle. Patronage continued to decline, so that by 1958 it was only one-

half of the 1946 level. To reverse this trend would require drastic action, but the railroad felt that it was possible because of the interurban-like nature of the route: Baltimore, Wilmington, Philadelphia, and Trenton, located between the financial and political capitals of America, were on-line traffic generators. Faster schedules and new equipment were an old formula that had worked in the past and might just succeed again. The railroad's interest was reinforced by the public statements of Senator Claiborne Pell, who voiced concern over the growing congestion of the Northeast corridor and suggested that it could best be relieved by upgrading existing rail service rather than building more highways and airports.

In 1958 the railroad ran some high-speed tests with its new Pioneer III electrics which indicated that M.U. equipment would be suitable for long-distance travel.[246] It was viewed as the best choice on both engineering and economic grounds. New electric locomotives would be costly to design and build, and comparable cars would still be needed. By combining the loco-

Figure 8.80   In 1971 the Illinois Central began to replace its standard-era steel cars with new bilevel cars produced by St. Louis. (Illinois Central Gulf Railroad)

Figure 8.81   The Metroliners are the ultimate in electric self-propelled rail cars. Budd began deliveries in 1967. (Herbert H. Harwood, Jr.)

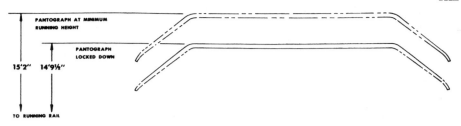

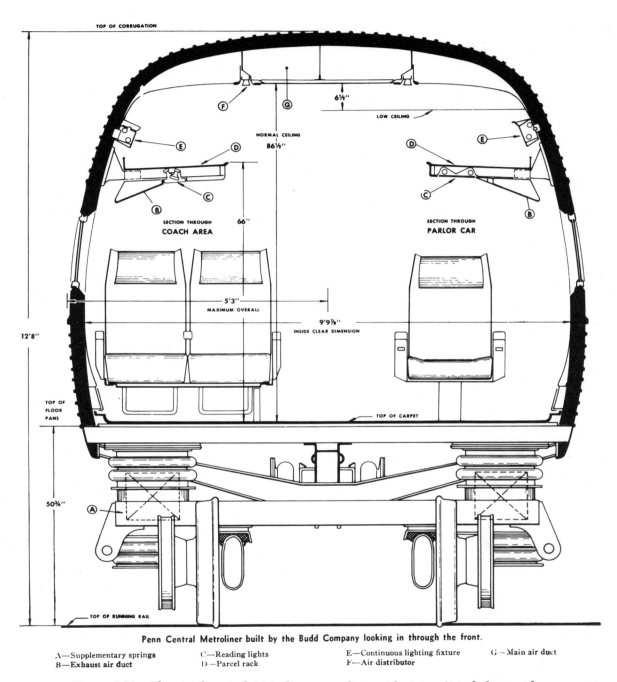

Penn Central Metroliner built by the Budd Company looking in through the front.

A—Supplementary springs   C—Reading lights   E—Continuous lighting fixture   G—Main air duct
B—Exhaust air duct   D—Parcel rack   F—Air distributor

Figure 8.82 The stainless-steel Metroliners are distinguished by their bulging sides, small windows, and airplane-like interiors. (Car and Locomotive Cyclopedia, 1970, p. 490)

motive and car, Pennsylvania engineers felt that faster deliveries and lower costs would follow. M.U. equipment offered better acceleration, full power no matter how many cars were in a train, and increased dynamic braking. Track maintenance would be eased by eliminating heavy locomotives. More studies followed, resulting in a detailed report that called for major track and

overhead wire improvements as well as new rolling stock. For 55 million dollars' worth of improvements, it was estimated that 150-mph schedules could be offered. At this speed the railroad could overwhelm its competition.

As an act of good faith, the High Speed Ground Transportation Act of 1965 pledged the Federal government to share in the costs of upgrading rail service in the Northeast corridor. The railroad, however, was expected to pay the major bills. The participants spoke of a renaissance of rail travel and optimistically announced that new service would begin in October 1967.

Fifty stainless-steel cars were ordered from the Budd Company in May 1966 at a cost of $435,000 each.[247] The coaches seated 80, the snack bar coaches 60, and the parlors 34. All measured 85 feet long by 10 feet, 6 inches wide over the curved sides (Figures 8.81 and 8.82). The coaches weighed 84 tons. Four 300-horsepower motors delivered an acceleration of 1.2 mph per second and were capable of propelling the cars at 160 mph. Because of the contemplated fast running, the underframes were covered over to protect electrical gear from flying ballast. The electrical system was complex and fragile. The controllers were minicomputers that automatically took over and precisely regulated the progression of power points, once the operator manually moved the lever to on. Deceleration was handled in the same fashion, and controls were so designed that dynamic braking (electric braking through the motors) would normally stop the train. The air brake was to be used only in emergencies. Although these elaborate systems were intended to ensure smooth operations, they were also prone to malfunction. The trucks were comparatively conventional, 8-foot, 6-inch wheelbase units with coil springs and equalizers produced by General Steel Industries. But even they proved troublesome in regular service.

The projected opening date came and passed with no Metroliners ready for service. Miles of wiring, resistors, and transistors proved fruitful sources of bugs. Most designs were untried and untested. No American firm had ever before attempted to build a high-speed electric rail car. The Metroliners worked satisfactorily at 120 mph, but developed serious problems in higher ranges.[248] Yet the specifications called for 150- to 160-mph operations, and the railroad was not about to accept the cars until they performed smoothly at that speed. It was unrealistic to expect such a complicated design to be worked out on the production line. More time and a prototype would have been the rational approach, but the Metroliner was not a simple matter of engineering. The Federal government had a hand in the project and was demanding an early and, in the opinion of some observers, unrealistic completion date. The axiom that politics and engineering rarely mix was proved in this case.

Finally on January 16, 1969, the first Metroliners were ready for service. By October enough cars were running to cover six round trips a day.[249] The public was pleased, and 700,000 passengers rode the trains during their first year. But the operating department remained unhappy. Breakdowns were frequent; 20 percent of the fleet was normally out of service. Sometimes as many as one-half of the cars were in the shop.[250] Even at the fastest regular operating speeds of 120 mph, the cars rode poorly. The speed was dropped back to 100 mph, which is the rate currently being observed. Track conditions were the primary cause for the orders to go slow. The multimillion-dollar track rebuilding program was abandoned after the Penn Central bankruptcy in June 1970. Limited maintenance did not help matters, but the cars' ride cannot be blamed entirely on the roadbed. The trucks were unequal to the job.[251] The coil springs were too stiff, and it was not possible to install anything much better within the confines of the side frames. Journal box wear compounded the problem. The secondary air springs caused the body to surge and did not cushion the ride.

In 1970 studies were begun to upgrade the electrical system and ride of the Metroliners. Several reworked cars were ready in 1974 that showed tangible improvements in both problem areas. Rewiring and roof-mounted exhaust grids for the dynamic brakes improved dependability. The original Metroliners could be expected to run 18,000 miles a month, while the improved model would cover 25,000 miles.[252] A monitor helped speed diagnoses of electric problems, many of which were simple enough to be corrected at the station platform rather than in the shop. But the major problem with the Metroliners remains how to find out what is wrong. Short of replacing the trucks, there seems no way to improve the ride. The most suitable substitute is a Swiss truck, the SIG type M, used successfully in Switzerland and France since the early 1960s. It features long, soft springs, bell crank axle boxes, and disc brakes. But these improvements are expensive: rewiring and new trucks would come to $500,000 per car, or more than the original purchase price.[253]

In retrospect, the Metroliner can be viewed as the greatest success and the greatest failure achieved by the multiple-unit cars. Metroliners proved a success in rebuilding rail passenger service, and they remain Amtrak's brightest star. Yet their large amount of downtime and high maintenance cost seem to have discredited the M.U. concept for high-speed train service. It is also ironic that so much money and effort was expended on special equipment that does not operate any faster than conventional trains. Amtrak's recent decision to purchase new electric locomotives and nonpowered cars is a demonstration of its disenchantment with the original Metroliners.

APPENDIX A: BIOGRAPHICAL SKETCHES OF CAR DESIGNERS AND BUILDERS

The lives of famous locomotive builders are comparatively well recorded. Anyone who has even sampled railroad literature recognizes the names of M. W. Baldwin, Richard Norris, and William Mason. But except for George Pullman and Webster Wagner, the names of car builders are unfamiliar and puzzling. Who was Benjamin Welch, for example, or James Leighton? The following sketches provide some answers, though this roster does not pretend to be definitive. In many cases the available information is incomplete or conflicting, but the notes offered here are at least a start and may encourage more investigation by other researchers.

In addition to the major figures in the design and production of the American passenger car, these selections cover less prominent men who were simply representative of their craft. Some of them devoted long years of service to their work although they made no spectacular contributions. In my opinion, material progress is not exclusively the creation of heroes and inventors; ordinary men are also necessary to the flow of the world's work.

Documentation is not offered in this section of the book, but the principal sources were the railroad trade journals. Among the standard biographical dictionaries that were consulted, the most helpful was the *Biographical Directory of Railway Officials of America* (first edition published in 1885; later titled *Who's Who in Railroads*). Obituaries in the *Master Car Builders Reports* were informative, and on occasion state and local histories provided data.

EVERETT E. ADAMS (1881–1954) was the designer of early and imaginative streamlined trains such as the City of Salina. He was born in Watertown, Massachusetts, and graduated from the University of California. He began railway service in 1905 on the Southern Pacific. Between 1913 and 1933 he worked for the Union Pacific, after which he joined the Pullman Car and Manufacturing Company. In addition to streamlined motor trains, he is credited with designing the duplex sleeping car.

FITCH D. ADAMS (1822–1904), one of the great personalities associated with American car building, was active in the trade for nearly fifty years and seemed to have had an opinion on every facet of it. At meetings of the Master Car Builders Association he could be expected to comment upon all papers given and generally seemed to delight in taking an opposite view from that of the speaker. He was a president of the organization and missed only one of its annual meetings.

Adams was born in Canterbury, Connecticut, and worked as a carpenter until 1848, when he entered the car-building shop at Norwich. In 1853 he joined the Buffalo Car Company as a subcontractor. He and his employer were wiped out by the 1857 panic. After a year in the lumber business, he reentered railway work at the Buffalo and Erie's car shop. He was soon made master car builder and stayed with the road until 1867. Then he became superintendent of the Ohio Falls Car Company, which failed in the spring of 1870. Adams joined the Boston and Albany and remained head of its car department at Allston, Massachusetts, until his retirement in 1896.

He built some of the lightest wooden cars ever constructed and was a firm believer in the arch truss. He was an equally outspoken advocate of the steel tired wheel, and seemed in general a man who was open to new ideas.

RICHARD N. ALLEN (1827–1890), inventor of the famous Allen paper car wheel, was born near Springfield, Massachusetts. His early years were spent as a locomotive engineer on lines in the South and Midwest. After the Civil War he began to try various occupations, such as inventing and well drilling in the Pennsylvania oil regions. Then his brother-in-law induced him to invest in and manage a strawboard mill in Pittsford, Vermont. The business was a poor one, and Allen began to seek new uses for his products. Somehow the unlikely idea of a paperboard-center steel tired wheel occurred to him. The first was produced in 1869. The invention was ridiculed at first, but Pullman endorsed Allen's design and the paper wheel became a success. Eventually two plants, in Hudson, New York, and Pullman, Illinois, were needed to fill the demand for Allen wheels.

WILLIAM C. ALLISON (1817–1891), who was born into a Quaker family of modest means in Chester County, Pennsylvania, built a great manufacturing business and fortune in Philadelphia. After apprenticing himself to a wheelwright, Allison opened his own wagon shop when he was only nineteen. He began building railroad cars around 1840 and developed his company into one of the largest car plants in the nation. In 1851 a blacksmith, John Murphy, who had supplied ironwork to Allison was taken in as a partner. The plant eventually became known as the Junction Car Works. Passenger car production was suspended in 1866, but the firm continued to build freight cars for many years.

WILLIAM P. APPLEYARD (1857–1905) was born in upstate New York and, unlike most early car builders, received a university degree. He graduated from Notre Dame's architecture school and worked in that trade for several years. In 1888 he joined Pullman as an inspector and became superintendent of repairs. Five years later he was made foreman of the New Haven's Old Colony division car department. In 1895 he was promoted to master car builder for the entire system. At about the same time he began work on his curious copper-sheathed passenger cars. In 1904 he returned to Chicago and the Pullman Company. When he was only forty-nine years old, he was run down and killed by a train at the Illinois Central Station.

WILLIAM C. BAKER (1828–1901), the originator of the renowned hot-water car heating system bearing his name, was born in Dexter, Maine. He went to New York City around 1850 and was soon attracted to the study of central heating. One of his earliest successful designs was for a hot-water heating system, patented in 1866, which was almost universally adopted for all first-class railway cars in North America. During his life Baker obtained over forty patents, most of which covered car heaters. He was also active in developing steam heating apparatus. He was accidentally killed by a train while visiting his daughter in the New York suburbs.

CHARLES E. BARBA (1877–19?), an important figure in steel passenger car design, was born in Freemansburg, Pennsylvania. In 1902 he joined the Pennsylvania Railroad engineering staff and worked with William Kiesel on designs for the P-70 class of passenger cars. He left Altoona in 1915 and later worked for the Midvale Steel Company and the Watertown Arsenal. Between 1922 and 1925 he was superintendent of the Osgood Bradley Car Works. During the following five years he served as mechanical engineer with the Boston and Maine Railroad. Barba prepared two valuable papers on passenger car construction. The first appeared serially in the 1907 *American Railroad Journal*; the second was published by the American Society of Mechanical Engineers in 1932.

JOHN H. BARKER (1844–1910) made the firm of Haskell and Barker into a major car builder. Born in Michigan City, he was the son of one of the founding partners of the firm and became president in 1878 after his father's death. Under his leadership the plant expanded until it was the largest single manufacturing establishment in the state of Indiana. It eventually became part of Pullman's car-building empire.

ELIAM E. BARNEY (1807–1880), one of the founders of the Dayton Car Works (better known by its later title of Barney and Smith), was born in Sackets Harbor, New York. He started his career as a teacher of classical literature and managed several schools in the East and Midwest. He settled in Dayton, Ohio, in 1843 to head the local academy, but his health failed after a few years and he resigned. Oddly, he decided to enter the sawmilling business as a cure. During these years he met Ebenezer Thresher, another intellectual turned industrialist. The two liked one another and determined to form a partnership in 1849. The result was the car works that was renamed Barney and Smith in 1864. The firm succeeded beyond the greatest hopes of its founders. For many years it was the second-largest car builder in North America and at one time employed 2,000 men. Barney's sons took an active part in the plant's operations. In 1892, however, the family turned the management over to outside interests, and the business went into a decline. Production ended in the early 1920s, and the firm was liquidated in 1923.

JACOB N. BARR (1848–1904), renowned for his contributions to cast-iron wheel technology, was also active in many other areas of rolling-stock design. He was born in Lancaster, Pennsylvania, and graduated from Lehigh University. He worked in the mechanical departments of several railroads, including the B & O and the Erie, but spent much of his career with the Milwaukee Road. Barr held sixteen patents, including a plan for a contracting cast-iron wheel mold and a design for a passenger car vestibule.

CHARLES BILLMEYER (?–1876) was a railroad car manufacturer of York, Pennsylvania, and a major partner in the Billmeyer & Small Company, proprietors of the York Car Works. (See David E. Small for more details.)

THOMAS A. BISSELL (1835–1902), one of the masters of wooden passenger car construction, was born near Erie, Pennsylvania. He was educated for the ministry but decided to go west and work on the railroad. He started as a bridge builder on the Burlington and soon graduated to the Aurora repair shops as a draftsman. In 1872 he joined Pullman and helped reorganize the car plant in Detroit. The following year he succeeded C. F. Allen as Pullman's mechanical superintendent and continued in that post until 1881, when he became superintendent of Barney and Smith. Five years later he took charge of the Wagner Sleeping Car Company's main shops in Buffalo. Declining health forced his retirement in 1895.

Bissell, a gifted car designer, did much to perfect wooden car framing. He had several patents covering vestibules, platforms, and other features. He was active in the Master Car Builders Association and served as a vice president. As the *Railroad Gazette* said at the time of his death, "His influence on the art of passenger car building has been of the first importance."

WILLIAM F. BIXBY (1857–1931) is credited with forming the American Car and Foundry Company. He was born in Adrian, Michigan, and began his railway career in 1873 as a baggageman. Twelve years later he entered the car-building trade as a lumber agent for the Missouri Car and Foundry Company of St. Louis. Within two years he was promoted to vice president and later took the presidency. As head of the company he became merger-minded. First he engineered a consolidation with the Michigan Peninsular Car Company of Detroit; then in 1899 he directed the combination of eighteen car builders into A.C.F. He became the first president of the new industrial giant, but was soon made chairman of the board and retired altogether in 1905.

During his retirement years Bixby earned two college degrees, helped to organize the American Red Cross, acted as a receiver for the Wabash Railroad, and took part in many educational and philanthropic enterprises.

JOHN E. BRADLEY (1860–1938), a grandson of Osgood Bradley, brought the Bradley Car Company successfully into the modern steel age. He left college in his second year and in 1882 joined the family business. After his uncle's death in 1896 he was made general manager, and when his father died five years later he became president and owner of the company. Recognizing that the firm must have a modern plant if it were to survive, he reorganized it as a joint stock company. The new shop was completed in 1910. Twenty years later, Pullman took over the business and Bradley retired.

Osgood Bradley (1800–1884), founder of the great car works in Worcester, was born in Andover, Massachusetts. His English ancestors came to the area during the time of the first Puritan settlements. After he completed high school, he was apprenticed to a carriage maker in Salem. In 1822 he moved to Worcester and opened his own shop. Like other carriage builders, he was drawn to railroad car construction, and by 1837 he was specializing in that line.

Despite the fact that his business survived to be handed on to his children, Bradley was in financial difficulties all his life. He failed several times and was considered a poor credit risk, if not a poor businessman, by Dun and Bradstreet. The company's fiscal affairs appear to have been managed more carefully after his passing.

Albert Bridges (1811–1881) was a partner in the car-building firm of Davenport and Bridges between about 1837 and 1851. For the next quarter century he operated a large New York City railway supply house, which specialized in all varieties of railroad-related hardware. (More on his car-building career is given in the sketch on Charles Davenport.)

John G. Brill (1817–1888) founded the Philadelphia-based J. G. Brill Company, which became the largest streetcar builder in North America. Brill was born near Cassel, Germany, and came to the United States in 1840 after learning the cabinetmaker's trade in Bremen. He found employment at Murphy and Allison's Philadelphia car works, where he rose to the position of foreman of the streetcar shop. In 1868 he left Murphy and Allison and started his own car plant. Within a few years he was building railroad and city cars.

The firm continued to grow and after 1900 began to acquire other car plants, including the Wason Car Company of Springfield, Massachusetts. After the coming of steel construction, Brill appears to have abandoned the manufacture of equipment for main-line service, except for the self-propelled rail cars described in Chapter 8. It continued as an important streetcar fabricator until 1941. Motor bus production continued until 1954, when the firm suspended the manufacture of transportation equipment in any form.

Edward G. Budd (1870–1946) did much to advance the lightweight passenger car and stainless steel fabrication. He was born in Smyrna, Delaware, received a high school education, and worked in various metal-manufacturing plants in the Philadelphia area. In 1912 he opened a small shop to make automobile parts and soon began turning out complete auto bodies for such major firms as the Dodge brothers. Later he went into the automotive wheel business. By the time Budd decided to enter railway car manufacturing in the early 1930s, he was a major industrialist. Within a few years he was accepting substantial orders for railway cars, much to the indignation of the established builders. In 1945 a huge plant built in Red Lion, Pennsylvania, by the government for war production was taken over by Budd's company for passenger car manufacture. Budd died the following year, but his firm produced such famous postwar trains as the California Zephyr. Car orders fell off dramatically after 1960, and in 1972 Budd announced its intention to close the Red Lion plant. However, new orders have since reversed that decision.

Charles Davenport (1812–1903), an important pioneer passenger car manufacturer, had a plant in Cambridgeport, near Boston. He was born in Newton, Massachusetts, and became apprenticed to a carriage maker. In 1832 he opened his own shop under the firm name of Kimball and Davenport to produce coaches and carriages. He entered the car-building trade about two years later. Within a few years his company was probably the largest single car manufacturer in the United States. In 1837 or 1839 Davenport's boyhood friend, Albert Bridges, became a partner. Some ten years later Lewis Kirk of Reading, Pennsylvania, joined the firm to supervise the construction of locomotives, a new sideline. This venture was not successful, however, and the business was reorganized with Charles Davenport as the sole partner. The firm recovered, but many other car plants were now active. In 1855 or 1856 Davenport decided to close the plant and retire with a comfortable fortune. He claimed to have produced 4 million dollars' worth of cars for fifty different railroads.

William A. Davenport (1831–1888) operated a large car plant in Erie, Pennsylvania. He was born in Schuyler County, New York, and began life modestly as a clerk. In 1866 he opened a car wheel foundry and two years later established the Erie Car Works. The plant was a major freight car builder but did not, so far as is known, make any outstanding contributions in the area of passenger car design. Davenport took an active part in the affairs of the Master Car Builders Association, an interest which was curiously rare among his rivals in the car building field.

Otho C. Duryea (1880–1941) was the inventor of the cushion draft gear underframe. He was born in Wyoming, Illinois, the younger brother of J. Frank and Charles E. Duryea, the American automotive pioneers. He began work on the cushion underframe around 1925. The device was used on passenger equipment as early as 1935, but its great importance came in more recent years, for jumbo freight cars could not have been built without it.

Frederick W. Eames (1843–1883) created the vacuum brake bearing his name. He was born in Kalamazoo, Michigan, the son of an inventive mechanic-farmer, Lovett Eames. Because Frederick named an experimental locomotive, which he used to demonstrate the vacuum brake, after his father, some careless writers have ascribed the brake to the father rather than the son. Eames received his first patent in 1874 and began to market the apparatus about two years later. (Other details are given in Chapter 7).

Eames was shot and killed while attempting to enter his Watertown, New York, factory during a strike. The firm survived and was reorganized in 1890 as the New York Air Brake Company.

ORSAMUS EATON (1792–1872) was a partner in the prominent car-building firm of Eaton and Gilbert, proprietors of the Troy Car Works, which was for many years a major passenger car producer. Eaton opened a carriage shop in 1820, and Uri Gilbert became a partner ten years later. The railroad cars that they added to the business a few years afterwards soon became their main product. Eaton retired in 1862, but the business continued under the able management of Gilbert.

JULIUS E. FRENCH (1836c.–1910), a prominent railway supply manufacturer, was born in Perry, Ohio. In the early 1870s he helped establish the Winslow Car Roofing Company in Cleveland. In 1882 he organized another firm in the same city to produce a steel tired wheel designed by W. H. Paige, and fifteen years later he helped consolidate the major manufacturers of tired wheels into the Steel Tired Wheel Company. The firm was reorganized in 1902 as the Railway Steel Spring Company.

LEANDER GARY (1827–1886) was a founder of the Master Car Builders Association and the longtime head of the New York Central's car department. He was born in York County, Maine. He followed his father into the carpentry trade, and when he was about twenty-four entered the service of the American Car Company in Seymour, Connecticut. He worked briefly for the Naugatuck Railroad and in 1855 was made master car builder for the New York and Harlem Railroad. In 1873 he was appointed the head of all car shops on the New York Central, a position calling for unusual administrative ability. This quality Gary demonstrated as president of the M.C.B. between 1874 and 1885. Although he was an inventive mechanic, he was not given to patenting his work and it is difficult to document his innovations. Because of illness, Gary retired early in 1885 and died in November of the following year.

J. TAYLOR GAUSE (1823–1898) was born in Chester County, Pennsylvania. He went to work at Harlan and Hollingsworth in 1843 as an errand boy and filled many minor positions there before beginning to receive advancement. Finally in 1858 he could afford to buy an interest in the firm. In 1867 he was made a vice president, and he became president with the death of Samuel Harlan in 1883. He was a careful manager, and the business maintained its leading position under his direction. He retired in

1896 but was recalled to head the plant two years later, just months before his death.

GEORGE GIBBS (1861–1940) is best remembered in railway car circles as a pioneer in the introduction of steel passenger cars. His designs for the first New York subways did much to convince steam railways that all-metal cars were a sound investment. Although he considered electrical engineering his major field, he was understandably proud of his pivotal role in passenger car design.

Gibbs was born in Chicago into a prosperous mercantile family. He graduated from the Stevens Institute of Technology and worked for Thomas A. Edison for several years. Then he took a position with the Milwaukee Road, where he did important work in train-board electric lighting and steam heating. In 1897 he joined the Baldwin Locomotive Works, which recognized the potential of heavy electric traction. In 1912 he established the firm of Gibbs and Hill, Consulting Engineers, and continued his work in the area of railway electrification.

URI GILBERT (1809–1888) was the major figure in the important Troy, New York, car-building firm of Eaton and Gilbert, later known as the Gilbert Car Works. A carriage maker by trade, Gilbert also proved to be a gifted manager, and after 1862 he became the main partner in the firm. The plant produced a great variety of cars, both freight and passenger. It had a large export business, and it took special pride in the elegant equipment it produced for the Wagner Sleeping Car Company. After Gilbert's death the firm went into a decline. The panic of 1893 was the final blow, and two years later the great plant closed down.

EDWARD E. GOLD (1847–1931), a key figure in the introduction of steam car heating, was born in Waverly, Illinois. He began manufacturing central heating appartaus in 1867, and in 1882 he invented a system of steam car heating that was widely used in North America. Gold took out 100 patents during his lifetime. (Whether he was related to Egbert H. Gold, inventor of the vapor system of car heating, has not been determined.)

JAMES GOOLD (1790–1879), who built some of the first passenger cars produced in this country, was born into a farming family in Granby, Connecticut. He worked briefly as a bookbinder but found carriage making more to his liking. In 1813 he opened his own shop in Albany, New York, where he produced the first cars for the Mohawk and Hudson Railroad in 1831. Though carriages were his principal product, he continued to build railroad cars until about 1870. He also constructed streetcars between about 1865 and the time of his death. The business was continued for many years by his descendants but was finally closed in 1913.

CHARLES A. GOULD (1849–1926) was a manufacturer of couplers and electric train lighting apparatus. He was born in

Batavia, New York, and worked for several years as a merchant. Apparently Gould was not a mechanic, but he had the knack of recognizing marketable items for the railway supply trade and developed an immense business in this area. He bought into a steam forge business in 1884 and began to turn out car couplers. Several years later he acquired rights to the platform and vestibule patented by Thomas A. Bissell of the Wagner Car Company. Next he opened a steel plant in the Midwest, and he built a giant factory for the production of car parts, locating it adjacent to the New York Central shops at Depew, New York (near Buffalo). When electric car lighting came in, Gould began to manufacture batteries. His properties were taken over by the Symington Company of Rochester, New York.

THOMAS A. GRIFFIN (1852–1914) created the huge car wheel manufacturing combine that remains the largest single producer in that line today. His father, Thomas F. Griffin, emigrated from Ireland as a boy and started a foundry in Rochester, New York, where he began making car wheels in 1847. In 1870 the father moved the business to Detroit. Thomas A. Griffin broke away from his father, opened an independent wheel foundry in Chicago, and in time bought out his father's company. He took over or established new wheel factories in St. Paul, Kansas City, Denver, Tacoma, Boston, and Los Angeles. Griffin was a pioneer in profit sharing and helped organize the Association of Manufacturers of Chilled Car Wheels in 1908.

JOHN M. HANSEN (1873–1929), a gifted mechanical engineer and businessman, was born in Sandy Lick, Pennsylvania, and attended the Western University of Pennsylvania. He took a job in the engineering department of Schoen Pressed Steel Company, later the Pressed Steel Car Company, where he worked on the designs of early steel freight cars. In 1902 with the backing of James B. (Diamond Jim) Brady and the Mellon family, he organized the Standard Steel Car Company, with a plant at Butler, Pennsylvania. In 1906 a second shop was opened in Hammond, Indiana. Within a few years Hansen acquired control of the Middletown (Pennsylvania) Car Company, the Baltimore Car & Foundry Company, the Osgood Bradley Car Company, and the Keith Car Company. He also opened plants in France and Brazil. Hansen retired in 1923, and in the year of his death his car-building empire was taken over by Pullman.

SAMUEL HARLAN, JR. (1808–1883) was a partner in the car-building firm of Harlan and Hollingsworth of Wilmington, Delaware. He was born in Glen Cove, Long Island, and spent his early manhood in Wilmington learning the trade of cabinet-making. In 1836 he became superintendent of Betts and Pusey, owners of a new railroad car shop—one of the first companies to organize with this specific product in mind. A year later Harlan became a junior partner, and by 1849 he was the main partner in a business that had flourished since its founding. It expanded to include a shipyard, and a new partner, Elijah Hollingsworth, gave the firm its mellifluous corporate name that was to become

a standard-bearer in the field. By 1860 Harlan and Hollingsworth employed over 600 men and had become one of the biggest car builders in America.

After Harlan's death the firm continued to be well managed, but it seems to have concentrated more on its ship-building business. In 1904 Bethlehem Steel acquired Harlan and Hollingsworth. Car production was suspended in 1939, but ship building continued for many years at Bethlehem's Wilmington, or "Harlan," Works.

WILLIAM T. HILDRUP (1822–1909), longtime superintendent of the Harrisburg Car Works of Pennsylvania, was born in Middletown, Connecticut, and began his career as a carpenter. Around 1842 he entered the employ of Osgood Bradley in Worcester, Massachusetts, and learned the car-building trade. Ten years later he transferred to the Elmira, New York, car plant, but during the following year he helped establish a shop in Harrisburg with the financial backing of eight other partners. Although the plant originally produced all styles of cars, around 1865 it began to specialize in freights. The works prospered for several generations, then failed in 1890 and does not appear to have reopened.

ELIJAH HOLLINGSWORTH (1806–1866) was born of a Quaker family near Wilmington, Delaware. He learned the machinist's trade in his native city, and when he was only in his middle twenties he was made foreman of Baldwin's new locomotive shop in Philadelphia. Returning to Wilmington in 1841, he became a partner in the car-building firm of Harlan and Hollingsworth.

RICHARD IMLAY (1784–1867), one of the first names associated with American passenger car construction, was born in Hartford, Connecticut. His father was in the importing business. How Imlay entered the carriage- and coach-building trade is uncertain, as are most facts about his life. Around 1828 he opened a shop in Baltimore and two years later began producing cars for the B & O. He appeared to have relocated in Philadelphia during the following year. In 1835 he built an eight-wheel car, the Victory, which included several remarkable features, including a buffet and a water closet. Two or three years later he built for the Cumberland Valley Railroad what is probably the first sleeping car ever produced. Despite these innovations, Imlay's business failed during the depression following the panic of 1837. His shop apparently closed in 1840, and he seems to have had no subsequent connection with railway cars except for a patent suit involving his alleged invention of the truck center plate.

JOB H. JACKSON (1833–1901) was a founding partner of Jackson and Sharp, operators of the Delaware Car Works in Wilmington. This plant, in combination with the Harlan and Hollingsworth factory, made Wilmington the car-building center of America before Pullman opened his great Chicago plant in 1881.

Jackson was born into a family of modest means in Chester County, Pennsylvania. He quit school at an early age and worked as a tinsmith and mechanic. In 1863 he and Jacob Sharp opened a small car shop, in which Jackson managed the office while Sharp, an experienced car builder, supervised the shop. In 1870 Sharp retired, and a joint stock company was formed. Jackson, a shrewd businessman, acquired most of the shares within a few years and developed the Delaware Car Works into a formidable enterprise. He added a shipyard in 1875 and became involved in the management of the Woodruff Sleeping Car Company. By the late 1880s the Jackson and Sharp works was producing 400 passenger cars a year. It also built streetcars and produced sash work.

The works was taken over by the American Car and Foundry Company in 1901. Within a few years A.C.F. concentrated its export work at the Wilmington plant and produced hundreds of wooden narrow-gauge cars. The plant closed around 1945.

ELI J. JANNEY (1831–1912), who invented the present-day car coupler, was born into a farming family in northern Virginia near Washington, D.C. He served in the Confederate Army and returned home so poor after the war that he gave up his farm and went into the dry-goods business. Apparently he became interested in coupler design when he heard about injuries suffered by a local trainman. Although Janney knew little of mechanics, he began whittling models and thinking about the problem. In 1868 he obtained a patent for a design that led nowhere, but five years later he received a second patent that clearly foreshadowed the hinged, knuckle type of coupler. In time the Janney coupler (sometimes erroneously called the "Jenny" coupler in the popular literature) was perfected and accepted by the industry.

JOSEPH JONES (1818–1879), recognizing the need for an association of car builders, called together a group of car masters in 1864 to discuss problems of mutual interest. He suggested that a more formal organization could be beneficial, and from this beginning the Master Car Builders Association emerged in 1867.

Born in South Wales, Jones came to this country when he was twelve years old. He was apprenticed to a carpenter and went to work for the Utica and Schenectady Railroad in 1846. He became head of the New York Central's main car shops at West Albany in 1865. Two years later he was given the honor of affixing the first signature to the constitution and by-laws of the M.C.B. He retired from railroad work in 1875 but remained active in the carpentry trade until a week before his death.

WILLIAM F. KIESEL, JR. (1866–1954), an important figure in the design of steel passenger cars, was born in Scranton, Pennsylvania. In 1887 he graduated from Lehigh University as a mechanical engineer, and in 1888 he began a forty-eight-year career with the Pennsylvania Railroad's mechanical department. In 1899 he was made chief draftsman and within ten years was promoted to mechanical engineer.

He was active in locomotive and freight car design, but it was his work in developing the P-70 steel coach that made him a notable contributor to the development of the American passenger car. Kiesel held 135 patents for improvements in railroad rolling stock. He retired in 1936.

JOHN KIRBY (1823–1915) served as master builder on the Lake Shore and Michigan Southern for twenty-two years. He was born in Oxfordshire, England, and was trained to be a coach maker. He emigrated to the United States in 1848 and worked in the car shop of the Syracuse and Utica Railroad for six years. Then he went west to take a job in the Michigan Southern's shops at Adrian, Michigan, where he became foreman after two years and head of the car department in 1870. In addition to repair work these shops constructed many new cars, including some elaborate palace sleepers.

Kirby was an early member of the Master Car Builders Association. He served one term as its president just before his retirement in 1892 and was elected treasurer for several terms between 1900 and 1909.

JAMES T. LEIGHTON (?–1892) was a second-echelon car builder whose wandering career typifies that of many railroad men. No record of his early life can be discovered, except for the fact that he was born in Portland, Maine. In 1873–1874 he took out several patents for improvements in sleeping cars. He and his brother founded the New Haven Car Company, where they built cars of all types. They also operated their sleeping cars for Wagner and for at least one Southern line until about 1879. Leighton ran his own line of sleeping cars in the New Haven area until about 1885, when he went to work for Jackson and Sharp. In later years he was employed by the Baker and Sewall car heating companies.

JOHN S. LENTZ (1847–1935) was employed by the Lehigh Valley Railroad's engineering department for a record seventy years. He was born in Mauch Chunk, Pennsylvania, and began work on the Lehigh Valley in 1865. He was promoted to the position of master car builder in 1881 and held the job under various titles until 1928. After this time he acted as a consultant to the car department. Lentz was also active in the Master Car Builders Association and served as its president for the 1894–1895 term.

JOHN LIGHTNER (1811–1896) was employed in the car shops of the Boston and Providence Railroad for fifty-five years. He was born in Baltimore and learned the trade of carriage making. After working briefly for the B & O, he was hired in 1833 by the Boston and Providence as a car repairman. In 1835 he was promoted to master car builder at the road's Roxbury shops, a position he held until retiring in 1888. Lightner made many mechanical contributions, the most important of which was a journal bearing box patented in 1848.

GEORGE G. LOBDELL (1817–1894) was a principal partner in the prominent car wheel firm of Bush & Lobdell. He was born in New York City and was apprenticed in 1832 to his uncle, Jonathan Bonney, a pioneer car wheel maker. When Bonney died in 1838, Lobdell succeeded him as a partner in the Wilmington, Delaware, firm. He became a leading manufacturer and a recognized authority in his field.

EDWIN LOCKWOOD (1810–1881), master car builder for the Camden and Amboy Railroad, was born in Orange County, New York, and began his working life as a wheelwright. In 1832 he was hired by the Stevens family, who owned the C & A, to work on the Hoboken ferry. A year later they transferred him to the C & A shops at Bordentown, New Jersey. In 1837 he was made foreman of the car department, a position he held until his retirement in 1875. The cars of the Camden and Amboy differed from those of most other early U.S. lines because of the exposed truss framing beneath their bodies and the use of six-wheel trucks many years before they were common elsewhere.

WILLIAM McCONWAY (1842–1925) was a leading figure in the introduction and manufacture of the Janney coupler. He was born in Ireland but came to the United States as a boy. He entered the malleable iron casting business and by 1866 was a junior partner in a Pittsburgh iron firm. Four years later he and John J. Torley established their own company, which still produces materials for the railroad industry. McConway became involved with Janney during the formative years of the automatic coupler, and in 1884 he incorporated the Janney Car-Coupling Company. All the shares were owned by McConway and his associates; ironically, Janney was not a member of the firm. Without the technical expertise of McConway and Torley, however, the Janney coupler might not have become a practical reality.

WILLIAM R. McKEEN, JR. (1869–1946) did outstanding work in developing early gasoline-propelled rail passenger cars. He was born in Terre Haute, Indiana, the son of a prosperous banker. He received degrees from Rose Polytechnic Institute and the University of Berlin. Entering railway service in 1891, he worked for several Midwestern lines until 1898, when he joined the Union Pacific. In 1908 he was made superintendent of machinery. Three years earlier he had built a gasoline rail car which proved so promising that it was put into commercial production. Over the next twelve years about 150 of his motor cars were manufactured. McKeen was unquestionably a mechanical genius; one account claims that he was granted 2,000 patents.

EZRA MILLER (1812–1885) invented a car coupler and platform that was popular on American passenger cars in the 1870s and 1880s. He was born in Pleasant Valley, New Jersey, and studied and practiced civil engineering. He joined the New York State Militia and rose to become a colonel—a title he enjoyed using for the rest of his life. In 1848 he moved to Wisconsin to survey Western land. A few years later he began to study the car coupler problem, an interest presumably aroused during the extensive rail travel required by his work. His invention slowly matured, and he obtained patents for it in 1863 and 1865. The device proved successful and was widely used before the introduction of the Janney coupler. Miller's outgoing, politically minded personality was an asset in achieving the rapid and widespread acceptance of his device.

KARL F. NYSTROM (1881–1961) made important advances in the design of lightweight passenger cars. He was born in Sweden and worked his way through school, earning a degree in mechanical engineering in 1904. The following year he came to the United States. He worked as a draftsman and engineer for Pressed Steel, Pullman, A.C.F., and several railroads before joining the Milwaukee Road in 1922. By this time he was concentrating on car design. His work in developing lightweight, all-welded cars is described in Chapter 2. He retired in 1949, a recognized leader in his specialty.

WILLIAM H. PAIGE (1842–1885), for many years superintendent of the Wason Car Company, held several patents for sleeping-car improvements that were used by Wagner. Around 1881 he resigned his post at Wason to devote his time to a firm organized to produce steel tired wheels of his design.

GEORGE M. PULLMAN (1831–1897), the only name connected with railroad cars that is known to the general public, is best remembered for something he did not invent—the sleeping car. As the success story of a self-made man, his life can hardly be rivaled; it was filled with achievement and reward. He did much to upgrade and civilize railway travel. He created a vast and intricate business empire that built, serviced, and operated a huge fleet of sleeping cars. He was a man of incredible vision, energy, and judgment. It is curious that his fame rests on his modest mechanical skills, while his great gifts as an organizer and capitalist receive scant attention.

Pullman was the third of ten children born in Brocton, New York, to a house builder and mover of small means. George left school when he was fourteen and worked as a carpenter and cabinetmaker. He started his own house-moving business when he was 22 and did well in it. A few years later he transferred the business to Chicago, where he entered the sleeping-car trade and began the brilliant career that is described in Chapter 3.

E. J. W. RAGSDALE (1885–1946) was chief engineer of the Budd Company's railway car division. Born in San Francisco, he traveled widely in the Far East and led an adventurous life before joining Budd shortly after World War I. He developed the Shotwelding process vital to the manufacture of rivetless stainless-steel cars. Ragsdale was an advocate of disc brakes and did much to encourage a departure from traditional passenger car design.

WILLIAM E. RUTTER (1812–1882) operated a car plant in Elmira, New York, for a brief period and worked as a car builder for several railroads. He was born in Providence, Rhode Island, and learned the carpentry trade in Baltimore. In the early 1830s he took a job with the Boston and Providence and in later years became master mechanic on the Stonington Line. In 1851 he moved to Elmira and opened a car repair shop to service the newly constructed New York and Erie Railway. A year later he began to build cars. In 1858 he sold out to the Erie but stayed on as the manager. In 1870 the works was leased to the Erie and Atlantic Sleeping Car Company, an Erie-sponsored corporation that was quickly absorbed by Pullman. The Elmira facility was used by Pullman for repairs until 1886.

Rutter's son, James H. (1836–1885), became a favorite of the Vanderbilts and served as president of the New York Central.

CHARLES T. SCHOEN (1844–1917), who was prominent in the development of steel freight cars, also made a notable contribution to passenger equipment by introducing wrought-steel wheels. Schoen was born in Concord, Delaware, and worked with his father in the cooper's trade. He went to Philadelphia at the end of the Civil War to learn the metal-fabricating business. In 1892 he established a plant in Pittsburgh that made pressed-steel car parts, and soon he began to manufacture freight cars that revolutionized freight car design. The business was reorganized as the Pressed Steel Company in 1898 and became an important producer of steel passenger equipment in later years. Schoen was forced to resign from the car plant in 1901. He opened a wheel plant at Butler, Pennsylvania, but sold out to U.S. Steel in 1907.

CHARLES A. SELEY (1856–1939) drafted the basic specifications for the railway post office car. Their aim was to set minimum-strength standards that would safeguard railway mail clerks, but the specifications were eventually adopted by the railroad industry for all passenger cars.

Seley was born in Wapella, Illinois, and entered railway service as a draftsman in 1876. By 1899 he was a mechanical engineer with the Norfolk and Western. He was elected president of the Western Railroad Club in 1907. In later years he went into the supply trade.

HENRY H. SESSIONS (1847–1915) is popularly (but incorrectly; see Chapter 5) remembered as the inventor of the vestibule. He was born in Madrid, New York, the son of Milton Sessions, who was master car builder for the Central Vermont Railroad and later superintendent of the Taunton Car Works. The son served an apprenticeship under his father and then struck out on his own, working for several railroads in New York, Iowa, and Texas. In 1885 he was made superintendent of the Pullman plant in Chicago and served in this post until 1892. For several years he worked as a consultant to Pullman, then began to devote full time to the affairs of the Standard Coupler Company.

Although he was not the originator of the vestibule, Sessions made important contributions to wooden passenger car design during its last decades. His ingenious metal reinforcing scheme did much to strengthen this form of construction.

JACOB F. SHARP (1815c.–1888) was a partner in the car-building firm of Jackson and Sharp. He was born in Hunterdon County, New Jersey, and worked as a carpenter and bridge builder before taking a job in 1840 at the Harlan and Hollingsworth car plant in Wilmington, Delaware. In 1863 he and Job Jackson founded the Delaware Car Works. However, Sharp left the firm in 1870, several years before it emerged as a major car builder.

DAVID E. SMALL (1824–1883) was the leading partner in the York, Pennsylvania, car-building firm of Billmeyer and Small. His father was in the construction and lumber business. It was not a major transition to expand the sash mill into a car shop, and in 1852 Charles Billmeyer was taken in to help organize the new firm, originally called the Etna Car Works but later better known as the York Car Works. The plant grew rapidly into a major car builder, a position it held until about 1885. For many years it specialized in narrow-gauge cars, and members of both founding families remained active in its affairs. Car production was suspended around 1910.

CALVIN A. SMITH (1822–1903) may have been the inventor of the side-panel truss for wooden passenger cars. Born in Newfield, Maine, Smith first worked as a carpenter. Around 1840 he found employment at a car shop in Lawrence, Massachusetts. In 1851 he began to work for the Erie and was soon made foreman of the main car shop at Piermont, New York. He is said to have built one of the first parlor cars, the *Metropolis*, sometime before leaving the Erie in 1878. Between that date and his retirement in 1902, he served as superintendent of the Union Tank Car Line.

PRESERVED SMITH (1820–1887) was a partner in the Dayton Car Works, also called Barney and Smith, between 1864 and 1877. Like the other major partners in this important firm, Smith was an educated man rather than a practical mechanic. A native of Warwick, Massachusetts, he was active in railroad management before joining the car building of Barney and Smith.

WILLIAM W. SNOW (1828–1910), a prominent car wheel maker, was a founder of the famous Ramapo Car Wheel Works. He was born in Heath, Massachusetts. Snow considered becoming a minister, instead went into the book-binding trade, then in 1848 joined his brother as an iron founder. Soon he was deeply involved in the mysteries of car wheel making. In 1859 he became general manager of the Union Car Wheel Works in Jersey City, which was a major wheel manufacturer. Six years later he opened his own plant at Ramapo, New York. The business was a success, and it grew into one of the largest independent car wheel makers in North America. Since Snow was dedicated to the manufacture of a first-class product, he turned to making track materials when competitors drove down the price of wheels so that it was no longer possible to offer top-quality merchandise. Snow was a founder of the Steel Tired Wheel Company and the American Brake Shoe and Foundry Company.

JOHN STEPHENSON (1809–1893), who is best remembered as a major producer of streetcars, also manufactured steam railroad cars earlier in his career. He was born in northern Ireland and was brought to this country two years later. His parents had him educated for a mercantile position, but he persuaded them to apprentice him to a carriagemaker when he was nineteen years old. Two years later he opened his own shop, specializing in the construction of omnibuses, and this naturally led to the manufacture of railroad cars. Around 1837 he introduced a peculiar body style that used a lattice frame.

The long financial depression following the panic of 1837 forced Stephenson to close his shop in 1842. He blamed his ruin on the worthless notes issued by his railroad customers, and vowed that he would never again build a railroad car. Technically he remained true to his word, but he did eventually reenter the trade by producing streetcars.

WALTER V. TURNER (1866–1919), an air brake expert who made many contributions in this field, was born in England. He worked first in the woolen textile trade, then emigrated to the United States in 1888 to take a job with a New Mexico ranch. Turner went into ranching for himself but failed, and he found employment as a car repairman with the Santa Fe Railway. From this unlikely beginning he rose to the top of his profession as a leading authority on pneumatics. He became fascinated with the air brake, quickly mastered its theory and mechanics, and began to patent improvements. His ideas were so impressive that in 1903 George Westinghouse hired him and within seven years made him a chief engineer. After about 1890 Westinghouse himself was no longer active in air brake design; it was associates like Turner who devised improvements in this area. For example, Turner developed the K triple valve, a longtime standard for freight cars. He also perfected the LN passenger car brake and the universal valve that replaced the triple valve. In all, Turner acquired some 400 patents.

WILLIAM VOSS (18?–19?) was the only practical car designer to write a book on the subject. He was born in Germany and graduated from a technical school. Later he came to the United States and worked as master car builder on the Burlington, Cedar Rapids, and Northern Railroad. In 1889 he became an assistant engineer with the Fox Pressed Steel Company. Next he served as assistant superintendent of the Barney and Smith Car Works, and in 1898 was appointed superintendent of Jackson and Sharp, where he worked until at least 1908. Little more information seems to be available on his life and career, even though his book, *Railway Car Construction* (1892), was the only one published in this field for many years.

WEBSTER WAGNER (1817–1882), sleeping-car pioneer and founder of the second-largest sleeping-car company in America, was born into a farming family in Palatine Bridge, New York. He learned the wagon-making trade from his elder brother, with whom he formed a partnership. Their business failed, however, and Wagner became a station agent in his home town. In 1860 he quit that job to devote more time to some sleeping cars that he had begun to manage on the New York Central Railroad. He became a protégé of Cornelius Vanderbilt, soon took over the management of the sleeping cars on the huge Vanderbilt system, and within a few years became a wealthy man.

In 1870 Wagner was elected to the New York State Assembly, and later he gained a seat in the State Senate. He looked the part, since he was "a white-haired, white whiskered but fresh and vigorous man," according to one description. Ironically, he was killed aboard one of his own cars in a terrible rear-end collision at Spuyten Duyvil, New York.

NATHAN WASHBURN (1818–1903) designed the standard pattern of the double-plate cast-iron car wheel. He was born on a farm in Stafford, Connecticut, and worked as a carpenter and iron founder in his early years. He organized his own foundry in 1840 and operated it in several locations. By 1860 Washburn was a well-established manufacturer with facilities in Worcester, Massachusetts, Troy and Schenectady, New York, and Toronto, Canada. These works produced pulleys, cotton machinery, railroad rails, and gun barrels, in addition to car wheels. After 1865 he decided to concentrate on car wheels. Later he acquired property with iron ore on it and developed a steel tired wheel.

THOMAS W. WASON (1811–1870) was born in Hancock, New Hampshire, the son of a carpenter and farmer who was too poor to give his children many years of schooling. Wason found work in Massachusetts as a carpenter, but in 1845 he resolved to start his own business in Springfield. His brother Charles joined him, and together they began to build freight cars with their own hands. The enterprise grew slowly. In 1848 the brothers leased a portion of the defunct Springfield Car and Engine Company. Charles left the firm in 1851 to open his own car plant in Cleveland, and Wason took in several new partners. His business expanded until it had become a major car builder by the time of Wason's death. The Wason Car Company continued as a major producer for many years but appears to have largely dropped out of steam railroad car building by 1900. Brill purchased the firm in 1907, and production was concentrated on street railway vehicles until the plant closed in 1932.

BENJAMIN WELCH (1827–1913), a pioneer West Coast car builder, carried the traditions of the East to the workshops of California. He was born in Peaks Island, Maine, and entered railway service in 1846. He worked briefly for the Portland Company, a maker of locomotives and cars, before setting out for the California gold fields. Welch failed as a miner and was glad to find a position in the small car shop of the Sacramento Valley

Railroad. In 1863 he took the job of car builder for the Central Pacific, whose Sacramento shops became the largest manufacturing establishment in the Far West. Scores of passenger cars, including some elegant private cars, and hundreds of freight cars were built under Welch's direction. He retired in 1902.

GEORGE WESTINGHOUSE, JR. (1846–1914) did not invent the air brake so much as introduce and develop a workable version of it. He was born in Central Bridge, New York, of German and Dutch-English parents. His father operated an agricultural-tool factory in Schenectady, and George received a practical schooling in the shop. He was a remarkably large, strong, and intelligent young man with an unmistakable gift for mechanics. When he was only nineteen years old he received the first of nearly 400 patents that he would take out during his lifetime. The early success of the air brake financed Westinghouse's explorations in electricity, natural gas, railway signaling, and steam turbines. He made significant contributions in all these areas.

Westinghouse was not only an imaginative and original engineer but also a skillful businessman. His great manufacturing empire failed during the panic of 1907, and a few years later he lost control of two major divisions. However, Westinghouse remained active in the air brake and signal divisions until the time of his death.

ASA WHITNEY (1791–1874), a major manufacturer of car wheels, was born in Townsend, Massachusetts, the son of a blacksmith. He followed his father's trade and worked in various small shops and factories in the Northeastern states. Around 1830 he entered the employ of the Mohawk and Hudson Railroad to make machinery and cars. Within three years he became superintendent of the line, and his subsequent progress was swift. He was appointed to the office of canal commissioner for the state of New York. In 1842 he became a partner with the Philadelphia locomotive manufacturer, M. W. Baldwin. Four years later he left Baldwin to open a car wheel plant, which prospered and became for a time the largest in North America. He served briefly as president of the Philadelphia and Reading Railroad but was forced to retire in 1861 because of failing health.

WILLIAM H. WOODIN (1868–1934) was for many years an official of the American Car and Foundry Company. His grandfather had been a founding partner of the old car-building firm of Jackson & Woodin of Berwick, Pennsylvania. After this plant merged with A.C.F. in 1899, Woodin worked for the parent corporation and in 1916 became its president. He also headed the American Locomotive Company, and he remained on the boards of both concerns for many years. Woodin, a prominent figure in business and financial circles, served under President Franklin Roosevelt as Secretary of the Treasury.

JONAH WOODRUFF (1809–1876), the elder brother of the sleeping-car pioneer Theodore T. Woodruff, also played a leading role in sleeping-car development. He was born in Watertown, New York, left school at the age of ten, and spent his early years on the family farm. He took up portrait painting with some success, and appears to have made a living at it. In 1838 he served as a colonel in the Canadian Rebellion's "patriot army" and was often called by this title afterwards. During the late 1850s he joined his brother's sleeping-car business and eventually showed a greater talent for the management of the firm than its founder. He helped organize the Central Transportation Company, which for a time seemed destined to dominate the sleeping-car trade in North America. After Pullman leased the Central Transportation Company in 1870, Woodruff decided to start over again and organized his own line. He had already been involved in sleeping-car design, and during his career he obtained several patents in the field. He is best remembered for devising the Silver Palace and Rotunda sleeping car designs.

Woodruff's new company, while it was not a threat to Pullman's empire, was expanding when his health gave way in the mid-1870s. Early in 1876 he went to Bermuda hoping to recover, but he died there a few months later.

THEODORE T. WOODRUFF (1811–1892) is, after Pullman, the popular choice as inventor of the sleeping car. Born in Watertown, New York, of New England parentage, he stayed on the family farm until he was sixteen years of age. Things mechanical were his interest, and he was pleased to leave the land to his parents' care and become a wagonmaker's apprentice. Around 1830 he took work in a local foundry. Expanding his knowledge of the mechanical trades, he demonstrated an inventive bent by designing a mowing machine. When the railroad was built as far as Watertown, Woodruff found a new outlet for his skills. Eventually he gained a practical knowledge of car building and constructed his first sleeping-car model. He went west in 1856 to take a job as master car builder for a railroad in Illinois. His real interest was in promoting his sleeping-car design, however, and within a year he left his new position. He secured patents, built a test car, and organized his company, as described in Chapter 3.

After retiring from the sleeping-car business in 1864, Woodruff lived the life of a wealthy gentleman for a time and built a gorgeous mansion in Mansfield, Ohio. But the old life was more to his taste, and in 1870 he returned to Philadelphia to take over a foundry in Norristown. The panic of 1873 ruined the business, however, and Woodruff was forced into bankruptcy. He bore his failure heavily and spent his last years in a flurry of patents and business schemes, vainly attempting to remake his fortune. A white-haired old man of eighty-five, he was on a business trip to promote a ship's propeller when a train struck and killed him in New Jersey.

APPENDIX B: STATISTICS

The most frustrating problem encountered during the research for this book was the lack of reliable statistics on the American railway passenger car. A scarcity of figures covering the early cars was to be expected, but discrepancies in the statistics available for the modern period make it very difficult to trace the history of special types of cars, such as the parlor. Certainly a large body of statistical data is available for the years after 1880; sources include the Census reports, Interstate Commerce Commission reports, *Poor's Manual of Railroads,* the railroad trade press, and the American Railway Car Institute. Yet they rarely agree, and many of the discrepancies are not small (see Table B.2). For example, there is a difference of 85,000 between the 1880 Census report of 455,000 freight cars in the U.S. fleet and the 1880 *Poor's* report of 540,000. Presumably *Poor's* included all service, work, and privately owned freight cars, while the Census counted only revenue cars owned by operating railroads. There is no way to be certain, however. Passenger car totals are likewise confused by various sources, inclusion or exclusion of Pullman, baggage, express, and mail cars. Also, some sources incorporated Canadian cars in U.S. totals; *Poor's* often did so. In general, governnment statistics (the Census and the I.C.C.) showed smaller totals than private reporters such as *Poor's* or the *Railroad Gazette.* This might be explained by the government agencies' placid acceptance of "information received," whereas the private publications, having more contacts within the industry, apparently compiled more complete reports. By the early years of the twentieth century, more general agreement prevailed. The discrepancies among sources were still noticeable, particularly in regard to figures on new cars produced. For example, in 1946 the A.A.R. reported delivery of 481 new passenger cars, while the figure given by the American Railway Car Institute was 1,329.

No official statistics are available for the early years of the railroad industry. Table B.1's totals of 900 passenger cars for 1840, 3,000 for 1850, 10,000 for 1860, and 13,000 for 1870 are simply estimates projected from known railroad mileages. During this period several crude reckonings were attempted, but they rarely agreed. The 1855 New York State Railroad Commission Report estimated that there were 10,120 passenger cars and 41,786 freight cars. *The American Railway Review* of September 18, 1859, claimed that there were 5,900 passenger cars and 114,500 freight cars. On April 18, 1861, the same journal revised the figures to 12,272 passenger and 114,750 freight cars, based on a rule of thumb of about one-third of a passenger car and four freight cars per mile of track. Perhaps this rule was reasonably accurate for the early period, but it does not agree with later mileages and known car figures. Compared with 1880 figures, for example, the formula overestimates passenger cars by 13,000 and underestimates freight cars by nearly 200,000! Albert Fishlow attempted to resolve the problem in his economic history of American railroads before 1860.[*] His totals are based on the passenger revenue figures presented in the published reports of the Massachusetts and New York railroad commissions for 1849. He concludes that there was a national total of 500 passenger cars in 1839, 1,400 in 1849, and 4,900 in 1859. But these totals seem too low, particularly when compared with the mileage formula given above. And both methods may produce data wide of the actual totals.

Perhaps some more accurate measure of our pioneer fleet can be gained from using a representative trunk line as an example. The Baltimore and Ohio's annual reports gave far more data on its rolling stock, including breakdowns by types of cars, than other roads did. Following are some statistics from selected years:

B & O Report, 1836, with about 120 miles of line (p. 74).
1,062 freight cars (82 of which are on 8 wheels)
46 coaches (27 of which are on 8 wheels)

B & O Report, 1847, with about 220 miles of line (pp. 46–47).
983 freight cars
54 passenger cars (23 coaches, 9 ladies' cars, 3 smokers, 7 baggage, 1 sleeper, 3 mail, 2 emigrant, 6 express 6-wheelers)

B & O Report, 1851, with about 250 miles of line (table H).
1,443 freight cars
78 passenger cars (34 coaches, 11 ladies', 7 smokers, 12 baggage, 7 mail, 7 excursion)

B & O Report, 1865 (p. 56).
3,398 freight cars
164 passenger cars (130 coaches, 24 smokers, baggage, & mail, 8 office, 1 Post Office, 1 pay car)

B & O Report, 1876 (p. 58).
12,072 freight cars
353 passenger cars (237 coaches, 58 baggage, 13 Post Office, 6 palace [parlor?], 6 sleepers, 10 office, 9 emigrant, 2 pay cars, 12 combines)

The tables given in this appendix are only general indicators; their figures should not be taken as absolute. Because of the conflicting nature of the evidence, no effort was made to determine exact numbers. Instead, the highest figure was generally accepted and rounded off to the nearest tenth or hundredth. Besides the fact that the numbers are approximations, the tables may contain statistical defects, for the author does not pretend to be an expert in this area. As they stand, the tables represent the most satisfactory arrangement of figures that could be assembled by a layman threading his way through "lies, damn lies and statistics."

[*] *American Railroads and the Transformation of the Ante-bellum Economy,* Cambridge, Mass.: Harvard University Press, 1965, pp. 419–421.

Table B.1 American Passenger Cars by Type, 1840–1970

YEAR	COACH	PARLOR	DINER	BAGGAGE	MAIL°	SLEEPING	COM-BINE	OFFICE	SELF-PRO-PELLED	OTHER	TOTAL
1840											900e
1850											3,000e
1860											10,000e
1870	9,000e			3,500e		400†e				100	13,000e
1880	12,500	130‡	40	4,000		1,300†				430	18,400
1890	19,000	370‡	160‡	6,700	3,000e	2,100†	350			720	32,400
1900	20,000	500‡	350	6,000	3,500e	3,500†	4,400	600		250	38,200
1910	25,800	1,350†	950	8,220	1,500e	4,500†	5,700	780		3,000	52,300
1920	29,300	1,570†	1,350	12,100	1,240	7,100†	5,680	950		4,610	63,100
1930	25,100	1,760†	1,760	12,900	950	8,500†	5,270	880	3,660	2,320	63,900
1940	15,200	1,370†	1,530	12,000	1,500	6,400†	2,500	570	4,100	30	45,200
1950	14,200	730†	1,800	13,600	1,750	6,100†	2,000	560	2,650	110	43,500
1960	10,250	300	1,170	11,400	830	2,600†	750	400	2,900		30,600
1970	4,030	210	570	3,710	320	800	80	210	2,630		12,560

° Mail car figures are confused by designation of R.P.O. and combined mail-baggage cars as mail cars.
† Includes Pullman
‡ Less Pullman
e Estimate

Source: Data from I.C.C. statistical reports, Association of American Railroads yearbooks and statistics on class 1 railroads, *Poor's Manual of Railroads*, and other sources mentioned in the opening statements of this appendix.

Table B.2 American Railroad Car Statistics from Various Sources, 1880–1960

YEAR	U.S. CENSUS	POOR'S	AM. RLY. CAR INST.	A.A.R.	I.C.C.
1880	16,805 P	17,575 P			
	455,450 F	539,255 F			
1890	26,820 P	30,211 P			
	918,491 F	1,061,970 F			
1900	34,713° P	34,995° P			34,713° P
	1,365,531 F	1,350,258 F			1,416,125 F
1910	47,179° P	51,158 P	47,179° P		47,095° P
	2,148,478 F	2,297,620 F	2,135,121 F		243,236 F
1920	56,102° P		56,102° P	53,501 P	56,102° P
	2,388,424 F		2,417,999 F	2,322,122 F	2,388,424 F
1930	53,584° P		53,584° P	52,130 P	53,584 P
	2,322,267 F		2,352,046 F	2,276,867 F	2,322,267 F
1940	38,308° P		38,308° P	37,817 P	38,308° P
	1,684,171 F		1,706,387 F	1,653,663 F	1,706,387 F
1950	43,372 P		37,359° P	37,146 P	37,359° P
	1,745,778 F		1,769,983 F	1,721,269 F	1,769,983 F
1960	25,746 P		25,746° P	25,655° P	26,947° P
	1,690,000 F		1,709,549 F	1,676,629 F	1,690,396 F

P—Passenger cars
F—Freight cars
° Less Pullman

Source notes: U.S. Census figures are from *Historical Statistics of the United States: Colonial Times to 1957*, 1960 (see also supplement to 1962). Figures offered disagree from those in original published reports of the Census. (In 1890, for example, the original reports claim 27,653 passenger cars.)

Poor's Manual of Railroads figures for 1880 are from 1892 edition; figures for 1870, 1900, and 1910 are from 1911 edition.

Table B.3   Passenger Car Prices, 1831–1960

TYPE	YEAR	R.R. OR BUILDER	SOURCE	PRICE OR COST[*]
4-wheel coach	1831	B & O	Annual report	$ 700
4-wheel coach	1840c.	All	Von Gerstner†	800–900
8-wheel coach	1840c.	All	Von Gerstner†	2,000
8-wheel coach	1843	All	Baldwin papers, Hist. Soc. of Penna	1,700
8-wheel coach	1845	Eastern	A.R.R.J., Oct. 9, 1845	2,200
Coach	1848	Louisa	R.&L.H.S. Bulletin 65 (1945)	1,350
Coach	1850	P R R	J. E. Thomson letter book, P R R Library	1,850
Sleeper	1860	Wagner	Am. Rly. Rev., May 31, 1860	3,500
Sleeper	1863	Pullman	West. R.R. Gaz., July 18, 1863	6,500
Sleeper	1866	Central Transportation	R.R. Record, Jan. 18, 1866	8,000
Parlor	1867	Lake Shore	R.R. Car Jour., Nov. 1894	17,500
Private	1875	D. O. Mills	N.C.B., Jan. 1876	15,000
Coach	1875	All	N.C.B., May 1875	4,800
Baggage	1879	All	N.C.B., Sept. 1879	2,200
Coach	1880	P R R	Rly. Age, Nov. 11, 1880	5,500
Sleeper	1882	Pullman	Rly. Rev., Aug. 12, 1882	14,000
Coach	1882	Pullman	Rly. Rev., Aug. 12, 1882	7,500
Diner	1886	All	R.R.G., Jan. 1886	12,000
Coach	1887	Southern Pacific	N.C.B., Sept. 1887	4,500
Coach	1892	All	R.R.G., Aug. 5, 1892	5,000
Drawing Rm.	1892	All	R.R.G., Aug. 5, 1892	12,000
Coach	1906	All	R.R.G., Dec. 28, 1906	8,000
Coach	1907	All	R.R.G., Dec. 27, 1907	8,500
Coach	1915	All	Rly. M.E., June 1915	12,800
Mail	1915	All	Rly. M.E., June 1915	11,000
Sleeper	1915	All	Rly. M.E., June 1915	22,000
Business	1915	All	Rly. M.E., June 1915	15,000
Motor	1915	All	Rly. M.E., June 1915	20,000
Baggage	1915	All	Rly. M.E., June 1915	8,500
Coach	1920	A C L	Rly. Age, Jan. 7, 1921	34,036
Baggage	1920	Ill. Cent.	Rly. Age, Jan. 7, 1921	26,650
Diner	1920	Ill. Cent.	Rly. Age, Jan. 7, 1921	51,005
Postal	1920	C & N W	Rly. Age, Jan. 7, 1921	27,500
Coach	1925	N Y C	Rly. Age, Jan. 2, 1925	27,625
Diner	1925	F E C	Rly. Age, Jan. 2, 1925	48,500
Baggage	1925	M P	Rly. Age, Jan. 2, 1925	18,168
Business	1925	Reading	Rly. Age, Jan. 2, 1925	84,104
Coach	1930	All	Military Eng., Sept.–Oct., 1930	27,000–33,000
Diner	1930	All	Military Eng., Sept.–Oct., 1930	43,000–51,000
Baggage	1930	All	Military Eng., Sept.–Oct., 1930	18,000–20,000
Rebel motor train	1936	G M & O	Rly. Age, Aug. 22, 1936	184,175
Rail bus	1936	—	Rly. Age, Aug. 22, 1936	24,611
Coach	1936	Milwaukee	Rly. Age, Aug. 22, 1936	24,649
Coach	1940	All	Am. Rly. Car Inst., 1940	53,905
Diner	1940	All	Am. Rly. Car Inst., 1940	68,851
Baggage	1940	All	Am. Rly. Car Inst., 1940	19,945
Sleeper	1940	All	Am. Rly. Car Inst., 1940	98,642
Combine	1946	—	Rly. M.E., May 1946	84,541
Diner	1946	—	Rly. M.E., May 1946	98,076
Tavern-Lounge	1946	—	Rly. M.E., May 1946	96,936
Coach	1947	P R R	Wash. News, April 17, 1947	90,000
Coach	1950	All	Am. Rly. Car Inst., 1950	112,413
Diner	1950	All	Am. Rly. Car Inst., 1950	138,275
Baggage	1950	All	Am. Rly. Car Inst., 1950	43,632
Sleeper	1950	All	Am. Rly. Car Inst., 1950	143,240
Dome	1955	Great North.	Trains, Dec. 1959	243,738
Bilevel commuter	1954	—	R.R. Mag., March 1954	140,000
Coach	1958	All	I.C.C. Report, 1958	235,934
Sleeper	1958	All	I.C.C. Report, 1958	307,497
Diner	1958	Great North.	Trains, Dec. 1959	242,569
Pioneer III Coach	1956	Budd Co.	Trains, Nov. 1956	90,000
Coach	1960	All	Am. Rly. Car Inst., 1960	154,137
Diner	1960	All	Am. Rly. Car Inst., 1960	184,668
Baggage	1960	All	Am. Rly. Car Inst., 1960	36,355

[*] Most of these figures are prices, but in cases where the cars were built by the railroads, the amount given represents the cost.

† F. A. Ritter Von Gerstner, *Die innern communication der vereinigten Staaten von Nord-America*, Vienna, 1842–1843.

Table B.4 Cars Owned by the Pullman Company and Its Competitors

YEAR	SLEEPING	DINING	PRIVATE RENTAL	PARLOR	MISC.	TOTAL PULLMAN	WAGNER, WOODRUFF, ETC.	RAILROAD-OPERATED SLEEPERS
1870						300e	100e	
1880						700e	300e	277
1890						2,135	600e	516
1900	3,213	24	24	466	152	3,879		393
1910	4,094	25	34	629	65	4,847		490
1920	6,559	0	27	1,019	121	7,726		558
1930	8,263	18	22e	1,186	312	9,801		223
1940	6,320	1	17	479	93	6,910		110
1950	5,993	0	0	227	6	6,226		148
1960						2,650		523
1970						0		677

e Estimate

Source: 1936 I.C.C. Report on the Pullman Company, published annual reports of the Pullman Company, and data in *Poor's Manual of Railroads*. Figures for 1900 are from a Pullman publication of Feb. 1, 1901, entitled *A Descriptive List*.

Table B.5 Passenger Car Production, 1871–1960

YEAR	NO. BUILT	NO. ORDERED	YEAR	NO. BUILT	NO. ORDERED
1871a	1,507		1920	751	1,613
1872–1879	(No data)		1921	1,162	308
1880b	395		1922	908	2,425
1881	407		1923	1,888	2,191
1882	458		1924	2,332	2,704
1883	587		1925	2,312	2,140
1884	374		1926	2,798	1,987
1885	259		1927	2,007	1,542
1886	456		1928	1,549	2,287
1887	663		1929	2,419	2,315
1888c	2,471		1930	1,520	679
1889d	(No data)		1931	243	11
1890e	1,654		1932	39	39
1891	1,640		1933	9	16
1892	2,195		1934	275	421
1893	1,986		1935	202	126
1894	516		1936	189	448
1895	430		1937	621	564
1896	474		1938	420	268
1897f	494		1939	273	319
1898g	669	1,046	1940	250	370
1899h	1,201	1,590	1941	363	546
1900i	1,515		1942	393	4
1901j	1,949	2,249	1943	675	1,720
1902	1,948k	2,510	1944	995	740
1903	2,007k	2,153	1945	928	2,993
1904	2,144k	2,055	1946	1,329	1,238
1905	2,551k	2,947	1947	858	316
1906	3,167k	2,716	1948	951	506
1907	5,457k	1,475	1949	939	109
1908	1,645k	1,065	1950	1,078	102
1909	2,698k	3,891	1951	183	261
1910	4,136k	3,073	1952	195	520
1911	3,362	2,466	1953	329	164
1912	2,509	2,888	1954	351	610
1913	2,654	3,535	1955	412	1,078
1914	3,589	1,494	1956	391	269
1915	1,513	1,878	1957	232	78
1916	1,344	2,006	1958	143	63
1917	1,684	1,118	1959	70	193
1918	750	6	1960	251	264
1919	126	290			

a 1871 figures from *Railroad Gazette*, Oct. 4, 1873.

b 1880–1887 from *R.R.G.*, Jan. 6, 1888 (Est. of major builders and R.R. shops).

c 1888 from *R.R.G.*, Jan. 4, 1889.

d For 1889 *R.R.G.*, Jan. 10, 1890, reported on passenger car production but gave no figures.

e 1890–1896 from *R.R.G.*, Jan..1, 1897 (9 major commercial builders).

f 1897 from *R.R.G.*, Dec. 31, 1897.

g 1898 from *R.R. Car Journal*, January 1899, and *R.R.G.*, Dec. 30, 1898.

h 1899 from *R.R. Car Journal*, January 1900.

i 1900 from *R.R.G.*, Dec. 28, 1900 (exclusive of R.R. shops).

j 1901–1960 from *Railroad Car Facts*, 1944 and 1961 eds. (Data include express refrigerator and milk cars.)

k Figures include foreign output. (From *Railway Age*.)

APPENDIX C: AMERICAN RAILROAD CAR CHRONOLOGY

1830	A typical four-wheel passenger car, seats 15, weighs 2 tons, and measures 15 feet long.
1831	Mail is carried on a car of the South Carolina Railroad. The *Columbus*, the first eight-wheel or double-truck passenger car, begins service.
1833	The B & O is using reversible seat backs. The need for separate baggage cars is recognized.
1834	J. S. Kite patents the safety beam as a safeguard against broken axles.
1835	The passenger car *Victory*, made by Richard Imlay, features a toilet, a buffet, and a clerestory roof.
1837–1838	Richard Imlay builds the first sleeping car, the *Chambersburg*, for the Cumberland Valley Railroad.
1838c.	The double-truck passenger car is now standard in America.
1838	Bonney, Bush, and Lobdell patent the double-plate, cast-iron wheel.
1839	John Tims patents a journal bearing box. The Western Railroad (of Massachusetts) provides a special car for its director. A sixteen-wheel passenger car is built by Davenport and Bridges.
1840	The typical coach is 35 feet long and 8½ feet wide; it weighs 8 tons and seats 50.
1841	The swing bolster truck is patented by Davenport and Bridges.
1845c.	Twelve-wheel cars appear on the Camden and Amboy and are adopted by several other lines within two years.
1845	Equalizers are used for passenger car trucks. The Eastern Railroad is operating parlor cars.
1848	John Lightner patents a journal bearing box.
1849	At least eight railroads have operated one or more sleeping cars.
1850c.	Double-acting brakes are introduced. Side-panel truss permits long bodies and lightweight construction. Some lines begin using iron body bolsters.
1850	Nathan Washburn patents a double-plate car wheel that will remain popular until 1928. Rubber springs have become popular for passenger car trucks.
1851	The Hudson River Railroad tests gas lighting for cars.
1852	Charles Waterbury patents a platform canopy that foreshadows the vestibule.
1853	The Hudson River is operating stateroom cars.
1856	The Illinois Central begins running stateroom cars.
1857	T. T. Woodruff builds his first sleeping car on patents received the year before. The Spear hot-air car stove is patented.
1858	Webster Wagner enters the sleeping-car business.
1859	George M. Pullman enters the sleeping-car business. An all-iron passenger car is built after a design patented in 1854 by B. J. La Mothe. Webster Wagner reintroduces the clerestory roof.
1860	Primitive lunch counter cars appear on a few lines.
1862	The Central Transportation Company, the first large-scale sleeping-car firm, is organized. A Railway Post Office car fitted for en route sorting appears on the Hannibal and St. Joseph Railroad.
1862–1863	The Pittsburgh, Fort Wayne, and Chicago acquires about twenty-four all-iron passenger cars.
1864–1865	Pullman's famous sleeper, the *Pioneer*, enters service.
1865	Dark exterior colors replace the yellow-cream tints formerly favored for passenger cars. Kerosene lamps are growing in popularity for car lighting. The Baker hot-water car heater is introduced. It will be in wide use within a decade.
1866	The New York Central Sleeping Car Company is formed by Webster Wagner.
1867	The Pullman Palace Car Company is incorporated. Wagner begins parlor car service. Limited food service is offered aboard Pullman's hotel cars.
1868	The first full-sized dining car, the *Delmonico*, enters service.
1869	Allen's paper car wheel is introduced. The Master Car Builders adopt a standard bearing. Westinghouse receives his first patent for an air brake (patents had been issued to other inventors for air brakes in earlier years).
1870	Pullman leases the Central Transportation Company and begins to gain control of the nation's sleeping-car business.

1870	Soft white-metal bearings become popular for wheel journals.
1873	G. F. Chalender patents an iron-plate, side-panel truss for wooden cars.
	Eli Janney's second patent embodies the basic plan of the modern car coupler.
	The Westinghouse automatic air brake is introduced.
1875c.	Parlor cars are now relatively common on many U.S. railroads.
	The vacuum brake and the air brake compete for the power brake market.
1875	The Miller coupler is widely used on passenger cars.
1879	The Central Pacific introduces emigrant sleeping cars.
	An air-activated water system for cars is patented, though it will rarely be used until about 1890.
1880	Serious attempts begin to develop a practical steam car heating system.
1881	Pintsch gas lighting for cars is introduced into the U.S.
	Pullman opens a giant car works south of Chicago.
1882	The Pennsylvania Railroad tests electric car lighting. Webster Wagner is killed aboard one of his palace cars.
	The Pennsylvania begins lounge car service.
1883	The Mann Boudoir Car Company is incorporated to reintroduce the stateroom sleeping car to the American public.
	Buffet cars appear on the West Shore railway.
1885c.	Dining cars are becoming common on most U.S. lines.
1885	A patent is issued for the Forney car seat—an early attempt to fit the seat to its occupant.
1887	The first Sessions vestibule enters service.
1888	The quick-action air brake is introduced.
1889	The Robbins steel passenger car is completed after a design patented in 1868.
1890	Iron and steel reinforcing is devised to strengthen wooden car construction.
1893	The New York Central introduces a class of wooden cars remarkable for their 72-foot-long bodies.
	Pullman patents the wide vestibule.
1894	Railway Mail Service has grown enormously since 1869. Over 3,000 cars are carrying the mail.
	The high-speed air brake is introduced.
1897	George M. Pullman dies.
1899	The American Car & Foundry Company (A.C.F.) is formed from a merger of seventeen smaller firms. The Pullman Company takes over the Wagner Sleeping Car Company.
1900	Steam heating is now standard on first-class cars. Steel beam end platforms are becoming common for the better class of passenger car.
1904	The first steel baggage car is delivered to the Erie. Vapor (low-pressure steam) car heating is developed.
	Cast-steel trucks are tested on the Big Four railroad.
1905	The Garland car ventilator is introduced. The first steel mail car enters service.
1907	The first production-model steel passenger car is built.
	The first steel Pullman, the *Jamestown*, is completed.
1910	Mass production of steel sleeping cars begins with Pullman's *Carnegie*.
	Steel axles are now standard; the iron axle has reluctantly been abandoned.
	Large-scale production of wrought-steel wheels begins.
	Electric car lighting is finally accepted as practical.
1911	The Santa Fe tries an ice cooling system on its deluxe train.
	The Post Office establishes minimum strength standards for its cars.
	The type D Janney-style coupler is adopted—the first coupler with interchangeable parts.
1913	The last wooden cars are ordered by a U.S. railroad.
1920	In the entire passenger car fleet, 60 percent of the cars are still wooden.
1921	The Pennsylvania begins using roller bearings for passenger cars.
1922	Pullman and Haskell & Barker merge their car plants.
1923	The first large order of aluminum passenger cars is completed for the Illinois Central Railroad by Pullman.
1927	Pullman tests an electromechanical air conditioner on one of its sleepers.
1929	Pullman takes over the Standard Steel Car Company.
1930	Heat-treated (hardened) car wheels are introduced.
1931	The B & O inaugurates the first fully air-conditioned train.
	The type E Janney-style coupler is adopted.
1934	*City of Salina*, the first streamline motor train, enters service on the Union Pacific.

1934	An all-welded passenger car is produced by the Milwaukee Road shops.
1935c.	Lightweight arched-roof passenger cars are now accepted as the new standard design. Triple bolster trucks are introduced by Pullman.
1936	There are 5,800 air-conditioned cars in service.
1937	The Southern Pacific begins to purchase articulated, standard-sized passenger cars. The first Roomette sleeping car enters service.
1938	The tightlock style of coupler is adopted.
1938–1939	Budd introduces disc brakes, though they will not be widely used until about 1955.
1939–1940	The Southern Pacific purchases triple-unit dining-car sets.
1940c.	Floating bolsters for car trucks are introduced.
1942	The private car *Ferdinand Magellan* is remodeled for the use of President Franklin D. Roosevelt.
1945	The Sleepy Hollow car seat is developed. The first Vista Dome car begins service on the Burlington. Roller bearings are now standard for new passenger cars.
1945–1946	Railroads place huge orders for new cars—3,000 in all.
1947	Pullman sells its sleeping-car division to an association of U.S. railroads.
1949	The first all-electric dining car appears on the Illinois Central.
1950	Only about 15 percent of the passenger fleet consists of modern lightweight cars. Gallery, or double-deck, cars are adopted for suburban trains in the Chicago area.
1956	Budd produces a cheap production-line coach, the Pioneer III, in a vain effort to revive new car orders. The first Slumber Car—an economy sleeper—enters service.
1963	The Railway Mail Service, which had already been much reduced, is cut heavily.
1965	An order of new coaches is delivered to the Kansas City Southern. No new conventional passenger cars will be ordered for nearly a decade.
1966	Metroliners are ordered from the Budd Company.
1968	The Pullman Company, which has been owned since 1947 by an association of railroads, suspends sleeping-car operations altogether.
1969	Metroliners begin regular service.
1971	Nearly all passenger cars are acquired by Amtrak, a government-sponsored corporation, organized to operate long-distance passenger service in the United States.
1973	Amtrak orders 57 new cars from Budd—the first conventional passenger equipment purchased in nearly ten years.
1974–1975	Amtrak adds another 435 units to its 1973 order.
1975–1976	Amtrak orders 284 bilevel cars from Pullman.
1977	The last railway post office car operates between New York and Washington.

NOTES

CHAPTER FIVE

1. *Ross Winans vs the Eastern Railroad Company*, U.S. Circuit Court, Massachusetts District, October 1853 term, p. 473.
2. *Illustrated London News*, April 10, 1852, p. 285.
3. *Railroad Advocate*, Sept. 20, 1856, p. 2.
4. E. T. Freedley, *Leading Pursuits and Leading Men*, Philadelphia, 1856, p. 320.
5. *George M. Pullman and Pullman's Palace Car Company vs. the New York Central Sleeping Car Company and Webster Wagner*, U.S. District Court, Northern District of Illinois, 1881, p. 80.
6. *American Railway Review*, April 26, 1860, p. 241.
7. *New England Railroad Club*, April 1924, p. 87.
8. *Railroad Car Journal*, November 1896, p. 279; see also *Railroad Gazette*, Dec. 18, 1885, p. 802, and 1888 *Car Builders' Dictionary*, fig. 1139.
9. 1888 *Car Builders' Dictionary*, figs. 1123–1130.
10. *R.R.G.*, June 4, 1886, p. 377, and Dec. 18, 1885, p. 801; *American Railroad Journal*, March 1887, p. 107.
11. 1928 *Car Builders' Dictionary*, p. 645.
12. *Railway Mechanical Engineer*, February 1930, p. 87.
13. Ibid., April 1936, p. 163.
14. Ibid., July 1947, p. 340.
15. Ibid., August 1945, p. 330.
16. *Report of the Mechanical Advisory Committee to the Federal Coordinator*, Washington, D.C., 1938, p. 645.
17. *Rly. M.E.*, September 1938, p. 338.
18. *Car Builders' Dictionaries* of all dates offer information on types of seat coverings.
19. Edward Hungerford, *Men and Iron*, New York: Crowell, 1938, p. 143.
20. *R.R. Adv.*, July 19, 1856, p. 3.
21. *R.R.G.*, July 7, 1876, p. 302.
22. Data on physical qualities of mohair were taken from *Car Builders' Dictionary*, 1922, p. 487, and 1928, p. 653.
23. *Pullman News*, April 1935, p. 105.
24. *National Car Builder*, March 1890, p. 46, and *R.R.C.J.*, March 1893, p. 133.
25. *N.C.B.*, August 1870, p. 3.
26. Alvin F. Harlow, *Steelways of New England*, New York: Creative Age Press, 1946, p. 380.
27. *A.R.R.J.*, Jan. 30, 1836, p. 49. McWilliams's patent was issued Nov. 29, 1832.
28. David Stevenson, *Sketch of the Civil Engineering of North America*, London, 1838; F. A. Ritter Von Gerstner, *Die innern communication der vereinigten Staaten von Nord-America*, Vienna, 1842–1843.
29. Von Gerstner, vol. 1, p. 122.
30. August Mencken, *The Railroad Passenger Car*, Baltimore: Johns Hopkins, p. 102, quoting a British travel account of 1839.
31. *R.R. Adv.*, Feb. 7, 1857, p. 4.
32. Ibid., June 2, 1855, p. 4.
33. *Am. Rly. Rev.*, March 27, 1862, p. 398, and *National Cyclopedia of American Biography*, vol. 2, p. 416.
34. From Spear's advertisement, dating about 1875, in Reading Company papers, Historical Society of Pennsylvania, Philadelphia.
35. *Master Car Builders Report*, 1908, p. 340.
36. *Railway Age*, Nov. 27, 1884, p. 731.
37. *A.R.R.J.*, Dec. 12, 1846, p. 793.
38. *R.R.G.*, July 5, 1873, p. 272.
39. *N.C.B.*, January 1878, p. 5.
40. *Engineering News*, April 5, 1890, p. 328.
41. Data on Baker's life and early years are drawn from obituaries appearing in 1901 engineering journals and a sketch published by *Locomotive Engineering*, January 1890, p. 45.
42. *Western Railroad Gazette*, March 21, 1868, p. 1.
43. E. Lavoinne and E. Pontzen, *Les Chemins de Fer en Amerique*, Paris, 1882, vol. 2, p. 26.
44. This description of the Baker system was compiled from *Appleton's Cyclopedia of Applied Mechanics*, 1886 vol. 2, p. 116; *Proceedings of the Institute of Civil Engineers*, 1878, vol. 53, p. 49; and *Car Heating*, an International Library of Technology manual on car appliances, 1912, part I, pp. 1–19.
45. *Pullman News*, April 1954, p. 8.
46. *R.R. Car Jour.*, January 1891, p. 46; *R.R.G.*, March 18, 1892, p. 201; *Railway Review*, Jan. 18, 1890, p. 31.
47. Lavoinne and Pontzen, vol. 2, p. 28.
48. *Rly. Rev.*, July 15, 1882, p. 410.
49. Ibid., April 2, 1887, p. 202.
50. *Railway Master Mechanic*, January 1892, p. xi.
51. *Rly. Rev.*, Dec. 13, 1890, p. 753.
52. *R.R. Car Jour.*, March 1891, p. 72.
53. *Rly. Rev.*, Jan. 15, 1887, p. 31.
54. *A.R.R.J.*, March 11, 1882, p. 163.
55. Ibid., Feb. 8, 1868, p. 145.
56. *R.R.G.*, March 18, 1887, p. 178.
57. *N.C.B.*, November 1875, p. 168.
58. *R.R.G.*, April 1887, p. 54.
59. *Rly. Rev.*, Dec. 13, 1890, p. 753.
60. Eugene S. Ferguson (Ed.), *Early Engineering Reminiscences (1815–1840) of George Escol Sellers*, Washington, D.C.: Government Printing Office, 1965, p. 12, and Hamilton Ellis, *Railway Carriages in the British Isles from 1830 to 1914*, London: Allen & Unwin, 1965, p. 45.
61. *Master Mechanics Magazine*, January 1888, p. 193.
62. *R.R.G.*, March 4, 1887, p. 149.
63. *American Railway Times*, Feb. 11, 1865, p. 46.
64. *Journal of the Franklin Institute*, 1877, vol. 103, p. 196.
65. *American Artisan*, September 1875, p. 229.
66. *Railway World*, Dec. 22, 1877, p. 1214.
67. *Locomotive Engineering*, January 1890, p. 45.
68. *Master Mechanics Mag.*, November 1888, p. 167.
69. *Rly. Age*, May 11, 1882, p. 257.
70. *R.R.G.*, March 8, 1889, p. 162.
71. *R.R. Car Jour.*, May–June 1891, p. 95.
72. *R.R.G.*, Jan. 21, 1887, p. 39.
73. *Engineering Magazine*, vol. 9, 1895, p. 716.
74. *R.R.G.*, Jan. 20, 1888, p. 45.
75. *N.C.B.*, August 1889, p. 115.
76. *The World* (New York City newspaper), Oct. 28, 1888.
77. Testimonial letter by F. K. Hain, printed in 1899 Gold catalog.
78. *Rly. Rev.*, Jan. 15, 1887, p. 38.
79. 1906 *Car Builders' Dictionary*, rear advertising section, p. 33. The figure 40,000 undoubtedly includes street railway cars.
80. L. H. Haney, *A Congressional History of Railways*, vol. 2, Madison, Wis., 1910, pp. 313–318.
81. *A.R.R.J.*, July 3, 1869, p. 737.
82. *N.C.B.*, April 1892, p. 53.
83. *R.R.G.*, Aug. 5, 1887, p. 517.
84. *N.C.B.*, May 1894, p. 75.
85. *R.R.G.*, March 11, 1887, p. 157.
86. Ibid., Dec. 2, 1887, p. 783.
87. Ibid., April 17, 1891, p. 263.
88. *N.C.B.*, November 1893, p. 171.
89. *R.R. Car Jour.*, January 1895, p. 13.
90. *Rly. Age*, Dec. 10, 1921, p. 1163.
91. *Western Railroad Club*, November 1892, p. 4.
92. Ibid., May 1895, p. 382.
93. Ibid.
94. *R.R.G.*, Oct. 12, 1894, p. 707.
95. *Rly. Rev.*, Dec. 5, 1891, p. 790.

96. *M.C.B.*, 1912, p. 365.

97. *Scientific American Supplement*, April 9, 1887, p. 9389.

98. "The Actual Efficiency of a Modern Locomotive," *Record of Recent Construction, No. 60*, pamphlet published by the Baldwin Locomotive Works, Philadelphia, 1907, p. 6.

99. *R.R.G.*, June 12, 1890, p. 410.

100. Ibid., May 19, 1905, p. 566.

101. Marshall M. Kirkman, *The Science of Railways*, vol. 4, Chicago: World Railway Publishing Company, 1904, p. 159.

102. *Rly. M.E.*, April 1934, p. 112.

103. For the Gold vapor system, see *R.R.G.*, June 24, 1904, p. 81, and June 16, 1909, p. 1262; *Rly. Rev.*, Dec. 19, 1908, p. 1026.

104. *Moody's Industrials*, 1920, p. 1936.

105. *Rly. M.E.*, March 1951, p. 52, and *Heating, Ventilating and Air Conditioning Guide: 1956*, 34th ed., American Society of Heating and Air Conditioning Engineers, p. 1089.

106. *Rly. M.E.*, December 1948, p. 712.

107. *Car Builders' Cyclopedia*, 1928, p. 732.

108. *Rly. M.E.*, March 1951, p. 52.

109. *Federal Coordinator's Report*, 1935, p. 640.

110. *Car Builders' Cyclopedia*, 1953, p. 723.

111. *New York Railroad Club*, Jan. 20, 1949, pp. 61–74.

112. Conversation with W. D. Edson, September 1973.

113. *New England Railroad Club*, Oct. 10, 1894, pp. 20 and 24.

114. *A.R.R.J.*, July 8, 1854, p. 428, and June 6, 1857, p. 361; *R.R. Car Jour.*, November 1894, p. 243.

115. *R.R.G.*, Sept. 20, 1895, p. 622.

116. *Railroad Record*, Oct. 5, 1854, p. 538.

117. Fredrick Moné, *An Outline of Mechanical Engineering*, New York(?), 1851, no page numbers; see description of the locomotive *Croton*.

118. *A.R.R.J.*, March 20, 1852, p. 188, and *Rly. Age*, Sept. 6, 1889, p. 587.

119. Harlow, *Steelways of New England*, p. 391.

120. *R.R.G.*, March 29, 1898, p. 309.

121. *R.R. Adv.*, June 13, 1856, p. 2.

122. *R.R.G.*, Oct. 1, 1870, p. 1, and *R.R. Car Jour.*, November 1894, p. 243.

123. *R.R. Car Jour.*, November 1894, p. 243.

124. Lavoinne and Pontzen, p. 24.

125. Lucius M. Beebe, *Trains We Rode*, vol. 2, Berkeley, Calif.: Howell-North, 1966, p. 798.

126. Mencken, p. 47.

127. *R.R. Car Jour.*, October 1894, p. 208.

128. *R.R. Adv.*, July 19, 1856, p. 1.

129. *Dictionary of Canadian Biography*, 1972, vol. X, p. 636.

130. *A.R.R.J.*, Sept. 9, 1854, p. 565.

131. Ibid., Feb. 7, 1857, p. 92.

132. Ibid., Aug. 20, 1859, p. 534.

133. Ibid., June 20, 1863, p. 576.

134. *Scientific American*, July 19, 1862, p. 40.

135. *A.R.R.J.*, Feb. 8, 1868, p. 145.

136. *R.R.G.*, Jan. 4, 1873, p. 6.

137. Ibid., June 22, 1877, p. 282.

138. *R.R.G.*, Dec. 14, 1883, p. 819, and 1888 *Car Builders' Dictionary*, p. 191.

139. *R.R.G.*, Aug. 1, 1884, pp. 562 and 568.

140. *A.R.R.J.*, Dec. 15, 1840, p. 354.

141. *R.R.G.*, Jan. 4, 1873, p. 8.

142. *M.C.B.*, 1875, p. 11.

143. *R.R.G.*, Jan. 21, 1887, p. 40.

144. *New England Railroad Club*, Oct. 10, 1894, p. 6.

145. *M.C.B.*, 1875, p. 14.

146. *Rly. Age Gaz.*, Aug. 25, 1911, p. 362.

147. *M.C.B.*, 1908, p. 362.

148. Ibid., p. 340.

149. *New England Railroad Club*, Oct. 10, 1894, p. 24.

150. Ibid. and 1888 *Car Builders' Dictionary*, fig. 1566.

151. *Western Railroad Club*, January 1890, p. 92.

152. *R.R.G.*, March 24, 1876, p. 131.

153. Dudley's ventilator is described in the following sources: *Jour. Franklin Inst.*, July 1897, p. 1; *A.R.R.J.*, June 1901, p. 77; *R.R.G.*, Sept. 30, 1904, p. 392; *M.C.B.*, 1908, p. 341.

154. *R.R. Car Jour.*, May 1894, pp. 80 and 85.

155. Data on Crowder is from *R.R. Age Gaz.*, Aug. 25, 1911, p. 362; *Harper's Weekly*, Feb. 7, 1914, p. 21; J. C. Welliver, Pullman Progress (proof sheets, dated 1929, held by the Chicago Historical Society), pp. 55–57.

156. *Rly. Rev.*, Feb. 22, 1908, p. 147.

157. *R.R. Age Gaz.*, June 20, 1908, p. 244.

158. *Rly. M.E.*, March 1924, p. 158.

159. *Scientific American*, May 21, 1881, p. 323.

160. *R.R.G.*, Dec. 21, 1900, p. 841.

161. *N.C.B.*, September 1891, p. 131.

162. *Loco. Eng.*, August 1898, p. 397.

163. *R.R.G.*, June 20, 1908, p. 245.

164. *Rly. M.E.*, March 1924, p. 158.

165. *N.C.B.*, March 1873, p. 59.

166. *R.R.G.*, April 5, 1889, p. 232.

167. *Rly. M.E.*, November 1930, p. 638.

168. *R.R. Age Gaz.*, Jan. 5, 1912, p. 23.

169. R. S. Henry, *This Fascinating Railroad Business*, Indianapolis, 1942, p. 258.

170. Margaret Ingels, *Willis H. Carrier: The Father of Air Conditioning*, New York: Doubleday, 1952, p. 78. Hereafter referred to as Ingels, *Carrier*.

171. William W. Kratville, *Passenger Car Catalog: Pullman Operated Equipment, 1912–1949*, Omaha: Kratville Publications, 1968, p. 149.

172. *American Society of Mechanical Engineers Transactions*, 1937, p. 733.

173. Ingels, *Carrier*, pp. 78–81, and *Rly. M.E.*, September 1930, p. 508.

174. *Rly. M.E.*, February 1931, p. 108, and November 1931, p. 556.

175. Ibid., May 1932, p. 214.

176. Ibid., September 1930, p. 515, and December 1930, p. 696.

177. Ibid., January 1932, p. 47.

178. Ibid., August 1933, p. 306.

179. Ibid., June 1934, p. 211.

180. Ibid., October 1931, p. 521.

181. Kratville, *Passenger Car Catalog*, p. 149.

182. *M.C.B.*, 1936, p. 254.

183. *Car Builders' Cyclopedia*, 1937, p. 550, and *Fortune*, January 1938, p. 96.

184. *Rly. M.E.*, February 1931, p. 63.

185. *Railroad Magazine*, August 1946, p. 55.

186. *A.S.M.E. Transactions*, December 1937, p. 733.

187. *Rly. M.E.*, October 1931, p. 521.

188. *A.S.M.E. Transactions*, December 1937, p. 737.

189. *Rly. M.E.*, March 1934, p. 76.

190. Ingels, *Carrier*, p. 80.

191. *Rly. M.E.*, October 1931, p. 515, and March 1937, p. 124.

192. *A.S.M.E. Transactions*, December 1937, p. 733.

193. *Trains*, August 1951, p. 13.

194. Statement by W. D. Edson, former car superintendent of Amtrak, January 1971.

195. *Rly. M.E.*, May 1932, p. 173.

196. Ibid., February 1946, p. 90, and March 1947, p. 111.

197. Notes on early lighting were drawn from *Early Lighting: A Pictorial Guide*, Rushlight Club, 1972; L. S. Russell, *A Heritage of Light*, Toronto: University of Toronto, 1968; H. F. Williamson and A. R. Daum, *American Petroleum Industry*, Evanston, Ill.: Northwestern University Press, 1958. The last two works are scholarly studies which have been used elsewhere in this section without specific credit.

198. Mencken, p. 107.

199. Mencken, p. 173.

200. *M.C.B.*, 1875, p. 9.

201. *N.C.B.*, February 1880, p. 27.

202. *R.R.G.*, May 19, 1905, p. 563.

203. *Western Railroad Club*, February 1891, p. 161.

204. *R.R. Car Jour.*, November 1891, p. 15.

205. *Journal of the Franklin Institute*, 1831, vol. 2, p. 76.

206. *N.C.B.*, May 1880, p. 77.

207. Ibid., February 1878, p. 21.

208. *Engineering News*, April 26, 1890, p. 396.

209. *A.R.T.*, Feb. 15, 1868, p. 55.

210. Ibid., March 7, 1868, p. 79.

211. *N.C.B.*, May 1880, p. 77, and *Rly. Age*, Aug. 11, 1881, p. 5.

212. *A.R.R.J.*, April 5, 1834, p. 197.

213. *R.R. Adv.*, Sept. 6, 1856, p. 3.

214. Ibid., Aug. 16, 1856, p. 4, and *A.R.R.J.*, June 11, 1859, p. 377.

215. A. Bendel, *Aufsätze Eisenbahnwesen in Nord-Amerika*, Berlin, 1862, p. 37, and *A.R.R.J.*, October 1899, p. 329.

216. *N.C.B.*, December 1877, p. 180.

217. *Am. Rly. Rev.*, Jan. 17, 1861, p. 21.

218. *R.R.G.*, March 2, 1883, p. 137.

219. *N.C.B.*, June 1871, p. 1.

220. James Dredge, *The Pennsylvania RR.*, London, 1879, p. 151.

221. Lavoinne and Pontzen, vol. 2, p. 19.

222. *N.C.B.*, March 1879, p. 38, and *Engineer*, Aug. 29, 1884, p. 164.

223. A. M. Wellington, W. B. D. Penniman, and C. W. Baker, *The Comparative Merits of Various Systems of Car Lighting*, New York: Engineering News Publishing Company, 1892, p. 3.

224. *Rly. Rev.*, Sept. 17, 1887, p. 543.

225. *R.R.G.*, Feb. 29, 1884, p. 164; *Report of International Railway Congress*, September 1900, p. 104.

226. *Rly. Age*, July 24, 1884, p. 5 advertisement.

227. *Rly. Rev.*, Jan. 17, 1882, p. 14, and July 8, 1882, p. 395.

228. *R.R.G.*, April 6, 1883, p. 209.

229. *National Cyclopedia of American Biography*, 1904, vol. 12, p. 189.

230. *R.R.G.*, Jan. 30, 1891, p. 84.

231. *Loco. Eng.*, January 1893, p. 15, and February 1893, p. 52.

232. *Rly. Rev.*, May 7, 1892, p. 293.

233. *A.R.R.J.*, October 1899, p. 329.

234. *Rly. Rev.*, Feb. 10, 1900, p. 70.

235. *R.R.G.*, Oct. 30, 1908, p. 1269.

236. *R.R.G.*, March 26, 1886, p. 210.

237. *N.C.B.*, July 1889, p. 98, and *Western Railroad Club*, February 1891, p. 165.

238. *R.R.G.*, June 7, 1889, p. 371.

239. *Rly. Rev.*, Nov. 5, 1892, p. 699.

240. Ibid., Aug. 25, 1900, p. 476.

241. Ibid.

242. *R.R.G.*, March 29, 1901, p. 219.

243. *R.R.G.*, April 28, 1905, p. 134.

244. *R.R. Age Gaz.*, July 9, 1909, p. 59.

245. *Western Railroad Club*, Feb. 17, 1891, p. 180.

246. *N.C.B.*, February 1879, p. 18.

247. George Behrend, *Pullman in Europe*, London: Ian Allen, 1962, p. 87,

248. *Rly. Age*, Dec. 29, 1881, p. 731.

249. *R.R.G.*, May 31, 1889, p. 355, and Sept. 22, 1882, p. 587.

250. C. W. T. Stuart, *Car Lighting by Electricity*, New York, 1923, p. 2.

251. *R.R.G.*, March 1, 1889, p. 143, and Nov. 29, 1889, p. 786.

252. Ibid., Nov. 1, 1889, p. 718.

253. *Loco. Eng.*, September 1899, p. 394, and *Western Railroad Club*, February 1891, p. 167.

254. *R.R.G.*, May 31, 1889, p. 356.

255. *Rly. Rev.*, Feb. 6, 1886, p. 62.

256. *R.R.G.*, June 13, 1890, p. 410.

257. *R.R. Age Gaz.*, July 9, 1909, p. 56.

258. *R.R.G.*, Feb. 12, 1904, p. 116.

259. Ibid., April 3, 1896, p. 237.

260. Ibid., May 20, 1898, p. 351.

261. Ibid., March 17, 1882, p. 168, and *A.A.R.J.*, September 1884, p. 184.

262. *R.R. Car Jour.*, July 1893, p. 260.

263. *R.R.G.*, June 19, 1896, p. 439.

264. Ibid., April 30, 1897, p. 301, and *A.R.R.J.*, August 1899, p. 275.

265. *Rly. Rev.*, Aug. 25, 1900, p. 471.

266. *R.R.G.*, March 4, 1904, p. 164.

267. *Rly. Rev.*, June 8, 1907, p. 503.

268. *A.R.R.J.*, December 1912, p. 614.

269. Stuart, *Car Lighting by Electricity*, pp. 5 and 55.

270. *Railway Electrical Practice, Part 2*, a handbook published by the American Technical Society, 1942, p. 151.

271. *Rly. Rev.*, Oct. 12, 1907, p. 888.

272. *Rly. M.E.*, August 1944, p. 385, and August 1945, p. 358.

273. Ibid., September 1941, p. 338.

274. *R.R. Age Gaz.*, Jan. 5, 1912, p. 15.

275. *Rly. Age Gaz. M.E.*, October 1915, p. 522.

276. *Rly. Age*, July 29, 1939, p. 175.

277. Illinois Central Railroad papers, Newberry Library, Chicago.

278. Von Gerstner, vol. 1, p. 129.

279. *A.R.R.J.*, June 15, 1842, p. 380.

280. *Harper's Weekly*, Feb. 7, 1914, p. 21.

281. *R.R.G.*, Aug. 17, 1877, p. 375.

282. *N.C.B.*, October 1882, p. 110.

283. *R.R. Car Jour.*, July 1893, p. 256.

284. *International Textbook Library of Technology*, vol. 27, 1912, p. 61.

285. *N.C.B.*, September 1884, p. 121.

286. *Harper's Weekly*, Feb. 7, 1914, p. 20.

287. Railway and Locomotive Historical Society, *Bulletin 65*, October 1944, p. 62.

288. *R.R. Record*, Aug. 11, 1853, p. 377.

289. *Loco. Eng.*, June 1898, p. 287.

290. *R.R. Adv.*, Sept. 20, 1856, p. 2.

291. *A.R.R.J.*, July 1909, p. 299.

292. *Railroad Man's Magazine*, January 1917.

293. *Lippincott's Magazine*, May 1873, p. 521.

294. Bendel, p. 38.

295. *A.R.R.J.*, Sept. 19, 1835, p. 577.

296. *R.R. Car Jour.*, June 1896, p. 109.

297. *History of the Delaware and Hudson Company*, Albany, 1925, pp. 658–659.

298. This statement was reprinted in *Railway and Locomotive Engineering*, May 1910.

299. *New York Tribune*, Sept. 24, 1893.

300. *R.R. Car Jour.*, December 1896, p. 303.

301. R. Haynes, *Troy and Rensselaer County*, New York, 1925, p. 357.

302. *R.R. Car Jour.*, November 1897, p. 349.

303. *New England Railroad Club*, Oct. 11, 1898, p. 10.

304. William Chambers, *Things as They Are in America*, London, 1853; *R.R.G.*, Nov. 11, 1881, p. 631.

305. *R.R. Car Jour.*, November 1892, p. 30.

306. *A.R.R.J.*, Aug. 17, 1850, p. 527.

307. *A.R.T.*, March 16, 1867, p. 86.

308. *R.R. Adv.*, Sept. 20, 1856, p. 1.

309. *Manufacturer and Builder*, September 1870, p. 271; *R.R.G.*, July 5, 1873, p. 272; *N.C.B.*, June 1873, p. 136, and December 1877, p. 178.

310. *N.C.B.*, March 1881, p. 30.

311. *Pullman News*, December 1927, p. 287, and July 1939, p. 6.

312. *N.C.B.*, November 1878, p. 163.

313. *Lippincott's Magazine*, May 1873, p. 521.

314. *Rly. Rev.*, Sept. 3, 1887, p. 516.

315. R. J. Wayner (Ed.), *The Pullman Scrapbook*, New York: Wayner Publications, 1971, p. 34.

316. *N.C.B.*, January 1876, p. 6.

317. *Rly. Rev.*, March 3, 1888, p. 115.

318. *N.C.B.*, November 1892, p. 165.

319. *R.R.G.*, Aug. 5, 1881, p. 431; *Rly. Rev.*. Nov. 8, 1884, p. 576; *A.R.R.J.*, December 1884, p. 292.

320. *R.R.G.*, Jan. 13, 1882, p. 17.

321. *N.C.B.*, June 1884, p. 73.

322. *Rly. World*, May 15, 1880.

323. *R.R.G.*, Aug. 19, 1881, p. 451; William Voss, *Railway Car Construction*, New York: Van Arsdale, 1892, p. 110.

324. *Rly. Rev.*, Aug. 13, 1881, p. 453.

325. Ibid., Nov. 19, 1887, p. 672.

326. Stanley Buder, *Pullman: An Experiment in Industrial Order*, New York: Oxford, 1967, p. 73.

327. *N.C.B.*, June 1884, p. 71.

328. *R.R.G.*, June 8, 1894, p. 397; Cincinnati Historical Society *Bulletin*, Spring 1973, p. 58.

329. *R.R. Car Jour.*, November 1898, p. 339.

330. Russell Lynes, *The Tastemakers*, New York: Grosset and Dunlap, 1949, pp. 95–96.

331. Ibid., p. 96; *R.R. Car Jour.*, November 1900, p. 317.

332. *R.R.G.*, October 22, 1897, p. 754.

333. Ibid., June 30, 1905, p. 773.

334. *Rly. M.E.*, February 1925, p. 97.

335. Ibid., February 1924, p. 91.

336. Ibid., June 1933, p. 191.

337. R. C. Reed, *The Streamliners*, San Marino, Calif.: Golden West Books, 1974.

338. *R.R. Car Jour.*, October 1896, p. 253.

339. *N.C.B.*, January 1881, p. 9.
340. Ibid.; *R.R.G.*, April 5, 1878, p. 178.
341. *Car Builders' Dictionary*, 1888, p. 122.
342. *R.R. Car Jour.*, January 1900, p. 20.
343. Ibid., November 1891, p. 12.
344. *N.C.B.*, September 1891, p. 142.
345. *R.R. Car Jour.*, October 1898, p. 304.
346. *R.R.G.*, Nov. 11, 1881, p. 631.
347. *N.C.B.*, March 1881, p. 30.
348. *Rly. Rev.*, Feb. 6, 1897, p. 79; *A.R.R.J.*, November 1898, p. 387.
349. *R.R.G.*, Sept. 12, 1884, p. 672.
350. *R.R. Car Jour.*, March 1897, p. 75.
351. Ibid., Aug. 21, 1903, p. 598, and March 4, 1905, p. 318.
352. Railway and Locomotive Historical Society, *Bulletin 103*, October 1960, p. 59.
353. *New England Railroad Club*, Oct. 11, 1898, p. 46.
354. *Rly. M.E.*, December 1925, p. 757.
355. Ibid., October 1914, p. 527.
356. Ibid., June 1931, p. 292.
357. Ibid., October 1935, p. 435.
358. On early vestibule devices, refer to *R.R. Car Jour.*, November 1891, p. 18; *R.R.G.*, March 2, 1894, p. 152; *Glasgow Practical Mechanics Magazine*, 1847, vol. 2, p. 204.
359. *R.R. Record*, June 16, 1853, p. 247, and Sept. 22, 1853, p. 471; *A.R.R.J.*, July 1, 1854, p. 406.
360. A rare photograph of a train fitted with Atwood and Waterbury's vestibule is reproduced in L. M. Beebe and Charles Clegg, *Hear the Train Blow*, New York: Grosset and Dunlap, 1952, p. 44. This photograph agrees with the line drawing shown in Figure 5.86.
361. *R.R.G.*, Aug. 26, 1887, p. 557.
362. Ibid., May 25, 1888, p. 341.
363. *N.C.B.*, June 1887, p. 80; 1879 *Car Builders' Dictionary*, p. 194.
364. *Rly. Age*, Sept. 28, 1882, p. 540.
365. *Rly. Rev.*, April 16, 1887, p. 226.
366. *R.R.G.*, Dec. 9, 1887, p. 796.
367. Ibid., Aug. 24, 1888, p. 557.
368. *R.R. Car Jour.*, June 1899, p. 151.
369. *Rly. Rev.*, Oct. 18, 1890, p. 627.
370. *R.R.G.*, Nov. 28, 1889, p. 791, and *N.C.B.*, September 1888, p. 142.
371. *R.R.G.*, March 2, 1894, p. 153.
372. Ibid., Oct. 17, 1890, p. 723.
373. Ibid., March 9, 1894, p. 173.
374. *Rly. Master Mech.*, July 1893, p. 111.
375. *R.R. Mag.*, October 1949, p. 35.
376. *Rly. Age*, Jan. 19, 1946, p. 151.

CHAPTER SIX

1. A. M. Earle, *Stage Coach and Tavern Days*, New York: Macmillan, 1900, p. 236.
2. Memorandum on the history of baggage service dated February 8, 1929, by W. M. Skinner of the American Association of General Baggage Agents.
3. Ibid.
4. *Locomotive Engineering*, February 1895, p. 92.
5. F. A. Ritter Von Gerstner, *Die innern communication der vereinigten Staaten von Nord-America*, Vienna, 1842–1843, vol. 2, p. 96.
6. August Mencken, *The Railroad Passenger Car*, Baltimore: Johns Hopkins Press, 1957, p. 125.
7. A. Bendel, *Aufsatze Eisenbahnwesen in Nord-Amerika*, Berlin, 1862, pp. 38–39.
8. Karl Ghega, *Die Baltimore-Ohio Eisenbahn*, Vienna, 1844, p. 195.
9. E. H. Mott, *Between the Ocean and the Lake: The Story of the Erie*, New York, 1901, p. 397.
10. Alvin Harlow, *Steelways of New England*, New York: Creative Age Press, 1946. The Old Colony stock certificate is reproduced on the endpapers of this volume.
11. *New York Railroad Club*, November 1922, p. 6817.
12. *National Car Builder*, January 1881, p. 5.
13. *Railway Age*, July 6, 1882, p. 372.
14. *Railway Review*, March 2, 1902, p. 146.
15. *Railroad Car Journal*, November 1899, p. 335.

16. *Rly. Rev.*, July 18, 1909, p. 573.
17. *Master Car Builders Report*, 1908, pp. 307 and 322.
18. F. Gutbrod, "The Construction of Iron Passenger Cars on the Railways in the United States of America," *International Railway Congress Bulletin*, vols. 26 and 27, 1912–1913, part 3, p. 64. Also *Railroad Gazette*, June 18, 1908, p. 133.
19. J. W. Shine and F. M. Ellington, *Passenger Train Equipment of the Atchison, Topeka & Santa Fe*, vol. 1, Colfax, Iowa: Railcar Press, 1973, p. 80.
20. Bendel, p. 38.
21. Von Gerstner, vol. 1, pp. 79, 238, 247, and 250.
22. Lucius M. Beebe and Charles Clegg, *The Central Pacific and the Southern Pacific Railroads*, Berkeley, Calif.: Howell-North, 1963, pp. 26 and 29.
23. *Loco Eng.*, July 1898, p. 353.
24. Von Gerstner, vol. 1, p. 121.
25. Illinois Central Papers, Newberry Library, Chicago. Letter from the Buffalo Car Company, May 10, 1855.
26. Michigan Central annual report, 1855, p. 37.
27. *Railroad Magazine*, September 1952, p. 66.
28. *N.C.B.*, April 1879, p. 55.
29. *Rly. Rev.*, Sept. 18, 1883, p. 532.
30. *Rly. Age*, March 1, 1889, p. 144, and *R.R.G.*, Feb. 15, 1889, p. 115.
31. *Rly. Rev.*, May 3, 1884, p. 232.
32. A copy of "Santa Fe Route Tourist Sleeper" of 1900 is in the collections of the Library of Congress.
33. *Railway Mechanical Engineer*, October 1933, p. 380.
34. *Rly. Age*, Nov. 15, 1947, p. 234.
35. *Trains*, June 1947, p. 7.
36. General comments made here and elsewhere about the Railway Mail Service are drawn largely from existing histories on the subject, especially B. A. Long and W. J. Dennis's *Mail by Rail*, New York: Simmons-Boardman, 1951 (hereafter *Mail by Rail*), and Carl Scheele's *A Short History of the Mail Service*, Washington, D.C.: Smithsonian Institution, 1970. Scheele's work (hereafter Scheele) is particularly valuable for its extensive bibliography.
37. *Mail by Rail*, p. 96.
38. *Rly. Age*, July 9, 1938, p. 36.
39. Hamilton Ellis, *Railway Carriages in the British Isles from 1830 to 1914*, London: Allen & Unwin, 1965, p. 42.
40. *Quarterly Review*, vol. 84, 1849, p. 44.
41. Von Gerstner, vol. 1, p. 95.
42. *Rly. Age*, Aug. 6, 1938, p. 223, and C. H. Corliss, *Main Line of Mid America*, New York: Creative Age Press, 1950, p. 74.
43. U.S. Post Office Department, *History of the Railway Mail Service*, Washington, D.C.: G.P.O., 1885, p. 40. Hereafter *History R.M.S.*
44. *American Railroad Journal*, July 28, 1855, p. 470.
45. *R.R. Car Jour.*, November 1894, p. 243.
46. Biographical file on W. A. Davis in the Smithsonian's Division of Postal History.
47. *Beginnings of the True Railway Mail Service and the Work of George B. Armstrong in Founding It*, Chicago, 1906, pp. 74–84.
48. *Railway World*, May 28, 1881, p. 510.
49. *A.R.R.J.*, Sept. 10, 1864, p. 894.
50. *Rly. World*, May 28, 1881, p. 510.
51. *Scribner's Monthly*, June 1873.
52. *Chicago Magazine*, January 1912, p. 350.
53. J. E. White, *A Life Span and Reminiscences of the R.M.S.*, Philadelphia, 1910, p. 12. Hereafter White, *Reminiscences*.
54. *Scribner's Monthly*, June 1873.
55. *R.R. Car Jour.*, August 1892, p. 200.
56. *N.C.B.*, October 1870, p. 5.
57. *Scribner's Monthly*, June 1873.
58. *Rly. World*, Dec. 1, 1877, p. 1142.
59. *R.R.G.*, Sept. 18, 1875, p. 387.
60. *The American Railway*, 1892, p. 318.
61. *N.C.B.*, October 1875, p. 154.
62. *R.R.G.*, Sept. 18, 1875, p. 387.
63. *The American Railway*, 1892, p. 319.
64. White, *Reminiscences*, p. 15.
65. *Scribner's Monthly*, June 1873.
66. *History R.M.S.*, p. 74.
67. *N.C.B.*, August 1882, p. 89.
68. *Rly. Age*, April 30, 1885, p. 283.

69. Post Master General's annual report, 1880–1881.
70. *N.C.B.*, November 1894, p. 173.
71. *Mail by Rail*, p. 136. *Railway Age Mechanical Engineer*, June 1915, p. 299, gives a total of 5,866 mail cars, only 1,566 of which were full R.P.O.s as of December 31, 1914.
72. *N.C.B.*, October 1883, p. 109.
73. E. Lavoinne and E. Pontzen, *Les Chemins de Fer en Amerique*, Paris, 1882, atlas, plate VI.
74. *N.C.B.*, October 1880, p. 164.
75. *Rly. Rev.*, Dec. 19, 1891, p. 816.
76. White, *Reminiscences*, pp. 168–170.
77. *Engineer*, Aug. 31, 1883, p. 163, and *R.R.G.*, Jan. 29, 1886, p. 70.
78. M. McKenna, *History of Fond Du Lac County*, Chicago, 1912, vol. 1, p. 324.
79. *N.C.B.*, May 1885, p. 64.
80. *R.R. Car Jour.*, July 1895, p. 158.
81. *Railway Mechanical Engineer*, April 1913, p. 212.
82. *History R.M.S.*, p. 201.
83. *Mail by Rail*, pp. 141–142.
84. 69th Congress House Hearings (No. 1904), Jan. 31, 1927, p. 3.
85. *N.C.B.*, September 1894, p. 133.
86. White, *Reminiscences*, p. 171, and *Engineering News*, Feb. 20, 1892, p. 173.
87. White, *Reminiscences*, p. 173.
88. *Rly. Age*, Feb. 3, 1911, p. 190, and Feb. 10, 1911, p. 291.
89. *Southern & South Western Railroad Club*, May 1916, p. 11, and *Rly. Age*, Nov. 24, 1911, p. 1049.
90. 69th Congress House Hearings (No. 1904), Jan. 31, 1927, p. 2.
91. *R.R.G.*, June 22, 1906, p. 687, and *Rly. Rev.*, June 15, 1907, p. 522.
92. White, *Reminiscences*, p. 178.
93. *Rly. M.E.*, June 1915, p. 299.
94. Ibid., December 1922, p. 737.
95. Shine and Ellington, *Passenger Train Equipment of the Atchison, Topeka & Santa Fe*, vol. 1, p. 8.
96. House Hearings on Steel Railway Post Office Cars, 69th Congress, First Session, H.R. 4475, April 30, 1926.
97. House Hearings on Steel Railway Post Office Cars, 70th Congress, First Session, H.R. 9678, May 16, 1928.
98. *Car Builders' Cyclopedia 1928*, pp. 515–528.
99. *Railroad Car Facts*, American Railway Car Institute, New York, 1961, p. 73.
100. *Car Builders' Cyclopedia 1953*, p. 472, and *Rly. M.E.*, March 1915, p. 123.
101. *Mail by Rail*, p. 340.
102. Scheele, p. 169.
103. Ibid., pp. 176–177.
104. *Trains*, February 1971, p. 31.
105. *Trains*, October 1960, p. 12.
106. *Year Book of Railroad Facts*, 1972, p. 12, and 1974, p. 10.
107. Shine and Ellington, *Passenger Train Equipment*, vol. 1, p. 10.

CHAPTER SEVEN

1. *Railway Mechanical Engineer*, April 1930, p. 218.
2. *Railroad Advocate*, June 2, 1855, p. 2.
3. *Ross Winans vs. The Eastern Railroad Company*, U.S. Circuit Court, Massachusetts District, October 1853 term (Boston 1854), p. 913.
4. *Railroad Magazine*, June 1942, p. 103.
5. *New York Railroad Club*, April 19, 1900, p. 22.
6. Ibid.
7. *National Car Builder*, January 1881, p. 8, and *Rly. M.E.*, January 1916, p. 27.
8. William Voss, *Railway Car Construction*, New York, 1892, p. 112.
9. *Rly. M.E.*, November 1915, p. 569.
10. *Rly. M.E.*, March 1945, p. 110.
11. *R.R. Mag.*, June 1942, p. 98.
12. *Institution of Civil Engineers Proceedings*, vol. 28, 1869, p. 380.
13. *American Railway Review*, Jan. 19, 1860, p. 33.
14. *Pullman News*, October 1929, p. 191.
15. *N.C.B.*, September 1876, p. 134.
16. Ibid., June 1879, p. 84.
17. *The Engineer*, Philadelphia, Oct. 11, 1860, p. 68.
18. *N.C.B.*, January 1891, p. 4.

19. *Scientific American*, Sept. 11, 1845, p. 10.
20. *New York Farmer & Mechanic*, May 29, 1845.
21. *N.C.B.*, March 1884, p. 32.
22. *Rly. Rev.*, Sept. 7, 1889, p. 524, and *Railroad Gazette*, June 6, 1890, p. 394.
23. *R.R.G.*, May 15, 1896, p. 334.
24. Ibid., Sept. 16, 1898, p. 661.
25. Ibid., June 17, 1904, p. 16.
26. *A.R.R.J.*, March 1906, p. 96.
27. *Rly. M.E.*, November 1915, p. 569.
28. *Car Builders' Cyclopedia*, 1928, fig. 2181.
29. *Rly. M.E.*, March 1945, p. 108.
30. Ibid., March 1946, p. 122.
31. *Railway Age*, July 30, 1949, p. 207.
32. *Car Builders' Cyclopedia*, 1943, p. 1117.
33. Data from W. D. Edson, former car superintendent for Amtrak.
34. *A.R.R.J.*, Oct. 25, 1834, p. 658.
35. William Whiting, *Arguments of William Whiting, Esq., in the case of Ross Winans vs. Orsamus Eaton et al.*, Boston, 1853, p. 126.
36. *Rogers Locomotive and Machine Company*, 1876 catalog, p. 26.
37. *Report of the Entire Testimony in the Case of W. C. Fuller vs. F. M. Ray*, pamphlet published by F. M. Ray, New York, 1850.
38. *A.R.R.J.*, Jan. 5, 1850, p. 11.
39. Ibid., Feb. 28, 1852, p. 136.
40. D. A. Wells (Ed.), *Annual of Scientific Discovery*, Boston, 1857, p. 54.
41. *Car Builders' Dictionary*, 1879, fig. 118.
42. *Journal of the Franklin Institute*, vol. 6, 1843, p. 168.
43. *A.R.R.J.*, Oct. 30, 1847, p. 689.
44. *American Railway Times*, Jan. 19, 1861, p. 24.
45. *Rly. Age*, Dec. 1, 1969, rear cover advertisement.
46. John H. White, Jr., *James Millholland and Early Railroad Engineering*, Contributions from the Museum of History and Technology, paper 69, U.S. National Museum Bulletin 252, Washington, D.C., 1967, p. 7.
47. *Railroad Car Journal*, August 1894, p. 167.
48. *R.R.G.*, June 30, 1876, p. 290.
49. Nicholas Wood, *A Practical Treatise on Rail-Roads*, London, 1832 edition, p. 544, mentions the use of oil cellar boxes on the B & O.
50. *Winans vs. Eastern Railroad*, p. 553, and *Niles Register*, Jan. 8, 1829, p. 299.
51. *Niles Register*, Feb. 21, 1829, p. 432.
52. Ibid., Aug. 29, 1829, p. 2.
53. *R.R.G.*, April 7, 1876, p. 149.
54. B & O annual report, 1831, p. 108.
55. *R.R.G.*, April 7, 1876, p. 149.
56. *A.R.R.J.*, Dec. 8, 1855, p. 777.
57. South Carolina Railroad annual report, 1840, p. 6, and 1843, p. 8.
58. *R.R. Adv.*, June 2, 1855, p. 4.
59. John Cochrane (Ed.), *Railway Patent Claim Reporter*, vol. 1, No. 2, July 1860, p. 37.
60. Ibid., p. 47.
61. *Master Car Builders Reports*, 1875, p. 78.
62. *A.R.R.J.*, May 1899, p. 147.
63. Ibid., December 8, 1855, p. 777.
64. *M.C.B.*, 1883, p. 85.
65. *N.C.B.*, September 1885, p. 113.
66. *A.R.R.J.*, June 1900, p. 171.
67. F. A. Ritter Von Gerstner, *Die innern communication der vereinigten Staaten von Nord-America*, vol. 1, Vienna, 1842–1843, p. 298.
68. *Am. Rly. Rev.*, July 14, 1859, p. 2.
69. *A.R.R.J.*, September 1843, p. 254.
70. Ibid., Nov. 8, 1856, p. 716.
71. Ibid., Sept. 1884, p. 162.
72. *N.C.B.*, December 1874, p. 182.
73. *M.C.B.*, 1875, p. 160.
74. *N.C.B.*, August 1891, p. 122.
75. James Dredge, *The Pennsylvania R.R.*, London, 1879, p. 90.
76. *Rly. Rev.*, June 16, 1888, p. 351.
77. *New York Railroad Club*, Jan. 18, 1900, p. 31.
78. Ibid., p. 12.
79. *M.C.B.*, 1875, p. 76.
80. Ibid., 1875, pp. 76 and 159; 1888, p. 37.

81. *Rly. M.E.*, May 1953, p. 56.
82. *Car Builders' Cyclopedia*, 1961, p. 851.
83. *Rly. M.E.*, September 1950, p. 494.
84. Railway and Locomotive Historical Society, *Bulletin 12*, 1926, p. 48. (Hereafter R.&L.H.S.)
85. *N.C.B.*, September 1870, p. 2.
86. *Rly. M.E.*, November 1943, p. 567.
87. *R.R. Mag.*, October 1953, p. 122.
88. *N.C.B.*, August 1891, p. 122.
89. *New York Railroad Club*, Jan. 18, 1900, p. 5.
90. *Rly. M.E.*, November 1934, p. 409, and May 1946, p. 260.
91. *Rly. M.E.*, March 1943, p. 120.
92. *R.R.G.*, Aug. 11, 1905, p. 121.
93. *M.C.B.*, 1889, pp. 45–52.
94. *N.C.B.*, July 1890, p. 97. Fletcher was not the inventor but rather the advocate of this form of journal box lid.
95. *R.R.G.*, Aug. 11, 1905, p. 121.
96. *Rly. M.E.*, September 1936, p. 405.
97. Ibid., November 1934, p. 407.
98. *Rly. Age*, March 9, 1946, p. 504.
99. *R.R. Mag.*, November 1950, p. 65.
100. *Rly. Age*, Sept. 15, 1969, p. 83.
101. H. T. Morton, *Anti-Friction Bearings* (published by the author), 1954; also *Cassier's Magazine*, May 1897, p. 60.
102. *Cassier's Magazine*, August 1899, p. 392.
103. *R.R.G.*, Nov. 22, 1889, p. 765.
104. *N.C.B.*, June 1892, p. 99.
105. *R.R.G.*, Aug. 30, 1889, p. 575.
106. Ibid., Oct. 5, 1888, p. 653.
107. *N.C.B.*, June 1894, p. 98.
108. *Master Mechanics Magazine*, August 1888, p. 120.
109. *Rly. M. E.*, January 1914, p. 19.
110. Ibid., December 1946, p. 688.
111. Ibid., November 1926, p. 667.
112. *Report of the Mechanical Advisory Committee to the Federal Coordinator*, Washington, D.C., 1935, p. 655.
113. *Railroad Car Facts*, 1961, p. 74.
114. *New York Railroad Club*, Oct. 20, 1949, p. 208.
115. R.&L.H.S. *Bulletin 23*, November 1930, p. 9.
116. *A.R.R.J.*, September 1843, p. 267.
117. *R.R. Adv.*, Sept. 20, 1856, p. 2.
118. *A.R.R.J.*, Feb. 18, 1854, p. 105.
119. *R.R.G.*, Feb. 13, 1883, p. 107.
120. *A.R.R.J.*, Sept. 28, 1850, p. 609.
121. Edward Hungerford, *Men and Iron: The History of the New York Central*, New York: Crowell, 1938, p. 244.
122. *Report on the Causes of Railroad Accidents and the Means of Preventing Their Recurrence*, New York Legislature, January 1853.
123. *A.R.R.J.*, Oct. 15, 1839, p. 256.
124. Ibid., Nov. 15, 1839, p. 289.
125. Charles Paine, *Elements of Railroading*, New York, 1895, p. 113.
126. *M.C.B.*, 1875, p. 134.
127. *R.R.G.*, Sept. 20, 1889, p. 616.
128. *R.R. Adv.*, Sept. 20, 1856, p. 2.
129. *A.R.R.J.*, July 16, 1853, p. 460; *Rly. Rev.*, April 9, 1887, p. 205; *R.R.G.*, April 2, 1886, p. 227; Paine, *Elements of Railroading*, p. 113.
130. Marshall Kirkman, *The Science of Railways*, vol. 4, Chicago: World Railway Publishing Company, 1904, p. 222.
131. *A.R.R.J.*, Aug. 15, 1857, p. 522.
132. Marshall Kirkman, *The Science of Railways. Cars: Their Construction, Handling and Supervision* (supplementary volume on cars), New York and Chicago, 1908, p. 97.
133. Robert S. Henry, *This Fascinating Railroad Business*, Indianapolis: Bobbs-Merrill, 1942, p. 268.
134. *N.C.B.*, February 1879, p. 25.
135. Ibid., November 1872, p. 9; November 1876, p. 165; February 1879, p. 25.
136. Peter Temin, *Iron and Steel in Nineteenth-Century America*, Cambridge, Mass.: M.I.T. Press, 1964, pp. 218–222.
137. John H. White, Jr., *American Locomotives 1830–1880*, Baltimore: Johns Hopkins, 1968, pp. 178–179.
138. *R.R. Adv.*, Sept. 20, 1856, p. 2.

139. *A.R.T.*, Jan. 19, 1861, p. 24.
140. *R.R.G.*, Jan. 20, 1893, p. 47 and Paine, *Elements of Railroading*, p. 113.
141. *Rly. Rev.*, Feb. 11, 1894, p. 106.
142. *R.R.G.*, Dec. 13, 1878, p. 599.
143. *N.C.B.*, August 1875, p. 121.
144. *R.R.G.*, Dec. 13, 1878, p. 599, and *N.C.B.*, July 1878, p. 104.
145. *N.C.B.*, March 1889, p. 43.
146. *Western Railroad Club*, May 1904, p. 270.
147. Ibid., p. 276.
148. Marshall Kirkman, *The Science of Railways. Cars: Their Construction, Handling and Supervision*, New York and Chicago, 1908, p. 94.
149. *R.R. Adv.*, Sept. 20, 1856, p. 2.
150. *Proceedings, A.A.R. Mechanical Division*, 1941, p. 356; and *Car Builders' Cyclopedia*, 1949–1951, p. 1052.
151. Data from W. D. Edson, former car superintendent for Amtrak.
152. *M.C.B.*, 1885, pp. 16–27.
153. *Rly. M.E.*, August 1939, p. 307, and J. E. Muhlfeld, *The Railroad Problem and Its Solution*, New York: Devin-Adair, 1941, p. 193.
154. *A.R.R.J.*, June 15, 1833, p. 369.
155. M. J. T. Lewis, *Early Wooden Railways*, London: Routledge & Kegan Paul, 1970, p. 195.
156. R.&L.H.S. *Bulletin 61*, May 1943, p. 49.
157. *Baltimore Engineer*, August 1928, p. 7.
158. Kirkman, *Science of Railways*, vol. 4, 1904, p. 217.
159. Muhlfeld, *Railroad Problem*, p. 192.
160. *Car Builders' Dictionary*, 1888, fig. 2150.
161. George H. Burgess and M. C. Kennedy, *Centennial History of the Pennsylvania Railroad*, Philadelphia, 1949, p. 324.
162. *N.C.B.*, September 1877, p. 131, and November 1877, p. 169.
163. Memorandum written by Isaac Dripps in 1885. Now in papers of J. E. Watkins, Smithsonian Archives.
164. Petersburg Railroad annual report, 1835, p. 14.
165. *Budget* (Troy, N.Y., newspaper), Sept. 15, 1835.
166. Baldwin mss. collection, Historical Society of Pennsylvania, Philadelphia, letter dated Feb. 17, 1837, to Joseph Johnson.
167. Lewis, *Early Wooden Railways*, p. 195; C. F. Dendy Marshall, *A History of the Railway Locomotive Engine*, London: Locomotive Publishing Company, 1953, p. 126.
168. R.&L.H.S. *Bulletin 65*, October 1944, p. 49.
169. *N.C.B.*, September 1891, p. 140.
170. R.&L.H.S. *Bulletin 32*, October 1933, p. 57.
171. *A.R.R.J.*, June 19, 1847, p. 392.
172. *Baltimore Engineer*, August 1929, p. 5; *Business History Review*, Summer 1968, p. 171.
173. White, *American Locomotives*, pp. 280–286.
174. B & O annual report, 1834, p. 27.
175. *A.R.R.J.*, July 9, 1853, p. 434.
176. *Institution of Civil Engineers*, vol. 28, 1869, p. 383.
177. *U.S. Centennial Reports and Awards*, vol. VI, 1880, p. 41; *A.R.R.J.*, June 1896, p. 108, and September 1896, p. 222.
178. *Anson Atwood vs. Portland Company, U.S. Circuit Court, District of Maine, Sept. 23 and 24, 1879*. Printed copy of pleadings, proofs, briefs, and opinions, in Library of Congress. Hereafter *Atwood vs. Portland*.
179. Zerah Colburn, *Recent Practice in the Locomotive Engine*, London, 1860, p. 53.
180. *Atwood vs. Portland*, p. 42.
181. *Rly. Rev.*, Feb. 2, 1889, p. 64.
182. *A.R.R.J.*, Nov. 6, 1847, p. 705.
183. *Cassier's Mag.*, March 1910, p. 440.
184. *Railway and Locomotive Engineer*, December 1924, p. 367.
185. *R.R.G.*, April 20, 1872, p. 175.
186. Ramapo Wheel & Foundry Company, Ramapo, N.Y., catalog 1887c., p. 14.
187. *Institution of Civil Engineers*, 1869, p. 402.
188. Ibid., 1878, p. 37.
189. *Rly. Rev.*, Feb. 11, 1882, p. 89.
190. *R.R.G.*, April 20, 1872, p. 175.
191. Ibid.
192. Ibid.; *U.S. Centennial Reports and Awards*, vol. VI, 1880, p. 49; *Scientific American Supplement*, July 29, 1876.
193. *R.R.G.*, April 29, 1898, p. 304.

194. Ibid., Feb. 10, 1911, p. 268.
195. *M.C.B.*, 1883, p. 32.
196. *Southern and South Western Railroad Club*, November 1939, p. 37.
197. *Institution of Civil Engineers*, 1878, p. 69.
198. *R.R.G.*, May 22, 1896, p. 364.
199. *Engineer*, Dec. 28, 1883, p. 499.
200. *Report of the Committee Investigating the Affairs of the Boston and Providence R.R.*, Boston, 1856, p. 26.
201. *Rly. M.E.*, July 1947, p. 347.
202. Ibid., August 1933, p. 296.
203. *R.R.G.*, April 20, 1872, p. 175.
204. Bush and Lobdell Catalog, 1851, p. 6. Copy in A.A.R. Library as of 1968.
205. *American Society of Civil Engineers Transactions*, vol. 20, 1899, p. 615.
206. *Baltimore Engineer*, August 1929, p. 9.
207. *R.R.G.*, April 29, 1898, p. 304.
208. Baldwin Locomotive Works, *Record of Recent Construction*, no. 52, 1905, p. 13; *Baltimore Engineer*, August 1929, p. 9.
209. Dredge, *The Pennsylvania R.R.*, p. 83. Data gathered in 1875 and 1876.
210. *Engineer*, Dec. 28, 1883, p. 499; *Cassier's Mag.*, March 1910, p. 440.
211. *R.R.G.*, July 5, 1889, p. 439, and June 19, 1908, p. 211.
212. *Jour. Franklin Inst.*, September 1843, p. 212; October 1845, p. 273; February 1848, p. 119, and September 1848, p. 210.
213. Alvin F. Harlow, *Road of the Century*, New York: Creative Age Press, 1947, p. 146; quotes from a newspaper account dated May 26, 1849.
214. *A.R.T.*, Jan. 19, 1861, p. 24.
215. Zerah Colburn, *The Locomotive Engine*, Philadelphia, 1854, p. 72.
216. *Report on the Causes of Railroad Accidents and the Means of Preventing Their Recurrence*, New York Legislature, January 1853, pp. 51 and 59.
217. *A.R.T.*, Jan. 19, 1861, p. 24.
218. *Western Railroad Gazette*, April 24, 1869, p. 1.
219. *A.R.T.*, April 10, 1869, p. 118.
220. *Rly. Rev.*, Oct. 18, 1890, p. 629.
221. *R.R.G.*, June 3, 1887, p. 373.
222. *N.C.B.*, March 1872, p. 4.
223. Ibid., February 1871, p. 4.
224. *Leslie's Illustrated Newspaper*, Aug. 25, 1877.
225. *Rly. Age*, Sept. 1, 1881, p. 494.
226. *A.R.R.J.*, April 22, 1882, p. 255, and *R.R. Car Jour.*, June 1893, p. 232.
227. Advertising leaflet, 1883c., published by the Allen Paper Wheel Company.
228. *R.R. Car Jour.*, June 1893, p. 232.
229. *R.R.G.*, Jan. 26, 1883, p. 53.
230. *N.C.B.*, July 1880, p. 124.
231. Ibid., July 1882, p. 73, and *Rly. Rev.*, Feb. 10, 1883, p. 83.
232. *R.R.G.*, June 11, 1886, p. xviii.
233. *R.R. Car Jour.*, June 1893, p. 232.
234. *R.R.G.*, June 11, 1886, p. xviii.
235. *M.C.B.*, 1875, p. 140; *N.C.B.*, August 1877, p. 117.
236. *R.R.G.*, Sept. 4, 1891, p. 613.
237. Conversation with George Krambles, general manager of the Chicago Transit Authority, May 1972.
238. *M.C.B.*, 1875, p. 140.
239. Ibid., 1883, p. 58.
240. *Engineer*, Dec. 28, 1883, p. 499.
241. *R.R.G.*, Dec. 13, 1878, p. 599.
242. *N.C.B.*, June 1888, p. 85.
243. *Rly. Age*, Sept. 1, 1881, p. 494.
244. Ibid.
245. *N.C.B.*, November 1879, p. 166.
246. *R.R.G.*, Jan. 26, 1883, p. 53.
247. *N.C.B.*, January 1893, p. 1.
248. *Rly. Age*, Aug. 18, 1881, pp. 473 and 476.
249. *Locomotive Engineering*, April 1897, p. 320.
250. *Pullman News*, April 1923, p. 364.
251. *Engineering News*, July 1, 1915, p. 11.
252. *Rly. & Loco. Eng.*, March 1917, p. 79.
253. *R.R.G.*, May 24, 1901, p. 352.

254. *Southern & Southwestern Railway Club*, March 1917, p. 4.
255. L. S. Marks, *Mechanical Engineer's Handbook*, New York, 1930, p. 1483.
256. *R.R. Mag.*, March 1969, p. 18.
257. *Sci. Am.*, January 1924, p. 35.
258. *A.R.R.J.*, Dec. 27, 1862, p. 1027.
259. *A.R.T.*, Jan. 18, 1868, p. 23.
260. Ibid., July 31, 1869, p. 246.
261. *A.S.M.E. Trans.*, 1881, p. 300.
262. *Rly. Rev.*, June 9, 1888, p. 330; *R.R.G.*, Nov. 21, 1902, p. 889.
263. *R.R.G.*, Feb. 17, 1888, p. 110.
264. *Year Book of the American Iron & Steel Institute*, 1913, p. 367.
265. *R.R.G.*, March 1, 1912, p. 369.
266. *Report of the Mechanical Advisory Committee to the Federal Coordinator*, 1935, p. 650.
267. *Wheel and Axle Manual: Standards, Recommended Practices, and General Information*, 9th ed., Chicago: Association of American Railroads, 1968, pp. 142–145.
268. Ibid.
269. Michael Reynolds, *Continuous Railway Brakes*, London, 1882, p. 1.
270. *Rly. Age*, Dec. 21, 1882, p. 720.
271. Von Gerstner, vol. 2, p. 255.
272. F. C. B. Bradlee, *The Boston and Maine Railroad*, Salem, Mass., 1921, p. 39.
273. *N.C.B.*, April 1890, p. 58.
274. S. M. Whipple, *Appendix. Documents and Depositions . . . in behalf of Respondents before the Senate and House Committees on Patents, in Answer to the Memorial A. G. Batchelder, et. al.* A pamphlet published in 1874 by the Eastern Rail Road Association, Springfield, Mass. Hereafter Whipple.
275. Seymour Dunbar, *A History of Travel in America*, vol. III, Indianapolis: Bobbs-Merrill, 1946, p. 987.
276. *Railroad Record*, Oct. 27, 1853, p. 551.
277. Ibid.
278. Von Gerstner, vol. 1, p. 261.
279. *R.&L.H.S. Bulletin 12*, 1926, p. 48.
280. *N.C.B.*, July 1881, p. 84.
281. George Bradshaw, *Prevention of Railroad Accidents*, New York, 1912, p. 18.
282. H. G. Prout, *A Life of George Westinghouse*, New York, 1922, p. 33.
283. *Report of the American Institute*, New York, 1855, p. 548.
284. *R.R. Adv.*, June 7, 1856, p. 3.
285. Jonathan Knight, *Report upon the Locomotive Engines*, Baltimore, 1838, pp. 4 and 10.
286. *Report on Causes of Railroad Accidents*, N.Y. Legislature, 1853, pp. 43, 67, 88.
287. *M.C.B.*, 1875, p. 24.
288. Voss, *Railway Car Construction*, p. 77.
289. Whipple, p. 44.
290. Ibid., p. 24.
291. Ibid., pp. 1–140; *Rly. Rev.*, Jan. 27, 1883, p. 45.
292. *Rly. Age*, April 7, 1881, p. 188.
293. *Jour. Franklin Inst.*, October 1835, p. 228.
294. *A.R.R.J.*, Sept. 15, 1841, p. 168, and Sept. 28, 1848, p. 691.
295. S. A. Howland, *Steamboat Disasters and Railroad Accidents*, Worcester, Mass., 1840, p. 262.
296. *Sci. Am.*, Dec. 9, 1854, p. 100.
297. *Am. Rly. Rev.*, Oct. 20, 1859, p. 3.
298. M. Chevalier, *Histoire et Description Des Voies De Communication*, vol. 2, Paris, 1841, p. 515; *Hazard's Register of Pennsylvania*, Aug. 2, 1838.
299. *A History and Description of the Baltimore and Ohio Railroad*, no author, Baltimore, 1853, p. 48.
300. Hamilton Ellis, *Railway Carriages in the British Isles from 1830 to 1914*, London: Allen & Unwin, 1965, p. 194.
301. *R.R. Adv.*, Feb. 28, 1857, p. 4.
302. *Am. Rly. Rev.*, July 27, 1859, p. 13.
303. *West. R.R. Gaz.*, Jan. 19, 1861.
304. *Report American Institute*, 1855, p. 552.
305. Paul Synnestvedt, *Evolution of the Air Brake*, New York, 1895, 15.
306. Reynolds, *Continuous Railway Brakes*, p. 41.
307. *R.R.G.*, Dec. 31, 1870, p. 314.

308. *A.R.R.J.*, July 14, 1855, p. 449.
309. *R.R. Adv.*, June 7, 1856, p. 4.
310. *Am. Rly. Rev.*, May 23, 1861, p. 312.
311. Ibid., June 1862, p. 181.
312. *Engineering*, July 12, 1867, p. 32.
313. *R.R.G.*, March 4, 1871, p. 532.
314. *Decisions of the Commissioners of Patents*, Washington, 1870, p. 133.
315. *Master Mechanics Report*, 1872, p. 114.
316. *Master Mechanics Magazine*, May 1898, p. 70.
317. Synnestvedt, *Evolution of the Air Brake*, p. 10.
318. *A.R.R.J.*, March 24, 1855, p. 190.
319. Pamphlet on steam brakes by T. W. Smith, 1856, in Association of American Railroads Library, Washington, D.C.
320. *R.R.G.*, Aug. 14, 1875, p. 338; *N.C.B.*, December 1874, p. 181.
321. G. Weissenborn, *American Locomotive Engineering*, N.Y. 1871, p. 174.
322. *N.C.B.*, August 1890, p. 121.
323. Ibid., October 1871, p. 5.
324. Ibid., March 1873, p. 60.
325. Weissenborn, p. 174.
326. *N.C.B.*, February 1877, p. 24.
327. *R.R.G.*, April 14, 1876, p. 160.
328. *N.C.B.*, October 1876, p. 153.
329. Ibid., February 1879, p. 17.
330. Ellis, *Railway Carriages in The British Isles*, p. 197.
331. *A.R.R.J.*, August 1892, p. 355; *Rly. Rev.*, April 2, 1881, p. 179.
332. *N.C.B.*, June 1881, p. 66.
333. *Rly. Rev.*, July 4, 1885, p. 323.
334. *The New York Air Brake Company Fiftieth Anniversary, 1890–1940*, 30-page pamphlet published by the N.Y.A.B. Co. Hereafter *N.Y.A.B. History*.
335. J. H. White, Jr., *American Single Locomotives*, Smithsonian Studies in History and Technology No. 25, Washington, D.C.: Smithsonian Institution, 1973, p. 7.
336. *Rly. Rev.*, Oct. 9, 1880, p. 528.
337. Ibid., Sept. 30, 1882, p. 561; *Railway Magazine*, London, 1907, p. 468.
338. E. A. Forward, *Handbook of the Science Museum: Land Transport*, vol. 3, part I, London, 1931, p. 89. Hereafter *Science Museum Handbook*.
339. Reynolds, *Continuous Railway Brakes*, p. 92.
340. *N.Y.A.B. History*, p. 3.
341. *R.R.G.*, April 27, 1883, p. 271.
342. *Rly. Age*, Nov. 4, 1886, p. 611; *Rly. Rev.*, March 10, 1886, p. 179.
343. *Car Builders' Dictionary*, 1888, p. 190.
344. *R.R.G.*, July 8, 1898, p. 501.
345. *A.R.R.J.*, Aug. 1892, p. 355.
346. *Moody's Industrials*, 1920, p. 1218.
347. *Fortune*, March 1937, p. 115.
348. John Bourne, *Ensamples of Steam, Air and Gas Engines*, London, 1878, p. 171.
349. *Science Museum Handbook*, p. 89.
350. *Sci. Am.*, Dec. 25, 1852, p. 116.
351. *R.R.G.*, April 10, 1903, p. 272.
352. Prout, *Life of Westinghouse*, p. 24.
353. *Loco. Eng.*, February 1900, p. 75.
354. A.S.M.E. Presidential Address Dec. 6, 1910. George Westinghouse's recollection of origins of the air brake. Hereafter Westinghouse, *Recollection*.
355. *Loco. Eng.*, February 1900, p. 75.
356. *Von Nostrand's Engineering Magazine*, September 1870, p. 327.
357. *Engineering*, May 24, 1872, p. 344.
358. Dredge, *The Pennsylvania R.R.*, p. 170.
359. Garrett papers, Maryland Historical Society, Baltimore.
360. C. Clark, "The Railroad Safety Movement in the United States" (doctoral dissertation, University of Illinois, 1966), p. 74.
361. Clark, "Railroad Safety Movement," 1966, p. 74.
362. *Jour. Frank. Inst.*, April 1874, p. 245.
363. Prout, *Life of Westinghouse*, p. 34.
364. *R.R.G.*, July 23, 1880, p. 403.
365. *N.C.B.*, April 1880, p. 66.
366. *Rly. Age*, Jan. 28, 1882, p. 55.
367. Ibid., April 1887, p. 51.

368. Ibid., Nov. 1887, p. 148.
369. Ibid., March 1883, p. 30.
370. *Rly. Rev.*, Sept. 22, 1883, p. 563.
371. Ibid., Jan. 1, 1881, p. 10.
372. *Fortune*, March 1937, p. 118.
373. *R.R.G.*, Nov. 11, 1898, p. 816.
374. M. Wilkins, *Emergence of Multinational Enterprise*, Cambridge, Mass.: Harvard, 1970, p. 59.
375. Prout, *Life of Westinghouse*, pp. 53–62.
376. *R.R.G.*, Dec. 29, 1893, p. 950.
377. W. V. Turner, *The Air Brake as Related to Progress in Locomotion*, Pittsburgh, 1910, pp. 19–22.
378. Ibid.
379. *R.R. Car Jour.*, August 1897, p. 240.
380. *Rly. & Loco. Eng.*, August 1924, p. 250.
381. *Federal Coordinators Report*, 1935, p. 668.
382. *Sci. Am.*, June 20, 1914.
383. Turner, *The Air Brake*, p. 38.
384. *Sci. Am.*, June 20, 1914.
385. *Car Builders' Cyclopedia*, 1943, p. 951.
386. S. K. Farrington, Jr., *Railroads of Today*, New York: Coward-McCann, 1949, p. 36.
387. *Rly. M.E.*, February 1949, p. 66.
388. L. K. Sillcox, *Mastering Momentum*, New York, 1941, p. 69.
389. *Rly. Age*, Nov. 17, 1969, p. 7.
390. *Loco. Eng.*, October 1891, p. 191.
391. *M.C.B.*, 1874, p. 93.
392. *R.R. Car Jour.*, June 1894, p. 132.
393. *Western Railroad Club*, October 1889, p. 33.
394. *M.C.B.*, 1875, p. 29.
395. Ibid., p. 16.
396. *Rly. M.E.*, Feb. 1913, p. 109.
397. *Car Builders' Dictionary*, 1909, p. 523.
398. *R.R.G.*, Dec. 9, 1887, p. 793.
399. *Rly. Age*, June 12, 1912, Aug. 16, 1912, pp. 1267 and 279. See also R.&L.H.S. *Bulletin 97*, October 1957, p. 65.
400. *N.Y.R.R. Club*, 1949, p. 247.
401. *Rly. M.E.*, December 1947, p. 719.
402. Ibid., March 1946, p. 121.
403. Ibid.
404. *N.Y.R.R. Club*, 1949, p. 247; *Rly. Age*, March 14, 1960.
405. *Car and Locomotive Cyclopedia*, 1966, p. 759.
406. Frank W. Stevens, *Beginnings of the New York Central Railroad*, New York: Putnam's, 1926, p. 42.
407. *A.R.R.J.*, September 1843, p. 285.
408. *Rly. Age*, June 19, 1884, p. 390.
409. *M.C.B.*, 1875, p. 115.
410. *N.C.B.*, June 1876, p. 85.
411. *R.R.G.*, Nov. 21, 1884, p. 829.
412. Ibid.
413. Wilson, Walker, & Company, Pittsburgh. Catalog 1875c., in DeGolyer Library, Dallas, Texas.
414. *N.C.B.*, March 1880, p. 39.
415. *N.C.B.*, December 1880, p. 182.
416. *Rly Rev.*, Dec. 25, 1886, p. 700.
417. Ibid., April 10, 1897.
418. *R.R.G.*, Oct. 4, 1873, p. 403; Sept. 18, 1885, p. 598; March 12, 1897, p. 190; and *R.R. Mag.*, January 1930, p. 233.
419. *R.R. Mag.*, May 1947, quotes the Baltimore *Clipper* for July 31, 1847. See also *A.R.R.J.*, March 18, 1848, p. 177.
420. *Am. Rly. Rev.*, July 27, 1859, p. 15.
421. *N.C.B.*, December 1875, p. 186.
422. Ibid., Nov. 1871, p. 3.
423. *R.R. Adv.*, Dec. 27, 1856, p. 1; *Great Railway Celebration*, Baltimore, 1857, rear advertising section, p. 24.
424. *Industrial America, or Manufacturers and Inventors of the United States*, New York, 1876, pp. 405 and 412.
425. *Dictionary of American Biography*, vol. 12, 1933, p. 625.
426. *A.R.R.J.*, September 1887, p. 401.
427. *R.R. Car Jour.*, November 1894, p. 243.
428. *R.R.G.*, Oct. 2, 1885, p. 630.
429. *A.R.R.J.*, Sept. 11, 1869, p. 1043.

430. Ibid., June 4, 1870, p. 644.
431. Ibid., Nov. 13, 1875, p. 1441.
432. *Industrial America*, 1876, p. 401.
433. *A.R.R.J.*, Dec. 26, 1874, p. 1633.
434. *Rly. Age*, July 16, 1885, p. 456.
435. *Miller's Trussed Platforms*, 1868 published catalog in A.A.R. Library, Washington, D.C.
436. *A.R.R.J.*, June 1887, p. 250.
437. Miller contract is in the collection of the Newberry Library, Chicago.
438. Letter to John W. Garrett, president of the B & O, Sept. 14, 1877.
439. Voss, *Railway Car Construction*, p. 93.
440. *R.R.G.*, April 11, 1890, p. 251.
441. Ibid.
442. I.C.C. Annual Report, 1893, p. 28.
443. Ibid., 1893, p. 28, and 1899, p. 35.
444. *N.C.B.*, November 1874, p. 165.
445. Clark, "Railroad Safety Movement," p. 129.
446. *R.R.G.*, Oct. 2, 1885, p. 630.
447. E. Wilson (Ed.), *History of Pittsburgh*, Chicago, 1898, p. 1048; *The Evolution of Car Couplings*, pamphlet published in 1905 by McConway & Torley, Pittsburgh.
448. *Engineering*, July 13, 1883, p. 33.
449. *R.R.G.*, Oct. 26, 1888, p. 710.
450. *M.C.B.*, 1888, pp. 100–101.
451. *R.R.G.*, Oct. 26, 1888, p. 710.
452. *Eng. News*, May 8, 1890, p. 422.
453. Ibid.
454. *R.R.G.*, March 10, 1899, p. 166.
455. Ibid., Nov. 13, 1891, p. 794.
456. Ibid., Oct. 16, 1896, p. 719 and *R.R. Car Jour.*, July 1899, p. 208.
457. Henry, *Fascinating Railroad Business*, p. 231.
458. *Baldwin Mag.*, July 1928, p. 30.
459. *Rly. M.E.*, Aug. 1919, p. 473.
460. Ibid., August 1920, p. 521.
461. Ibid., May 1932, p. 207.
462. Ibid., September 1938, p. 359.
463. Ibid., March 1931, p. 126.
464. *Car Builders' Cyclopedia*, 1953, p. 738.
465. *Trains*, October 1968, p. 16.
466. *R.R.G.*, Sept. 16, 1904, p. 345.
467. *Trains*, January 1969, p. 31.
468. *Trains*, April 1975, p. 6.
469. Henry, *Fascinating Railroad Business*, p. 233.
470. *R.R.G.*, Nov. 17, 1899, p. 794; *Rly. M.E.*, October 1913, p. 520.
471. *Rly. M.E.*, September 1914, p. 453.
472. Ibid., March 1930, p. 125.
473. *A.R.R.J.*, Jan. 1, 1839, p. 4.
474. Ibid.
475. *R.&L.H.S. Bulletin 23*, November 1930, p. 11.
476. Zerah Colburn, *Locomotive Engineering and the Mechanism of Railways*, London, 1872, p. 97.
477. *M.C.B.*, 1875, p. 108.
478. Ibid., p. 107.
479. *Rly. M.E.*, September 1952, p. 86.
480. *Car Builders' Dictionary*, 1879, figs. 252–258; 1888, figs. 409–643; 1895, figs. 1936–2354.
481. Prout, *Life of Westinghouse*, p. 77.
482. *Eng. News*, March 1, 1890, p. 207.
483. *R.R.G.*, Jan. 25, 1889, p. 61.
484. Ibid., Oct. 13, 1899, p. 705.
485. *Rly. M.E.*, July 1914, p. 341.
486. Ibid., December 1943, p. 591.
487. Ibid., August 1930, p. 445.
488. *Railroad Mag.*, February 1948, p. 68.
489. *Rly. M.E.*, May 1935, p. 171.
490. *R.R.G.*, June 2, 1893, p. 406.

CHAPTER EIGHT

1. R. W. Kidner, *Development of the Railcar*, South Godstone, England: Oakwood Press, 1958, ed. p. 3.
2. J. H. White, Jr., *American Single Locomotives and the Pioneer*, Washington, D.C.: Smithsonian Institution, 1973, p. 10.
3. *American Artisan*, April 19, 1865, p. 394.
4. *Western Railroad Gazette*, June 9, 1860, p. 1.
5. *American Railway Review*, Nov. 15, 1860, p. 293.
6. *Scientific American*, Nov. 24, 1860, p. 340.
7. *American Railroad Journal*, Nov. 24, 1860, p. 1039.
8. George B. Abdill, *Civil War Railroads*, Seattle: Superior, 1961, pp. 139 and 162.
9. *A.R.R.J.*, May 26, 1860, p. 442.
10. *Am. Rly. Rev.*, March 8, 1860, p. 144.
11. Ibid., Dec. 5, 1861, p. 168; *American Artisan*, April 19, 1865, p. 394; *Locomotive Engineering*, January 1898, p. 16.
12. *Am. Rly. Rev.*, Aug. 8, 1861, p. 26, and Aug. 15, 1861, p. 42.
13. Railway and Locomotive Historical Society, *Bulletin 95*, October 1956, p. 61; *Loco. Eng.*, March, 1911, p. 101.
14. John H. White, Jr., "Grice and Long: Steam-Car Builders," in Jack Salzman (Ed.), *Prospects: An Annual of American Cultural Studies*, vol. 2, New York: Burt Franklin Company, pp. 24–39.
15. *Western Railroader*, June 1970, p. 16.
16. *Loco. Eng.*, October 1891, p. 205.
17. *American Railway Times*, May 5, 1866, p. III.
18. *A.R.R.J.*, Sept. 23, 1865, p. 917.
19. *A.R.T.*, Aug. 11, 1866, p. 254.
20. *Scientific American Supplement*, June 17, 1876, p. 399.
21. *National Car Builder*, October 1872, p. 1.
22. Ibid., January 1890, p. 14.
23. *Railroad Gazette*, June 13, 1879, p. 323.
24. *New England Railroad Club*, Dec. 14, 1897, pp. 6–54.
25. *A.R.R.J.*, November 1897, p. 367.
26. *Loco. Eng.*, November 1897, p. 817.
27. *Engineering News*, Feb. 6, 1902, p. 102.
28. *Railroad Magazine*, June 1945, p. 129.
29. *R.R.G.*, April 23, 1897, p. 295.
30. *Railway Review*, June 11, 1898, p. 315.
31. *R.R.G.*, April 8, 1898, p. 255.
32. *Eng. News*, Feb. 6, 1902, p. 102.
33. *R.R.G.*, June 24, 1898, p. 463.
34. *Sci. Am.*, June 24, 1899, p. 407.
35. *Eng. News*, Feb. 6, 1902, p. 104.
36. *Rly. Rev.*, June 11, 1898, p. 315.
37. *Eng. News*, Feb. 6, 1902, p. 104.
38. *Rly. Rev.*, April 8, 1899, p. 196.
39. *Eng. News*, Feb. 6, 1902, p. 104.
40. *R.R.G.*, April 13, 1900, p. 238.
41. Ibid., Dec. 30, 1904, p. 688.
42. *New York Railroad Club*, April 19, 1907, p. 688.
43. *R.R.G.*, March 29, 1907, p. 451.
44. *International Railway Congress Bulletin*, 1909, p. 1322.
45. *Rly. Rev.*, July 11, 1908, p. 564.
46. *Car Journal*, September 1909, p. 17.
47. *R.R.G.*, Sept. 17, 1909, p. 511.
48. *R.R. Mag.*, April 1966, p. 16.
49. *New York Railroad Club*, April 19, 1907, p. 685.
50. *Rly. Rev.*, March 24, 1906, p. 213, and Oct. 5, 1907, p. 872.
51. Alvin Staufer and Fred Westing, *Erie Power*, Medina, Ohio: Alvin Staufer, 1970, p. 360.
52. E. D. Worley, *Iron Horses of the Santa Fe Trail*, Dallas: Southwestern Railroad Historical Society, 1965, p. 385.
53. *New York Railroad Club*, May 1922, p. 6743; *Car Review*, October 1909, p. 13.
54. *Car Review*, September 1909, p. 18.
55. *Rly. Rev.*, Sept. 29, 1906, p. 748.
56. *R.R.G.*, June 16, 1905, p. 672.
57. *Car Review*, September 1909, p. 17.
58. R. B. Jackson, *The Steam Cars of the Stanley Twins*, New York: Henry Z. Walck, Inc. 1969, p. 54.
59. *Electric Railway Journal*, Sept. 16, 1916, p. 506.
60. *New York Railroad Club*, May 1922, p. 6745.
61. *Mechanical Engineering*, February 1937, p. 120.
62. *Railway Age*, Oct. 14, 1922, p. 713.
63. *R.R. Mag.*, March 1942, p. 77.
64. *Railway Mechanical Engineer*, December 1927, p. 768.
65. *Brotherhood of Locomotive Firemen and Engineers Magazine*, October 1929.

66. *Rly. M.E.*, May 1931, p. 268.
67. *Trains*, June 1948, p. 52.
68. New Haven Railroad Technical Information Association, Standing Data Sheet 6–9210, 1974.
69. *Rly. M.E.*, November 1936, p. 485.
70. *Railway Gazette* (England) March 12, 1937, p. 495.
71. *Street Railway Journal*, May 1889, p. 112.
72. *Railroad History No. 129*, Autumn 1973, pp. 86–93.
73. *R.R.G.*, May 5, 1899, p. 313.
74. Julius S. Morton (Ed.), *Illustrated History of Nebraska*, vol. 3, Lincoln, Neb., 1913, pp. 473–477.
75. *Trains*, July 1960, p. 30.
76. *New York Railroad Club*, April 1907, p. 669.
77. *A.R.R.J.*, August 1905, p. 294; *Rly. Rev.*, Nov. 23, 1907, p. 1008.
78. *R.R. Mag.*, January 1951, p. 17.
79. *A.R.R.J.*, November 1905, p. 420; *Rly. Rev.*, Oct. 7, 1905, p. 720.
80. *Railway Age Gazette*, Nov. 12, 1909, p. 946.
81. *Master Mechanics Report*, 1909, p. 87; *International Railway Congress Bulletin*, 1909, p. 1358; *New York Railroad Club*, April 1907, p. 669.
82. *Rly. Rev.*, Nov. 23, 1907, p. 1008.
83. *Rly. Age*, Dec. 22, 1911, p. 1267.
84. *Atlantic Monthly*, July 1937, p. 542.
85. *Rly. Age*, April 19, 1924, p. 977.
86. *Trains*, July 1960, p. 39.
87. A. D. Dubin, *More Classic Trains*, Milwaukee: Kalmbach Publishing Company, 1974, p. 350.
88. *American Society of Mechanical Engineers Transactions*, December 1929, p. 29.
89. *R.R. Mag.*, January 1951, p. 23.
90. *Trains*, November 1973, pp. 39–49. All data on G.E. cars are from this source unless otherwise noted.
91. *Rly. Age Gaz.*, June 23, 1909, p. 1451.
92. *N.Y.R.R. Club*, January 1912, pp. 2653–2682.
93. *Cyclopedia of Engineering*, vol. 3, Chicago: American Technical Society, 1926, pp. 385–510. The text appears to have been written for the 1915 edition of this work.
94. *Rly. Age*, Sept. 27, 1924, p. 536.
95. *Rly. Age*, April 6, 1906, p. 625.
96. *Int. Rly. Cong. Bulletin*, 1909, p. 1335.
97. *A.R.R.J.*, November 1909, p. 460.
98. *Rly. Age*, June 19, 1914, p. 1525.
99. Electric Railway Historical Society, *The St. Joseph Valley Line*, Bulletin 16, 1955, p. 24.
100. *Rly. Age*, June 22, 1929, p. 64.
101. *Colorado Rail Annual*, no. 9 (1971), Golden, Colo: Colorado Railroad Museum, pp. 3–49.
102. *Rly. Rev.*, Nov. 14, 1925, p. 769.
103. *Rly. Rev.*, May 27, 1922, p. 741.
104. *Engineers and Engineering*, Philadelphia Engineer's Club, May 1923, p. 133.
105. *Car Builders' Cyclopedia*, 1922, p. 355.
106. *Car Building in Philadelphia*, 1926, p. 11. (Pamphlet published by the J. G. Brill Company.)
107. Edwards Railway Motor Car Company Catalog, 1928. Reprinted in 1970 by Old Line Publishing Company, Milwaukee—contains a short history of Edwards.
108. *Loco 1: The Diesel*, Ramsey, N.J.: Railroad Model Craftsman, 1966, p. 13.
109. *Rly. Age*, July 22, 1922, p. 170.
110. Ibid., Aug. 5, 1933, p. 219.
111. *Motor Coach Age*, December 1969, p. 21.
112. R. L. Kulp (Ed.), *History of Mack Rail Motor Cars and Locomotives*, Allentown, Pa.: Lehigh Valley Chapter, Nat. Rly. Hist. Soc., 1959, pp. 9–11.
113. *Extra 2200 South*, February–March 1969, p. 16.
114. *Motor Coach Age*, April 1968, p. 5.
115. *Car Builders' Cyclopedia*, 1925, p. 357.
116. *Rly. Age*, Dec. 27, 1924, p. 1155.
117. *Railway and Locomotive Engineer*, April 1926, p. 93.
118. Brill production data was furnished by Harold E. Cox of Philadelphia through Louis A. Marre of Dayton, Ohio.
119. *Rly. & Loco. Eng.*, April 1926, p. 93.
120. *Rly. Age*, Aug. 19, 1922, p. 352.

121. *Car Builders' Cyclopedia*, 1928, fig. 1039.
122. E. H. Wagner, Jr., *The Colorado Road*, Denver: Intermountain Chapter, Nat. Rly. Hist. Soc., 1970, p. 260.
123. *Rly. Rev.*, Sept. 6, 1921, p. 343; F. M. Reck, *On Time: The History of Electro Motive Division*, La Grange, Ill.: EMD, 1948.
124. *Rly. Age*, Aug. 30, 1924, p. 371.
125. Ibid., and *Rly. Rev.*, Sept. 6, 1924, p. 343.
126. Wagner, *The Colorado Road*, p. 261.
127. *Our GM Scrapbook*, Milwaukee: Kalmbach Publishing Company, 1971, p. 14.
128. *Rly. Age*, Nov. 5, 1949, unnumbered adv. page.
129. *Rly. & Loco. Eng.*, December 1926, p. 337.
130. *Rly. Age*, Jan. 29, 1927, p. 353.
131. Ibid., Jan. 1932, p. 68.
132. *A.S.M.E. Transactions*, December 1929, p. 30.
133. *Rly. Age*, Oct. 17, 1925, p. 695.
134. *Extra 2200 South*, March–April 1971, p. 21.
135. *Rly. Age*, July 7, 1928, p. 17.
136. *Rly. Age*, Jan. 5, 1929, p. 110.
137. *Rly. Age*, May 26, 1934, p. 773; the car was built in 1932.
138. *Rly. Age*, Jan. 2, 1932, p. 68.
139. *Rly. Rev.*, May 27, 1922, p. 741.
140. *Rly. Age*, Nov. 7, 1931, p. 700.
141. Ibid., April 16, 1926, p. 1195.
142. *Wall Street Journal*, April 24, 1934.
143. *Rly. Age*, Oct. 7, 1933, p. 502.
144. Ibid., Sept. 13, 1924, p. 447.
145. Ibid., Jan. 26, 1929, p. 247.
146. Dubin, *More Classic Trains*, p. 200.
147. *A.S.M.E. Transactions*, December 1929, p. 30.
148. *Atlantic Monthly*, July 1935, p. 535.
149. Art S. Genet, "Profile of Greyhound," Newcomen Society in North America, 1958, pp. 17–18.
150. J.F. Stover, *Life and Decline of the American Railroad*, New York: Oxford, 1970, p. 156.
151. *Rly Age*, Jan. 27, 1934, p. 166.
152. *Pullman News*, October 1933, p. 35.
153. J. H. Lemby, *The Gulf Mobile and Ohio*, Homewood, Ill.: Richard D. Irwin, 1953, p. 137.
154. *Rly. Age*, Dec. 19, 1955, p. 29.
155. H. E. Ranks and W. W. Kratville, *The Union Pacific Streamliners*, Omaha: Kratville Publications, 1974, p. 40. Hereafter *U P Streamliners*.
156. J. E. Muhlfeld, *The Railroad Problem*, New York, 1941, p. 186.
157. *U P Streamliners*, p. 45.
158. Ibid., p. 123.
159. *Rly. M.E.*, May 1936, p. 185.
160. Coverdale & Colpitts, *Report on Streamline, Light-Weight, High-Speed Passenger Trains*, New York, 1938, p. 52.
161. *New York Times*, April 30, 1935.
162. *Rly. M.E.*, May 1935, p. 185.
163. Ibid., February 1939, p. 43.
164. Ibid., July 1935, p. 306.
165. *Trains*, April 1962, p. 14.
166. *Rly. M.E.*, April 1935, p. 136.
167. *Rly. Age*, Jan. 18, 1936, p. 140.
168. *Rly. M.E.*, August 1940, p. 303.
169. Ibid., April 1932, p. 137.
170. *Society of Automotive Engineers Journal*, Feb. 1933, p. 54.
171. *Trains*, April 1973, p. 25.
172. *Modern Transport*, Aug. 7, 1937.
173. *Baldwin Magazine*, January 1934, p. 36.
174. *New York Times*, Jan. 17, 1934.
175. *Rly. Age*, April 14, 1934, p. 533.
176. *Mech. Eng.*, June 1937, p. 402.
177. *Atlantic Monthly*, July 1937, p. 540.
178. *Rly. M.E.*, March 1940, p. 107.
179. *R.R. Mag.*, October 1972, p. 61.
180. *Rly. M.E.*, December 1941, p. 503.
181. Dubin, *More Classic Trains*, p. 291.
182. *Investors Reader*, Oct. 6, 1952, p. 9.
183. *Trains*, March 1953, p. 20.
184. Ibid., May 1950, p. 11.
185. Ibid., March 1953, p. 17, and December 1968, p. 38.

186. Ibid., March 1953, p. 20.
187. *Business Week*, July 7, 1951.
188. *Trains*, January 1957, p. 42.
189. Walker made these statements in a conversation with the author in November 1973.
190. J. R. Day, *More Unusual Railways*, New York: Macmillian, 1960, p. 168.
191. *Rly. M.E.*, July 1949, p. 369.
192. *Trains*, September 1958, p. 22.
193. *Wheels*, May–June 1955, p. 19.
194. *Trains*, July 1956, p. 6.
195. R. J. Wayner, *Car Names, Numbers and Consists*, New York, 1972, p. 246.
196. R. C. Reed, *The Streamlined Era*, San Marino, Calif.: Golden West Books, 1975, p. 255.
197. *Trains*, July 1956, p. 6.
198. Ibid., August 1958, p. 6.
199. Ibid., March 1956, p. 10.
200. *U P Streamliners*, p. 391
201. *Trains*, March 1956, p. 10.
202. Ibid., March 1970, p. 32.
203. *Star News*, April 5, 1974.
204. *7th Report High Speed Ground Transport*, Washington, D.C.: Department of Transportation, 1973, p. 35.
205. *Trains*, June 1971, p. 29.
206. W. D. Middleton, *When the Steam Railroads Electrified*, Milwaukee: Kalmbach Books, 1974. Hereafter cited as Middleton.
207. *Business Hitstory Review*, Summer 1962, p. 153.
208. *R.R.G.*, Sept. 10, 1880, p. 487.
209. Edward Hungerford, *The Modern Railroad*, Chicago, 1911, p. 315.
210. *Railroad Herald*, November 1922, p. 31.
211. *Trains*, December 1957, p. 15.
212. Middleton, p. 266.
213. Ibid., p. 268.
214. *Rly. Age*, August 1932, p. 221.
215. *R.R. Mag.*, December 1971, p. 45.
216. *Trains*, December 1957, p. 26.
217. Ibid., p. 27.
218. *G.E. Review*, November 1914, p. 1025.
219. *Master Car Builders Report*, 1928, p. 283.
220. *Rly. M.E.*, October 1950, p. 580.
221. *Trains*, August 1965, p. 48.
222. Middleton, p. 73.
223. *Electric Railway Journal*, March 26, 1910, p. 518.
224. Middleton, p. 79.
225. *Elect. Jour.*, February 1916.
226. *Railroad Herald*, Nov. 1922, p. 34.
227. *Middleton*, p. 314.
228. J. F. Stover, *History of the Illinois Central*, New York: Macmillian, 1975, p. 295.
229. *R.R.G.*, May 21, 1880, p. 276.
230. *Rly. & Loco. Eng.*, August 1926, p. 239.
231. Ibid., January 1928, p. 25.
232. Middleton, p. 292.
233. *Rly & Loco. Eng.*, December 1927, adv. p. 8.
234. *G.E. Review*, April 1927, p. 212.
235. *Rly. M.E.*, February 1949, p. 89.
236. *Headlights*, December 1968, p. 4. (Published by the Electric Railroaders Association, New York City.)
237. *N.Y.R.R. Club*, March 16, 1950, p. 154.
238. Middleton, p. 334.
239. *Rly. M.E.*, June 1954, p. 78.
240. *Rly. Age*, June 30, 1958, p. 18.
241. Ibid., May 6, 1963, p. 34.
242. Ibid., March 11, 1974, p. 34.
243. Ibid., Dec. 2, 1968, p. 18.
244. *Rly. Age*, April 12, 1971, p. 22.
245. *New York Times*, Oct. 31, 1972.
246. *Trains*, November 1967, p. 21.
247. Ibid., p. 27.
248. *Rly. Age*, June 10, 1968, p. 12.
249. *Trains*, July 1969, p. 27.
250. From a conversation with W. D. Edson, former Amtrak car superintendent.
251. *Rly. Age*, June 11, 1973, p. 20.
252. *Loco. Eng.*, April 11, 1973, p. 3.
253. National Railway Historical Society *Bulletin*, vol. 40, no. 5, 1975, p. 54.

BIBLIOGRAPHY

Manuscript source material was nearly nonexistent for this book. A few original papers covering the eight-wheel car case and the Pullman Company are cited in Chapters 1 and 3.

The trade journals offered the most concrete and detailed information, particularly for the period 1880 through 1915. The book could not have been written without them.

The following books, reports, and pamphlets are not a complete list of all works consulted. General reference books, individual railroad corporate histories, biographies, and local histories have not been specified, although most of them are cited in the footnotes. None of the publications in this bibliography are scholarly works, at least in the formal sense. Few offer more than the most elementary notation of the sources consulted. Nearly all are the product of enthusiasts whose expertise, style, and reliability vary considerably, yet they preserve and collect data that are indispensable to the production of a general survey such as this book. My debt to the Dubins, Wayners, and Kratvilles who pursue railroad car history for pleasure is immense.

BOOKS

Arnold, Ian. *Locomotive, Trolley and Rail Car Builders*. Railway History Quarterly, vol. II, no. 1, January 1965.

A superficial summary that repeats Charlton's data on the subject of car builders in a more condensed form. Of modest value.

Association of Manufacturers of Chilled Car Wheels. *Chilled Tread Wheels*. Chicago, 1928.

An illustrated technical booklet which explains the properties and manufacture of cast-iron wheels. Produced by a trade association, it is obviously prejudiced, but the data provided are valuable.

Beebe, L. M. *The Overland Limited*. Berkeley, Calif.: Howell-North Books, 1963.

An elegant pictorial testimonial to a long-lived Union Pacific–Chicago North Western train. This Chicago–San Francisco train's equipment is well illustrated from the late 1880s to the modern era.

———. *Mr. Pullman's Elegant Palace Car*. Garden City, N.Y.: Doubleday & Company, 1961.

A long, informal history of Pullman and first-class rail travel in America. The author's grandiloquent prose is equal to the subject, but readers are cautioned not to depend uncritically on all the factual data offered. Pictures are abundant and well reproduced.

———. *Mansions on Rails*. Berkeley, Calif.: Howell-North Books, 1959.

Accurately subtitled "the folklore of the private railway car." A delightful book for which the Pullman glass-plate negatives were made available for the first time to any modern author by the custodians of that valuable collection.

Behrend, George. *Pullman in Europe*. London: Ian Allen, 1962.

A long, rather disorganized and rambling account. The focus is on the British Pullman operation, and British Pullman cars are listed and illustrated.

Charlton, E. Harper. *Railway Car Builders of the United States and Canada*. Los Angeles: Interurbans Special 24 (Ira L. Swett), 1957.

A hobbyist-oriented survey. Emphasis is on street railway car builders. Valuable for quick reference, although dates do not always agree with contemporary sources. The author depended on information supplied by correspondence. Early builders are neglected.

Cockle, G. R. (ed.). *Car and Locomotive Cyclopedia*. New York: Simmons-Boardman Publishing Corp., 1974.

The third edition of this work, which is a continuation of the *Car Builders' Dictionary* (see M. N. Forney).

Combes, C. L. (ed.). *Car and Locomotive Cyclopedia*. New York: Simmons-Boardman Publishing Corp., 1966, 1970.

The first and second editions of this work which combined the locomotive and car cyclopedias using the same format established in 1879 by M. N. Forney.

———. *Car Builders' Cyclopedia*. New York: Simmons-Boardman Publishing Corp., 1957, 1961.

The twentieth and the twenty-first editions of this work.

Commault, Roger. *Georges Nagelmackers*. Uz s, France: La Capitelle, 1970c.

A small, admiring biographical sketch of the founder of the Wagon-Lits company, written by an employee of the firm. Illustrated. The text is in French.

Dorin, P. C. *Coach Trains and Travel*. Seattle: Superior Publishing Company, 1975.

The author falls short of his plan to trace the development of coach travel for the past 175 years. His comments are limited almost entirely to the twentieth century, with heavy emphasis on the post-1930 period. Pictures and text largely repeat what Dubin and Kratville have already published.

———. *The Domeliners*. Seattle: Superior Publishing Company, 1973.

A pictorial history of dome cars in America that offers a useful, nontechnical survey.

Dubin, A. D. *More Classic Trains*. Milwaukee: Kalmbach Publishing Company, 1974.

A sequel to *Some Classic Trains* that deals with lesser-known trains.

———. *Some Classic Trains*. Milwaukee: Kalmbach Publishing Company, 1964.

A pictorial tribute to the great name trains of America. The Broadway Limited and other luxury trains are shown in excellent photographs and described in trade literature of the period. The focus is on developments between 1890 and 1950. The text is useful for dating equipment and establishing the chronology of name-train service. The author's long interest in the subject and his extensive collection of memorabilia are well displayed in this large volume.

Ellis, Hamilton. *Railway Carriages in the British Isles from 1830 to 1914*. London: George Allen & Unwin, 1965.

A survey history written for the enthusiast. Scholarly readers will criticize the lack of documentation and depth of coverage, but the book remains a valuable and useful introduction to the subject. Well illustrated.

Forney, M. N. (ed.). *The Car Builders' Dictionary*. New York: Railroad Gazette, 1879.

The first volume of this basic series, which was revised and republished every three to four years. It begins with definitions of standard terms and provides engravings of general car types with assembly drawings and details of components. This general format was followed in subsequent editions. The 1879 edition has been reprinted three times since 1949.

Fowler, G. L. *The Car Wheel*. Pittsburgh: Schoen Steel Wheel Company, 1907.

A subsidized work of 160 pages intended to promote the sponsor's product. The author was a respected mechanical engineer and author. The text is technical.

Hitt, Rodney (ed.). *The Car Builders' Dictionary*. New York: Railroad Gazette, 1906.

The fifth edition of this work. Reprinted in 1971 by Newton Gregg of Kentfield, California.

————. *The Car Builders' Dictionary*. New York: Railroad Gazette, 1903.

The fourth edition of this work.

Husband, Joseph. *The Story of The Pullman Car*. Chicago: A. C. Mc-Clurg, 1917.

A popular and not remarkably objective or accurate account. It is useful as a starting point, and for many years it was one of the few books available on the subject.

Kidner, R. W. *The Development of the Railcar*. South Godstone, England: Oakwood Press, 1958.

A modest but valuable British publication which compresses considerable data on the subject into a small space.

Kirkman, Marshall M. *The Science of Railways*. Chicago: World Railway Publishing Company, 1904 ed.

A multivolume series written by a railway official and a prolific writer on railroad management and engineering. An unnumbered volume contains a broad but inaccurate history of cars. Volume IV has sections on heating, lighting, wheels, and axles.

Kratville, W. W. *Passenger Car Catalog: Pullman Operated Equipment, 1912–1949*. Omaha: Kratville Publications, 1968.

A descriptive catalog of heavyweight Pullman cars organized by type of car. Names, lot, and plan numbers are given. Photographs and floor plans are included. Part II offers a brief illustrated overview of mechanical equipment that includes a good summary of air-conditioning systems used by Pullman.

————. *Steam, Steel & Limiteds*. Omaha: Kratville Publications, 1962.

A comprehensive pictorial survey of passenger train travel in America during the heavyweight era. A celebration of the wonders of train travel, intended for the enthusiast. The variety and scope of service offered is clearly represented but not analyzed in this 400-page volume.

Kratville, W. W., and H. E. Ranks. *Union Pacific Equipment*. Omaha: Kratville Publications, 1969.

Another textless scrapbook of photographs and diagram drawings that contains a treasury of uncoordinated raw data.

Kulp, R. L. (ed.). *History of Mack Rail Motor Cars and Locomotives*. Allentown, Pa.: Lehigh Valley Chapter, National Railway Historical Society, 1959.

A concise but complete record, well illustrated. An example of railroad-buff literature at its best.

Lister, F. E. (ed.). *The Car Builders' Dictionary*. New York: Railway Age Gazette, 1909.

The sixth edition of this work.

Lucas, W. A. *100 Years of Railroad Cars*. New York: Simmons-Boardman Publishing Corp., 1958.

A popular book of plans and photographs aimed at the model builder. Useful for quick reference. The author-editor was an expert on the subject and served for many years as associate editor of the *Car Builders' Cyclopedia*.

McConway & Torley Company, *The Evolution of Car Couplings*, Pittsburgh, 1905.

An illustrated pamphlet issued by a leading manufacturer of Janney couplers. The work is actually limited to the Janney style of coupler.

McKee, L. A., and A. L. Lewis. *Railroad Post Office History*. Pleasantville, N.Y.: Mobile Post Office Society, 1972.

A hardbound scrapbook of lists, photographs, and cancellation stamp imprints. Many pictures are not captioned, and few are dated. The minimal text consists of reprints of popular articles on the subject. The book is of modest value to the historian.

Master Car Builders Association Report of the Proceedings, 1867–1918.

The earliest reports (1867–1872) were actually not published until 1885. After 1873 the reports reproduced papers and related discussions that offer fascinating insights into the art of railroad car design. However, very little space is devoted to passenger car construction. After 1918 these reports were published as the Association of American Railroads Mechanical Section in combination with the former Master Mechanics Association.

Mencken, August. *The Railroad Passenger Car*. Baltimore: The Johns Hopkins Press, 1957.

This pioneering history of the American passenger car was for many years the only work available on the subject. The modest text concentrates on nineteenth-century developments. No formal documentation is offered, and readers are cautioned not to accept the information uncritically. Reproductions of contemporary travel accounts, which occupy one-half of the pages, form the most valuable portion of the book.

Middleton, W. D. *When the Steam Railroads Electrified*. Milwaukee: Kalmbach Publishing Company, 1974.

An informed and complete survey of main-line electric operations in America, which includes valuable information on multiple-unit operations.

Peck, C. B. (ed.). *Car Builders' Cyclopedia*. New York City, Simmons-Boardman Publishing Company, 1949 and 1953.

The eighteenth and nineteenth editions of this work.

Railway Age Company, *The Modern Railway Car*, Chicago, 1883.

An illustrated booklet describing the private car *Railway Age*. It has an introduction on the history of the passenger car.

Randall, W. D. *From Zephyr to Amtrak*. Park Forest, Ill., 1972.

A listing of lightweight cars and streamlined trains which offers basic data on builders, consists, and service. Illustrated mainly with builders' photographs. Minimal text. See Wayner's *Car Names, Numbers and Consists*, which covers the same subject.

Randall, W. D., and A. R. Lind. *Monarchs of Mid-America*. Park Forest, Ill.: P.O. Box 191, 1973.

A collection of diagram drawings, photographs, and schedules covering Illinois Central Railroad passenger equipment—no text beyond the captions. The book concentrates on twentieth-century developments.

Ranks, H. E., and W. W. Kratville. *The Union Pacific Streamliners*. Omaha: Kratville Publications, 1974.

A remarkably complete study that covers design, construction, and service in both pictures and text. One of the most valuable works yet published on American railroad equipment.

Reed, R. C. *The Streamlined Era*. San Marino, Calif.: Golden West Books, 1975.

A thoughtful analysis of the streamline phenomenon, its origins and applications. The book treats automobiles, ships, and furnishings but the emphasis is on rail vehicles. A good balance of text and illustrations.

Ross Winans vs. The Eastern Railroad Company in Equity. Boston: Moore & Crosby, 1854.

A massive printed transcript that reproduces the testimony and dispositions of the major and minor pioneer master car builders.

This volume is an indispensable source for the early history of the American passenger car.

Rush, R. W. *British Steam Railcars*. South Godstone, England: Oakwood Press, 1971.

A comprehensive, detailed, but undocumented work. Illustrated by modern line drawings.

Scribbins, Jim. *The Hiawatha Story*. Milwaukee: Kalmbach Publishing Company, 1970.

An in-depth study of the Milwaukee Road's premium passenger trains in the streamline era. An informed, readable text, reinforced by excellent pictorial coverage.

Shine, J. W., and F. M. Ellington. *Passenger Train Equipment of the Atchison Topeka & Santa Fe. Volume 1: Head End Cars*. La Mirada, Calif.: Railcar Press II, 1973.

A folio-sized record book containing drawings, photographs, and a brief service history from 1880 to the present time.

Sillcox, L. K. *Mastering Momentum*. New York: Simmons-Boardman Publishing Corp., 1941.

A sophisticated engineering treatise written by an official of the New York Air Brake Company that also treats wheels, axles and truck theory, materials, and design.

Stöckl, Fritz. *Comfort on Rails: Sleepers, Diners, and Saloon-cars on the European Railways*. Basel, Switzerland: Claude Jeanmaire, 1970.
Much on the order of Stöckl's *Rollende Hotels*.

———. *Rollende Hotels*. Part 1. Vienna: Rudolf Bohmann, 1967.

An illustrated history of the International Sleeping Car Company—a solid, well-illustrated popular work. Text in German.

Stuart, C. W. T. *Car Lighting by Electricity*. New York: Simmons-Boardman Publishing Company, 1923.

A textbook written for the practical mechanic. The principal commercial systems are explained and illustrated. A brief historical chapter exhibits more careful preparation than is usual for works of this nature.

Synnestvedt, Paul. *Evolution of the Air Brake*. New York: Locomotive Engineering, 1895.

Treats the development of individual components rather than the general system itself. A small, valuable work, illustrated with line drawings.

Turner, W. V., and S. W. Dudley. *Development in Air Brakes for Railroads, with a Brief Review of Past and Present Operating Conditions*. Pittsburgh: Westinghouse Air Brake Company, 1909.

A learned, clear account written by top experts in the field. The fine narrative is illustrated by numerous tables, drawings, and halftones.

Wait, J. C. (ed.). *The Car Builders' Dictionary*. New York: Railroad Gazette, 1895.

The third edition of this basic work, in which the page size was again enlarged.

Wayner, R. J. *Car Names, Numbers and Consists*. New York: Box 871, Ansonia Station, 1972.

A listing of streamlined and lightweight passenger cars built and operated in the United States. Dates of service and dispositions are generally included in notes. Illustrated by floor plans and diagram drawings. (See Randall's *From Zephyr to Amtrak*.)

———. *New York Central Cars*. New York: Box 871, Ansonia Station, 1972.

A pictorial scrapbook of photographs and captions, mainly twentieth-century views. Wayner has produced similar booklets on the Santa Fe and the Southern Pacific.

———. *Pullman and Private Car Pictorial*. New York: Box 871, Ansonia Station, 1972.

A collection of exterior photographs and captions covering the period 1890–1960.

———. *Rio Grande Car Plans*. New York: Box 871, Ansonia Station, 1969.

Reproduces diagram drawings with caption identifications. Covers the wooden and steel eras. The same author has produced a similar work on the Pennsylvania Railroad.

———. *Pullman Panorama*. Vol. I. New York: Box 871, Ansonia Station, 1967.

A listing of sleeping and lounge cars built at the Calumet shops (Chicago) between 1910 and 1931. Plan and lot numbers, construction dates, car names, and dispositions are given in the book. No illustrations are offered, nor have the subsequent volumes that are described in the introductory text been published so far.

Wellington, A. M. (ed.). *The Car Builders' Dictionary*. New York: Railroad Gazette, 1884. (1888).

A revised and enlarged edition of the first dictionary. The page size was much enlarged. The material was assembled in 1884 but not actually published until four years later. Reprinted in 1971 by Newton Gregg, Kentfield, California.

Wellington, A. M., W. B. D. Penniman, and C. W. Baker. *The Comparative Merits of Various Systems of Car Lighting*. New York: Engineering News Publishing Company, 1892.

A detailed study of car lighting as it was late in the nineteenth century. Theory and practice are carefully studied. Good line drawings are included.

Whittlesey, G. P. *Abridgements of the United States Patents Pertaining to Continuous Car Heating*. New York: Railroad Gazette, 1888.

A specialized booklet that is essentially a listing and summary of patents.

Wright, R. K. *Southern Pacific Daylight Train 98–99*. Vol. 1. Thousand Oaks, Calif.: Wright Enterprises, 1970.

An incredible graphic record of an important West Coast train. Within 650 pages are drawings and photographs that provide construction and in-service views. Both locomotives and cars are included. The book reproduces a complete specification for Pullman-built cars dating from 1936.

Wright, Roy V. (ed.). *The Car Builders' Dictionary*. New York City: Simmons-Boardman Publishing Company.

The following editions were compiled by the same editor: 1912 (seventh), 1916 (eighth), 1919 (ninth), 1922 (tenth); note title change to *Car Builders' Cyclopedia*; 1925 (eleventh), 1928 (twelfth), 1931 (thirteenth), 1937 (fourteenth), 1940 (fifteenth), 1943 (sixteenth), 1946 (seventeenth).

Voss, William. *Railway Car Construction*. New York: R. M. Van Arsdale, 1892.

The finest work ever published on American railway car construction. The text explains concepts, materials, and proportions. It is not a history and was meant to be a handbook for the practical car builder. Both freight and passenger cars are included. Extremely fine engineering drawings and engravings illustrate the work. This book goes beyond *The Car Builders' Dictionary*, which only illustrates what was done; Voss explains why it was done.

Zimmermann, K. R. *The Story of the California Zephyr*. Starrucca, Pa.: Starrucca Valley Publications, 1972.

An intelligent appreciation of this splendid train that includes a long text and abundant pictures.

PERIODICALS

American Journal of Railway Appliances. Vol. 1, 1883–Vol. 21, 1901. Space was devoted to rolling stock declines during 1890s; then the journal evolved into a commentary on machine tools and related matters. Title changed to *Railway Machinery* in 1901. Not a major source.

American Railroad Journal. Vol. 1, 1832, to 1975. The earliest years contain only scattered notes on cars, but more information was included after about 1890. A rich source from that time forward. In 1913 the title changed to *Railway Mechanical Engineer.* In 1953 it changed again to *Railway Locomotive and Cars.* This journal continued to be published under the original volume sequence until January 1975.

National Car Builder. Vol. 1, 1870-Vol. 26, 1895. The earliest years are incomplete; there were few illustrations until after 1880. The name was changed 1886 to *National Car and Locomotive Builder.* It is the single most valuable source for information on the subject.

Railroad Car Journal. Vol. 1, 1890–Vol. 12, 1902. Started as *The Journal of Railroad Car Heating and Ventilating.* Changed to *Railroad Car Journal* in October 1891. Name again changed to *Railroad Digest* in 1901, and the periodical expired during the following year. Well illustrated.

Railroad Gazette. Vol. 1, 1870–Vol. 44, 1908. Always concerned with engineering matters, this paper devoted frequent, if not abundant, space to railroad car items. Occasional drawings were published on the subject. The paper merged with *Railway Age* in 1908. M. N. Forney was the editor for many years.

Railway Age. Vol. 1, 1876 to present. A journal originally devoted to corporate matter; little space was given to rolling stock until about 1890. Then the subject was well covered until about 1955, when it was deemphasized.

INDEX